Experimental Physics

Principles and Practice for the Laboratory

Experimental Physics

Principles and Practice for the Laboratory

Edited by Walter Fox Smith

CRC Press
Taylor & Francis Group
Boca Raton London New York

CRC Press is an imprint of the
Taylor & Francis Group, an **informa** business

CRC Press
Taylor & Francis Group
6000 Broken Sound Parkway NW, Suite 300
Boca Raton, FL 33487-2742

© 2020 by Taylor & Francis Group, LLC
CRC Press is an imprint of Taylor & Francis Group, an Informa business

No claim to original U.S. Government works

Printed on acid-free paper

International Standard Book Number-13: 978-1-4987-7847-3 (Hardback)

**Visit the Taylor & Francis Web site at
http://www.taylorandfrancis.com**

**and the CRC Press Web site at
http://www.crcpress.com**

Contents

Part III   Fields of Physics

Preface

To the Student

You are holding the most important textbook ever written. I'm being humorous, of course, but only partly. Science is the most important field. Physics is the most important science; in addition to the amazing, world changing discoveries made in physics itself, physics underlies all the other sciences and fields of engineering, and is responsible directly or indirectly for the biggest advances in them. Physics is based on experimental results, so the most important physics book is about experimental physics. This is by far the best experimental physics book ever written. (For example, it's the only one written by a team of experts in each subfield of experimental physics, each of us having a deep devotion to undergraduate education.)

If you're using this book as part of a course, buckle up! You will find yourself challenged in ways you've never been before. You may find that the course is harder or less hard than other advanced physics courses you've taken, but it will definitely be very different. You'll be asked to master techniques that are used in cutting-edge research labs, to think very deeply about what you're doing, and (as you progress through the course) to become more independent as an experimental physicist, to design your own experiments, troubleshoot them yourself, and develop your own models to explain what you see.

Your approach to this course must be different from what you've done before. You truly have to do the pre-lab reading before lab, and far enough before that you can ask your instructor or classmates about anything that is unclear before lab. Otherwise, you will find yourself totally lost in lab, understanding only a small fraction of what you're supposed to, using more of the instructor's time than is your fair share, and having to stay later than the other students.

If you put in the effort needed to prepare properly, you will find much delight in the labs. You'll also find plenty of frustration – that's part of experimental physics. But solving the problems and puzzles that arise will be deeply satisfying.

Perhaps you will come up with a question that you and your instructor think many other students would have. Please email it to me, Walter F. Smith, wsmith@haverford.edu. I will add it to the FAQ on the book's website, with credit to you and your instructor. (You can ask not to be credited if you prefer.) Also, please email me with any other suggestions for the book.

For each chapter, please look at the resources on the book's website, ExpPhys.com. There, you will find important updates, additional instructional materials, links to related research, and occasional amusing anecdotes and other content.

Happy experimenting!

To the Instructor

Advanced lab courses have two central goals that are in tension with each other. One is concrete: to teach students about the most important experimental and data analysis techniques. This is most efficiently accomplished with a series of well-defined, programmed exercises. However, there is also an essential and more abstract goal: to transform students, at least partly, into independent experimentalists. Students should learn how to formulate worthwhile research questions, and how to discover what is already known about the area of research. They should begin to understand the strengths and weaknesses of the range of experimental techniques that might be applied. They should become more confident in their ability to design, assemble, and debug experimental apparatus, to use it to take meaningful data, and to analyze the data in a way that most clearly shows its meaning.

The advanced lab course prepares students for research. It does so in a way that is more efficient for both the instructor and the students than simply throwing them into actual original research. Instead, it combines structured experiences that teach fundamental skills, challenges to design experiments to accomplish well-defined goals, and more open-ended explorations in carefully chosen areas that are likely to produce interesting results in a limited time.

As an experimentalist and a professor, I've been enormously frustrated by the books available for advanced lab courses. There are few, they are outdated, they lack many of the content I consider essential, and they are all written by one or two people.

Experimental physics is far broader than any other science. It is simply impossible for one or two people to write authoritatively about ideas that are central to cutting-edge research in fields as diverse as particle physics; atomic, molecular, and optical physics; condensed matter physics; photonics; etc. Therefore, each of the advanced chapters in this book is written by a research-active expert in the area. **Extensive instructor's manuals are available for every lab, so that an instructor with a background in, say, condensed matter, can confidently teach the labs in particle physics.**

The experiments included emphasize connecting the students with current research, rather than reproducing historic experiments. I think this is an essential goal of the advanced lab course. In some cases, it may necessitate upgrading the equipment used in the course. However, equipment costs have been kept as low as possible, experiments are designed to use equipment you may already own whenever possible, and upgrades can be made over several years. Further, most major pieces of equipment are used for several labs. This is helpful budgetarily, but also serves a teaching goal, since students remember and understand better when they use the same equipment in several contexts.

The development of this book was guided in part by recommendations in a 2014 report from the American Association of Physics Teachers.* This emphasizes six focus areas: constructing knowledge; modeling; designing experiments; developing technical and practical laboratory skills; analyzing and visualizing data; and communicating physics. "Constructing knowledge" and "modeling" refer to teaching students how to use their own data to create their own sets of equations and/or computer simulations that describe the system being studied; the goal is that students should construct "knowledge that does not rely on an outside authority."

An important part of the "designing experiments" focus area is troubleshooting: "students should be able to troubleshoot systems using a logical, problem-solving approach." A common definition of "expert" is someone who has made all the mistakes that can possibly be made in a particular area. So, it is clearly quite important for students to learn how things can go wrong, how to recognize what has gone wrong, and how to correct it.

Unfortunately, this aspect of experimental science is usually only taught when a problem arises at random as students work through a lab. Therefore, some students may get through a lab course with much less experience in troubleshooting than others. In contrast, some of the experiments in this book include at least one explicit troubleshooting exercise. These parts of the labs safely and efficiently develop the students' troubleshooting skills.

The book is divided into three parts. "Part I: Fundamentals" covers ideas and analysis techniques that are common to many areas of experimental physics. We suggest that Chapters 1–3 be assigned over the first two weeks of the course, since these cover ideas that are essential for every lab. Chapter 4, on uncertainty analysis, should be assigned before the first lab assignment that requires it; ordinarily, this will be the first lab from Part III. If possible, the reading of Chapter 5, on scientific ethics, should be combined with one or more discussion sessions.

"Part II: Tools of an Experimentalist" covers experimental skills that are fundamental to most physics research. The emphasis in Chapter 6, on analog electronics, is on proper use of commonly available commercial equipment. Individual components are used sparingly, however, there are two labs devoted to op amps. This approach reflects the reality in most physics research labs. Part II also includes chapters on digital electronics (including Arduino and FPGAs), optics, vacuum, and particle detection. Except as noted, each experiment in Part II takes approximately three hours.

"Part III: Fields of Physics" includes an overview and experiments in each of the major fields of physics. Each chapter is written by an active researcher in that field. Some of these chapters include a high-intermediate experiment (which takes approximately six hours). All of them include an advanced

---

* "AAPT Recommendations for the Undergraduate Laboratory Curriculum,"
  https://www.aapt.org/Resources/upload/LabGuidlinesDocument_EBendorsed_nov10.pdf

experiment with an open-ended component (which takes approximately 12 to 24 hours of lab time). These labs were selected by each chapter author using the rule, "What do I wish my students already knew when they started work in my research lab?"

Throughout the sequence, we reuse the same equipment elements as much as possible, so that when they reach the advanced experiments, the students have a reasonable understanding of the uses and limits of these elements.

The experiments include learning goals, required pre-lab reading, pre-lab questions, and discussion of any safety concerns. Bold-face questions embedded in many of the labs are designed to get students to think more deeply. Instructors have the option of making reports consist only of the answers to these bold-face questions.

There are far more labs in this book than it is possible for a single student to work through in one course. We've structured the book this way so that you can tailor things to match your preferences and the needs of your students.

The focus of this book is experimental physics. However, in many cases it's important for students to understand theoretical aspects before starting an experiment. (In other cases, students are expected to develop theoretical models on their own.) This book includes summaries of relevant theory, together with intuitive qualitative models. We refer the students to specific other sources (including web addresses or specific ranges of page numbers) for additional detail on the theory, and we include homework problems that allow instructors to probe whether the students have actually consulted and at least somewhat understood these other sources. We suggest that instructors assign these readings and problems in the week preceding the lab, so that the students begin the lab with a reasonable level of understanding of the theory, and can focus on the experiment.

An instructor's manual is available from the publisher via the instructors' area of the book's website, ExpPhys.com. This includes full details on how to do each experiment, how long it's likely to take, the most common sources of trouble for the students, sample data and analysis, and answers to all pre-lab and bold-face questions. **It is essential to consult the instructor's manual before teaching any of the labs.** The instructor's area also includes difficulty ratings and full solutions for the end-of-chapter problems, suggested exam questions, and solutions, updated suggestions for equipment suppliers, and other hints on how to get the most instructional value from each chapter.

The student's area of ExpPhys.com includes errata, instructional videos, and links to helpful animations and related current research.

Every experiment in this book has been tested with students, revised, tested again, and revised again. Most of the experiments have gone through four or more cycles of testing and revision. We've taken full advantage of this experience to create labs that create lively learning experiences (you'll often hear, "Ahhhh! I get it!" during lab), get students to think deeply, and give students a real understanding of experimental physics.

The book includes three approaches to computer data acquisition and control: LabVIEW, Python, and Arduino. You can choose the one you prefer, or teach more than one. In the future, we plan to add support for data acquisition and control with MATLAB®; check ExpPhys.com to see if this has happened yet.

To reduce the cost of this book, we've chosen to move some of the content to ExpPhys.com.

You will find that much of the book, especially Parts I and II, is helpful for your research students.

If you have any questions or suggestions, please contact Walter F. Smith at wsmith@haverford.edu.

Acknowledgments

The authors of this book deeply thank the following people, each of whom has put in many hours specifically for this book, testing experiments, organizing groups of student testers, providing insightful feedback, helping with editing, and creating figures. Without their help, this book would not have been possible: Eric Beery, Charles Bene, Ayesha Bhika, Travis Crawford, Jackson Davis, John Dusing, Sara Fadem, Michael Fernandez, Moira Ferrer, Helena Frisbie-Firsching, Rebeckah Fussell, Shamira Gonzalez, Lou Han, Charles Holbrow, Sunxiang Huang, Merritt Jacob, Meredith Jones, Lucas Kasle, Julia Kotler, Brian Kroger, Sasha Levine, Kate Matthews, Oliver Maupin, Aidan McGuckin, Marian McKenzie, Dani Medina, Saul Medrano, Alan Mendez, Richard Morash, Axel Moreen, Carly Press, Kristina Qualben, Margarita Rivers, Adeeb Saed, James Silva, Jonathan Simpson, Jeanette Smit, Greg Soos, Emily Swafford, Elizabeth Teng, Ryan Terrien, Kelly Tornetta, James Vickery, Jonam Walter, Longqi Wu, Shufan Xia, and Hongwen Yu.

We are also grateful to the many students in our classes over the last several years who have done the labs in this book, and because of their hard work and enthusiasm have given us critical feedback.

The development of the labs in Chapter 20 was funded by grants to Prof. Enrique Galvez from the divisions of Undergraduate Education and Physics of the National Science Foundation.

Part I

Fundamentals

1

Introduction

Walter Fox Smith

If (based on the poem by Rudyard Kipling)

If you can take your data, when all about you
Are losing theirs, and blaming it on you,
If you can trust your understanding, when others doubt you,
But make allowance for their doubting, too,
If you can persevere, and not give up,
And insist on good controls, when others say you're wasting time,
Or think and plan instead of diving in,
And yet not plan too long, nor let your thinking stop your daring,

If you can dream, and not make dreams your master,
If you can theorize, and not make theories your aim,
If you can meet with defeat after defeat,
And keep on trying 'till you find the reason,
If you can bear to hear the truth you've spoken
Questioned by referees who don't know what they're saying,
Or watch the apparatus you gave your life to broken,
And build it better than it was before,

If you scorn to publish tiny variations,
But striving for results that change the world,
Yet knowing almost always you will fail,
Not breathe a word about your disappointment,
If you can fill the lab with joy not bitterness,
And wordlessly remind your mates why they chose science,
And finish taking data when it's boring,
Refusing to do other than your best,

If you can talk at conferences and fire the imagination,
Or walk through forests thinking of science,
If your collaborators know your virtue,
If journal editors value your review,
If you can read the manual and phone the expert,
If you can take a risk, and yet stay prudent,
Yours is the lab, and everything that's in it,
And – which is more – you'll be an experimentalist, my student!

Physics is an experimental science. That means that, although we greatly value the insights of theorists, the determination of whether a new idea is true or not is done experimentally. In addition to this important role of checking theories, experimentalists also often discover exciting, unexpected phenomena. (Unlike portrayals in the media, such discoveries are usually the result of carefully planned explorations and months or years of hard work.) In a third role, experimental physicists frequently make the first steps toward applying recent theoretical and experimental discoveries toward the creation of important new devices and technologies.

Even though experiments are so important, the vast majority of your physics time so far has been spent learning theory. This is partly because you need a solid understanding of fundamental theoretical ideas to design and interpret experiments. Nevertheless, since this may be the only full-credit course you take that's devoted to experimental physics, it's clearly the most important course you will ever take.

The experiments you'll do are chosen carefully to allow you to learn efficiently. You will not be expanding humanity's understanding of the universe, but the experiments will give you a genuine taste of the joys and frustrations of original research, and the experience should provide important guidance about careers you may wish to pursue.

You will improve your ability to understand a complicated undertaking at many levels, including theoretical background, planning, deep understanding of apparatus, optimal ways of connecting components, pilot testing, and rational debugging. It will require a great deal of patience and perseverance. You will likely need to change your approach, compared with earlier physics labs you've done. You will need to prepare more, focus more during the lab, be more ready to find outside resources in addition to the assigned reading, and think more deeply about your results.

If you put in the effort needed, you'll find that this is one of the most rewarding and memorable courses of your career. What you learn will be essential in the research lab, but will also serve you well in areas from improving industrial processes to home repair. In fact, the ideas of debugging an apparatus and carrying out an experiment in the most efficient way can be applied to almost all areas of your life!

Note on Part II: Tools of an Experimentalist

It is *absolutely essential* that you complete the pre-lab reading before each lab in this section. If you don't, you won't be able to finish the lab in a timely fashion, you'll understand little of what you're supposed to, you'll take more than your fair share of the instructor's time, and you'll have to stay later than the other students. It's possible that in some previous lab courses you were able to make do without careful preparation; that approach truly will not work here.

You should expect to understand everything in the labs in this part fairly thoroughly, and you should get fairly close to the expected results, except when we tell you explicitly that something unexpected may happen. (Nevertheless, you'll need to think hard!) If you complete a section of a lab and aren't reasonably sure that you understand what's going on, you should consult with a classmate or your instructor before proceeding to the next section.

Note on Part III: Fields of Physics

The labs in this part are designed to push you to become more independent as an experimentalist. You are expected to find related journal articles and relevant content in other textbooks. Much of the design of the experiment and the apparatus will be up to you. However, it's smart to discuss your ideas with your instructor to avoid wasting time.

Walter Fox Smith is the Paul and Sally Bolgiano Professor of Physics at Haverford College. His research centers on self-assembling electronics. The eventual goal of this field is to start with several beakers of carefully chosen chemicals, and by mixing them in the appropriate sequence and applying appropriate stimuli (such as changing temperature or applying a voltage) cause the molecules to assemble into 10^{23} identical circuits, with complex three-dimensional structures. Currently, the field's focus is on understanding the principles that guide the self-assembly and the fundamental physics of charge transport through individual self-assembled circuit elements such as wires and transistors. Prof. Smith

FIGURE 1.1 Walter Fox Smith on a family trip to Belle Ile, France.

is also the author of "Waves and Oscillations: A Prelude to Quantum Mechanics" (Oxford University Press). He enjoys role-playing games, board and card games, jiu jitsu, and singing. He writes songs about physics and collects historical physics songs; you can find some of each at his website, PhysicsSongs.org (Figure 1.1).

2

Planning and Carrying Out Experiments

Walter F. Smith

CONTENTS

There's a tremendous amount of work building the apparatus, getting the experiment to work. But sitting there late at night in the lab, and knowing light is going at bicycle speed, and that nobody in the history of mankind has ever been here before - that is mind-boggling. It's worth everything.

– Lene Hau (physicist at Harvard who used a Bose-Einstein condensate to slow the speed of light to 17 m/s)

2.1 Literature Research

An essential part of science is understanding the state of your particular field of interest. What are the most important, well accepted fundamentals? What are the areas of current inquiry and controversy? What is possible experimentally with currently available instruments? Whether you are planning an experiment for research or conducting one of the more advanced experiments in this book, you should do some delving into journal articles (collectively referred to as "the literature").

There are hundreds of research journals relating to physics. One of the most important for labs in this book is the *American Journal of Physics*, which focuses on instruction at the undergraduate level. Be aware that the papers are written for faculty, so it will be challenging to understand every part of them. However, you can understand more than half the content with only a little work, and all of it if you choose to persevere with that particular article.

There are two main types of articles (also known as "papers") in the literature: research articles and review articles. Review articles summarize the state of a particular field of research, so they are especially helpful. Some journals, such as *Reviews of Modern Physics*, are dedicated to review articles, but many other journals include them occasionally.

Luckily, it is now fairly easy to find papers in a particular field of interest using databases available through your library and electronic journal subscriptions. You should be thankful that you're alive today;

what you can accomplish researching the literature in 30 minutes at any location with internet access used to literally take days of full-time effort in the library.

A conventional internet search has its uses, but also limitations. It's especially helpful for finding websites of researchers in the area of interest; often, these include easily understandable summaries. You may also find some useful papers, e.g., through Google Scholar. However, this is in no way a thorough search of the literature; the fact that you find some good papers doesn't mean you've found all the good papers.

For such a search, you need to use a database that is available through your library. The best ones for physics are Web of Science and Scopus. Both of these cover almost all physics journals, as well as almost all journals in chemistry, biology, and other related fields. However, their coverage of engineering is quite limited. The database Inspec covers all physics journals and all engineering journals but has limited coverage of chemistry and biology.

You can search on these databases the same way you would with an internet search, including using quotes to indicate phrases that must be matched exactly. However, you can also do complex Boolean searches, such as "silver AND (resist* OR conduct*) NOT (photo* OR halide)." The * character is a "wildcard"; for example, resist* will yield results that include any of the words resist, resistor, resistance, resistivity, etc.

You can choose the "field" you want to search in. For example, you can choose to search for words in the titles of the articles or for authors. By default, the field is usually set to "topic," which means that you're searching the title, abstract, and the keywords that have been entered by the authors or editors.

You can search on multiple search fields. Normally the results are combined with "AND," meaning that all criteria must be satisfied. For example, if you search for "silver nanowires" in the Topic field, and "review" in the Document Type field, the results will be review articles that contain the words "silver" and "nanowires" in the title, abstract, or keywords.

Often, you need to do some exploring to figure out the best words to search on. For example, say you're interested in the effect of a magnetic field applied to a crystal on the polarization of light passing through it. You might begin by entering the following in the search for the topic field: "magnetic field" crystal polarization.

This would yield results that contain the phrase "magnetic field" as well as the words crystal and polarization in the title, abstract, or keywords. You would find that most of these results are not really what you're interested in, since magnetic fields, crystals, and polarization are used in many different ways in physics papers. However, after looking through the titles of the first ten papers, you would come across the term "Faraday effect," which more precisely identifies what you're after. You would then re-do the search using this better term.

The search results are given as a list of titles, with the most recent articles at the top. By clicking on the title, you can see the abstract. There should be a button you can click to go to a page where you can download a pdf of the article. Find out from your professor or librarian which button to click; it's usually not the one that says, "Full Text from Publisher." If your library doesn't subscribe to the journal, and if you're confident the article is important, you can order it through interlibrary loan and normally get it within a few days.

There is a second, completely different mode of searching which is just as important as the keyword search described above: citation searching. Say you've found an important early paper in the field of interest, and you think most later papers will include it in their references. Find the early paper (usually it's easiest to do this by searching for the exact title), then click on the number of "Times cited" or "Cited by" which appears on the right. This brings up a list of all the articles written more recently which reference the early article. This is a tremendously useful tool and should be part of every serious literature search.

Finally, of course, as you read each article of interest, you will see references to other articles. From the context of the reference, you can often judge what the article is about.

2.2 Reading Scientific Papers

Professional scientists must keep abreast of a daunting volume of new papers. So, all of us use shortcuts. We begin, of course with the title, and if that seems sufficiently interesting, we skim the abstract. If the paper still seems worthwhile, we do not simply start reading from the beginning. Instead, we look at the figure and captions. In most cases, that's as far as the reader goes, because that's usually enough to give a good sense of the work. If further depth is needed, we read the conclusions. Finally, in some cases we look at the introduction, and read the full results and discussion section. Only if the work is very closely connected with our own research do we read the methods section.

Even PhD scientists don't understand a paper fully on the first reading. So, if the paper seems really important, expect to need to go through it a few times, and even then, don't be surprised if there are parts that are confusing. If it's important to understand those parts, don't hesitate to ask someone, such as your professor, for help.

For any paper that seems reasonably interesting, you should download a pdf, and mark it up with highlighting, comments, and questions. You might want to assign a score from 1–10; thinking about the appropriate score will help you to remember the paper better. Save the pdf with the filename "Zhou 2003 Nanomasks for molecular conductivity," where Zhou is the last name of the first author, and the phrase that follows the year is something that will remind you of why the paper matters to you; this might or might not be a condensed version of the title. You may want to enter the paper into a database/citation system; this makes it easier to enter the citation when you're writing your own paper that will reference it. (Zotero.org offers such a system for free.)

If you think there's a good chance you'll want to reference the paper, then as soon as you're done reading it, you should write a sample sentence or short paragraph that you might use to reference the paper. Either save this in your database or put it into a document that lists all such sample sentences and includes a reference section at the end. This requires a small extra effort at the time of reading the paper but saves an enormous amount of time when you are writing your own paper and gathering references.

2.3 Experimental Design

When designing an experimental study, the most important thing is the "research question," i.e., what question or small number of questions do you wish to answer? The focus brought by having a clear research question in mind makes your work much more efficient. The research question should be rather tightly focused, preferably quantitative, and preferably tied to a theoretical prediction.

For example, "What happens when you create a plasma from hydrogen gas?*" is much too vague. "What does the emission spectrum of hydrogen plasma look like?" is better. However, a more focused and quantitative version would be even better, such as: "For hydrogen plasma, what wavelengths of visible light are emitted?" Finally, by tying to a theoretical prediction, we can make an excellent research question: "For hydrogen plasma, are the wavelengths of visible light that are emitted the same as predicted by the Bohr model?"

In research, you are often exploring phenomena for which there is no theory. In those cases, it is of course fine not to connect the research question to theory. It's still important in such exploratory research to have a rather focused research question, but also to notice unexpected effects, and be ready to follow up on them.

Many experiments consist of varying one quantity (call it x), and measuring a second quantity (y). For example, one might observe a beam of electrons in a vacuum tube traveling in a circle under the influence of a magnetic field, and measure the radius of the circle as a function of the strength of the magnetic field. In such experiments, often you want to determine the functional dependence of y on x, e.g., does the data follow $y = Ax^2$ or $y = Bx^3$? If the functional dependence is already known, you often want to determine the coefficient involved (A or B in our examples). For both purposes, you want to extend the range of the data you take as far as possible. For example, as shown in Figure 2.1, if you only take data from $x = 0$ to $x = 0.5$, it would be hard to tell the difference between the two possible behaviors, whereas if you extend the range up to $x = 1$, the difference is obvious.

---

* It's fairly easy to create a plasma (in which one or more electrons are separated from each atom, creating a cloud of electrons and ions) by applying a large voltage across a glass tube filled with low-pressure gas.

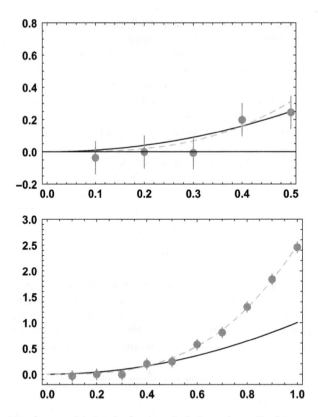

FIGURE 2.1   A comparison of two possible theories for a hypothetical experiment. For the range $x=0$ to $x=0.5$, the difference between the two is small, and so it might be hard to distinguish between them experimentally. However, by extending the range of x up to 1, the difference becomes obvious.

So, as an experimental physicist, you should always ask, "What limits how high I can go in x? What limits how low I can go? Are there any easy ways around those limits?" Often, the limits come from the available instruments. In that case, you might ask, "Can I get other instruments?" If not, you should determine whether it is safe to push the instruments you have as far as they can go. There can be safety concerns for you as the experimenter, for other instruments that are being used in the experiment, and for the sample you're measuring. Try to find out as best you can what these real limits are.

Once you have determined those real maximum and minimum limits, they should be the first two data points you take, since they will usually show the range of behavior.

There is a design cycle in experimental physics. You try an experiment and get some preliminary results. Based on these, you revise your experimental apparatus or technique to get more reliable or more precise results, or to investigate something unexpected that you found. You do another experiment, get more results, then refine further.

It's important to go through this cycle quickly, so that you can get a lot of iterations in the time available and get to the point where you're taking really good data. (Of course, don't go so quickly that you compromise safety or your understanding of what's going on!) When possible, do multiple experiments in parallel so as to speed things up.

In y vs. x experiments, you may ask, "How many data points should I take?" It wastes time to take a huge number, but if you take too few you may miss something important. If you can run through the maximum range of x fairly quickly, do so, and observe the variation of y qualitatively. Then, cluster your points in the regions where y varies the quickest.

The actual number of points to take depends on how long it takes to acquire each point, on how much noise there is, and on whether the process is automated. If it takes only a few seconds per point, and the process is automated (so that you don't have to manually take each point) then usually about 200 points will work well. If you have to manually take each point, and/or it takes 30 seconds to a couple of minutes per point, and your noise level is low (compared with actual variation of y with x), then take about eight data points per hump or dip. (If you were taking data for a research article and know that this particular figure will be part of that article, you might take somewhat more.)

If the functional form is completely known, see if you can rearrange it into a form that will yield a linear plot. For example, if $y = Ax^2$, then a plot of $\sqrt{y}$ vs. x gives a straight line. If you can make such a linear arrangement, and the only thing you need to determine is the coefficient, then really you only need two data points. (Of course, you will space them as far apart as possible!)

When you change x and observe a change in y, you are always also changing something else: time. This can affect your experiment in many ways, such as oxidation of your sample, slow drift in one of your instruments, and changes in room temperature. So, to be sure the observed variation in y is really due to your intentional variation of x and not to something else, when you complete your variation of x (e.g., starting at the lowest possible value and going up to the highest possible value), you should go back to the first value of x and re-take that data point. This point also applies to more qualitative experiments. For example, if turning off the room light apparently causes your measured voltage to decrease dramatically, you need to turn the light back on to make sure the voltage goes back up before drawing any conclusions.

Think carefully about whether your experiment requires one or more "controls" to be convincing. A control is an experiment that is run specifically to check that the intentional variation of one variable is really what caused the observed change in the measurement. The example above of going back to check the original value of x is a simple type of control. Often, more involved controls are called for. For example, if you're testing solar cell materials, and introducing a small amount of selenium into the sample apparently improves efficiency, you need to do a control in which the sample is treated in exactly the same way (e.g., being immersed in a beaker of water and then dried), but without the selenium. The control is just as important as the main experiment, since without it the main experiment is not compelling. (Many experiments in physics do not require controls that are more involved than going back to the original value of x, or simply repeating the experiment. In the example of measuring the emission spectrum of hydrogen plasma, no other control is called for.)

2.4 Modeling

The development of "models" is an essential component of every experiment. A model is a simplified description that captures what you believe to be the important behavior of the system you're studying, but ignores what you believe to be less important aspects. A good model helps you to understand the system and makes testable predictions. The model is usually represented by one or more equations, diagrams, and/or computer programs.

You are familiar with theoretical modeling from your previous physics courses, though you may not have used that name for it. When you describe the trajectory of a thrown ball, you usually treat it as a point particle. That's a model – really, the ball is made from a gigantic number of atoms, each of which is vibrating due to thermal motion. However, we can usually neglect those facts if we only care about the motion of the center of mass.

To understand the outcome of an experiment, you need to model both the *physical system* you're studying, and the *apparatus* you're using to study it*.

In some cases, the theoretical model for the physical system is given in the lab manual, or in the case of research is given by previous theory papers. In other cases, you need to develop the model for the physical system on your own, inspired by the results of your experiments. Such models are normally based on accepted physics principles, such as Maxwell's equations and Ohm's Law.

You already know that to understand a system deeply, you need a theoretical model, and that this must be developed in an iterative process: experiments are performed, their results are compared with the

---

* Many ideas in this discussion are drawn from B. M. Zwickl, D. Hu, N. Finklestein, and H. W. Lewandowski, "Model-based reasoning in the physics laboratory: Framework and initial results," *Phys. Rev. ST Phys. Educ. Res.* **11**, 020113 (2015).

theory, and in some cases this leads to changes in the theory. You also know that theoretical models are based on simplifying assumptions, and that to understand the regime over which the theory is applicable, you need to know those assumptions.

The idea of modeling the apparatus may be new. However, it's just as important as the modeling of the physical system if you are to really understand your results. In fact, you've been doing apparatus modeling to some extent without thinking about it. For example, when you use a voltmeter, you assume that it's correctly calibrated, so that when it displays a reading of 1.000 V, the input voltage really is 1.000 V. However, in fact the calibration may have drifted somewhat since the last time it was checked (probably when the voltmeter was manufactured), so that a reading of 1.000 V corresponds to an actual voltage of 0.991 V. For many applications, the model is good enough, but it's important to consider explicitly whether it is adequate for your particular experiment.

As another example, consider a signal generator, a device that creates a sinusoidal voltage, with frequency and amplitude controlled by knobs. You might assume that if you leave the amplitude knob at a fixed setting, then the amplitude will be constant even when you vary the frequency. This is a model, not the full reality. In fact, for most signal generators there is a small frequency dependence to the amplitude, but for many applications it is so small that it can be ignored.

It's better to be more intentional about your apparatus modeling, instead of doing it subconsciously as in the above examples. Being explicit will help you to realize the limits of the model. It is absolutely essential that you deeply understand your instruments. Another way of phrasing this is that you need to develop a model for your instruments that is accurate under your experimental conditions. Without such understanding, it's inevitable that you will misinterpret your experiment. This is more important in physics and astronomy than in other sciences, because physics researchers are always pushing the limits of their instruments.

As a final basic example, say you are measuring how the brightness of an LED depends on the current flowing through it. You use a sensor that measures the power of light that strikes it, and position it close to the LED. You then begin varying the current through the LED, starting at 0, and going to a maximum of 20 mA. As you increase the current up to 13 mA, the power increases steadily from 0 up to 1.07 mW. As you go higher in current, the power remains constant at 1.07 mW. If you were very naive, you would conclude: "The more current that flows, the more light is produced, until we reach the maximum amount of power the LED can produce, 1.07 mW." However, if you understood the power meter more deeply, you would realize that it has a maximum power it can read, which is nominally 1 mW. (The actual maximum is often slightly higher than the nominal value.) So, as you increased the current beyond 13 mA, the power really was increasing, but the power meter had "saturated" at its maximum value.

The starting point for developing a model for each instrument in your apparatus is the manual or datasheet. Whenever possible, you should read the manual*. This is not always practical in the context of a course, but it's quite important for research.

You need to understand how your instrument departs from ideal behavior. For example, the output voltage of every instrument ever made has a "zero offset," i.e., a small non-zero voltage that is added to the output, so that when the instrument should be reading 0 V it reads (say) 3.4 mV, and when it should be reading 1 V it reads 1.0034 V. Often, the manual specifies how big such departures from the ideal are likely to be, but often you need to do a few simple experiments to characterize them yourself.

Often, your model may consist of ignoring these departures, but it's important to remember that this is a model, and there will be some circumstances in which it's not justified. As with theoretical models, be sure you understand the simplifying assumptions that your model is based on, and the resulting range of circumstances over which the model is valid.

As with the theoretical model, the apparatus model should be developed iteratively. You do an experiment, and use your apparatus model to interpret the results, and then compare them with predictions of the theoretical model of the physical system. If the comparison yields a discrepancy outside your

---

* The manual can often be found easily on the internet by typing in the manufacturer, model number, and the word "manual." It's fine to skip the parts having to do with programming or remote control, unless you're using those functions.

uncertainty, then you modify the model of the physical system, or the apparatus model, or both. Often, you'll realize that you need to gain a deeper understanding of the background theory, or a deeper understanding of your apparatus. The latter often requires carrying out simple experiments to characterize the parts of your apparatus.

2.5 Important Guidelines for Conducting Experiments

Preparation

Before each lab, make sure you've completed the pre-lab reading, read through the lab itself, and answered all the pre-lab questions. You will not be able to complete the lab in the time allotted without this preparation.

Safety

Before you actually begin the experiment, be certain you understand all relevant safety precautions. **If you have any questions at all about the safety aspects, do not proceed without consulting with your instructor.** (The same goes for research and in fact all situations; make sure always to pause before doing anything that is at all hazardous and think through the precautions you're taking. Consult with more experienced people if you have any doubts at all.)

Pilot Testing

It's almost always a good idea to run through an experiment quickly first, to get an idea of what's going to happen, and make sure your equipment is adjusted correctly and working the way you expect. This is called "pilot testing." After this quick check, you will go back and take your "real" data more carefully. In many experiments, you're changing one thing (such as the temperature) and measuring the resulting change in something else (such as the pressure). We'll refer to the thing you change as the "independent variable" and the thing you measure as the "dependent variable." In the pilot test, you should go to the extremes of the independent variable, to get an idea of what will happen and the appropriate settings for your equipment.

Pilot testing often reveals flaws in your experimental design. For example, perhaps you observe in pilot testing that the quantity being measured shows no appreciable change between the two extremes. This might indicate that you're doing something wrong, or that you need to change things so that you can vary between much wider extremes of the variable you're controlling.

Taking Data

The phrase "taking data" evokes the idea of changing an independent variable and measuring a dependent variable. However, it also includes single measurements, such as measuring the value of a resistance. Think about what you're doing. Take care with the data taking; it really makes a difference! Plot your data as you take it or immediately after; this allows you to determine if something is going wrong while there's time to fix it easily. Think carefully about all the ancillary measurements you need. For example, do you need the mass of your sample? The room temperature? The exact value of a particular resistance? Think about how you'll do the uncertainty analysis and what you'll need for that.

2.6 Lab Notebooks

Once you've completed pilot testing and are ready to take real data, it's essential that you have a system prepared to record it and to keep it well organized. Set yourself a goal for this course to learn how to keep a good lab notebook. This skill is helpful in many areas of life, but is absolutely indispensable for serious science. It's not easy. It takes time, and it's tempting to fool yourself into thinking you don't need to be so scrupulous, because you'll remember everything you need when it comes time to write up your results. However, the fact is that human memory is much less precise than it appears. You're a very busy person, and details that really are important for interpreting your experiment will slip from your mind when you get distracted by other work.

Even though the notebook you create for this course may be used only for yourself, you should develop the habit of writing it so that a competent physicist who has never met you can fully understand what you've done by reading your notebook. This *doesn't* mean that you need to write down procedures in your notebook that are clearly recorded elsewhere (such as this book), but you do need to include clear references to such procedures.

You should illustrate your notebook copiously. The entry for every experiment should include a schematic diagram for the apparatus, unless such a diagram is included in the procedure you've referenced, or unless it would be absolutely clear to any competent physicist how the apparatus is constructed, and how things are connected together. Note that photographs don't fill this role, because critical details are often obscure, and unimportant features are often distracting. When you create a schematic diagram, you decide what should be included. This gives a more comprehensible picture, and also enhances your understanding.

Do use photographs where appropriate. For example, to record a rough version of a graph, sometimes it's easier to take a photo of the computer screen and paste that into your notebook than to use the computer to do a screen capture. (For completeness: Alt-PrtSc copies the active window to the clipboard on Windows computers. Control-Shift-Command-3 copies the entire screen to the clipboard on Macs.) Of course, you will also save the data for the graph. It's also nice to include occasional photos of yourself with your apparatus.

Choose names for data and image files thoughtfully, so that they will remind you of the contents while not being too long. Include full details in your notebook about the filenames and where each file is stored. It is smart to put a box around each filename entry in your notebook, to make it easy to find later. You should include a brief description of what each file contains, e.g., "Voltage vs. time for decade box resistor, with 3 ohm increase in the middle. Surprisingly high noise level after the increase."

One of the most difficult habits to keep, and also one of the most important, is to paste all the graphs and images you take into your notebook. This takes an extra step, and you'll be tempted to skip it because you know the file is available on the computer, so you can access it any time you want. However, if you don't do the pasting on the same day you take the data, you'll never actually get around to it, and as a result, your notebook will be of little value.

Your instructor may direct you to keep your notebook in either paper or electronic form. Each has advantages and disadvantages. It is easier to insert computer-generated graphs and images into an electronic notebook, easier to make a back-up copy, easier to share with your partners, and easier to search for a particular phrase or filename. It's harder to have it handy at all times (since you really need a laptop computer or at least a tablet), and perhaps harder to insert hand-drawn figures. (If you have a touch screen, you can create hand-drawn figures directly on your computer. If not, you can take a photo of such a figure and paste the photo.) Although not relevant for this course, in later research you may be required to use either a paper notebook or special software to document priority for patent matters.

If you work with partners, you should decide whether each of you will keep your own notebook, or instead if one person will act as record keeper. If the latter, make sure to rotate from one lab session to another, so that everyone gets practice.

Please visit the entry for this section on ExpPhys.com for examples of good and bad notebooks.

2.7 Troubleshooting

Walter F. Smith and Melissa Eblen-Zayas

It's better to avoid trouble than to have to troubleshoot. *The key to avoiding trouble with an experimental apparatus is to test each stage as you assemble the apparatus*, rather than building the whole thing first and then hoping it will work*. However, even if you're careful to test as you build, things may stop working at some point after the construction is complete. One of the reasons that so many companies are happy to hire experimental physicists is that they are expert troubleshooters.

When the apparatus doesn't work, the instinct of most students is to carefully check each part, including each wire for the case of a circuit, against a diagram that shows how things were meant to be set up. This is a poor way to start. Not only will it take longer than the strategies described below, but it will result in little or no learning. In contrast, using the following strategies will make you a stronger experimentalist, not only helping you to fix the current problem, but also giving you understanding that will improve your designs of future experiments.

There are three good strategies for troubleshooting: split-in-half, topographic, and function/discrepancy[†].

In the split-in-half approach, you divide the problem in half, and test each part. Once you determine which half is responsible for the issue, you divide in half again, proceeding in this way until the particular source becomes clear. *This is usually the most efficient approach*, and it applies well to all areas of life, not just physics experiments.

Let's consider the example of a circuit that is supposed to count pulses from a subatomic particle detector; a schematic is shown in Figure 2.2. The detector emits a voltage pulse each time a particle is detected, but the pulses are low in amplitude and very short in duration. So, the next stage amplifies each pulse, and the one after that stretches it out in time. Next, the pulse goes through a "discriminator," which determines if the pulse is large enough to be counted as a "real" event; if it is, the discriminator sends a cleaned-up square-shaped pulse to a "counter" chip, which increments the count at its output. The output of the counter is shown on an LED display.

You built this apparatus on a breadboard, using integrated circuits ("chips") for each of the stages. It had been working fine, but after moving it to a new location, the display always reads zero counts. Following the split-in-half approach, you begin by connecting an oscilloscope ("scope") to point A in the figure. The voltage there is always zero, so the problem is probably in the components to the left of point A. (Note that so far you didn't need to literally divide the circuit into two separate circuits by disconnecting the output of the stretcher from the input of the counter; instead, you mentally divided the problem.) Next, you connect your scope to point B, and you do see pulses. So, the pulse stretcher could be the problem. However, instead, the problem could be with the input of the discriminator; if that input is shorted to ground, then even if the pulse stretcher is working correctly, the voltage at point A would be zero. So, to divide the problem in two again, you remove the connection between the pulse stretcher

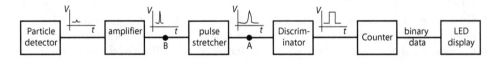

FIGURE 2.2   A hypothetical circuit for a particle detector.

---

* I have to relate a story from my days in graduate school to support this edict. When I first started in my graduate research group, there was an undergraduate who'd been assigned the task of building a complex piece of electronics to create various pulse sequences. Instead of soldering, he'd chosen to use "wire wrapping" as a way of connecting the components together. In this technique, you use a special tool to tightly wrap the wire around a square cross-section post. The tool automatically strips the insulation off the wire, and the sharp corners of the post bite into the wire, making a good electrical connection. I noticed that the more senior grad students would ask him each day how it was going, and whether he had tested what he'd built so far. He would always reply, "I'm being really careful. I'm sure it's all correct." In fact, he wasn't testing each stage. Finally, the big day came. He had applied nicely printed labels to all the front panel controls. He called us all in to view the instrument as he powered it up for the first time. He flipped the power switch and nothing happened! Eventually, he figured out that the wire wrapping tool he had used was defective, and hadn't been properly stripping the insulation off the wires. Every one of the hundreds of connections he'd made was bad. At that point, it was the end of the academic year, and he was graduating. No one else wanted to re-do his work, so we just discarded the instrument he'd worked on for months.

† D. H. Jonassen, and W. Hung, "Learning to troubleshoot: A new theory-based design architecture," *Educ. Psychol. Rev.*, **18**, 77–114 (2006).

output and the discriminator input. You still see a zero voltage at the pulse stretcher output, so the problem is with the pulse stretcher. Again, you divide the problem in two: is the problem with the wiring of the pulse stretcher chip (probably), or with the chip itself? The first thing to check on the wiring is the power supply connections. So, you touch your scope probe to the two power supply pins to verify that each is at the right voltage, and you find that the one that should be at +15 V is actually at about 0 V. From there, you quickly find that the far end of the wire that carries that voltage popped out of the breadboard when you moved the apparatus.

Note that in some cases, the problem is caused by the *interaction* between two parts of your apparatus, rather than by one of the parts. In the above example, perhaps the pulse stretcher can only provide 1 mA of output current, but the discriminator requires 5 mA of input current. Both of them work on their own, but they won't work when you connect them together.*

Also, be aware of the effect of your test equipment on your apparatus. For example, your scope requires a very small but non-zero current at the input to make measurements. In most cases this can be ignored, but not if you're testing a stage that can only provide a very small output current. (For the special case of circuitry, problems caused by interactions or the effect of test equipment are best understood using the ideas of input and output impedance; see Sections 6.2 and 6.3, and Lab 6A.)

We've used an example of a circuit, but this same approach can be applied to optics (e.g., checking the beam halfway through the apparatus), to computer code (checking the output of the middle subroutine in a sequence of subroutines), or really to anything.

The second troubleshooting strategy is topographic. This works best for linear chains of instruments, such as the example above, where something is passed from one unit to the next along a single path. In this troubleshooting approach, you either trace forward one stage at a time in the path starting from a point where things are working, or you trace back starting from a point where things aren't working. This requires a bit less thinking than the split-in-half approach, so often once you've used that to narrow things down to a small range, you switch to the topographic approach.

The final approach is function/discrepancy detection. This can be applied in situations where the split-in-half approach is difficult to use, perhaps because it is not easy to separate the experimental set-up into two parts. The idea is to find a mismatch between expected operation and actual behavior, and then fix it. There are four key steps[†] in this troubleshooting approach:

1. **Formulate the problem description.** Begin by cataloging what the system does wrong and what it does right. Make measurements that define the nature and scope of the problem. Are your mental models correct? At what points is the structure/function relationship what you expect and where is it not?

2. **Generate causes.** Once you've identified the symptoms, think about things that could give rise to them. (Or is the problem with the model that underlies your theoretical understanding? In that case, the apparatus might be working fine, but the results will not match your expectation.)

3. **Test.** If the faults are due to apparatus problems, and not with your theoretical understanding, then what measurements can you make to decide which of the problem sources from step 2 is the right one?

4. **Repair and evaluate.** Make changes to the system based on the diagnostic results, and then check if it's fixed. If not, was there a misdiagnosis of the original fault? Or were there multiple faults? If the initial repair doesn't resolve the issue, repeat the process from the beginning.

Although the four steps might imply a linear path to troubleshooting, the process often includes cycling back to previous steps depending on what you learn along the way.

---

* In this case, the circuit would never have worked; you would have noticed this problem as soon as you added the discriminator.
† Dimitri R. Dounas-Frazer, Kevin L. Van De Bogart, MacKenzie R. Stetzer, and H. J. Lewandowski, "Investigating the role of model-based reasoning while troubleshooting an electric circuit," *Phys. Rev. Phys. Educ. Res.*, **12**, 010137 (2016). https://doi.org/10.1103/PhysRevPhysEducRes.12.010137

A final piece of advice for the special case of troubleshooting circuits: if things are totally screwy, and literally don't add up, the problem is probably with grounding. Make sure all parts of your circuit are connected to the same ground. Be aware that some instruments, such as voltmeters, have a "floating ground," meaning that the negative terminal is not connected to the ground of the AC power line. Sometimes this is desirable, e.g., for measuring the voltage difference across a component, but sometimes it causes problems, e.g., if you use a signal generator with a floating ground and fail to connect that ground to the ground of your scope. (See Section 6.6 for more details about grounds.)

3

Presenting Your Results

Walter F. Smith

CONTENTS

The thing I find most compelling is when an experiment has seen something interesting, and I want to figure out: did they make a mistake, or is nature actually telling us something new?

– Janet Conrad (particle physicist at MIT)*

3.1 The Process of Scientific Communication

There are three main types of formal scientific communication: papers, talks, and posters. Each serves a distinct purpose in disseminating, checking, and improving research.

In physics, the main record of progress in research and the accepted standard for priority of discovery are papers published in scientific journals.† A scientist's record of journal publications is a central part of decisions in hiring, promotions, and grant awards. Different journals carry different levels of prestige; generally, the broader the audience, the greater the prestige. The journals *Nature* and *Science* are usually recognized as the highest prestige journals for physics, in part because they are read by scientists in all disciplines. The next tier includes *Physical Review Letters* and *Nature Physics* (both covering all areas of physics), as well as *Proceedings of the National Academy of Sciences* (covering all science disciplines). Next are journals with a more specialized focus, too numerous to list here. Many scientists prefer to publish in journals sponsored by a national or international professional society (e.g., the various flavors of *Physical Review*, which are published by the American Physical Society), as these are subscribed to by the libraries of most colleges and universities.

For many scientists, free public access to the publication is important, since this reduces inequities in access based on economic means, and since most research is supported by tax dollars. On the other hand, creating a journal requires a great deal of effort, and so requires money. In some cases, the costs

---

* As quoted in "MIT News," November/December 2018, p. 14.

† In many subfields, pre-publication versions are posted on arxiv.org, and in some subfields the papers posted there are more widely read than the final version that is published in a journal. Be aware that papers on arxiv.org are not yet peer reviewed, and so are more likely to have omissions or errors, occasionally significant ones. However, in many subfields, it's the best place to get information about the bleeding edge of research.

are absorbed at least partially by the scientists who publish in the journal, but this has its own ethical issues. The situation regarding free access journals is changing rapidly.

When a paper is submitted to a journal, one of the editors makes a decision as to whether it is suitable for peer review. (For most journals, the strong majority of papers received are sent out for review.) The paper (officially called a "manuscript" at this stage) is then sent to one, two, or three scientists with expertise in a relevant subfield. These "referees" carefully review the manuscript for the quality and importance of the science it contains, as well as the clarity of presentation. In some cases, they recommend rejection. In others, they give a list of required improvements. These lists are sent back to the authors, with the referees' identities kept secret. The authors revise the paper, and resubmit, along with a letter in which they may dispute some of the requirements of the referees. If the authors feel the referees have been truly unfair, they can request new referees, but this is rarely done. The referees examine the revised paper and make a final recommendation as to whether it should be published. Peer review is an important mark of approval from the broader research community.

Talks serve primarily as a way of bringing your research to the attention of other scientists. They can be critically important for building your reputation and for networking. In some cases, authors present results in talks that are quite new, and may not yet quite be ready for publication in a paper. Talks come in two variants: contributed and invited. For a contributed talk, you typically submit the abstract to a conference. Depending on the conference, your abstract may or may not be peer reviewed, and then may or may not be approved for inclusion in the conference. Contributed talks themselves are much lower prestige than published papers. However, for some conferences if the talk is accepted, then the authors are expected to submit a paper that will be published in the conference proceedings; these are often published as part of a regular journal.

Scientists who have recently made exciting discoveries may be invited to give a talk at a conference and/or to give a talk at a colloquium or seminar series at a college or university. Invited talks help to boost the reputations of the invited scholars.

Posters are usually not peer reviewed. However, they give much more opportunity for one-on-one interactions than talks do, and so many consider them to be more useful for networking. Further, you are likely to get more feedback, and feedback that is more frank, than from a talk. It's typical to get some important ideas to improve or expand your research when you present a poster. When you give a talk, you usually only get those ideas during conversations after the talk.

3.2 Data Visualization

Graphs

Graphs are the main way that scientists tell stories. So, when you prepare a graph, put a lot of thought into it. Think carefully about the axis scales. In some cases, in order to tell the full story of the data, you may need to present the same data twice, with the axes scaled in two different ways. (Usually this is done by including a small graph as an inset to the main graph.)

Frequently, you need to compare two datasets. It's usually better to do this by putting them on the same graph, rather than on stacked graphs (see Figure 3.1). Distinguish between the different datasets by notations on the graph, rather than in the caption; it is strongly preferable to do so by labeling the curves directly with color-coded labels, rather than by using a key, as shown in Figure 3.1. (However, when there are a large number of curves, you may need to use a key to make things clear, and in rare cases it may be necessary to explain the labeling in the caption.)

Frequently, you need to present two datasets that share the same horizontal axis, but have different vertical axes. Use the left vertical axis for one dataset, and the right for the other, with color coding of the axes and arrows near the datasets to indicate which axis goes with which dataset; see Figure 3.2. (Also specify the axes in the caption.)

Use arrows and other annotations on the graph to clarify things as needed, without ever obscuring the data, as shown in Figures 3.1 and 3.2. The more your reader can understand from the graph itself, rather than having to consult the caption or the main text, the bigger the impact will be. See the entry for this section on ExpPhys.com for more examples.

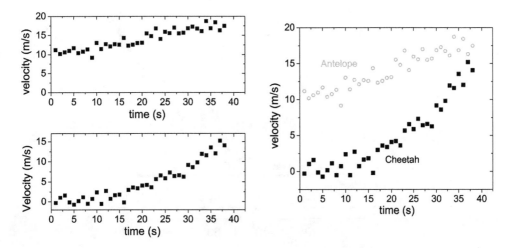

FIGURE 3.1   It is usually better to present datasets that are being compared in a single graph (right) rather than stacked graphs (left).

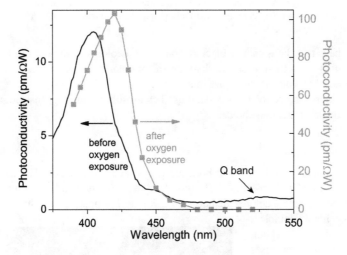

FIGURE 3.2   Example of a graph with two different vertical axes. The caption might be, "The black trace (left axis) shows the photoconductivity before oxygen treatment, and the gray (blue in ebook) trace (right axis) shows photoconductivity after treatment."

When comparing data and theory, use points for the data and a solid line for the theory, as shown in Figure 3.3. Also, you should ordinarily include a residuals plot (data – theory) at the bottom, as shown in Figure 3.3. This highlights whether there are any systematic deviations between theory and experiment.

Unfortunately, the default font size and line weight in most graphing software is too small. The smallest lettering in your graph (usually the tick labels) should have a height of about 3.5% of the overall height of the graph. Plot points and curves should have a thickness of about 2% of the overall graph height. Make the thickness of the axes and ticks nice and bold; typically, this means making the thickness twice the default. Making your fonts and symbols this large ensures that you can use the same figure for papers, talks, and posters.

If you use color, make sure you're consistent between figures. For example, if you use black for sample A and red for sample B, follow that convention for all your figures.

Bear in mind that 8% of men and 0.5% of women have some form of color blindness, most commonly a reduced ability to distinguish reds, greens, browns, and oranges. So, you should use other elements in addition to color to distinguish different datasets (e.g., different plotting symbols or line dash styles). If you search for "Coblis" (COlor BLIndness Simulator) on the internet, you can view a jpg or png version of your graphic as it would be seen by a color blind person.

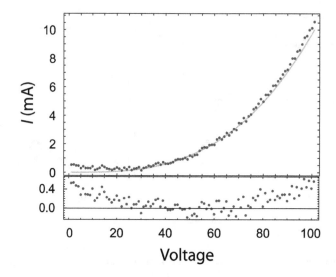

FIGURE 3.3   One common way to present residuals. The main graph is at the top, and the residuals below. Although the theory (solid line) appears to be a pretty good fit in the main graph, the residuals show a parabolic deviation from zero, indicating that the theory is incomplete.

For graphs that have two datasets, use black for one, and a color for the other, rather than using two different colors. This makes them easier to distinguish for color blind people, and also helps with the overall color scheme when you use the figure in a poster.

Avoid using jpg figures, since the compression used can result in blurring. Instead, use eps format if possible, with png being the second choice.

Images

Images *must always* include indications of both the lateral and color scales. The lateral scale can be indicated either by a scale bar (somewhat preferred) or in the caption (e.g., "500 nm image of …"). The color scale is best indicated by a color scale bar used as an inset or presented to the side of the image, as shown in Figure 3.4. It's acceptable to indicate the color scale in the caption, e.g., "z-scale: 50 nm." In a few cases, the color is not quantitatively important, and so the scale can be omitted; examples include photographs and scanning electron micrographs.

Diagrams

For papers written for a course, you should include at least one diagram of your apparatus. Diagrams convey the key information much more clearly than photos, which always include extraneous objects, and for which it's hard to get the lighting correct.

Diagrams are often used, especially in talks and posters, to convey experimental processes. When possible, coordinate the colors used in the diagram with colors used for graphs (see

FIGURE 3.4   Image with lateral scale bar (lower left) and color scale bar (upper right).

example on ExpPhys.com.) As with graphs, the smallest text on a diagram should have a height that's about 3.5% that of the diagram.

3.3 Writing Scientific Papers

It's best to start the process of writing a paper by preparing the figures. These ordinarily convey the key results, and so you can build your paper around them. Then, make an outline, or at least a list of the main points you wish to make in presenting your results.

Papers follow a standard format, which has evolved for the most efficient communication. Each section plays a key role:

- **Title.** This should clearly convey what the paper is about. Avoid catchiness for its own sake. For example, "Photoconductivity of Self-Assembled Porphyrin Nanorods" is clear (at least if you understand the meaning of each word), while "Astonishing Photoelectronic Effects in Self-Assembled Organic Nanowires" is too hyperbolic and less informative.
- **Author list.** A list of everyone who made a substantial contribution to the work described, with the institution at which they work. In physics, the first author is typically the student or postdoctoral fellow who did the main experiments, while the last author is the leader of the research group. (In particle physics, where the groups are larger, authors are usually listed in alphabetical order.)
- **Abstract.** This is a summary of your paper, typically 200 words or less.
  - Start with a sentence or two describing what you investigated, and why it is important. Example: "Photoelectronically active nanostructures that are self-assembled from organic molecules hold the promise of tailored functionality with simple and inexpensive production. Comparison of nanowires assembled from related compounds can give important insights into the details of self-assembly and the conduction mechanisms."
  - In the next sentence or two, present the specific question you investigated. Example: "We report the photoconductivity of nanofibers that are self-assembled from a porphyrin with long alkane substituents, one of which is attached to a chiral center."
  - Next, describe your methods in one to three sentences. Example: "Nanofibers were deposited from solution onto substrates with interdigitated electrodes, then illuminated with a wavelength matching the absorption peak. Conductivity was measured as a function of illumination intensity and voltage."
  - Then, describe your principle results in two or three sentences, including the most important quantitative findings, with uncertainty. Example: "In contrast to previously studied porphyrin nanowires, the photoconductivity increases as atmospheric O_2 is increased, at a rate of 2.3 ± 0.2 nS m^2 kW per percent O_2. This can be explained using the same model used in the previous studies, by assuming a different lineup of the bands of the nanofilaments with the electrode Fermi level, leading to a different height of Schottky barriers."
  - Conclude the abstract with a sentence or two that explains the significance of the findings in a broader context. Example: "Schottky barriers at the interface between organic nanostructures and electrodes strongly affect conduction and photoconduction, and are strongly influenced by atmospheric gases such as O_2."
  - The abstract should not include references and is usually written in the third person.
- **Introduction.** This should begin with the "motivation," which explains (in more detail than the abstract) what the general area of the article is and why it's important. This is followed by the "background," which summarizes the previously published papers that are most closely related to the current work and clarifies the state of knowledge prior to the previous work. For papers written for a course, the background should include the most important figures from the previously published papers. The introduction finishes with a clear statement of the "research question," i.e., the main question that you set out to answer. However, you should phrase it as a statement. For example, "Here we report measurements on the effect of atmospheric O_2 on the

electronic properties of nanostructures self-assembled from chiral porphyrins." (In some cases, there may be two, or rarely three, research questions.)

- **Methods.** Describe the experimental or computational methods in enough detail that a competent scientist in the same subfield would be able to reproduce the experiment. Include the manufacturer and model number of all major equipment. Include the supplier and grade for any chemicals or other materials used. For papers written for a course, this should always include at least one diagram of your apparatus. Computer programs you've written should be described only briefly; for a paper written for a course, you should include a complete listing of the program as an appendix. (For a LabView program, include screen images of both the front panel and block diagram for the main program, and also for any sub-VIs you've written.)

- **Results and discussion.** Clearly present your results, and discuss what they mean, including quantitative comparison with expectations, using uncertainties (see Chapter 4). You need to walk your reader through all the results in a coherent way. Sometimes it's better to divide this section into separate Results and Discussion sections. This works well if your results can all be explained by a single framework, which you present in the Discussion section. At other times, it's better to combine into a single section, in which you present a result, discuss its significance, then move on to the next result.

  Figures are the most important part of this section. Every figure must have a clear, but brief, caption, which can mostly be understood without reading the main text; most readers will start by reading the abstract and then looking at the figures. Every figure must be referred to in the main text, e.g., "As shown in Figure 7, ..."; it's okay if there is some duplication between what is said about a figure in the main text and the caption.

- **Conclusion.** Here you should reinforce the most important findings that were presented in the Results and Discussion, and discuss their broader significance. It is fine if you wind up essentially repeating things you said in earlier sections. Most readers will read the conclusion after the abstract and figures, and before looking at earlier sections. If this is for a course, include a "future directions" subsection, in which you discuss ways the experiment could be improved, as well as additional avenues that would be worth exploring.

- **Acknowledgments.** Thank anyone who made a contribution that was important, but not significant enough to justify listing them as a co-author. For example, you might thank someone who explained a theoretical idea, a person who characterized some aspect of your sample with only a modest effort on their part, or someone (such as a machinist) who made an important suggestion regarding your apparatus. Also, if your work was funded by an external agency, you should include an acknowledgement with the grant number.

- **References.** List the papers referred to in your main text. If you're submitting to a journal, you should look at other papers published there for the citation format. For papers written for a course, I recommend the following:
  - Example of the reference given in the main text: "Zhou and co-authors (Zhou 2003) explain how to..."
  - Example of the corresponding citation at the end of the paper: Y. X. Zhou, A. T. Johnson, J. Hone, and W. F. Smith, "Simple fabrication of molecular circuits by shadow mask evaporation," *Nano Lett.* **3**, 1371–1374 (2003).
  - The volume number is in boldface, followed by the range of page numbers. If you are using this format, then the references at the end of the paper should be listed in alphabetical order by the last name of the first author.

- **Division of responsibility.** (Usually only included for papers written for courses.) If you have one or more co-authors, you should clarify who did what, including in the writing of the paper.

- **Appendices.** Here, you present details that are important, but are too detailed to be of interest to most readers. If you're writing a paper for a course, this is where you should present the details of your uncertainty calculations. Ask you professor whether they prefer for these to be done in a symbolic algebra/calculus program such as Mathematica, or in another format instead.

Some Nitty-Gritty Reminders

- Plan your work so that you'll be done writing the paper three days before it is due. Then, revise it at least twice before considering it ready for submission.
- Be as brief as you can.
- You should use first person when describing what you did. If you worked alone, it is fine to use "I."
- Use past tense to describe what you or others have done, and present tense for everything else. Avoid the future and conditional ("would") tenses.
- Each paragraph should have one central idea, and the reader should be able to infer that idea from the first sentence.
- Avoid the phrase, "as such," since it is usually misused; it does not mean "therefore."*
- The verb "affect" means to alter something, e.g., "The light affected the photoconductivity." The verb "effect" means to bring about or accomplish, e.g., "By adding ten points to each score, the professor effected the change in grading policy."
- Use "that" if the sentence would make no sense if what follows were omitted, e.g., "The sample that had been bombarded with ions was an electrical insulator. The sample that had not been bombarded was a conductor." Use "which" if what follows is optional, e.g., "The sample, which had been bombarded with ions, was an electrical insulator." In this example, the "had been bombarded" phrase is a description, where in the first example it is being used to distinguish between two samples. Another way to tell whether to use "that" or "which": if you could put a comma before it, use "which."
- *Every time you write a number, you must include units (unless the number is dimensionless).* You will only be taken seriously as an experimental scientist if you do so.
- Ordinarily, you should include uncertainties for your main experimental results. (Within a course, you may be directed not to calculate uncertainties for some labs, so that you can spend time in other ways. Also, scientists often do not quote uncertainties for ancillary quantities that are not of central importance, instead allowing a rough uncertainty to be inferred from the number of significant digits, e.g., "All measurements were carried out at 22°C.") You should clarify the confidence interval (see Chapter 4) the first time you quote an uncertainty. For example, "The measured voltage was 83.2 ± 0.3 nV. (All quoted uncertainties represent 95% confidence intervals.)" or equivalently, "The measured voltage was 83.2 ± 0.3 nV (95%)."
- It is usually clearer to use prefixes than scientific notation. For example, the above could have been quoted as "$(8.32 \pm 0.03) \times 10^{-8}$ V." This is not wrong, but it's clunky.
- Except in very high precision experiments, quote uncertainties to one significant digit (e.g., 0.3 nV), unless the leading digit of the uncertainty is a "1," in which case you should quote the uncertainty to two significant digits (e.g., ± 0.14 nV).
- Only quote the number of significant digits allowed by your uncertainty. For example, 83.25 ± 0.3 nV would be too many significant digits, while (in a more precise experiment), 83.25 ± 0.14 nV would be correct. If you have not determined the uncertainty, you should quote three significant digits.
- Numbers such as 0.3 should be written with the leading zero; if it were instead written as .3, someone might not notice the decimal point.
- Put a space between the number and the unit: 83.2 nV, not 83.2nV.

---

* The correct meaning is "in the exact sense" or "in that sense." For example, "I am not a tennis player as such, but I can hit the ball." Another example: "Diana was made the leader of the platoon. As such, she began issuing orders."

3.4 Preparing, Delivering, and Listening to Talks

Begin organizing your talk by deciding which figures to include. Your talk should include mostly the same sections as a paper, but will be less detailed. Omit the abstract, the division of responsibility, and the appendices. Consider using alternatives for the standard section titles to enliven things a little. For example, instead of "Methods" you could use "Making the samples," and instead of "Results and Discussion," you could use, "What did we find? What does it mean?" or "Resulting physics," or "Properties of our samples."

For a ten-minute talk, plan eight to ten slides. For a 30-minute talk, plan 20 to 25 slides. For a 60-minute talk, plan about 40 slides. (The longer the talk is, the more questions you will get during the talk, rather than at the end.) Don't use fancy transitions between slides or when you add new elements to a slide; use no transition, or a simple fade in. Avoid putting anything important in the bottom 15% of the slide, since some members of your audience may not be able to see.

Don't use complete sentences on any of your slides, except the one devoted to the research question (see below). Instead, use bullet points. This keeps your audience more engaged, and makes you sound more spontaneous. Make sure there are reminders on the slides about everything you want to say, either in the form of bullet points or figures. This way, you will not need to use separate notes when you give the talk.

The first slide should have the title, authors, and institution(s), and preferably include an image that will engage your audience, such as a microscope image of a sample, or a schematic representation of a possible application; don't use a graph.

The second slide should be devoted to motivation – why is this project important? This is the most critical slide of the whole talk. If you fail to engage audience members here, they won't be much interested in the rest of the talk. You must include a compelling image. If you can't find good images to use for both this and the title slide, this has higher priority.

Devote a slide to the research question, and phrase it in question form. This could come before or after the background summary of the state of the field.

Almost every slide should include an image, graph, or graphic. These graphic elements help you explain things and help to maintain audience focus. When you're summarizing the state of the field, include figures from the papers you're discussing. (Include the reference at the bottom of the slide.) It's okay to have a few slides that are text only, and better to do this than to include silly graphics.

Every time you present an equation, be sure to define every symbol (unless it's been previously defined).

Practice your talk on your own several times, then at least twice with peers. If it's too long, you need to cut material, rather than talking faster.

When presenting your talk, make frequent eye contact with members of the audience. Every time you show a graph, explain the axes, unless they're the same as the previous graph. Try to avoid "um" and "ah"; if you need a brief time to think, just pause. Be aware that the audience will always appear bored, even if they're not – that's the way most people look when their face is relaxed. If you're using a laser pointer, hold it with two hands – it will shake less. Attempt to answer all questions yourself; don't refer questions to your research supervisor. However, don't pretend to understand something that you don't – it's fine to say, "That's a good question, and I'll need to think more about the answer."

Listening to Talks

When you're listening to someone else's talk, take notes. This will keep you much more engaged, make the talk more interesting to you, and help you to remember the talk much better, even if you never refer to the notes again. (In fact, I know one professor who discreetly discards his notes right after each talk!) Challenge yourself to ask a question. This shows the speaker that you have invested yourself in the talk. If you can't think of anything else, a good question often is, "What are the bigger implications of this work?", unless the speaker has already addressed that point.

3.5 Preparing and Presenting Posters

In some ways, posters are a hybrid between talks and papers. They are more formal than talks, and less formal than papers. During the poster session, you will present your poster to one to four people at a time, like giving a mini talk. So, as in a talk, you should emphasize graphics as much as possible, and mostly use bullet points rather than paragraphs. However, it's nice to have a poster that is comprehensible when you're not there, like a paper. For example, in some conferences, the posters are up for long periods, but the authors are only expected to be present for a specified "poster session" time. Also, it is nice after the conference to hang your poster in the hallway, so that casual passersby can get an idea of what you did. So, it's acceptable to use complete sentences when you really need to, such as in figure captions and acknowledgments.

As with papers and talks, you should begin by deciding which figures to include.

The most common error in poster design is using too many words. The poster is a summary of your project, not a complete account. You can explain things in more detail to those who are truly interested. Keep your total word count to 500 or less, including figure captions, but not including acknowledgments and references. Using bullet points instead of complete sentences helps. Don't hesitate to use whitespace around your text; it makes your poster more approachable. Use a minimum of 24-point font size. Use left-aligned rather than justified text, since your text columns will be narrow.

Your poster should have the same sections as a paper (with much less detail), except that you should omit the abstract, the division of responsibility, and the appendices. As with talks, you should consider using alternate names for the standard sections, such as "Sample preparation" instead of "Methods." However, don't feel obliged to do so. Enclose each section in a frame, with the name of the section clearly labeled. Make the "research question," which should be at the end of the introduction, into its own section (with its own frame), make sure it is framed in question form, and put it in a larger font than other text.* (You may also put the research question before the introduction, which makes it more prominent. The disadvantage is that it probably won't be comprehensible until the viewer has gone through the introduction.) It's often helpful to have subsections within each frame, but don't use subframes.

Usually the methods section should be quite short; your audience will be more interested in your results and interpretation. However, it's not uncommon to be at a stage with a project where the main thing you've accomplished is to work out a good set of methods; in that case, you can be more generous with the space allocated to the methods section.

The references section should include only the papers you reference in the poster; if you're also writing a paper, the reference section in the poster will usually be shorter than that in the paper.

At the bottom right, add one more section with its own frame: contact information. List your own name and email address. Consider including a photo of yourself; this helps people remember you, especially if they take one of the miniature versions of the poster that you'll have available (see below).

Make your figures large – they're the part that your audience cares the most about. The bottom line is to make the most important content easily readable from 1.5 meters away.

Choose an overall dominant color for your poster, preferably the same one that appears in your data plots. Use this color for graphic elements such as the frames around each section. For your presentation, consider dressing in coordination with the dominant color of the poster.†

An easy way to include videos is to put a QR code for the video on your poster. Interested people can then use their phones to view the video.

For examples of good and bad posters, suggestions on where to print them, and other ideas, see the listing for this section on ExpPhys.com.

---

* Often when posters are judged, an important criterion is whether the research question is stated clearly.

† This idea is taken from Colin Purrington's excellent and detailed guide to poster design, colinpurrington.com/tips/poster-design/. Check out his great ideas about "How to make your poster more engaging," including adding video or 3D effects. I don't agree with his advice about logos – they are an important way of identifying your institution.

Before printing your poster in full size, first zoom in on each graphic to 100% magnification to make sure it still looks good; if it looks blurry in this view, it will also look blurry when you print it. Then, use the print to pdf capability of your software to make a pdf version, and print the pdf out on a standard sheet of paper (don't just look at it on the screen). Often, you will find that the colors come out differently than you expected, and that images have lower contrast. Also, looking at things on a printed page, especially graphics, causes you to notice different errors than looking at it on the screen. After correcting any problems, print a new copy on a standard sheet of paper, and ask at least two people not involved with your project for feedback.

Work out a roughly seven-minute mini talk that guides viewers through the most important parts of your poster, with you speaking at a normal pace. It's fine if you skip some parts of the poster in this talk; interested viewers can ask about them. Practice your talk on your own until you feel reasonably confident about it, then practice it with your peers at least twice.

Print out 10–20 copies of your poster on standard size paper. During the poster session, offer these only to people who seem genuinely interested in your work. This really helps in making a lasting impression.

During the poster session stay with your poster for the entire assigned time, even if there are periods (as there always are) when no one is viewing the poster. If visitors look at your poster for more than 30 seconds, ask if they have questions; usually this leads into your mini talk. Have a notebook ready to jot down thoughts as they are offered or sparked by viewers of your poster; very often, you will get important insights from these visitors.

4

Uncertainty and Statistics

Paul Thorman

CONTENTS

Although this may seem a paradox, all exact science is dominated by the idea of approximation. When someone claims to know the exact truth about anything, you are safe in inferring that this is an inexact person. Every careful measurement in science is always given with the probable error. Careful observers admit that they are likely wrong, and know about how much wrong they are likely to be.

              – Adapted from Bertrand Russell, *The Scientific Outlook*

Suppose that you calculate the output of a circuit and predict an output voltage of 6 volts. You then build the circuit in the laboratory and your digital voltmeter reads 5.98 V. Does this measurement support your prediction or contradict it? The answer depends on your *uncertainty.*

4.1 Random vs. Systematic Errors

Uncertainty, also known as *error*,* is how scientists communicate their expectations about how closely their experimental result will match the results from past or future experiments, or match a theoretical prediction. Uncertainty is an essential part of every scientific prediction or measurement – it defines what we mean by "agreement," making it possible for our experiments to confirm or disconfirm hypotheses.

Accuracy vs. Precision

There are two main ways that errors affect measurements: (1) by lowering the *precision* (statistical error) and (2) by lowering the *accuracy* (systematic error). Precision tells you how much your measurements look like one another; accuracy tells you how close they are to the correct value. No matter how careful you are in making your measurements, there will always be some unpredictable spread in your data due to random changes in variables beyond the experimenter's control (e.g., air currents), from variations in how the experimenter takes readings from analog devices (e.g., meters with needle pointers rather than digital displays), and sometimes from variations in quantities that are intrinsically random (e.g., the path a particular photon follows). Suppose that you track the motion of a ball through the air. Even if the air is at rest on average, there are always air currents on small scales that can deflect the ball. Since the currents are just as likely to push the ball up as down, if you repeat the experiment many times, your data will look scattered or

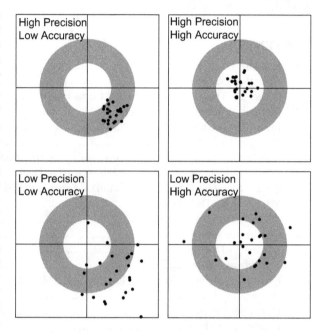

FIGURE 4.1 Results with different accuracy and precision. Note how improving the accuracy (left to right) produces a distribution that is better centered on the true value (center of the circle), while improving the precision (bottom to top) reduces the "scatter," but without necessarily drawing closer to the correct value

imprecise. By averaging the scattered data, you can hope to "average out" the air's random influence and lower your uncertainty. The more data you have to average, the more you can reduce this kind of uncertainty (Figure 4.1).

The scatter due to imprecision does not reflect any *systematic* errors, the type of errors in procedure or equipment that make repeated measurements consistently high or low. Systematic errors cause a lack of *accuracy* in your measurement – no matter how many repeated measurements you take, systematic errors will cause them all to differ from the true value. In our ball example, while the air can push the ball either left or right, it will always push the ball backwards and not forwards: air resistance will slow the ball down. If we were to repeat the experiment multiple times this effect would always act in the same direction, so all the data would be shifted in a systematic way and the

---

* Although uncertainty is sometimes referred to as error, its presence is not necessarily an indication of some experimental mistake (as in "human error"). When we use "error" to mean uncertainty, we are referring to its alternative meaning of "wandering," since the error is an estimate of how much our measured value will wander from our best estimate of its true value.

result will be *inaccurate*. Systematic errors can be much harder to quantify than statistical errors because you don't necessarily know that the data have been shifted, even if you repeat the experiment multiple times.

Where Do These Systematic Errors Come From?

They fall into a small number of general categories:

- *Offsets*: your measuring device reads an incorrect value that always differs from the true value by a fixed additive amount.
  - your voltmeter reads 0.2 V even when the probes are connected together
  - your meter stick is worn off at one end, reducing all your measurements
  - a pressure sensor came from the factory always reading 0.02 kPa below the actual pressure
  - your light sensor generates a small thermal current, even in the dark
- *Calibration errors*: your measurement device reads an incorrect value, with an error that depends on the input value.
  - your voltmeter reads 1% too high, so that it reads 1.01 V when you apply 1.00 V and 2.02 V when you apply 2.00 V
  - your meter stick expands slightly, reading 1 cm for every 1.01 cm of actual distance
  - a force sensor always reads $1.2 \times$ the actual force applied
  - a light sensor is designed to produce one digital count for every 3.6 photoelectrons, but instead produces one count for every 4.1 electrons

Note that calibration errors are often multiplicative, as in the above examples, but not always. For example, a current meter might read 1% too high for currents below 1 mA, changing gradually to 2% too high for currents above 5 mA.

- *Insertion errors*: your measurement affects the quantity to be measured in some way.
  - your thermometer heats up the sample as you measure its temperature
  - the addition of a force sensor changes the mass of your experiment
  - the electronics on your light detector give off infrared radiation that gets picked up by the light sensor
  - your current meter has a resistance that is significant compared with that of your circuit, so that when you put the meter into the circuit, the current flowing in the circuit is significantly reduced
- *Drift*: changes in experimental conditions not related to the phenomenon you are studying cause the experiment to change over time.
  - the lab room heats up during the day, changing your experiment's cooling rate
  - vibrations from vehicle and foot traffic disturb your experiment more during the day than at night
  - your radioactive sample decays during the experiment, inadvertently causing your detection rate to decline
  - impacts from cosmic rays degrade your detector over time, lowering its efficiency

Some of these systematic errors can be identified after the fact and their effects can be corrected or partially remediated, but others must be measured at the time of the experiment by frequently calibrating your measuring apparatus against a stable reference standard. This can be accomplished by checking your measurements against a separate measuring device that has been calibrated against an outside standard, or by applying your device to some well-understood physical system with standard properties;

for example, you might use the spectrum from a sodium emission lamp, which has a large number of well-measured spectral lines, to calibrate a spectrograph.

4.2 Methods of Determining Uncertainty

There are two basic methods for determining uncertainty: instrumental uncertainty and multiple trials.

Instrumental Uncertainty

Sometimes an experiment is so difficult or time-consuming that it can be performed only once, or your measuring apparatus makes such coarse measurements that repeated trials will return the same numerical value. In these cases, your uncertainty will be a direct result of the quality of your measuring apparatus, which you can either measure directly or obtain from some accepted source (such as other users or the manufacturer). If you need to make your own precision estimate, you can observe the variations of your measuring device while making some measurement that you expect to be stable; if you need to know your accuracy, you can apply your measuring device to a known standard. You can also get error values for instruments from the device manufacturer's specifications (which summarize their attempts to make the above observations). This method allows you to take both accuracy and precision into account when determining your uncertainty. We'll discuss this method in more detail in Section 4.9.

Multiple Trials

If the experiment can be repeated many times quickly, it is usually easiest to use the results of these multiple trials to estimate the precision. This method does not measure the accuracy; for example, it won't detect a calibration error. If you don't know anything about the accuracy of the instruments, you might proceed under the assumption that the uncertainty is dominated by the lack of precision, rather than the lack of accuracy. But in that case, if you get a statistically significant difference between your measurement and the expected value, that might be a true discrepancy (an important result!) or it might merely mean that your uncertainty is not primarily statistical, but systematic instead.*

4.3 Standard Error of the Mean and Probability Distributions

(Much of the following discussion is phrased in terms of the multiple trials method, but as we'll see it also applies to the instrumental uncertainty method.)

4.3.1 Sample vs Population and the Gaussian Distribution

When you go into the laboratory, you perform a finite number of measurements; this set of measurements is your "sample." You might be measuring a speed, a voltage, or some other physical property; this is your variable, x. However, the result contains some uncertainty – the readings are not always identical. Often, you'll report the mean $\mu_S = \bar{x}$ of your sample, i.e., the average of your multiple measurements. To characterize the uncertainty in this reported mean, you will use the "Standard Error of the Mean," or SEM. To understand this fully, you need a bit of statistics background and nomenclature.

Think of your sample as being drawn from an infinitely large pool of possible results; this pool is called the "population." The range and frequency of the values in the population are specified by a "probability distribution." The probability distribution is a function, $P(x)$, of your variable, and the integral of

---

* A related tool for estimating statistical uncertainty is *resampling*, in which you simulate multiple hypothetical samples (usually with the aid of a computer) from your one actual data set to gauge the effects of outliers and sample size.

$P(x)$ from x to $x + dx$ gives the probability that a value in that range will show up in a single measurement. Since every measurement returns some value, the probability is normalized to have integral 1 over all possible measurement values.

The vast majority of laboratory measurements that are not counting experiments (e.g., counting radioactive decays) follow a "normal" (or "Gaussian") distribution. The normal distribution has a *mean*, or average value, that is also its *mode*, or most popular value, and the likelihood of getting a different value falls off rapidly on either side of the mean according to

$$P(x) = \frac{1}{\sqrt{2\pi\sigma^2}} e^{-\frac{(x-\mu)^2}{2\sigma^2}} \tag{4.1}$$

where μ is the population mean and σ is the "standard deviation," which measures the breadth of the distribution (Figure 4.2).

The standard deviation describes the variation of values in the population, or equivalently it tells you how wide $P(x)$ is. It can be estimated from your measurements using*

$$\sigma = \sqrt{\frac{\sum_i (x_i - \mu_S)^2}{N-1}} \tag{4.2}$$

where $\mu_S = \bar{x}$ is the mean of your sample, a sample which is assumed to contain N independent measurements drawn from the population. This is only an estimate for σ because, due to your finite number of measurements, you don't know the true mean μ of $P(x)$, i.e., the mean of the population. Instead, you only

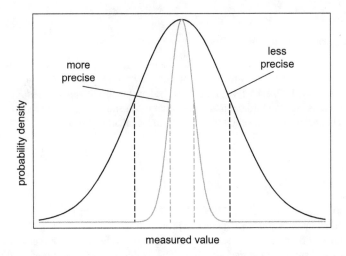

FIGURE 4.2   Two normal distributions with different widths. The darker line plots a distribution with a large standard deviation, while the lighter line shows a smaller standard deviation. The standard deviation is indicated in both cases by a pair of dashed lines enclosing 67% of the probability.

---

* This is simply a way of measuring the typical deviation from the average. Starting from the inside of the expression, we take the difference between each data point and the average. We square this so that both positive and negative differences from the average count in the same way. Then (with the summation and the division by $N-1$), we essentially take the average. (See the next footnote for an explanation of why we use $N-1$ rather than N.) Finally, we take the square root so that the result has the same units as x.

know the mean of your sample μ_S. However, if you have at least 15 measurements, it's usually a good enough estimate. (See Section 4.5 if you have fewer than 15 measurements.*)

4.3.2 Standard Deviation vs. Standard Error of the Mean

If you take 5000 measurements instead of 50 measurements, your estimate for σ will be about the same; you're just getting a more accurate idea of the variations in the population. However, your sample mean μ_S is a much better estimate of the true mean μ, so the uncertainty in your reported mean is much lower.

The uncertainty in μ_S is the SEM, which you can estimate using:[†]

$$\text{SEM} = \sigma_{\bar{x}} = \frac{\sigma}{\sqrt{N}} \tag{4.3}$$

The standard error of the mean now properly decreases with a larger sample size.

4.3.3 Other Distributions

The normal distribution does a good job of describing data sets in which the total error is the sum of many smaller contributing errors,[‡] which is why it is so common and useful in the physical sciences. Other common distributions include the *Poisson distribution*, which describes the uncertainty in a discrete count of events, and the *binomial distribution*, which describes the uncertainty in a set of independent binary observations, such as a series of coin flips. Poisson statistics are discussed in detail in Section 13.4.

4.3.4 Median and Mode

Example: Characterizing Voltage Data

We test a battery's voltage by hooking up a voltmeter to its terminals. After repeating this procedure 20 times, the results are as follows: 1.40, 1.41, 1.43, 1.43, 1.44, 1.44, 1.44, 1.44, 1.45, 1.45, 1.45, 1.46, 1.46, 1.46, 1.47, 1.47, 1.48, 1.48, 1.50, 1.51. What are the *mean*, *median*, and *mode* of these data? What is the standard deviation and what is the standard error on the mean?

Solution: The mean of an unweighted data set can be calculated from

$$\mu_S = \frac{\sum x_i}{N} \tag{4.4}$$

which for this data set gives us $1.4535 \approx 1.45$ V. This is the mean that is returned by Excel's AVERAGE function or *Mathematica*'s Mean[] function.

The median is the "middle" value of the data set, the value that is larger than half of the measurements and smaller than the other half. The median is sometimes preferred to the mean

---

* If $N-1$ were replaced with N, the quantity under the square root would be the average squared deviation of the measurements from the mean; by using $N-1$, we increase the estimate of σ for small samples. Since the members of your sample are likely to be closer to their own sample mean than to the true population mean (i.e., $\sum(x_i - \mu_S) \le \sum(x_i - \mu)$ where μ is the true population mean, not the mean of your particular sample), the factor of $N/(N-1)$ helps to correct what would otherwise be a low-biased estimator. Another way to think of this: because you had to use your sample to determine the mean, you have only $N-1$ independent measurements of how the sample is scattered around the mean.
[†] Please see the entry for this section on ExpPhys.com for the derivation of this formula.
[‡] This observation is known as the *central limit theorem*.

because it is less easily changed by *outliers*, data that lie at a great distance* from the mean. Since this data set has 20 values, the middle falls between the 10th and 11th values, so the standard procedure is to take the mean of values 10 and 11. Both of those values are 1.45 for this data set, so the median (like the mean) is 1.45 V.

The mode is the value that appears most frequently in the data. We can visualize the data using a histogram (Figure 4.3) and identify the most common value as 1.44, with four occurrences. No other value[†] appears more than three times, so 1.44 is our mode. For probability distributions that are very different from the normal distribution, the mode might do a better job than the mean of telling you what to expect in future experiments.

Finally, we should estimate the standard deviation and the standard error of the mean. We get the standard deviation from summing the squared differences from the sample mean and dividing by $N-1$. The sample mean is 1.454, so:

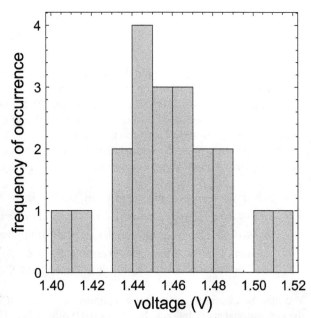

FIGURE 4.3 A histogram of the voltages from the example. The height of each bar represents the number of times that the corresponding value appeared in the data set. The mode can be identified by finding the value with the most appearances, which for these data is 1.44 V.

$$\sigma \approx \frac{(1.40-1.454)^2+(1.41-1.454)^2+2*(1.43-1.454)^2+\cdots}{(20-1)}=0.03$$

This standard deviation can be computed with Excel's **STDEV.S** function or *Mathematica*'s StandardDeviation function.

The standard error of the mean, on the other hand, is

$$\sigma_{\bar{x}}=\frac{\sigma}{\sqrt{n}}=\frac{0.027}{4.47}=0.006$$

which is much smaller than the standard deviation.

4.4 Confidence Intervals

You should report results as the mean of the measurements and the width of a specified *confidence interval*. This is defined as a range, Δ, of your measurement variable that contains the mean, μ_S, and has the property that

---

* Determining what constitutes a "great distance" and how to properly account for outliers without allowing our biases to overrule the actual data are difficult issues to be discussed in Section 4.11.

† This single most common value makes our data *unimodal*; if there were more than one widely separated value that dominated over its neighbors, the data would be *multimodal*.

$$\int_{\mu_S-\Delta}^{\mu_S+\Delta} P(x)\,dx = C \tag{4.5}$$

where C is a probability threshold that you must choose. Common values for C include 68%, 95%, and even as high as 99.9999%, which is typical for particle physics, where a premium is placed on avoiding false positive results. *It is critical that you specify the value of C the first time you quote a result with uncertainty. Alternatively, you can quote the value of C as a percentage in parentheses, as in the examples below.* You should use 95% in most cases.

For the specific case of the normal distribution, if the true population mean is μ and you make 15 or more independent observations and average them to get $\bar{x} = \mu_S$, then

- 68% of the time, your measured $\bar{x}$ will fall within the range $\mu \pm \sigma_{\bar{x}}$
- 95% of the time, your measured $\bar{x}$ will fall within the range $\mu \pm 2\sigma_{\bar{x}}$
- 99.7% of the time, your measured $\bar{x}$ will fall within the range $\mu \pm 3\sigma_{\bar{x}}$,

  where again $\sigma_{\bar{x}}$ is the standard error of the mean. So, if you're using a 95% confidence interval, then you will report your result as $\bar{x} \pm 2\sigma_{\bar{x}}$ (95%).

You must be careful to use the term "confidence interval" (CI) properly. For example, say you measure the gravitational acceleration to be (9.83 ± 0.02) m/s^2, where the ± 0.02 m/s^2 is the 95% CI. The expected result of 9.81 m/s^2 is right at the edge of the 95% CI. It would be correct to say, "The accepted value is within the 95% confidence range of my experiment, indicating a reasonable level of agreement." However, because the expected value is right at the edge of the CI, it would also be correct to say, "There is only a 5% chance that the discrepancy between my measurement and the expected value was caused by fluctuations and inaccuracies that are included in my uncertainty analysis." It would be **wrong** to say, "My results show with 95% confidence that the expected value is correct." After all, if you had chosen instead to quote a 99.7% CI and the expected value was just at the edge of this CI – i.e., the discrepancy is greater than above – you would certainly not say, "My results show with 99.7% confidence that the expected value is correct." In this case, there is 99.7% chance that the discrepancy was caused by a systematic error in your experiment that is not accounted for in your uncertainty analysis, or that the accepted value is wrong.

4.5 Student's t-Distribution

For fewer than 15 measurements, your confidence intervals should be broader than those quoted here, due to additional uncertainty caused by your small sample. If we assume that your measurements represent an independent sample randomly drawn from a normally distributed variable with a well-defined mean and standard deviation, then your sample means should follow "Student's t-distribution.*" This distribution has one parameter, ν, which specifies the number of "degrees of freedom":

$$\nu = (\text{\# of independent measurements}) - (\text{\# of fitted or calculated properties}) \tag{4.6}$$

The Student's t-distribution is symmetrical, like the normal distribution, but has more probability farther out from the mean, and thus a higher likelihood of producing samples with means that are very far from the underlying population mean (Figure 4.4). This extra width is more pronounced for lower values of ν. The mathematical form for the *t*-distribution is complex, but it can be accessed directly in *Mathematica* using the StudentTDistribution[ν] function. The range, measured in standard deviations, that must be included to produce a given % CI can be calculated in Excel using the function T.INV.2T(1−C.I., ν). Alternatively, you can use Table 4.1.

---

* This was developed by William Sealy Gossett, Head Brewer of Guinness and an important figure in modern statistics. He published under the pseudonym "Student."

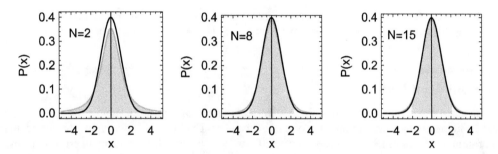

FIGURE 4.4 Normal (black line) and Student's t (gray line) distributions for 2, 8, and 15 degrees of freedom. The x coordinate is measured in standard deviations (σ). The shaded areas show the range enclosing 95% of the probability, which approaches 2σ as the sample gets larger.

TABLE 4.1

Factors for converting standard error to confidence interval using Student's t-distribution. Find the number of degrees of freedom in the first column and multiply your standard error by the factor under the desired confidence interval

	Probability contained		
η	67%	95%	99%
1	1.75	12.71	63.66
2	1.28	4.30	9.92
3	1.16	3.18	5.84
4	1.11	2.78	4.60
5	1.08	2.57	4.03
6	1.06	2.45	3.71
7	1.05	2.36	3.50
8	1.04	2.31	3.36
9	1.03	2.26	3.25
10	1.02	2.23	3.17
11	1.02	2.20	3.11
12	1.02	2.18	3.05
13	1.01	2.16	3.01
14	1.01	2.14	2.98
15	1.01	2.13	2.95

Example: Using Student's T-Distribution

Suppose that you are running an experiment to measure the resistance of a sample, collecting seven different values of current as you apply seven different voltages. You fit a line to the data and find that the best-fit slope corresponds to 50 ± 5 Ω, where 5 Ω is the standard error (the equivalent to the standard error of the mean, but the result of a fit) reported by your fitting software. How should you take your number of data points and fitted parameters into account if you want to report the 99% confidence interval?

Assuming that your seven data points are independent, the fit parameters should be distributed according to Student's t-distribution for the appropriate number of degrees of freedom. Here, we have seven data points minus two fitted parameters (slope and intercept of a line) = 5 degrees of freedom for our distribution. We wanted to report a 99% confidence interval, so we need to know how wide that is, in standard errors, for a Student's t-distribution with 5 degrees of freedom. In Excel, we use T.INV.2T(1 − 99%, 4) = 4.03. Then, $4.03\sigma_{\bar{x}} = 4.03 \times 5$ $\Omega = 20.15$ Ω. So for this data set, we can report that the resistance is $50 \pm 20 \ \Omega$ at 99% confidence, or simply $50 \pm 20 \ \Omega$ (99%). Why 20 and not 20.15? Because we don't want to overstate our confidence in the error value by reporting too many significant figures.

4.6 Significant Figures

We can use the confidence interval to find the correct number of significant figures for expressing our results without giving a false impression of our precision or accuracy. First, determine the correct number of significant figures for your uncertainty itself. Since we've seen that the standard error of the mean decreases as $\sqrt{n}$, where n is the number of measurements, then a safe assumption would be that

$$\frac{\text{uncertainty}_x}{x} \geq \frac{1}{\sqrt{n}} \tag{4.7}$$

So if we have tens or hundreds of measurements, we can justify rounding the uncertainty to the first significant figure, since that will correctly convey that we have $\approx 10\%$ error *on our uncertainty value*. The one exception would be if the uncertainty value itself begins with a 1, since rounding 14 to 10 (or 1400 to 1000) introduces an error larger than our 10% error estimate. So in the case where the first digit of our uncertainty is a 1, we can justify keeping one additional digit. If we are running a very large experiment that takes 10,000 data points or more, we can consider keeping two or more significant figures in our uncertainty, always rounding to the decimal place corresponding to $1/\sqrt{n}$.

Now that we have a properly rounded uncertainty value, we can let it guide us in how to round our actual measured value. There is no point in reporting a temperature of 22.56°C from a thermometer with an uncertainty of ± 2°C, and it might mislead future experimenters. To keep our value consistent with the uncertainty, we should round the value to the same decimal place as our uncertainty. In the thermometer example, the error is rounded to the units place, so the measured temperature should also be rounded to the nearest °C, for a reported value of 23 ± 2°C.*

4.7 Quantitative Comparisons, or How Not to Be Misled by Error Bars

Once we have determined the correct confidence range, we are ready to test for agreement between our result and other results or theory. For example, we might be testing whether a resistor has the correct value to match its label by applying different voltages and measuring the currents, then determining the slope of the line. If we find 550 ± 30 Ω (95%), but the theoretical value[†] is 500 ± 25 Ω (95%), is our value in agreement with the specified value?

In this example, the error bars from the measured and theoretical values overlap (see Figure 4.5), but that does **not** mean that our value agrees with the theoretical value at the 95% level. The correct way to check for agreement is to determine whether

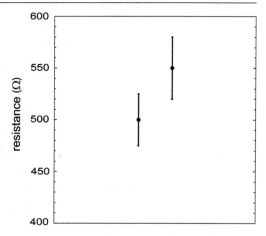

FIGURE 4.5 A graphical representation of the theoretical and measured resistances from the example, with uncertainties shown as error bars. Note that the error bars overlap in this example, with the higher end of the error range from the specified value exceeding the lower end of the error range for the measured value.

---

* We've chosen to report the central value, then the one-sided confidence interval, then the units. Although you might see other formats, this one has the advantage of communicating the value, error, and units without repetition or extra brackets.

† Theoretical values can have uncertainties, for example, because one of the parameters used in the theory is uncertain.

$$\left(x_{\text{measured}} - x_{\text{theoretical}}\right)^2 \leq \text{uncertainty}^2_{\text{measured}} + \text{uncertainty}^2_{\text{theoretical}} \qquad (4.8)$$

Using the example values, we find that the *discrepancy*, the difference between the measured and theoretical values, is 50 Ω, while the joint uncertainty on the discrepancy is* $\sqrt{25^2 + 30^2}$ Ω $\approx 40\,\Omega$, so despite those overlapping error bars, these two values do not agree at the 95% confidence level, meaning that we expect such discrepant values to indicate true disagreement at least 95% of the time.[†]

4.8 Propagating Errors

So far, we have found the errors in measured quantities, either by testing them ourselves or by referring to manufacturer's specifications (or other customary rules for estimating instrumental errors). But most experiments are not constructed so that your most interesting or useful quantity is directly measured in the experiment: you might measure speed when what you really want is kinetic energy, or measure the voltage across a resistor as a way of measuring the current. We need a way to get the error on that calculated quantity, based on the measured errors of the contributing quantities. This process is called *error propagation*.

Suppose that you have measured the speed of a 0.15 kg cart to be 0.30 ± 0.05 m/s (95%). Its kinetic energy, $K = \frac{1}{2}mv^2$, is plotted in Figure 4.6. The central value is $\frac{1}{2}(0.15\,\text{kg})(0.30\,\text{m/s})^2 = 0.0068$ J. There are a few valid ways to estimate how much your 0.05 m/s error will affect your determination of the kinetic energy.

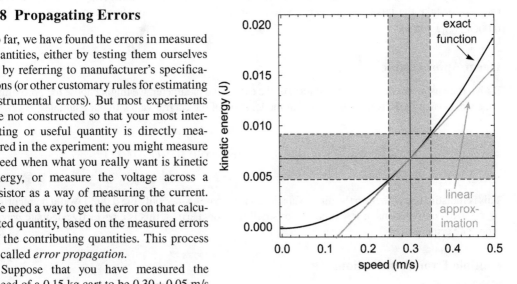

FIGURE 4.6 Kinetic energy of the cart as a function of speed. The gray areas span the error on the x-axis (0.05 m/s) and the error on the y-axis, as determined by evaluating the kinetic energy at the highest and lowest speed values given by the error bar. The straight line shows what we would obtain by using the derivative of the energy at the measured value (0.30 m/s) and extrapolating to those same error limits.

Direct Substitution

If you want to know how much the function $\frac{1}{2}mv^2$ changes when you change v by 0.05 m/s, you could simply plug in your new value and see how much the function changes, that is

$$\Delta K(v) = K(v + \Delta v) - K(v)$$
$$= K(0.30 \text{ m/s} + 0.05 \text{ m/s}) - K(0.30 \text{ m/s}) = 0.0024 \text{ J}$$

$$(4.9)$$

---

* Note that we can combine two uncertainties by adding in quadrature whether they represent $1\sigma_{\bar{x}}$ or $2\sigma_{\bar{x}}$ or any other confidence interval, so long as they both represent the same % confidence.

[†] The other 5% of the time, the apparent disagreement is accidental.

where ΔK is the uncertainty in the kinetic energy, with the same % confidence as for the uncertainty in velocity Δv. This method has the advantage that it correctly models the higher-order behavior of the function, which can be a substantial improvement for calculated quantities that are highly non-linear. It does not address a related problem: although adding 0.05 m/s to our measurement gives one estimated error, subtracting it from our measurement will give a slightly different error (0.0021 J, in this case) with no reason to prefer one result over the other. For our purposes, however, this is not an important distinction. The error is seldom known to better than 10% in most research, so regardless of whether we add or subtract the error, we would report 0.007 ± 0.002 J (95%), rounding the error to one significant figure and then rounding the central value to match that decimal place. If you find that the error estimate from adding differs from the one you get by subtracting, you may average the two, or take the larger of the two (to be conservative) or you may choose to report both, as in $0.0068^{+0.0024}_{-0.0021}$ J (95%).

Linear Approximation

Rather than using the exact functional form of the calculated parameter, we might choose to approximate the change using its derivative, measured at the location of the central value. In our example:

$$\Delta K = \Delta v \left| \frac{\partial K}{\partial v} \right|_{v=\mu} = (0.05\,\text{m/s})(0.15\,\text{kg})(0.3\,\text{m/s}) = 0.0023\,\text{J}$$

This is indistinguishable from the direct substitution estimates, since only the 0.002 J would be reported, so for errors on functions of low order or for small fractional errors, this is a perfectly adequate way to propagate errors.

Multiple Error Contributions

In the kinetic energy experiment, we had two measured quantities that contributed to the energy: mass and speed. So far, we've been neglecting the error in the mass. This is appropriate as long as the fractional error on the mass is at least three times smaller than the fractional error on the speed, that is

$$\frac{\sigma_m}{m} \bigg/ \frac{\sigma_v}{v} \leq 1/3 \tag{4.10}$$

An error three times smaller than the next largest error cannot affect the total error at the 10% level, which is the usual level of significance for lab exercises and most research results.

Suppose that the mass error were larger, comparable with the fractional error in the speed, for example, $m = 0.15 \pm 0.04$ kg (95%). How can we include this error in the error propagation?

Addition in Quadrature

For a function $y(x, z)$, if we assume that variations in z are uncorrelated with variations in x, then:*

$\sigma_{y,\text{total}} = \sqrt{\sigma_{yx}^2 + \sigma_{yz}^2}$, where σ_{yx} is the uncertainty in y due only to variations in x, and σ_{yz} is the uncertainty in y due only to variations in z.

Addition in quadrature

---

* Please see the derivation of this formula under this section at ExpPhys.com

This way of adding uncorrelated error sources (squaring, adding, then taking the square root) is called "addition in quadrature."

Returning to the kinetic energy example, we have

$$\sigma_{Kv} = K(m, v + \sigma_v) - K(m, v) = 0.0024 \text{ J}$$

$$\sigma_{Km} = K(m + \sigma_m, v) - K(m, v) = 0.0018 \text{ J}$$

$$\sigma_{K, \text{total}} = \sqrt{\sigma_{Kv}^2 + \sigma_{Km}^2} = \sqrt{0.0024^2 + 0.0018^2} \text{ J} = 0.003 \text{ J}$$

The resulting error bar of 0.003 J takes into account both the mass and speed errors, and is larger than either one alone.

In the linear approximation method, this would be

$$\sigma_{K, \text{total}} = \sqrt{\left(\sigma_v \frac{\partial K}{\partial v}\right)^2 + \left(\sigma_m \frac{\partial K}{\partial m}\right)^2}$$

More generally, for a function $y(x, z)$,

$$\sigma_{y, \text{total}} \cong \sqrt{\left(\sigma_x \frac{\partial y}{\partial x}\right)^2 + \left(\sigma_z \frac{\partial y}{\partial z}\right)^2} \tag{4.11}$$

You can show in problem 10 that if $y = Ax^n z^m$, then

$$\frac{\sigma_{y, \text{total}}}{y} \cong \sqrt{\left(n \frac{\sigma_x}{x}\right)^2 + \left(m \frac{\sigma_z}{z}\right)^2} \tag{4.12}$$

This shows that the importance of each measured error depends strongly on the power to which that quantity is raised. For our kinetic energy example, we get

$$\left(\frac{\sigma_{K, \text{total}}}{K_{\text{total}}}\right) = \sqrt{\left(\frac{\sigma_m}{m}\right)^2 + \left(2 \frac{\sigma_v}{v}\right)^2} = \sqrt{\left(\frac{0.04}{0.15}\right)^2 + \left(2 \frac{0.05}{0.3}\right)^2} = 0.43$$

Using this fractional error estimate gives $\sigma_{\text{tot}} \approx 0.43 \times 0.0068 \text{ J} = 0.0029 \text{ J} \approx 0.003 \text{ J}$, which is essentially identical to the 0.003 J estimate obtained from the direct substitution method.

4.9 More of the Instrumental Uncertainty Method, Including "Absolute Tolerance"

All the above discussion applies both to the multiple trials method of determining uncertainties and to the instrumental uncertainty method.* In the multiple trials method, the "population" from which your sample is drawn is the set of possible results in the absence of any systematic errors; this method gives you no information about systematic errors. So, when you find a significant discrepancy between theory and experiment (e.g., the theoretical value is outside the 95% confidence range), you can conclude that, very likely, there is either an error in the theory or a systematic error in the experiment. This should cause you to re-examine both the theory and the experiment for sources of the discrepancy.

---

* The one exception is the Student's t-distribution, which is only relevant for the multiple trials method and for quantities derived from fitting to a plot.

In contrast, the uncertainties for instruments normally do include allowance for calibration and offset errors,[*] so these types of errors are included in the population, which is represented by $P(x)$. The population and $P(x)$ represent an infinite set of measurements, each one taken with a different set of instruments (so that the calibration and offset errors are different). This population includes effects of random errors as well as calibration and offset errors. So, using this method, when you find a significant discrepancy between theory and experiment, you can be reasonably confident that it is not due to calibration or offset errors. It could still be due to other types of systematic errors in the experiment, or to problems with the theory.

From the information provided by the instrument manufacturer, we can estimate the standard deviation of the population, i.e., we can estimate σ. If you make only one measurement (which is typical for the instrumental uncertainty method), then the number of trials is $N = 1$, so for purposes of the preceding discussions, the "standard error of the mean" is $\sigma_{\bar{x}} = \dfrac{\sigma}{\sqrt{N}} = \dfrac{\sigma}{\sqrt{1}} = \sigma$. For example, if you use a balance with an uncertainty of $\sigma = 0.05$ g and you measure a mass of 50.00 g, then there is a 68% chance (including effects of typical random errors associated with the instrument, calibration errors, and zero offset errors) that the true mass is in the range 49.95 g to 50.05 g.

Typically, manufacturers quote an "absolute tolerance" for the accuracy. For example, a digital multimeter (DMM) may quote the accuracy of voltage readings as a "% of reading + counts," such as "0.5% + 2." The "counts" refer to the last digit displayed. For example, if the meter displays 1.56 V, the accuracy would be $0.5\% \times 4.56$ V $+ 2 \times 0.01$ V $= 0.0228$ V $+ 0.02$ V $= 0.0448$ V. This means that the manufacturer of the DMM guarantees that the reading is correct to within 0.0448 V, i.e., that the actual voltage is somewhere in the range 1.56 ± 0.0448 V.

How do we interpret this in terms of σ? The manufacturers usually achieve these accuracies by individual testing of each unit, with the failed ones being sent back for parts replacement and re-testing. This means that, if we make the same measurement with many different copies of the same DMM, we will get a non-normal distribution. The details of the distribution depend on the testing protocol of the manufacturer, which is proprietary. So, all we really know is that, to essentially 100% confidence, the true voltage is in the range 1.56 ± 0.0448 V.

We follow the advice of the National Institute of Standards and Technology (NIST) to estimate σ for such cases:

$$\sigma = \frac{\text{Absolute tolerance}}{\sqrt{3}}, \tag{4.15}$$

where the "absolute tolerance" is half the guaranteed range, i.e., the absolute tolerance is 0.0448 V in our example. So, in our example $\sigma = \dfrac{0.0448 \text{ V}}{\sqrt{3}} = 0.026$ V, where we've rounded to two significant digits.

You should try hard to find the manufacturer's specification ("spec") for the accuracy, which can usually be found in the manual on the "specifications" page. Often, you can find the manual on the web.

However, if you're unable to find the spec, as a second choice you can use the following rules of thumb.

For analog instruments (i.e., instruments without digital displays such as meter sticks, calipers, micrometers, volumetric flasks, needle and dial scales on pressure gauges), the total uncertainty (including effects of imprecision and inaccuracy), σ, can be estimated as half the smallest division. For example, if a meter stick is marked with 1 mm as the smallest division, then $\sigma = 0.5$ mm. Obviously, you can (and should) read the meter stick to a greater precision (you should be able to read it to within 0.1 mm); the 0.5 mm figure includes inaccuracies in the calibration of the instrument, effects of thermal expansion, etc., as well as the imprecision in your ability to read it.

---

[*] Be aware that uncertainties quoted by an instrument manufacturer are usually only guaranteed for one year after purchase. After this, the zero offset and calibration errors may increase. So, when possible, you should check the zero offset and calibration. If that is not possible, as long as the instrument is 15 years old or less, you can probably get away with using the quoted uncertainty.

For digital instruments, the accuracy of the instrument is often much lower than you might expect from the number of digits being displayed. However, it is usually safe to use $\sigma=$ (five times the value of the least significant digit place displayed). For example, if a voltmeter displays a reading of 36.93 V, you would use $\sigma = 5 \times 0.01$ V $= 0.05$ V.

In some cases (e.g., if you are forced to use an instrument in a less than optimal way, or if you are using a non-standard instrument), you will need to estimate the measurement uncertainty yourself, using a "seat of the pants" method. The best way of doing this is to estimate the 95% confidence interval (corresponding to $\pm 2\,\sigma$): what is the highest number that would be reasonable to record for the measurement? What is the lowest? This range corresponds approximately to the 95% confidence interval. The span of this interval, from highest to lowest, is equal to 4σ (i.e., $\pm 2\,\sigma$).

As described in Section 4.8, you can combine the errors from multiple instruments by adding them in quadrature, as long as you can assume the errors are not correlated with each other.

An important exception to this rule for combining errors occurs when two measurements have been made with the same instrument and they have given very similar results. For example, if you want to measure the pressure drop along a length of tube, you might choose to use the same barometer to measure both pressures, one after the other. The change might be relatively small, say from 101.5 kPa to 100.8 kPa. In that case, you might be justified in assuming that the systematic errors of the device are approximately the same for both measurements, in which case the difference $\Delta P = P_2 - P_1$ does not depend on those systematic errors, which cancel one another out. The only errors that would affect ΔP would be random errors, which will be less than the combined effects of random and systematic errors. If you encounter a situation like this in your lab, where you are using the same meter to make multiple similar measurements, you should consider trying to characterize the precision separately from the total instrument error.

4.10 Parameter Fitting

In many experiments, you will not be exploring in complete ignorance, but will have some set of expectations that you want to adduce evidence for or against. In those cases, you will test your data against a *model*, a set of assumptions based on previous observations that allow you to predict future observations. A model might include a functional form for how one variable depends on another or go farther and predict one or more constants of the fit based on known constants of nature, while leaving some parameters unknown. In order to determine whether your model has done a good job of explaining your data, it is often necessary to determine the *best fit* values, the numerical values that bring the predicted functional form into best agreement with your data set, and then compare those with other experimenters' best fit values or to values calculated using constants of nature.

The most common way to find these best fit values is by using *least squares* fitting. Your model should make a prediction for each observed case, whether it's a large number of repeated, identical measurements (like multiple observations of the same star's brightness) or a functional relationship (like the relation between current and voltage for a resistor). Subtracting your observed values from those predictions gives a set of differences that we call *residuals*. Least squares fitting changes your model parameters to minimize the sum of the squared residuals, or

$$r_{fit}^2 = \sum_{i=1}^{n} \left(x_i - m_i\right)^2 \tag{4.16}$$

where m_i is the model's prediction for observation i. This expression works well for data sets where each point is of equal quality, or where the errors for each data point are not well known. In these cases, a model's explanatory power can be expressed as an R^2 or *residual sum of squares* value. Taking the original data set, we can calculate the sum of squares using the mean of the data set (a zero-parameter fit, as it were):

$$r_{\text{mean}}^2 = \sum_{i=1}^{n} \left(x_i - \mu_S\right)^2$$

$$R^2 = 1 - r_{\text{fit}}^2 / r_{\text{mean}}^2 \tag{4.17}$$

The residual sum of squares ("R-squared") expresses how much of the variance of the original data set has been explained by the model. Values range from zero to 1, with values above 0.9 usually indicating a good model fit. However, R^2 tends to overestimate how much of the variance is explained by the model because it is built from two biased estimators for the variance. Using better estimators gives the "adjusted R-squared," given by

$$R_{\text{adj}}^2 = 1 - \left(1 - R^2\right)\frac{n-1}{\nu-1} \tag{4.18}$$

where ν is the number of degrees of freedom (Equation 4.6).

In *Mathematica*, least squares fitting is implemented by the LinearModelFit* command. This will return a FittedModel object that can be plotted directly or queried for parameters like R^2. In Microsoft Excel, least squares fitting can be obtained by inserting a scatter plot of your data, right-clicking on the data points, and selecting Add Trendline from the context menu. Option checkboxes in the Trendline dialog box will allow you to display R^2 and the fit parameters.

Example: Polynomial data: We can use linear least squares fitting to analyze the simulated data set plotted in Figure 4.7. If we were to approach this data set with no preconceived idea of the correct functional fit to use, we might still notice that there is a pronounced trend for values at the right of the plot to be higher than values at the left of the plot, suggesting a linear fit. We can attempt a linear fit in *Mathematica* using LinearModelFit[data,x,x], where the first x specifies that a linear term (and implied constant term) should be fitted, and the second x tells *Mathematica* which variable in the preceding expressions should be treated as the independent variable in the fit. The result is shown in Figure 4.8.

If we want to view the R^2 value in *Mathematica*, we can simply use the name we have assigned to our FittedModel object (*linearfit* was used in the example code) and use the desired parameter as function input, e.g., linearfit["AdjustedRSquared"]. In this

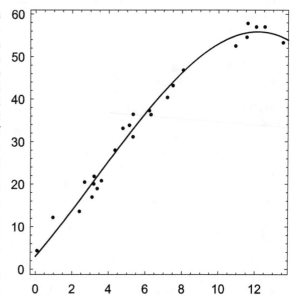

FIGURE 4.7 A pseudorandomly generated data set with normally distributed variations around a third-order polynomial, shown as the solid black line.

case, our R^2 value is 0.94, not dismal but not spectacular. But the real clue that we've missed some structure in our functional fit appears in the residual plot (Figure 4.8). The presence of second-order polynomial structure that is left over after our first-order fit strongly hints that the true order of the polynomial is ~3.

---

* Despite the name, LinearModelFit is not restricted to first-order polynomial fitting; the fits are linear in the sense that you provide *Mathematica* with a list of basis functions and it finds the best fit using a single coefficient for each one.

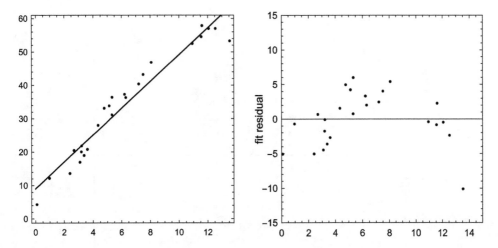

FIGURE 4.8 The first attempt to fit the data set using a line as our model. The left plot shows the line of best fit, while the right plots our residuals, relative to that linear model. Note that the residuals do not appear random, but instead follow a parabolic shape, arcing first upward and then downward.

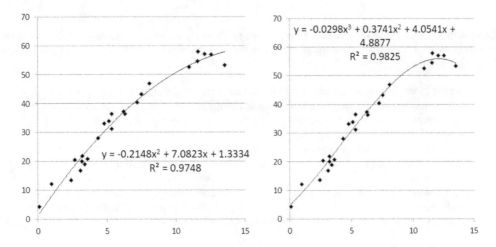

FIGURE 4.9 Excel plots showing second- and third-order fits to the same data set. Note the increasing R^2 values as we increase the order of the polynomial fit.

Using Excel, we can plot our data set and add trendlines using polynomials of orders 2 and 3, as shown in Figure 4.9. While the increasing R^2 is an encouraging sign, it's important to be careful of *overfitting*, or using more parameters than are necessary to describe your data set. After all, if you were given n parameters to fit a data set of n points, you could simply specify every point in your model, with no leftover variance at all – a perfect adjusted R^2 of one! But such a model would be undesirable in at least two ways: it would do nothing to briefly summarize the known data, since your model representation would be just as long as your data set; and it would be unable to predict the outcome of new experiments, unless they used precisely identical test conditions to earlier experiments. Even in the case where you use an overly complex mathematical function to fit your data, you will find that the more parameters you include to reproduce your existing data, the less stable your function will be outside the range of known data, often making it quite useless for extrapolation to new experiments.

How can you avoid overfitting? One method is to monitor your residuals as you continue to add parameters to your fit (Figure 4.10). If you continue to see non-random structure in your residuals, you might be justified in continuing to add fit parameters (or considering a different functional form that will better capture the shape of your underlying data). In physics we sometimes

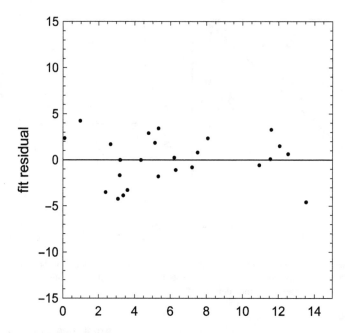

FIGURE 4.10 Residuals from the second-order polynomial fit. Checking residuals for structure is far from foolproof, especially in small data sets. Although the true form of the underlying function is cubic, it is difficult to detect any structure in this set of residuals.

make use of another strategy: using a fit function that is justified by some underlying theory. This has the advantage of guiding you in the choice of fitting function, but if your data are in genuine disagreement with the theory, restricting yourself this way may disguise that disagreement by forcing the data into the theoretical fit.

You can also test each parameter in your expression for overfitting by using a confidence interval test. What should your test hypothesis be? In most cases, a fit parameter equal to zero contributes nothing to the fit (but be careful with fitted exponents!) so you can consider discarding any parameter that lies closer to zero than its own confidence interval. For example, in our example data set above, where the true function was a third-order polynomial, the fit parameters and values for the first five polynomial orders are shown in Table 4.2.

TABLE 4.2

Best-fit parameters for the first five orders of polynomial fit to our test data, whose true functional form was a third-order polynomial. Under each parameter is the value divided by its standard error; values which differ from zero by at least two standard errors are shown in bold

Best-fit coefficients (value/standard error)	Constant	x	x^2	x^3	x^4	x^5
Polynomial order						
1	**9.02**	**4.02**				
	(5.95)	**(19.2)**				
2	1.33	**7.08**	**0.21**			
	(0.76)	**(12.1)**	**(5.39)**			
3	**4.89**	**4.05**	0.37	**0.030**		
	(2.58)	**(3.64)**	(1.91)	**(3.05)**		
4	**5.01**	3.86	0.60	0.067	0.0003	
	(2.30)	(1.94)	(0.74)	(0.56)	(0.12)	
5	**5.99**	0.64	2.20	−0.39	0.029	−0.0008
	(2.47)	(0.16)	(1.13)	(1.04)	(0.96)	(0.95)

Although nearly every polynomial fit in Table 4.2 has at least one coefficient that is indistinguishable from zero, it isn't until we reach order 4 that the majority of coefficients fail to be significant, a pattern which continues in order 5. While we cannot be definitively sure that the correct fit is the polynomial of order 3, there is a strong hint here that orders 4 and 5 might be overfitting.

4.11 Measurement Errors and χ^2 (also known as chi square)

So far, our fits have treated all of the input data as though they are of equal quality; but in a real experiment, we often want to combine measurements taken under different conditions, or with different instruments, which naturally have different errors of their own. These data should not be given equal weight, but instead more weight should be given to higher-quality data. To obtain the best weighted fit, each point should be weighted inversely by its *variance*, which for a normal distribution is σ^2. For example, a weighted mean would be given by

$$\mu_{\text{weighted}} = \frac{\sum_{i=1}^{n} \left(x_i / \sigma_i^2 \right)}{\sum_{i=1}^{n} \left(1 / \sigma_i^2 \right)} \qquad (4.18)$$

and the standard error on that weighted mean by

$$\text{SEM}_{\text{weighted}} = \left(\sum_{i=1}^{n} \frac{1}{\sigma^2} \right)^{-0.5} \qquad (4.19)$$

To extend this idea to our least-squares function fitting, we need to weight deviations from the fit less for less-reliable data. In un-weighted least-squares, the metric for how discrepant any given point is from the fit is the squared difference between the observed value and the model value; to add inverse variance weighting, we will divide each difference by the squared error on the observation:

$$\chi^2 = \sum_{i=1}^{n} \frac{\left(x_i - m_i \right)^2}{\sigma_i^2} \qquad (4.20*)$$

Minimizing this χ^2 will then have the effect of finding the best-fit parameters for our model. In *Mathematica*, weighted fitting is implemented by adding Weights $\rightarrow \{w_1, w_2, ..., w_n\}$ to the LinearModelFit command.

Interpreting χ^2

If you fitted a model to data and you included your error bars in the fitting process, then the value of χ^2 can tell you something about the quality of your fit and your error bars. Typically, we don't look at χ^2 directly, but at the *reduced* χ^2, which is χ^2/DOF, where DOF is the number of degrees of freedom from Equation 4.7. Since χ^2 is a sum over all data points, it requires this correction to keep χ^2 from being larger simply from having a larger data set. If no error bars were used in the fitting, then the best-fit model will still be the one with the smallest χ^2, but without error bars, we cannot predict what that smallest value will be.

On the other hand, if your error bars are a good representation of the actual scatter in your data, then you can expect a reduced $\chi^2 \approx 1$. If your error bars are too small, or your fitting function does a poor job of following your data, or both, then you may see reduced $\chi^2 \gg 1$, since your residuals will appear large relative to your expected residuals. If your error bars are too large, and your fitting function isn't

---

* In discussions of chi square fitting, you may run across a different definition of the test statistic:

$$\chi^2 = \sum_{i=1}^{n} \frac{\left(x_i - m_i \right)^2}{m_i}$$

This alternative definition is identical to Eq. 4.20 only for the special case of Poisson statistics, where m_i is equal to σ^2, and should only be used to determine the χ^2 for binned data sets (histograms), which naturally follow Poisson statistics.

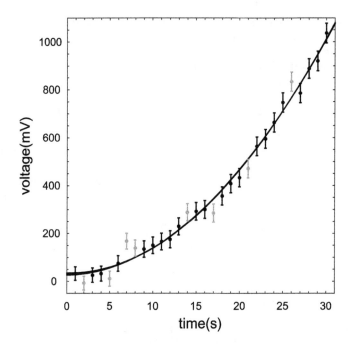

FIGURE 4.11 A simulated data set showing best-fit lines with no weighting, proper weighting, and incorrect weighting (solid black lines). While correctly weighting the fit makes only minor changes to the actual best-fit values in this case, note that the correctly calculated error bars (gray) disagree with the fit for eight of the 30 data points, very close to the expected number of 1σ disagreements (10).

completely incorrect, you may see reduced $\chi^2 < 1$, meaning that your data actually hew more closely to the fitting function than you would have expected, based on your error bars. You can also perform a quick visual check (Figure 4.11) to determine whether your error bars are capturing the true scatter in the underlying data: for normally distributed errors, we expect 33% of our data points to *disagree* with the fitted function by at least one standard error. In a data set of 30 points, there should be only a 1-in-165,000 chance that all of your data will lie within one standard error of the best-fit function, so if you find that your data are all or nearly all in agreement with your fit, it's an excellent bet that you've overestimated the error bars on your individual data points.

Fitting Routines and How to Make Them Work for You

When you need to find a model to fit your data, the most common thing to do is pick out a family of functions (polynomials, exponentials, etc.) and run a fitting algorithm to find the parameters that reproduce your data the best. If everything works according to plan, you'll get physically meaningful parameters (e.g., fitting the slope of a plot of force vs. acceleration to get a cart's inertial mass) with sensible confidence intervals. But just as often, you'll try to fit a model to your data and get parameters that are completely the wrong order of magnitude or get warnings that your fit "failed to converge." In order to understand these failures, it's worth taking a quick look at how model fitting is done.

Any model fitting routine will require a "goodness-of-fit" estimator to tell it whether it has found the best parameters for your data. The most traditional estimator is actually χ^2 from the earlier section on fit quality and error bar sizes. It has the desirable properties that for simple linear models it has a global minimum at the best-fit parameters and is everywhere continuous and differentiable, which makes it easy for automated routines to find its minimum – they can evaluate χ^2 and its gradient at a point in parameter space, then follow the downward gradient in a series of steps along the χ^2 surface to its lowest point (the method of steepest descent). But there's a possible failure in the plan: what we want is the best fit from all possible parameters, but the method of steepest descent is particularly susceptible to getting stuck in a merely local minimum, a set of parameters that is a better fit than all of its close-by neighbors in parameter space, but a worse fit than some very distant set of parameters.

A related technique that shares only some of the disadvantages of steepest descent is called downhill simplex. This method works better for functions whose derivatives might be unknown or difficult to calculate, so rather than use the calculated gradient at a location to determine which way the algorithm should take a step, we calculate the value of the goodness-of-fit parameter at a variety of small steps in every possible parameter direction (these points form a simplex in parameter space). We then take a step in the direction that reduces χ^2 by the largest amount before repeating the process, flowing downhill along the parameter surface like an amoeba. Unfortunately, this method has the same sort of problems with getting trapped in local minima that the steepest descent method has, although it can more easily adapt its step size and search radius in an attempt to escape from a local minimum and search for something even lower.

Regardless of your search method, any fitting algorithm can fail if the χ^2 surface is complex enough. Although a search for a best-fit line will always converge, as we move to more complicated parameterizations (exponential, sinusoidal, or even just polynomial), we have an increased chance that our search algorithm will miss the "correct" answer and just keep taking steps off into an unrealistic region of parameter space.

To have the best chance of finding good parameter fits:

- *Choose an appropriate fitting function.* You should plot your data before proceeding with a fit, and if the shape of the data suggests a particular functional form (linear, exponential, sinusoidal, etc.) then choose something appropriate, or use a fitting function based on theory. Even the best fitted parameters will provide a poor fit and worse predictions if they are calculated with a function that does not capture the behavior of your data.
- *Start with a good estimate.* Most iterative fitting programs will allow you to set the starting values for each search parameter. (In *Mathematica*, this can be done by placing each parameter in a list with its starting value, e.g., {a, 500} would start the search for parameter *a* at value 500.) If there is any way for you to estimate a realistic value for your parameter, you should include it as a starting point. This will dramatically decrease the odds that your fitting program will get caught in a local minimum or run off to infinity.
- *Lock down any parameters you are not using.* If you're using a standard expression for your fitting function, but you don't need certain parts of it, set those to be fixed values rather than allowing your program to fit unnecessary parameters. You might manually replace those parameters with numbers, or some fitting programs (such as Origin) allow you to specify certain parameters to have a fixed value rather than being fitted. This will prevent possible fitting errors from changes in those unwanted parameters.
- *Don't be afraid to experiment with alternative fits.* If you see something suggestive in the shape of your data or your residuals after a fit, try changing your fit function or the starting points of your parameters. Even if you have a strong theoretical basis for picking, say, a linear function for your fit, the quadratic outliers or sinusoidal pattern might be telling you something about systematic errors or unmodeled details of your real physical system.

Outliers and Outlier Rejection

Since we expect a certain level of disagreement between our data and our fitted model, how can we define an *outlier*, a point which, due to an error not present in the other data, lies anomalously far from the correct value? If there are truly outliers in the data, we would not wish to include them in the fit – since they are, in essence, drawn from a different underlying distribution of values, we cannot include them in our fit at the same weight as the other points without changing our fit in severe and often undesirable ways.

Sometimes, a simple statistical test can detect outliers and allow you to exclude them; for example, when combining multiple astronomical images taken of the same area of the sky, the pixels corresponding to one spot on the sky are often subjected to a "three-sigma clip," which removes any pixel whose value is more than three standard deviations away from the mean of the pixels at the same position in the other images. For normally distributed errors, we would expect this to be true of only 0.3% of non-outlier pixels, so for a stack of 20 images, there is a 94% chance that all 20 pixels at a given spot will meet this criterion, while it's almost certain to correctly reject pixels that have been hit by cosmic rays (giving a very high value) or pixels with unusually low response due to charge traps or other unrecognized issues with the camera. But bear in mind that if we have 1000 images to combine, then this criterion would be expected to eliminate three legitimate pixels out of that 1000. So there is a cost to using this clipping procedure to eliminate bad pixels, since it eliminates good pixels as well, but we might consider it worthwhile because the rate of bad pixels is relatively high and the inclusion of those bad pixels would impose a significant cost in the final image (by shifting the calculated mean to a distinctly incorrect value).

In another experiment, this type of clipping might be less appropriate: suppose that you were running a particle detection experiment where you were trying to measure how likely one rare decay was, compared with another more common decay. If you applied a sigma-based rejection criterion to each data point, you would find yourself rejecting more legitimate rare decays than legitimate common decays, simply because they would lie farther from the mean of the typical random sample.

These statistical methods are effective for identifying the occasional outlier whose cause is unknown. When running an experiment, you may consider treating data differently, or even discarding them, when there is a clear, articulable reason that they do not agree with your other measurements, for example:

- You discover that a piece of equipment was malfunctioning.
- The apparatus was warming up, cooling down, or otherwise unstable.
- Your records of that batch are flawed or incomplete.
- You opened a new box of supplies and find that all of your measurements have shifted.
- One series of data disagrees with an otherwise strong empirical trend.

In these cases, you must decide for yourself whether to eliminate the data from your report entirely, report them but exclude them from your analysis, include them in your analysis but at a lower weight, or just allow them to affect your results as they may. Excluding data points because they do not fit your expectations or your trend would be considered unethical, but real-world experimentation is full of tweaks to experimental setups, iterative improvements, bug fixes, and other realistic reasons to discard some or all of your data.

How can you make sure that you're taking a responsible approach to data collection, rather than cherry-picking your data to achieve your desired result?

- *Set data collection goals ahead of time and stick with them*: it's tempting to stop your data collection when you see the trend you were predicting, or keep going because you haven't seen anything yet, but this is one of the easiest ways to accidentally bias your data in favor of your desired outcome.
- *Conduct a blind analysis whenever possible*: in order to avoid confirmation bias (the natural tendency to weight new evidence by how well it confirms your preconceptions), many large collaborations use blind analyses, where a random offset is added to the data (and securely recorded elsewhere) so that the analysis team won't be able to tell whether their answer agrees with, say, the known speed of light, or the expected value of Hubble's constant or the gyromagnetic ratio of the electron. Since those numbers are so well known, it would be very tempting to keep seeking out new errors or conducting more complicated analyses until your answer matches the "right" answer, and then stop your work. This would make it nearly impossible for you to actually contribute any independent knowledge about the value you are purportedly measuring! Blinding allows an analysis team to truly put in its best effort without any bias toward the library value.
- *A new discovery, or at least a new kind of failure?* Consider that your "unusual" data points might be a genuine signature of a new unrecognized physical effect – possibly the most important clues that an experimenter can find! – or that they might point you to flaws in your experimental design or analysis that you can correct.

4.12 What to Do When Something Goes Wrong

Despite your best efforts, sometimes your experiment will not proceed according to plan – you'll get a result that you know to be incorrect, not merely because it disagrees with your expectations, but because you recognize, either during or after the experiment, that you did not carry the experiment out correctly. You might have used the equipment incorrectly, left out a key measurement that makes it impossible to analyze your results, or lost all record of your data points.

When this happens, the way out is simple: you must not report data that you know to be incorrect without commenting on them. If you can, replace the data by repeating the experiment correctly from the point where it went wrong, in consultation with your lab instructor or research supervisor. Sometimes this is a simple matter of measuring a forgotten parameter, or carefully setting up an experiment so that you can take a few more measurements; take care with your new data, since you might introduce systematic errors by carrying out the experiment at a different time, with different equipment, or simply by using the apparatus more than once.

If recovering the experiment is not possible, document the failure as clearly as you can and explain how it affects your results: which parts of the experiment can still be completed and which cannot, how you would have proceeded if the experiment had been successful, and what you would need to do differently to correct this issue in the future. A careful consideration of the failure points of an experiment can be a big help to the next person who attempts that measurement.

4.13 Homework Problems

1. For each experiment and effect, figure out whether it would contribute random error, systematic error, or both types:
   (a) Near a vibration-sensitive laser interferometry experiment: daily traffic rises and falls
   (b) Near a DC electronic experiment: 60 Hz electrical interference from nearby wires
   (c) Above a shielded cosmic ray detector: seasonal rise and fall in water table
   (d) During a sensitive light-measuring experiment: stray light leaking in from the room
   (e) During a photon-counting experiment: changing the gain setting so that one photon appears as ten counts (instead of one)

2. Why can't repeated trials reduce systematic error? How *can* systematic error be reduced?

3. In an experiment dominated by random error, you take automated readings from a highly accurate voltmeter and average them in groups of 20. If the settings are changed to average readings in groups of 80, how does this affect the mean, the standard deviation (σ), and the standard error on the mean (SEM), compared with the earlier measurements?

4. An experiment that you read about quoted a mean value for Hubble's constant of 71 km/s ± 3 (systematic) ± 4 (random). If you expanded the sample from that experiment by a factor of 25, how would you expect each of those error estimates to change? What would be the old and new total errors, if you assume that the systematic and random errors are uncorrelated?

5. What would be the standard way of reporting the following raw values and uncertainties, using the correct number of significant figures?
   (a) 75.8 ± 17.5
   (b) 110.7 ± 43
   (c) 20 ± 1.73
   (d) 3554 ± 75

6. For the following measured and theoretical values, determine whether they agree or disagree at the 95% level, given the attached uncertainties.
   (a) 10 ± 3 V (1σ, measured) and 11 ± 1.2 V (1σ, theoretical)
   (b) $1.6 \pm 0.5 \times 10^8$ (95%, measured) and $9 \pm 5 \times 10^7$ (95%, theoretical)
   (c) 150 ± 20 m/s (95%, measured) and 140 ± 20 m/s (95%, theoretical)
   (d) 4900 ± 300 Ω (1σ, measured) and 4550 ± 70 Ω (1σ, measured)

7. For many particle detection experiments, the signals are recorded using *photomultiplier tubes*, which can detect a single incoming photon by creating a cascade of electrons to amplify the

signal. Since the experiment counts single photons arriving during some time interval, the data should obey the *Poisson distribution*:

$$P(k) = \frac{\mu^k e^{-\mu}}{k!}$$

where μ is the mean number of events per interval (say, photons per second) and $P(k)$ is the probability that exactly k events will be detected in any particular time interval.

(a) If the mean number of photons detected per second is 5, calculate the detection probabilities for $k = 1, 2, 3, \ldots 10$. (The Excel EXP() and FACT() functions should be useful for this.)

(b) Repeat the previous calculation for $\mu = 25$ and $\mu = 100$, calculating k for each integer from 1 to 2μ. (Excel will suffer simultaneous under- and over-flow for $k > 150$, but you should be able to discern the pattern by then.)

(c) How does the plot of the Poisson distribution for $\mu = 100$ compare with a plot of the normal distribution for $\mu = 100$, $\sigma = 10$? When using Excel to calculate the values of the normal distribution, you may need to add extra parentheses to ensure that the argument of your EXP() function is always negative.

8. Suppose you are working on a radioactive decay experiment, measuring decays per minute over the course of ten minutes (ten data points). If your fitting function is

$$N(t) = N_0 e^{-at}$$

how many degrees of freedom does your fit have? How wide (measured in standard errors) should your 95% confidence interval be?

9. During an experiment to check Coulomb's Law $\left(F = \frac{kq_1 q_2}{r^2} \right)$, you measure the following values:

$$q_1 = +5 \pm 0.7 \, \text{mC} \qquad q_2 = -7 \pm 1.1 \, \text{mC} \qquad r = 0.05 \pm 0.01 \, \text{m}$$

You can assume the error on $k = 9 \times 10^9 \, \frac{\text{Nm}^2}{\text{C}^2}$ is much smaller than the other errors in the experiment. If your force gauge reads 0.63 μN, does that measurement support Coulomb's Law or contradict it?

10. For $y = Ax^n z^m$, derive equation 4.14.

Acknowledgment

I am grateful to Prof. Suzanne Amador Kane for her input into this section.

Paul Thorman is a Laboratory Instructor in the Departments of Physics and Astronomy at Haverford College. In addition to supervising dozens of students in lab, he gets to show up in lectures with demonstrations of astounding physics and introduce people to the wonders of the night sky. In his spare time, he enjoys building and overtaxing PC gaming systems and assembling mostly harmless robots (Figure 4.12).

FIGURE 4.12 Dr. Paul Thorman.

5

Scientific Ethics

Grace McKenzie-Smith

CONTENTS

To doubt everything, or, to believe everything, are two equally convenient solutions; both dispense with the necessity of reflection.

– Henri Poincaré

Now I am become death, the destroyer of worlds.

– J. Robert Oppenheimer quoting from the *Bhagavad Gita*

Scientific ethics is a broad topic, ranging in scope from the relatively straight forward, such as data fabrication, to the more ambiguous, such as including that extra author on a 100+ author paper. In this chapter, we will introduce the main areas of scientific ethics that you are likely to encounter in a career in physics or some adjacent field. The majority of ethical quandaries encountered by physicists in their day-to-day work are in a moral gray area where it is difficult to determine the truly "right" course of action.[1] In the few cases where there are clear answers, we will be sure to highlight them. However, in addition to ensuring you know the few knowable things in scientific ethics, the purpose of this chapter is to give you the starting place to develop a habit of noticing when there is something you need to think about.

Because of the ambiguity inherent in scientific ethics, it is always valuable to gain additional perspectives when thinking through an ethical problem. Thus, we strongly recommend working through this chapter with several peers or having a group discussion afterward.

This chapter includes numerous references and readings, most of which are easily available on the web. Consult ExpPhys.com for the links.

Finally, please note that while we will be discussing numerous cases of ethics violations, the strong majority of scientists are not committing these offenses and are dedicated to upholding scientific integrity.

---

[1] Johnson and Ecklund. "Ethical Ambiguity in Science." 2016.

5.1 A Brief Overview of Scientific Ethics

In the course of this chapter we will be examining the scientific ethics of physics – that is, the moral principles that govern the conduct of physicists in all the spheres to which that appellation applies: research, teaching, dissemination of information to other physicists and the public, etc. While this chapter is tailored toward those intending to pursue a career in physics, many of the quandaries presented here apply to other fields as well. It is also written from a US perspective with the US regulatory process in mind. Most countries with well-established scientific communities will have some set of similar regulations and agencies, so it is likely that the principles, if not the specifics, will be fairly universal.

Many of the topics we will touch on will also have ethical principles that are broadly applicable even outside of the sciences, such as treatment of subordinates, equality in the workplace, ethical use of funding, etc. However, there are three sources of ethical principles specific to science that shape modern thought on scientific ethics: the scientific method as the guiding principle behind the process of scientific research, the history of immoral human experimentation, and the development of nuclear weaponry. Francis Bacon's statement of the scientific method[2] can be thought of as the guiding light of the scientific community, and the source of much publishing and data ethics. However, the lessons on the danger of science are what shape the public and political view of scientific ethics and provide the scaffolding for principles of regulation and ethical choice of research subject.

Exercise 1: Go to the American Physical Society (APS) Guidelines on Ethics[3] and read the "Summary" and the "Introduction and Rationale" sections (we will be returning to this page many times, as it represents the most clear national code of physics ethics for the US physics community). What is the key interest this code is designed to promote and protect? Was there anything you found peculiar or surprising? If so, why do you think it was there?

5.2 FFP: The Cardinal Sins

As mentioned previously, the vast majority of scientific ethics are dominated by moral fuzziness and uncertainty. Yet, within this miasma there is an unholy trinity of research ethics violations: fabrication (making up results), falsification (altering results), and plagiarism (stealing results),[4] commonly referred to as FFP, and the foundation of the US Office of Science and Technology's Federal Research Misconduct Policy.[5] If you ever find yourself in a situation where you witness FFP, where you consider committing FFP, or where someone asks you to commit FFP, this is the clearest case where all of scientific ethics will tell you what to do: report it, don't do it, report it, respectively.

Much of the ethical discussion surrounding FFP focuses on the fact that it undermines the trust integral to the scientific institution. It is thus worthwhile to briefly explore why trust is so important to science, and how FFP violates that trust. Scientific progress relies on being able to stand on the shoulders of previous generations. Scientists gain new knowledge and a more accurate understanding of the universe by examining the results of previous studies and forming a new and exciting hypothesis – the famous step 1 of the scientific method. Without trust in previous work, a scientist would have to spend their whole career confirming all the basic knowledge their intended course of research rests on. Furthermore, science holds a special place in society. It is both the source of most modern dangers and most modern goods. It is imperative that governments and the public be able to trust scientists when they are advised on the efficacy of a new drug or the consequence of an environmental pollutant.

FFP violations damage the trust inherent to the scientific community because this type of scientific misconduct is clearly based on a desire to deceive for personal gain and cannot be excused as an honest mistake. Falsification and fabrication damage the scientific record and have the potential to waste the

---

[2] Bacon. *Novum Organum*. 1620.
[3] APS Statements. "Guidelines on Ethics."
[4] For a formal definition of these, go to the Office of Research Integrity's website.
[5] Office of Research Integrity. "Federal Research Misconduct Policy." 2000.

time of other researchers attempting to follow up on incorrect results. Plagiarism, while it does not damage the scientific record in the same way, harms the ethos of the field. Plagiarism removes proper credit, discourages good science, and undermines the core processes of advancement in a scientific career. It is imperative for the scientific community that these sorts of violations be heavily discouraged and severely punished.

Case Study 1: Read the executive summary of the Bell Labs report on the investigation into the scientific misconduct of Dr. Hendrik Schön,[6] one of the most blatant and extensive cases of its type. You may also find it interesting to look at the actual data in the appendices.
Questions:
(1) Why do you think the committee believes that wholesale substitution of data is a milder offense than the substitution of individual curves?
(2) Do you agree with the way the committee handled the question of coauthor culpability? What principles would you have adopted?
(3) What changes do you think Bell Labs could have made to make such misconduct less likely in the future? What changes could the scientific community implement?

While any breach of scientific ethics can be damaging to your career and reputation, an FFP case is likely to end the scientific career of the culprit and will taint anyone else associated. This highlights one of the things that makes scientific ethics so complicated: those most able to report a breach of ethics, i.e., those working most closely with the culprit, are also likely to have their own careers damaged. A student who notices questionable behavior from their advisor reports it at their own peril. Even in the best-case scenario where they are able to find another lab to finish their PhD in, they will inevitably lose time, and may have to continually answer questions about their original advisor for the rest of their career. This means that there is fairly strong motivation not to report FFP, and merely to remove oneself from an ethically questionable situation.

Case Study 2: Read this article detailing the experience of a student who joined the lab of a principal investigator (PI) embroiled in a scientific misconduct inquiry.[7]
Questions:
(1) Do you agree with the article's conclusion that research misconduct inquiries should be made more transparent?
(2) How would you react if you found yourself in this situation?
(3) Consider this case from the position of a potential future employer. What would you want to know about the grad student in question before making a decision on hiring them?
Alteration: consider the case where instead of a new student joining the lab, our subject was instead already in the lab, but working on a different project from the one in question. What level of responsibility for this misconduct does she bear?

Another issue with FFP is that its clear wrongness makes it a target for public shaming of the scientific community. As you may have gathered by now, the scientific community is very protective of its right to self-regulate, with good reason, and also of its good standing with the public and with the US government. Consequently, the scientific community and scientific regulators take a harsh stance against FFP not only because it damages the scientific process, but also because it damages the relationship between science and society. Yet, whistleblowers can't be publicly glorified, since this would draw more negative press. In fact, whistleblowing is often implicitly discouraged, since it can damage an institution's reputation.

[6] (Remember that links for most of the readings, including this one, are available at ExpPhys.com.) Lucent Technologies. "Report of the Investigation Committee on the Possibility of Scientific Misconduct in the Work of Hendrik Schön and Coauthors." 2002.
[7] Adam Ruben. "Sins of the Principal Investigator." 2016.

Exercise 2: Read this Retraction Watch article on gas lighting whistleblowers.[8] What career stages do you think would be most vulnerable to this sort of treatment? Oransky offers a suggestion on how to fight this sort of behavior. Do you think this would be effective? What other strategies could individual scientists or the scientific community pursue to protect whistleblowers?

5.3 Data Ethics

Our discussion of FFP leads us into dealing with data ethics more broadly defined. This includes things such as data omission, proper lab notebook upkeep, and data sharing and storage.

Since modern science is an empirical field, data is the foundation of all truly scientific knowledge, and must therefore be treated very mindfully. At every step in the process, the careful scientist must fully consider their handling of data. How is data being taken? Does the method of data collection introduce bias into the results? How is data being recorded? Will the method of recording mean the data is clear for future analysis? How is the data being stored? Will it be accessible to other researchers? How is the data being presented? Is this presentation true to the data?

In general, data handling should be guided by the basic principle of transparency: when taking data, potential errors and sources of bias should be noted. When presenting data, any processing should be explicitly mentioned and justified. In all cases, the scientific record should be preserved, the data should be presented clearly and fully, and the data should be made readily available to other researchers.

Exercise 3: Read the body of "On the Treatment of Data" from *On Being a Scientist* (you may skip the case study if you wish).[9] What surprised you about this text? Did it raise any new concerns for you? Can you think of any solutions to the problems raised by the authors about the modern age of digital information?

Within data ethics, one of the more controversial topics is that of data selection or omission. Unlike clear-cut cases of FFP, there are often nuances to how authors choose which data to present that skate lines of ethical conduct. If a representative image is published, is it truly "representative"? If outliers are omitted, do they need to be mentioned? Can the contrast be altered on a digital image to make some feature relevant to the results more prominent?

Exercise 4: Spend some time thinking about the following questions, and ideally talking them over with a peer. What are some valid reasons for excluding data from a published paper? How/when should excluded data be reported? Can you think of a circumstance where the most ethical course would be to exclude some points from a data set? Read sections 6.1 and 6.3 of Taylor's *An Introduction to Error Analysis*[10] and compare your conclusions with the ones presented therein.

Regardless of how data is reported, it is generally agreed that maintaining a permanent scientific record (often in the form of a lab notebook) is paramount. Lab notebooks are considered legal documents that may be used as evidence in patent cases, in proving primacy of discovery, and in scientific misconduct cases.[11] Poorly kept lab notebooks may lead to inaccurately reported results that damage you and your lab's credibility and send others in the scientific community on wild and expensive goose chases. As demonstrated clearly in the Baltimore Case (where a professor was accused by a student of scientific misconduct, only for it to be determined a mistake of record keeping after several painful years),[12] keeping a clear lab notebook is well worth the effort.

Case Study 3: Many of you will be familiar with Millikan's famous oil drop experiment to measure the charge of the electron. You are less likely to know that it is also a famous case of serious data omission and questionable research practices. Read the case explanation presented by the Online Ethics Classroom

[8] Oransky. "How Institutions Gaslight Whistleblowers – and What Can Be Done." 2018.
[9] National Academy of Sciences. *On Being a Scientist*, 3rd edition. 2009.
[10] Taylor. *An Introduction to Error Analysis*, 2nd edition. 1982.
[11] Ryan. "Keeping a Lab Notebook." Note that if you have not received training on how to keep a good lab notebook, you should definitely read this guide.
[12] *Nature*. "Margot O'Toole's Record of Events." 1991.

(OEC).[13] You may also find it interesting to review Millikan's original lab notebooks (particularly note the marginalia, such as on page 59 and page 71). [14]

Questions:

(1) Read the "Research Results" section of the APS ethics guidelines.[15] Does Millikan's handling of his data satisfy these requirements?

(2) Do you believe the fact that Millikan arrived at the correct result for the charge of the electron has any bearing on the ethical analysis of his data handling practices? Why or why not?

(3) What is a more ethical way Millikan could have handled the discrepancies in his data? How would this affect the strength of his conclusions in a published paper?

Alteration 1: what if Millikan had instead found the wrong result for the charge of the electron? How would this affect your analysis? How do you think it would affect the reaction of the rest of the scientific community?

Alteration 2: imagine you are a graduate student working closely with Millikan on these experiments, and you happened to see his lab notebook, noting his comments of "publish this" and "error high, will not use." How would you react?

An accelerating issue in data ethics is the use of digital image manipulation. As tools and techniques to alter images have become more accessible, the number of misconduct inquiries involving these manipulations has been on the rise.[16,17] One of the main issues with this type of data manipulation is that it is easy and cloaked in shades of the benign. Rather than going to the trouble of fabricating a whole dataset, a perpetrator merely alters a few pixels over the course of a few minutes. Since the action itself is smaller, the breach of ethics may seem less significant. Another facet that makes digital image manipulation such a complicated issue is that there is a large gray zone of what is acceptable or not. With increasingly complex data often matched by decreasing article length allowance, there is a strong push to make data easy to understand and space efficient. Researchers need to balance transparency of data with clarity of story, and most scientists aren't exposed to specific guidelines that delineate where acceptable adjustments end and fraud begins.

Exercise 5: What are some common sense rules that should be applied to digital images in scientific papers? Think about or discuss this question, and then go read this *Journal of Cell Biology* (*JCB*) article on image manipulation guidelines.[18] Did you come up with any of these rules on your own? What did you miss, or have additionally? Do you agree with the logic of this particular proposal?

Case Study 4 (hypothetical): You are getting ready to publish a paper about a process to fabricate a certain crystal structure, which is scientifically interesting but has no major commercial applications. Part of your data are atomic force microscopy images of your crystal. The process you use to produce this crystal also happens to create a second crystal, which you want to study in a future project. You believe this future project is likely to get you a publication in a high-profile journal, which would help you secure desirable positions down the road and give your current lab access to better funding. However, you know of another lab that has a very similar instrument set up to yours, and could easily replicate your process within a few months, potentially scooping this future project. Since you have already disclosed the fact that there are byproducts to your process, it would not change the main result of your paper to alter your AFM image to show blank substrate in place of this interesting crystal. Since this is not a cut-and-dry case of data fabrication or falsification, you are unsure about whether this is a permissible action.

Questions:

(1) Is this alteration within the realm of ethical acceptability defined by the *JCB* article discussed above? Would there be a way to protect this future project in a more acceptable way? If so, would you do it?

[13] Goldfarb and Pritchard. "Case Study 2: The Millikan Case." 1999.

[14] Millikan. Oil Drop Experiment Notebook 2. 1912.

[15] APS Statements. "Guidelines on Ethics."

[16] Parrish and Noonan. "Image Manipulation as Research Misconduct." 2009.

[17] Cromey. "Digital Images are Data: And Should be Treated as Such." 2012.

[18] Rossner and Yamada. "What's in a Picture? The Temptation of Image Manipulation." 2004.

(2) Consider your response to a friend who was in this situation. How would you advise them?

(3) What are the facets of science/publishing that make this situation possible?

Alteration 1: rather than yourself, it is your grad school advisor who suggests this image alteration. How would you respond?

Alteration 2: you believe that the second crystal may have potentially lucrative commercial applications. How does this affect your analysis of the ethics in this situation?

Alteration 3: rather than working on crystals, you are instead working in a biophysics lab studying the effects of drugs on metastatic cancer cells. The image you are potentially altering is a western blot showing the protein levels in these cancer cells, and you wish to remove a band on the blot that might lead one of your competitors toward a better treatment alternative. How does this affect your analysis of the ethics in this situation?

Data ethics also include the issues of proper storage and sharing of data. It seems self-evident that since data is the foundation of scientific progress, it must be stored carefully, and reasonably long term. Along with this, most published data must be shared upon request with other scientists to be used to facilitate future research or to verify results. Regulations on data storage and sharing are often enforced by funding agencies or journals, rather than specific national or international rules. For example, all National Science Foundation (NSF) proposals are now required to carry data management plans, and the results of NSF funded research must be shared with other researchers.[19]

Case Study 5: Read this case from the Committee on Publication Ethics (COPE) on the sharing of technical details from a published manuscript.[20]

Questions:

(1) How well do you think COPE and the journal editors handled this case? Is there anything you would have done differently?

(2) The COPE advice section makes reference to a potential "ominous reason" for the original authors' lack of cooperation. To what do you think they are referring?

(3) Consider the position of the journal editors. Suppose that in another few weeks they receive another email from the plaintiff stating that the authors have still not complied. How should they respond? What is the responsibility of the journal in this case?

5.4 Publishing and Credit

The main ways scientists publish their work are either by publication in a peer-reviewed journal or by presentation at a conference (some of which require no review for admission, such as the APS March and April Meetings). Publishing research in peer-reviewed journals is the most respected and accepted way of communicating your research to other scientists and claiming credit for your work. Publishing a paper can be a long and frustrating process with many difficult questions along the way. What is the scientific story your paper tells? Which journal should you submit to? Who should be included on the author list? Sometimes the answers to these questions will be straightforward, but often they will need to be carefully considered to come to an ethical decision.

Exercise 6: The publishing process can be a bit opaque to those who have not yet gone through it, so it will be useful to get a sense of what it entails. Read the introduction to this article by Wiley detailing their publishing process.[21] What steps in the process do you think have the potential to be abused? What step in this process do you think is the most important in preserving a pure scientific record?

Particularly integral to the concept of scientific publishing is the peer-review process. Peer-review is meant to control the quality of published results and to keep scientists accountable to each other. Unlike journal editors, peer-reviewers are experts in the field, and should be able to speak to whether the methods

---

[19] NSF. "Dissemination and Sharing of Research Results."

[20] COPE. "Non-compliance of Author with Request for Information." 2007.

[21] Ali and Watson. "Peer Review and the Publication Process." 2016.

of a paper up for publication are valid or problematic. Often, peer-reviewers will suggest additional experiments to validate the results of a study that make the overall conclusions stronger and more useful to the rest of the field. Peer-review is considered something of a gold standard of validity, and papers published without peer-review are generally not held as valid for consideration in hiring and other decisions. However, there are downsides to peer-review. It is a time-consuming process both for the reviewers and the authors. It is also a potential opportunity for abuse of the publishing system. A hostile reviewer can derail publication for months or even permanently, and good reviews can be used as bribes to secure citations for the reviewer in the published paper. While peer-reviewers are usually supposed to be anonymous, it is often possible for reviewers and reviewees to deduce each other's identity, especially if they work in a smaller field. Thus, a particularly unethical reviewer could delay publication of a competitor's paper in order to get one of their own papers published earlier. The anonymity of peer-review makes accountability more difficult, and it is usually up to the journals to be the final arbitrators on cases of potential reviewer abuse.

Case Study 6: Read this case from the COPE Forum discussing a reviewer breach of confidentiality.[22]
Questions:
(1) Why do you think there is an emphasis on the confidentiality of unpublished manuscripts?
(2) The journal editor and COPE counsel both agree that it is inappropriate to share the name of the reviewer with the paper author, since doing so would violate their anonymity. This might leave the author vulnerable to the same situation at some other journal. Do you think this is the right course of action?
(3) One of the issues in this particular case was the lack of institutional affiliation of the reviewer. Normally, an affiliated institution would be able to undertake an investigation and a potential punishment of misconduct. COPE appears to recommend against using unaffiliated reviewers. Do you think this is fair?

As alluded to in Wiley's article, not all publication outlets are created equal. A paper published in *Nature* or *Science* brings a lot more prestige than a paper published in *Physical Review E*. Journals are often judged on their impact factor – that is, the number of citations to papers published in that journal divided by the number of papers published by that journal in the past two years. This is not always an indication of the quality of the science; journals with higher impact factors tend to have higher retraction rates, possibly because they are intentionally publishing work that is perceived as more exciting and ground breaking.[23] Despite this, most fields have some concept of the "best" journals to publish in. There is a certain amount of gamesmanship that goes into which journal is selected for first submission, perhaps similar to the strategy used when applying early decision to an undergraduate institution. One does not wish to submit to a journal in which the paper is too unlikely to be published, such as a journal in the wrong field, or with a high rejection rate, because this will delay the publication date, and potentially jeopardize receiving credit for first discovery. However, some first submissions will be a "reach" journal, one that has a relatively low chance of accepting any given paper but that has a high impact factor. If the manuscript is rejected upon first submission, it can then be submitted to "safety" journals. This can, of course, lead to additional ethical complications.

Case Study 7: Read this case on the COPE Forum about potential resubmission of a paper in order to increase the impact factor.[24]
Questions:
(1) Do you believe that withdrawing and resubmitting a manuscript to a more prestigious journal is a "misuse of the impact factor"? Would it be considered misuse if the paper had been originally submitted to the high impact factor journal?
(2) The COPE Forum advocates for a much more lenient route than the journal editors originally intended to take in this case. What do you think accounts for this difference?
(3) The journal in question decided to permanently retire its blacklisting policy. Do you agree with this change? In what circumstances do you think blacklisting an author would be justified?

[22] COPE. "Online Posting of Confidential Draft by Peer Reviewer." 2013.
[23] Morrison. "Retracted Science and the Retraction Index." 2011.
[24] COPE. "Unethical Withdrawal after Acceptance to Maximize the 'Impact-Factor'?" 2017.

The publication record is one of the most important considerations in hiring decisions. It is therefore important to the scientific community that it is an accurate reflection of a scientist's accomplishments. There are various ways that one might game the system, such as redundant publications and salami publishing (breaking up science into least publishable units to get a larger number of papers) that harm our ability to accurately evaluate scientific achievement and merit.

Exercise 7: Consider why it is frowned upon to submit to more than one journal at once. Go read this article from the Office of Research Integrity.[25] Did it highlight any additional concerns?

Along with avoiding unethical paper submission practices, the scrupulous scientist must also contend with the careful allocation of credit: proper citation and proper authorship practices. Citation gives proper credit to those responsible for a discovery and helps to advance their careers. Citation is also important to the scientific process: any statement presented as fact that is not general knowledge or the direct conclusion of new results being presented in the paper at hand should be cited. This allows readers to trace back the evidence to its original paper and examine the data for themselves if they wish. This supports the general ethos of science as a discipline built upon its own history.

Authorship often presents a particularly thorny issue in scientific publishing. Authorship carries with it both credit and responsibility for the published work, and it is thus important to list as authors only those who have contributed significantly to the paper (not excluding any significant contributors). Authorship should not be gifted (e.g., listing someone influential in your department or field as an author to gain their approval) and authorship should not be withheld (e.g., leaving a junior student off of an author list to give a larger portion of the credit to more senior scientists). The situation is complicated by the fact that authorship traditions vary from field to field and lab to lab, such as whether a PI is always on a paper, what the author order signifies, and the amount of credit allocated to collaborators. Authorship number can also vary drastically, from single author theory papers to vast collaborations like CERN and LIGO that can have authorship numbers in the thousands.[26] The ambiguity surrounding authorship means it needs to be carefully considered and explicitly discussed before submission to a journal.

Exercise 8: Read the "Authorship" section of the APS ethics guidelines.[27] Do you believe these guidelines cover all the necessary bases? If not, what have they missed? Recall our previous case study on Hendrik Schön. Do you think it would have been harder for him to have committed scientific misconduct if he and his coauthors had followed these guidelines?

5.5 Academia

While much scientific research occurs outside of the academic context, academia still shapes scientific culture, including your current plight as students. It is therefore worthwhile to delve a little bit into the particular ethical issues in academic physics.

The academic job market is extremely competitive, with a very small number of positions for a very large applicant pool. If you are able to secure a job as a professor, then you are thrown from the stress of the job search to the stress of the push for tenure. The ethics of tenure are a story for a different book, but suffice it to say that most professors work harder to get tenure than they did to get their job in the first place. This brings the central conflict of academic science into focus: even though universities are nominally institutes of learning, the interests of the teachers are not always aligned with the interests of the students.

Exercise 9: Read the "Introduction" section of this essay on the "publish or perish" principle.[28] The authors raise arguments both for and against an emphasis on publication. Which do you find more convincing?

The degree to which professors focus on their students is frequently a function of the type of school at which they work. At a small liberal arts college, for example, teaching is often a highly weighted metric

---

[25] The Office of Research Integrity. "Avoiding Plagiarism, Self-plagiarism, and Other Questionable Writing Practices: A Guide to Ethical Writing." 2015.

[26] Mallapaty. "Paper Authorship Goes Hyper." 2018.

[27] APS Statements. "Guidelines on Ethics."

[28] De Rond and Miller. "Publish or Perish." 2005.

in both hiring and tenure decisions. At an R1 research university[29] on the other hand, hiring and tenure decisions are more focused on research prowess (primarily measured by publication record), and thus no strong external motivation exists to promote good teaching practices. Most professors in physics have no formal education in teaching techniques, and many have not led their own class before they reach assistant professor status. The long and short of the matter is that teaching is not a high priority for many professors at top research universities, and this is a problem of physics failing to re-invest in its future. Many current physicists can trace their interest in the science back to one particular class or professor who went above and beyond to provide an exciting introduction to the material. Traditional models of academic hiring and firing disincentivize this sort of effort.

Apart from the overarching pressure of a demanding publication schedule and a potentially equally demanding set of teaching requirements, professors at academic institutions also have to contend with the competitiveness inherent to science. Unfortunately, great minds often think alike, and in research that means that sometimes multiple labs are working on very similar projects. While this is actually a very good thing for science as a whole, since two independent experiments replicating each other is good evidence that the conclusions drawn are valid, it is seen as a bad thing for the individual researchers. Being "scooped" by another group means you often lose the majority of the credit for the discovery and are unable to publish in a high impact factor journal. This can have serious career implications, particularly for junior faculty seeking tenure.

Exercise 10: Read this article on how the online journal ELife proposes to handle scooping.[30] Do you believe ELife's strategy would make scooping less of a concern for a tenure-seeking assistant professor? What do you think of ELife's policy of allowing preprint posting of manuscripts?

The intense and pressurized environment of academia can sometimes push professors down unethical routes to improve their publication record (some examples of which we have discussed). They can also lead to the unethical treatment of students, technicians, and postdocs. While some abuse of power cases are clear-cut, such as pressuring students to commit research misconduct, others are less obvious. Is a grad student being over worked or just in a pre-publication crunch? Was the feedback at that lab meeting too harsh or just a specific mentoring style? Often the line between tough-but-fair and abuse only becomes clear after a pattern is established.

Case Study 8: Many of you have probably seen the infamous Caltech chemistry letter admonishing a postdoc for not coming in on the weekends.[31] Read this more modern case drafted by an astronomy program.[32]
Questions:
(1) Graduate school is considered a very demanding experience, and most graduate students expect to work more than 40-hour workweeks, particularly during periods of intense data collection or in the run-up to publication. What do you think defines the line between reasonable and unreasonable work expectations for graduate students?
(2) This letter highlights several different unhealthy aspects of this particular department. Which ones do you think would be the most difficult for you personally? Which ones do you think are most easily fixable?
(3) The author(s) of this letter claim to have worked "80–100 hours/week in graduate school." This is almost certainly false,[33] although perhaps is not an intentional deception. How do you think this unrealistic depiction of the grad school experience might contribute to the culture of the department? You may wish to review the phenomenon of impostor syndrome.[34]
(4) The author(s) of this letter do give some good (if rather exaggerated) advice on the usefulness of keeping up to date with current academic literature. They are also (perhaps laudably) frank about funding concerns. What other useful comments can you find? On the whole, how would receiving this letter from your department change your views or behavior?

---

[29] This is shorthand for a university with a substantial graduate program and emphasis on research. Examples include most state universities and most large private universities.
[30] Marder. "Scientific Publishing: Beyond Scoops to Best Practices." 2017.
[31] Carreira. "I Expect You to Correct Your Work Ethic Immediately." 1996.
[32] Charfman. "The Detection of Interstellar Boron Sulfide: A Motivational Correspondence." 2012.
[33] Robinson et al. "The Overestimated Workweek Revisited." 2011.
[34] Laursen. "No, You're Not an Impostor." 2008.

Regardless of the form abuse takes, the hierarchical structure of academia means that abuse often goes unreported and unaddressed. For example, once a grad student has found their thesis group, almost their entire future career rests in the hands of their advisor. At every step along their future career path they will be relying on their advisor to get reference letters. It is therefore considered risky to "get on your advisor's bad side." Thus, students find it difficult not only to report abusive situations for themselves and others, but also to speak out about scientific misconduct.[35] While it is possible to change advisors in most PhD programs, it often delays graduation and can result in a weaker reference letter, as your new advisor will have less time working with you.

Exercise 11: Read the "Treatment of Subordinates" section in the APS guidelines.[36] Can you think of any major points APS missed? What are the core principles they emphasize in this statement?

5.6 Equality and Equity

Historically, STEM fields have been dominated by white cis-gendered heterosexual-presenting men, due to a multitude of societal and historical factors. Currently, physics is one of the fields that has made the least progress on correcting historical underrepresentation in the field. Many programs are working to increase diversity with various amounts of success, such as women in physics groups or scholarships targeted at students from underrepresented ethnicities. The continued lack of diversity in physics is the responsibility of all members of the field, and we have an ethical obligation to attempt to correct it. But … why?

Exercise 12: Read the section in the APS ethics guidelines on "Explicit, Systemic, and Implicit Bias."[37] Think of an example, ideally something that you have experienced or witnessed, of each of the types of bias outlined in the guide. What is a way that you could respond to that situation if you are the victim? If you are acting as an ally?[38]

While the arguments in favor of increased diversity might be clear, it is worthwhile to outline them briefly. From a moral standpoint, many cultures hold equality and equity to be basic moral principles. A lack of representation implies some amount of discrimination and bias in the field; that is, members of underrepresented groups who would do well in physics are discouraged from pursuing careers in the field unfairly. Furthermore, if science is to be considered an enterprise of humanity, it only seems just that people of all races, genders, sexualities, etc. have the right to participate.

From a more pragmatic standpoint, diversity is healthy for the scientific institution as a whole and for the world at large.[39] Underrepresented groups in physics such as women and people of color contain large pools of untapped talent that could enhance discovery in the field and drive innovation. Diverse groups of people are better at analyzing and solving problems than less diverse groups,[40] and are able to give a wider range of potential interpretations of data than more homogeneous groups.

Case Study 9: Read Albert Einstein's essay on racial bias in America.[41]
Questions:
(1) What does Einstein point to as the primary source of racial bias in America? How does he propose to correct it?
(2) Imagine you are one of Einstein's intended readers. What part of his argument would you find the most convincing? How might this argument be adapted to apply to a different underrepresented group?
(3) Currently, many of the problems facing people of color in physics are less blatant than they were in the 1940s (although still very real and harmful). How do you think Einstein might update this essay to address a more modern audience?

---

[35] Schneider. "A Personal Tale of Scientific Misconduct." 2016. A particularly sad story.
[36] Hankel. "What to Do When Your Academic Advisor Mistreats You." 2014.
[37] APS Statements. "Guidelines on Ethics."
[38] Atcheson. "Allyship – The Key to Unlocking the Power of Diversity." 2018. A good review on allyship.
[39] Gibbs. "Diversity in STEM: What it is and Why it Matters." 2014. A good and detailed argument on this topic.
[40] Hong and Page. "Groups of Diverse Problem Solvers Can Outperform Groups of High-Ability Problem Solvers." 2004.
[41] Einstein. "The Negro Question." 1946.

From both a moral and pragmatic standpoint it is worthwhile to promote diversity in physics. While promoting diversity does not necessarily have to be your personal crusade, it should be something you think about and work toward whenever an opportunity presents itself. There are many ways that individuals can make life easier (or significantly harder) for members of underrepresented groups,[42] and it is certainly your responsibility to ensure that you are part of the solution rather than part of the problem.

Case Study 10: Review the slides from Alessandro Strumia's controversial 2018 talk on the discrimination against men in physics.[43] For a more complete picture, you may also want to listen to the audio and read Strumia's blog post.[44]

Questions:

(1) Are there points raised by Strumia that you find compelling or thought provoking? Can you think of a more helpful way to present these points?

(2) If you could challenge one of Strumia's pieces of supporting evidence for his "conservative" theory, which would you pick? How would you support your alternative explanation?

(3) This article includes several examples of various responses from the field to Strumia's talk.[45] Which of these responses do you think is the most helpful, and why?

(4) Imagine you were in the room where this was taking place and had the opportunity to debrief with other attendees afterward. How would you respond to the situation if you felt attacked by Strumia's comments? As an ally?

While most scientists these days agree that improving diversity within STEM is a worthwhile activity, the actual paths to achieving this goal can be controversial. Should diversity be a metric used in hiring decisions? Should extra effort be made to find mentors for underrepresented students? Should classes be changed in such a way as to appeal more to underrepresented students? Should departments undergo bias training? Strumia speaks to a fear many conservatives in the field feel: that pushes for diversity will cause an over-correction and become discriminatory against those who have traditionally succeeded ("reverse discrimination"). This mirrors the US debate around affirmative action in admissions decisions for college education.

Exercise 13: Read this APS article about affirmative action.[46] Do you think race and/or sex should factor into faculty hiring decisions? Why or why not? On the whole, did you find Baranger's argument convincing? Finally, why do you think reverse discrimination is often presented as more urgent and worthy of addressing than "traditional" discrimination?

5.7 Financial Considerations

As with most things, the ethical snares of science are further complicated by the addition of financial concerns. This involves the obtaining and responsible spending of grants, patents, commercialization, and conflicts of interest. This issue crosses barriers between scientists operating in the spheres of academia, industry, and government, as money is a constant concern everywhere.

Obtaining grants for research is becoming an increasingly fraught issue, especially as fewer federal grants are awarded each year. In an academic setting, a scientist's ability to find external funding is paramount, since it is the source of their students' and technicians' salaries, a significant factor in hiring and tenure decisions, and represents their ability to continue pursuing research in their field. The pressure for external funding drives professors to spend hundreds of hours grant writing (rather than teaching or researching), and to pursue research areas that match funding agencies values rather than those where

---

[42] Nelson. "Commentary: Diversity in Physics: Are you Part of the Problem?" 2017.

[43] Strumia. "Experimental Test of a New Global Discrete Symmetry." 2018.

[44] Strumia. "ENG – The Gender Talk at CERN." 2018.

[45] Banks. "Thousands of Physicists Sign Letter Condemning 'Disgraceful' Alessandro Strumia Gender Talk." 2018.

[46] Baranger. "Questioning Affirmative Action." 1996.

their personal interest lies.[47] It also encourages ethically questionable behavior, such as misspending of grant funds, over-hyping or fabricating preliminary data to secure grants, and publishing low-quality research.[48]

Case Study 11 (hypothetical): You are a pre-tenure professor writing a proposal for a grant that, if awarded, will allow you to continue supporting your current grad students for another three years. If you do not receive this grant, you will probably be unable to fund at least one of your students next year. You have what you believe are some promising preliminary results for improving a protein purification process. When the control and experimental means are graphed, it looks quite impressive … but the error bars are large and overlap to a degree that makes the difference between control and experimental means questionable. You think your results would be more compelling to the grants committee if you merely graphed the means without the error bars. Since grant proposals are very space-limited, it would not be notable that you included a simplified chart.

Questions:

(1) In your opinion, if you were to include the graph without error bars, would this be a case of scientific misconduct? Try to support your answer using some of the guidelines previously discussed.

(2) What are some of the pressures you feel that might compel you to act in a potentially unethical fashion? What would be one way to alleviate some of these pressures?

Alteration 1: imagine that your preliminary results have no issue with error bars, but are fundamentally uninteresting to your potential funding agency. It is possible (though you don't believe it likely) that this protein purification process could have applications in cancer immunotherapy. Would it be scientific misconduct to include these potential applications in your grant proposal? Would it be unethical?

Alteration 2: imagine that your grant application for the protein-purification process has been rejected, but you happened to get a different grant to study the properties of protein liquid crystals. There is enough money in the grant to support all three of your graduate students, but only two are actually working on that project while the third is continuing the shunned protein purification project. Would it be ethical to fund your third student with money from the liquid crystal grant? If not, is there a change to the scenario that would make it ethical?

As federal grants become harder to obtain, scientists are turning to alternative sources of funding, which can introduce additional conflicts of interest into the equation. For example, if your lab is funded by a fabrication company to test the tensile strength of their new material for use in bridge building, there will be a fair amount of pressure to find that the material is indeed strong enough to be used. These conflicts of interest can bring bias into the scientific record and must always be fully declared in published papers. It is not necessary to avoid private funding. However, if you are reading a privately funded paper, or working in a privately funded lab or within industry at all, this is a potential bias that you need to consider carefully.

Case Study 12: Read this Retraction Watch article on a case of a retraction involving undisclosed conflicts of interest.[49]

Questions:

(1) Consider the responsible parties referenced in this article: the reporting student, the journal editors, the peer reviewers, the other coauthors, and the university. Which of these parties do you think acted in an ethical matter? Where does the majority of the responsibility for the errors in the paper lie?

(2) As far as we can tell, the conflict of interest described here does not arise from funding sources, but rather from potential benefit to a commercial interest of the authors. Does this represent a misuse of funding from the primary funding source?

(3) Consider the position of the student in this situation. Does his institution owe him any compensation?

---

[47] Lilienfeld. "The Neglected Implications of Grant Culture." 2018.

[48] Horgan. "Study Reveals Amazing Surge in Scientific Hype." 2015.

[49] Oransky. "Stem Cell Retraction Leaves Grad Student in Limbo, Reveals Tangled Web of Industry-Academic Ties." 2012.

As referenced in the above case study, major ethics issues can arise surrounding the question of commercialization in science. Since the Bayh-Dole act was passed in 1980, there has been a large push in the US for universities to license scientific discoveries, either to already existing companies or to a university-sponsored startup. While this can lead to innovation and economic output, it also poses significant ethical issues.[50] A PI motivated by a promising startup may put students on projects that are potentially profitable rather than scientifically meritorious. Technologies and therapies that require more time in development may be fast-tracked to consumers based on economic pressures rather than solid evidence. Fundamentally, the goals of science and the goals of business are not the same.

Exercise 14: Read the abstract and the first part of the discussion of this article about commercialization of university research,[51] and then pick one of the "potential risks" sections to delve into more thoroughly. What ethical principles do the authors believe could be violated in this case? What are the potential negative effects on the scientific institution or on the public? Do you think the potential benefits (outlined at the beginning of the discussion section) of commercialization outweigh these negatives?

5.8 Safety

Safety in science seems like it ought to be relatively straightforward but, as with many of our topics, it grows more complex upon serious examination. We will consider safety within a research institution, safety in the translation of science to society (medicine, technology development, etc.), and disasters caused by the negligence of people within the scientific community. In each of these cases, there are conflicts between the pressure to carry out research efficiently and the pressure to hold safety paramount.

Safety is often regulated at the institutional level, with various mandatory trainings on proper disposal of hazardous materials, basic lab safety practices, and proper use of machine shop equipment. However, the letter of institutional law does not always match up with the actual practices of researchers. Many institutions have policies against working in a lab alone, and yet very few experimental scientists will manage to go a whole month without some solo lab time, even when working with potential hazards such as open flames or caustic chemicals. This can lead to a certain sense of casualness around safety regulations, heightened by the fact that the results of ignoring rules presented alongside each other may vary from mild skin irritation to death. This ambiguity can create hazardous research conditions with potentially fatal consequences.

As with several other topics, the arguments for laboratory safety standards can be presented from both a moral and pragmatic standpoint. Morally, it is clearly unfortunate when someone is maimed or killed in any preventable fashion. Pragmatically, casualties of laboratory research are potentially great losses to the scientific institution. It is generally considered the PI's responsibility to ensure that the practices followed in their lab are safe for all the workers therein, but if you ever notice an unsafe condition or practice you should consider it your personal responsibility to report it to your PI and to make sure it is dealt with. Note that following proper guidelines on the handling and disposal of hazardous materials protects not only yourself and your fellow researchers, but also the local environs of your institution and the cleaning staff of your research building.

Exercise 15: Read this ten-year retrospective on the death of Sheri Sangji.[52] What problems within the scientific institute did this article highlight? Do you think the solutions offered are likely to work?

For translational safety, ethical issues often arise in determining the timeline for public distribution of a new product. Whether it's the initial marketing of radium as a health aid[53] or the glacial pace of the introduction of self-driving cars, the potential for harm vs. the possible rewards must be carefully analyzed. The correct balance between conservative caution and progressive optimism is

---

[50] Kumar. "Ethical Conflicts in Commercialization of University Research in the Post-Bayh-Dole Era." 2010.

[51] Caulfield and Ogbogu. "The Commercialization of University-based Research: Balancing Risks and Benefits." 2015.

[52] Benderly. "A Decade after a Needless Lab Death, How to Strengthen Safety." 2019.

[53] Fröman. "Marie and Pierre Curie and the Discovery of Polonium and Radium." 1996.

rarely easily discernable, particularly in cases such as potentially life-saving treatments or innovative technologies.

Exercise 16: Read the "History" section of this article on the new cases of thalidomide-induced birth defects in Brazil.[54] Where do you believe the responsibility for this disaster lies? What do you think the main lessons on the ethics of translational safety are in this case? If you wish, take the time to compare your thoughts with those of Frances Kelsey, who was responsible for thalidomide not being approved for use in the US.[55]

Finally, scientific ethics meet safety in the form of preventable disasters. The demands within science often push researchers toward unsafe actions. The need to publish a paper earlier than a competitor might lead to lax lab safety standards. The need to gain access to a new market might lead to a drug being approved before all its side effects have been thoroughly studied. Likewise, in the broader world of applied sciences, particularly in engineering, the pressure to complete projects on time and under budget can lead to safety concerns and disaster on a large scale.

Case Study 13: Perhaps one of the most famous cases of a failure of safety ethics is the 1986 *Challenger* disaster, likely caused by the "go fever" of the US–USSR space race. Read the following case from the OEC (you may start reading at the "Recommendations" section if you wish).[56]

Questions:

(1) Review the engineering code of ethics.[57] Did the MTI engineers act ethically according to these guidelines? What pressures were they facing?

(2) What, if anything, could the MTI engineers have done differently to prevent the disaster from occurring?

(3) It is hard not to feel the pain of the narrator when reading this case. While this engineer clearly had strong ethical convictions about the right course of action, he was unable to prevail. What lessons can you take from this case?

5.9 Communication

Successfully communicating science to non-scientists is a large problem. While scientists have been trained to understand and respect the scientific method, to trust the work of other scientists, and to grasp the basics of quantitative and statistical arguments, the vast majority of the public has no such training. However, for both idealistic and pragmatic reasons, it is often worthwhile to attempt to communicate scientific results and scientific methodology to the public.[58,59]

Much of the communication between scientists and the public falls under the category of "scientific outreach." Scientific outreach includes any effort that increases public engagement with science – either broadly defined or your specific research. Funding organizations such as the NSF encourage this type of work (for example, through the "broader impacts" section that is required in grant proposals), and it is also prized by groups such as the APS. Outreach can include activities such as giving presentations at local schools about your research, participating in scientific demonstrations at science museums and festivals, and providing research experiences for interested high schoolers. Outreach is a way to give back to the public (who often pay for research through their taxes) by showing them the joys of science. Outreach is also a way to improve public perception of the scientific community and potentially encourage students to pursue scientific careers. In many cases, outreach efforts are targeted specifically at underrepresented groups, such as providing summer internships to first-generation college students or bringing scientific presentations to lower-income schools.

[54] Vargesson. "Thalidomide-Induced Teratogenesis: History and Mechanisms." 2015.
[55] Kelsey. "Thalidomide Update: Regulatory Aspects." 1988.
[56] Boisjoly. "Telecon Meeting (Ethical Decisions- Morton Thiokol and the Challenger Disaster)." 2006.
[57] National Society of Professional Engineers. "NSPE Code of Ethics for Engineers." 2019.
[58] Medvecky and Leach. "The Ethics of Science Communication." 2017.
[59] Safina. "Why Communicate Science?" 2012.

Exercise 17: Read this article on the effectiveness of scientific outreach.[60] What are the main issues the author has with more "traditional" forms of outreach? Do you believe his alternative suggestion would be effective? What are some potential pitfalls to carrying out "local" outreach?

Another facet of communicating with the public comes through the media, ranging from press releases about new discoveries to pop-sci articles about Nobel Prize winners. This can be a wonderful thing. Many of you are probably taking this course because you think physics is amazing, exciting, interesting, and worth sharing. Bringing science to the public can lead to great moments, like the shared joy of a successful moon landing or the broad excitement over the discovery of gravitational waves. However, it is difficult to effectively communicate the results of a complicated study in a single digestible sound bite, and this can lead to misunderstandings and exaggerations. Figuring out how to effectively communicate your science to reporters can be tricky, but it is also important that you do so in order to preserve both your own professional integrity and the standing of the scientific community at large.

Exercise 18: Read this article from *EMBO* on misleading scientific reporting in the media.[61] What lessons on how to communicate your research to the media can you draw from this article, which highlights the role that journals play in misreporting science? What steps could an author take to protect their research from this possibility?

Scientists must also be able to advise the public and policy makers so that they can make informed decisions about scientific topics. Advice can range from expert witness testimony to presenting reports to congress. Along with this advising, there is the possibility of specific policy advocacy. There is considerable debate on how much influence a scientist should try to exert over public policy. A scientist seen advocating for specific policies may be considered biased and may lose professional credibility. Furthermore, while scientists are often more aware of the details surrounding certain issues such as pollution, vaccination, and climate change, being more informed doesn't necessarily mean they have a monopoly on making the "right" decisions. On the other hand, some within the scientific community see it as an obligation of scientists to advocate for the policies they believe are most supported by the evidence. Some scientists choose to stay out of policy advocacy entirely,[62] while others make it a significant focus of their careers.[63]

Case Study 14: Read the case study from the OEC on the US Strategic Defense Initiative (SDI).[64]
Questions:
(1) Consider the questions suggested by the OEC. What evidence would you need to sign on to any of these statements? How certain would you need to be?
(2) Imagine that you are in cosmology theory (or some other field very unlikely to be involved in the SDI). Would these statements still be relevant to you? Would you feel qualified to take a firm position on them?
(3) The statements provide several arguments for and against the SDI. Which do you find most compelling, and why?
(4) Consider some matter of policy that could be influenced by scientists today. Do you think it should be? What sort of points might you include in a statement proposal to the APS on this topic?

When attempting to advise and inform the public and policy makers about scientific issues, it is important to be as accurate as possible while still being clear. As mentioned above, the statistical and quantitative arguments that most scientists are trained in are not as easy for the general public to parse. Furthermore, the philosophy of science discourages absolutism. Scientists have evidence that can support specific interpretations of the data, but most scientific statements should not be made with the same certainty that many policy makers state their beliefs. Nowhere is this more apparent, or more potentially

---

[60] Raman. "Science Outreach Is Great but Scientists Must Consider Who They're Reaching." 2018.
[61] Moore. "Bad Science in the Headlines." 2006.
[62] Revkin. "Melding Science and Diplomacy to Run a Global Climate Review." 2007.
[63] Gillis. "Climate Maverick to Retire from NASA." 2013.
[64] Thomsen. "The Strategic Defense Initiative." 2018.

harmful, than the discourse surrounding global climate change. As a scientist, you have some responsibility to society to find ways to communicate both your own research and more overarching concepts accurately, but also effectively.

Exercise 19: Read this article on how to effectively frame communication about climate change.[65] Intentionally framing an argument so as to push listeners toward a single conclusion might be considered unethical. Do you think it is ethical or unethical in the case of climate change? What do you think the personal responsibility of climate scientists is in communicating their research? What about the responsibility of a non-climate scientists discussing climate change with friends or colleagues? You may wish to compare your thoughts with the official APS stance on climate change.[66]

5.10 Regulations

Official sources of regulation are actually relatively sparse within physics, and the regulations that do exist are usually enforced by funding agencies, journals, and institutions rather than by governments or the international community. Journals and institutions are often involved in investigating and disciplining cases of scientific misconduct, as you will have seen in earlier examples. Institutions are also usually responsible for creating and enforcing safety regulations and carrying out proper disposal of hazardous research materials. Field organizations, such as APS, will usually have a set of guidelines that their members are encouraged/required to follow. There are also various watchdog and advisory agencies, such as COPE, Retraction Watch, and the US Office of Research Integrity. Regulation at the funding level essentially involves funding organizations choosing not to give money to projects they find intellectually or morally unworthy. They may also require additional trainings, such as the Responsible Conduct of Research training mandated by NSF grants.[67] Most of the public funding for physics research in the US will come from the NSF or the Department of Defense (DoD), each of which has its own specific mission (basic research and defense-oriented research, respectively). One notable (and perhaps obvious) exception to the general lack of national regulation on physics research is nuclear research, which often requires a DoD security clearance to get access to material.

Exercise 20: Read the "Summary" section of Vannevar Bush's report to President F.D. Roosevelt.[68] Bush was one of the driving forces behind the creation of the NSF. Bush posits several reasons why science is important for the nation. What other reasons can you think of? Think of some of the scientific papers you have read recently or go skim some abstracts from your favorite journal. How do these studies fit into the goals of their funding agency?

One of the main thrusts of Bush's letter was that science should be relatively free of external regulation, and physics has by and large succeeded in that goal. However, other fields are much more heavily regulated, and as science becomes increasingly interdisciplinary you may find yourself collaborating or working in an area with more oversight. The heaviest regulations exist, of course, around biology and medical science. Biologists must navigate regulation at the institutional, national, and international levels, largely because of an unfortunate history of human experimentation. While many of these regulations are universally agreed upon, others are more controversial. Some might argue that, rather than promoting good science, regulations stifle innovation and politicize basic research.[69] On the other hand, prudent regulations can protect the credibility of the scientific institution and potentially prevent great harm.[70] If you go into a field like biophysics, it is likely that you will brush up against these regulations, and possibly be in a position to influence future or current regulatory policy.

[65] Hendricks. "Communicating Climate Change: Focus on the Framing, Not Just the Facts." 2017.

[66] APS Statements. "Statement on Earth's Changing Climate."

[67] NSF. "Responsible Conduct of Research (RCR)."

[68] Bush. "Science the Endless Frontier." 1945. Note that this is a very interesting piece of scientific history, and you may find it worthwhile to read the full report.

[69] Wolpert. "Is Science Dangerous?" 1999.

[70] Intemann and De Melo-Martín. "Regulating Scientific Research: Should Scientists Be Left Alone?" 2007.

Exercise 21: The Belmont Report is the primary US guideline on the ethics of human subject research. Read the introduction to this article about the ethics of the Belmont Report.[71] What are some of the issues with the current ethical system of the report? Which of these issues do you think would be most likely to make carrying out effective research more difficult?

5.11 Choice of Research

Finally we come to the topic that has perhaps the least certainty and the most areas of gray. What responsibilities does a scientist have when choosing their research? Some believe that science ought to be an ivory tower soaring above the vagaries of the politics of the day. Science is the pure pursuit of knowledge, and whether that knowledge is used for good or ill is not the concern of those who discover it. It seems against the empirical philosophy of the field to avoid searching for knowledge in a particular area out of fear. On the other hand, science can be seen as a dangerous field, where scientists have to choose their research carefully so as not to harm the world with their findings. Many scientists carry out their research using public funds and are therefore obligated to ensure the safety and well-being of their society. Some might argue that had scientists been more careful or more forward thinking, modern society would be free of worries such as nuclear bombs and biological weapons. This point of view would posit that scientists must always consider their research through the lens of potential future harm.

Within physics these questions most often come up in the context of the scientists who were working on the Manhattan project that developed the nuclear bombs eventually dropped on Japan by the US during World War II. Of all the scientists on the Manhattan Project, only one, Joseph Rotblat, resigned due to ethical concerns. He then dedicated his life to attempting to mitigate the results of the discoveries he participated in making possible.[72]

Exercise 22: Consider what principles you think should drive how scientists choose their research. Read Dr. Rotblat's essay on the Hippocratic Oath for scientists.[73] How do his proposed guidelines match up with yours? Do you think his reasoning is well founded? Are there additional pieces of evidence you can think of that would either strengthen or weaken his argument?

Within biology, questions about responsible choice of research can extend from genetic engineering[74] to viral gain of function experiments.[75] In physics, the debate is most often focused on weapons research. A large goal of science funding in the US is to support military strategy and defense, and a significant portion of physics funding comes from the DoD. Hinting at military applications in grant proposals can increase the likelihood of funding, and even organizations like NASA have their roots and many continued interests in defense. Again, there is the question of how much responsibility a scientist has to consider the consequences of their research. For military applications this is additionally complicated by the scientist's own views on the foreign policy stance of their nation, and their particular values in regard to violence and pacifism. There is not necessarily a right or wrong answer here, but it is fair to say that any physicist must seriously reflect before undertaking research with direct military applications.

Case Study 15: Read the letters of Norbert Wiener, a scientist involved in weapons research during World War II who publicly refused to continue with any project that had military applications.[76]
Questions:
(1) Wiener seems to believe there are reasonable and unreasonable levels of scientific censorship. What level of censorship do you believe appropriate for his work? Do you believe science should be more censored during times of war?

---

[71] Shore. "Reconceptualizing the Belmont Report." 2006.
[72] The Norwegian Nobel Institute. "Joseph Rotblat Facts." 1995.
[73] Rotblat. "A Hippocratic Oath for Scientists." 1999.
[74] *Nature.* "Germline Gene-Editing Research Needs Rules." 2019.
[75] Ghose. "The Risks of Dangerous Research." 2012.
[76] Wiener. "A Scientist Rebels." 1947.

(2) Wiener seems to be more in favor of defense research than offense research. What do you think is the major difference he sees? What position do you think Wiener would have taken on the SDI?
(3) Read the first letter of response Wiener sent in October.[77] Why do you believe he chose to publish an extended version? Do you agree with his decision?
(4) Read the letter written to Wiener after his article was published. What do you believe would be the effect on someone's career today if they publicly stated they would not pursue weapons research?

Unfortunately, even the most cautious researcher most concerned with the consequences of their research on society may still end up contributing to an end product they are morally opposed to. Science is the search for new knowledge, and this means that it is often impossible to truly predict the consequences of a course of research. Could the first nuclear physicists have predicted the atomic bomb? Could Einstein have predicted the usefulness of GPS? In the end, whether you try your best to predict the consequences of your research or choose the projects that seem most interesting or lucrative, the unexpected may occur. It is worthwhile considering what you feel your responsibility is in those cases.

Case Study 16: Read the case study on Agent Orange from *On Being a Scientist*.[78]
Questions:
(1) Do you think Galston should have been able to predict the results of his research? If he had, do you think he should have continued?
(2) It appears that Galston's most pressing motivation for ending the use of Agent Orange was the personal responsibility he felt. What other good reasons might a scientist advocate for the same cause?
(3) Do you agree with Galston's statement that a scientist must "remain involved with it [their research] till the end"? It may be interesting to consider this in analogy to the legal and social concepts of parental responsibility for a child.

Grace C. McKenzie-Smith is a Graduate Student in the Princeton Physics Department. She has worked in research groups at several institutions, including the University of Georgia, the University of Pennsylvania, Bowdoin College, and Princeton University. Her current research is focused on the physics of social behavior in bumblebees, a particularly interesting example of a eusocial species. While pursuing her PhD, she still manages to find time for singing at the university chapel and spending time with her rabbits (named Strange and Charm, of course) (Figure 5.1).

FIGURE 5.1 Grace McKenzie-Smith and her pet snake Rorschach, at work on data analysis.

[77] Sage Publications. "Science, Technology, & Human Values." 1983. Has original letter drafts and a response to the article.
[78] National Academy of Sciences. *On Being a Scientist*, 3rd edition. 2009.

Part II

Tools of an Experimentalist

6

Analog Electronics

Walter F. Smith

CONTENTS

Distinguishing the signal from the noise requires both scientific knowledge and self-knowledge.

          – Nate Silver (statistician, host of FiveThirtyEight.com)

6.1 Introduction

Nearly every physics experiment begins with analog electronics, i.e., electronics that manipulate continuously varying voltages and currents. Sensors of all types, from pressure sensors to light detectors to viscosity meters, output analog voltages or currents. These signals must almost always be amplified before they can be converted to digital form, i.e., the form in which any voltage below an established value is interpreted as binary 0, and any voltage above that value is interpreted as binary 1. Much of the art of experimental physics comes in choosing the right sensor, connecting it in the right way to the best amplifier for the application, and using the right filtering to separate the desired signal from noise and interference. This chapter focuses primarily on the right way to combine commercial electronics instruments and to use them, since these are the skills most needed for research. Much of this depends on a very deep understanding of resistors and capacitors. You've encountered these before, but may not yet have the intuitive understanding you need to really be proficient with research level instruments.

    Occasionally one does need to build simple electronics from the chip level, so there is also content to teach the basics and the most important subtleties of op amps.

6.2 Input and Output Impedance: Part 1

Motivation, Voltage Dividers

You have presumably already seen the circuit shown in Figure 6.1, but let's review it quickly. It's called a "voltage divider" because the output voltage is a constant fraction of the input voltage, as we can easily show:

The current flowing in the circuit is $I = \dfrac{V_{in}}{R_1 + R_2}$, and the output voltage

(relative to ground) is just the voltage across R_2:

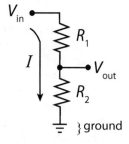

FIGURE 6.1 A voltage divider.

$$V_{out} = IR_2 = V_{in}\frac{R_2}{R_1 + R_2} \tag{6.1}$$

    So, the output equals the input times the fraction $R_2/(R_1+R_2)$. This is the basic equation for the output voltage of a voltage divider. Because voltage dividers (in various forms, including filters) are so very omnipresent, **you should memorize this equation**. In fact, it should be thoroughly ingrained into the very fiber of your being, so that if you grandmother shakes you awake at 4:30 am, points a gun at your head, and shouts, "What's the output voltage of a voltage divider?", you can respond without hesitation.

Now consider two voltage dividers, one after the other, as shown in Figure 6.2. What is V_{out}/V_{in} for this circuit? You might reasonably say, "Well, the voltage divider on the left has $R_1 = R_2$, so (using equation 1) the voltage at point P is half of V_{in}. This voltage then gets divided in half again by the voltage divider on the right, so $V_{out}/V_{in} = \frac{1}{4}$." However, if you were to check yourself experimentally, you would find instead that $V_{out}/V_{in} = \frac{1}{6}$!

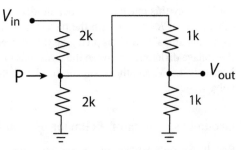

FIGURE 6.2 Two voltage dividers.

The reason for this reduction in the output voltage is that the divider on the right is "loading" the divider on the left. Loading occurs to some extent whenever *any* two circuits are connected together. So, we need to understand what causes loading, how to predict quantitatively the extent of loading, and how to reduce loading effects to acceptable levels. The key concepts for understanding loading are input and output impedance.

In this section, you'll learn the meaning and definitions for input and output impedance, how to measure them experimentally, how to calculate them theoretically from a circuit diagram, and how to apply these concepts to understanding, debugging, and designing circuits and systems.

[Note: the combination of two voltage dividers shown above is simple enough that you can easily analyze it correctly using what you know about resistors in series and parallel resistors, without reference to input and output impedance. However, most other circuits are not so simple, and for these it's essential to use the ideas of input and output impedance.]

Introduction

As you'll recall from a previous course, a real battery can be modeled as an ideal battery in series with a resistor, as shown in Figure 6.3. The value of the resistor is called the "internal resistance" of the battery. It can also be called the "output impedance" of the battery. The concept of output impedance can be extended to all electronic circuits. It is absolutely essential that you **thoroughly** understand this concept if you wish to understand either electronic circuits or the correct way to interconnect electronic instruments. In fact, after the ideas of voltage, current, and resistance, the matching concepts of input and output impedance are the most fundamental in understanding electronic systems.

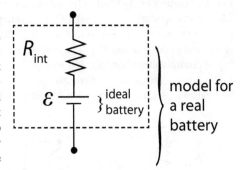

FIGURE 6.3 Model for a battery.

Before proceeding to the more general definition of output impedance, let's make sure you're solid on the simple case of a battery. Of course, if you sawed open a real battery, you would not find an ideal battery and a resistor inside! The combination of the ideal battery and resistor is a "model" for the actual battery. A "model" mimics most of the important behavior of the real object, but is much easier to understand.

What Is an Ideal Battery?

The purpose of a battery is to maintain a voltage difference between its two terminals. For example, an ideal D battery is supposed to maintain a voltage difference of exactly 1.5 V between its two terminals, no matter how much current it has to supply. For example, let's say we connect a resistor with $R = 10^{-9}$

Ω between the two terminals of an ideal battery, as shown in
Figure 6.4. By Ohm's law, the current flowing in this circuit is
equal to the voltage across the resistor divided by its resistance:
$I = V/R = (1.5 \text{ V})/(10^{-9} \ \Omega) = 1.5$ billion amps! No real battery
could supply this much current, but an ideal one would. On the
other hand, if $R = 10^9 \ \Omega$, then the battery only has to supply 1.5
nA ("nanoamps"), which is pretty easy. However, for either case,
the voltage difference between the terminals of the ideal battery
is exactly 1.5 V. Another term for an ideal battery is an "ideal
voltage source."

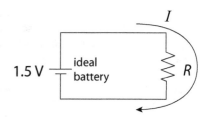

FIGURE 6.4 Ideal battery connected to external resistor.

Ground vs. Common, Behavior of Real Batteries with "No Load" vs. with R_{load}

Recall that only voltage *differences* are significant, for exactly the same reason that only *changes* in
potential energy are significant – the absolute value of the potential energy doesn't matter, since you can
always choose the point at which potential energy is defined to be zero. Similarly, for a given circuit, you
can always choose the point at which voltage is defined to be zero. This point is called the "common"
or "ground." The distinction between these two terms is simple: a point that is called "ground" must
actually be connected to the dirt, usually through a route that includes the third prong of an electrical
outlet. A "common" point need not be connected to the dirt.* However, for our present discussion, the
distinction is not important: since only relative voltages are important, connecting one point in the circuit
to the dirt does not affect the basic operation of the circuit. (The more subtle effects that can arise from
grounding relate to the pickup of electrical interference and will be dealt with in a later lab.) From this
point on, when we discuss voltages, it will always be implied that these are voltages relative to the common (or ground) point of the circuit.

Now let's consider a real battery driving a load resistor. For convenience, let's connect the negative
battery terminal to ground. Again, making a connection to ground at one point in a circuit does not affect
the operation of the circuit; it's just a way of choosing where we'll define $V = 0$. Assume that the voltage of
the ideal battery being used in the model for the real battery is $\varepsilon = 1.5$ V, and that the internal resistance
(or output impedance) is $R_{int} = 0.5 \ \Omega$. If nothing is connected to the battery and we measure the voltage
(relative to ground) at the positive output terminal, we would find 1.5 V; no current is flowing, so there
is no voltage drop across R_{int}. The output voltage when nothing is connected to the output is called the
"open circuit voltage" or "no load voltage."

Next, we connect an external resistor (often called the "load resistance") across the terminals of the
real battery (as shown in Figure 6.5) and measure what happens to the battery's output voltage. We see

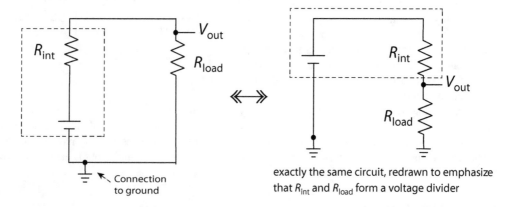

FIGURE 6.5 Real battery connected to external resistor, with ground connection.

---

* It's common for people, including me at times, to be sloppy with this language, and to use "ground" when we should say
"common."

that R_{int} and R_{load} form a voltage divider, so we can apply (6.1) to find the output voltage. For example, if $R_{load} = 1\ \Omega$, then

$$V_{out} = \varepsilon \frac{R_{load}}{R_{int} + R_{load}} = (1.5\ \text{V}) \frac{1\ \Omega}{(0.5\ \Omega) + (1\ \Omega)} = 1\ \text{V}$$

So, the output voltage has dropped by 0.5 V because of loading.

Definition of Output Impedance

The above example shows how to measure R_{int} (or, more generally, the output impedance of any device) experimentally: compare measurements made with and without the load resistor, and apply Ohm's law:

$$\left.\begin{array}{l} V_{no\ load} = \varepsilon \\[4pt] V_{load} = \varepsilon - I_{load} R_{int} \end{array}\right\} \Rightarrow \Delta V \equiv V_{no\ load} - V_{load} = I_{load} R_{int} \left.\begin{array}{l} \\ \\ \end{array}\right\}$$

$$\left.\begin{array}{l} \Delta I \equiv I_{no\ load} - I_{load} \\[4pt] I_{no\ load} = 0 \end{array}\right\} \Rightarrow \Delta I = -I_{load} \qquad \Rightarrow -\frac{\Delta V}{\Delta I} = \frac{I_{load} R_{int}}{I_{load}} = R_{int}$$

In fact, although we've made the comparison between a case where no current is flowing ("no load") and one where current does flow, you should convince yourself that

$$R_{int} = -\frac{\Delta V}{\Delta I}$$

holds if ΔV and ΔI are the difference in voltage and current measured for any two load resistances. (Note that the "−" sign is left "implicit" in most texts. Just recall that the internal resistance has to be positive, and you'll be fine.)

We will use this as the general definition of internal resistance, or, more generally, output impedance:

$$Z_{out} \equiv -\frac{\Delta \tilde{V}_{out}}{\Delta \tilde{I}_{out}} \tag{6.2}$$

Note that I've used the symbol Z, rather than R, and also used the complex versions of the current and voltage. (The actual voltage is given by $V_{out} = \text{Re}\,\tilde{V}_{out}$, and similarly for the current.) This allows us to discuss situations where the impedance includes a capacitor or inductor. Also, I've added the subscripts "out" on the right side, to prevent confusion with the rather similar definition of input impedance we'll encounter later.

How to Measure Output Impedance

In an actual experimental measurement of Z_{out}, instead of bothering to measure ΔI, it is easier just to use the two measured voltages and the known value for R_{load}:

$$Z_{out} \equiv -\frac{\Delta \tilde{V}_{out}}{\Delta \tilde{I}_{out}} = \frac{\Delta \tilde{V}_{out}}{\tilde{I}_{load}} = \frac{\tilde{V}_{no\ load} - \tilde{V}_{load}}{\tilde{V}_{load}/R_{load}} = R_{load} \frac{\tilde{V}_{no\ load} - \tilde{V}_{load}}{\tilde{V}_{load}} \tag{6.3}$$

Usually, Z_{out} is real, i.e., it does not have a capacitive or inductive component. In that case, all the voltages are in phase, and we can write

$$R_{out} = R_{load} \frac{V_{no\ load} - V_{load}}{V_{load}}, \tag{6.4}$$

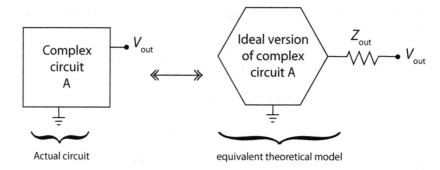

FIGURE 6.6   Output impedance.

where V_{load} is the DC voltage (for a measurement made at DC) or the amplitude of the voltage (for a measurement made with AC). Note that you will get the most accurate measurement if you use an R_{load} that's comparable with Z_{out}.

Generalization of Output Impedance, Perfect Buffers

As we've seen in the preceding discussion, the output voltage of a battery is highest under the no-load condition (i.e., no current flowing), and it drops as we draw more and more current. This behavior is typical not only of batteries, but of almost any electronic circuit. So, we can model almost any circuit as a combination of an ideal version of the circuit (i.e., a version which supplies constant voltage to the output no matter how much current is flowing) in series with a resistor, as shown in Figure 6.6. (The connection from "complex circuit A" to ground is simply there to remind us that V_{out} is the output voltage relative to ground.) Note that we've drawn Z_{out} using a resistor symbol, but in general it might be a complex impedance, such as the parallel combination of a resistor and a capacitor. We'll continue to use the resistor symbol to draw impedances that might or might not be complex.

To be more explicit about what is meant by a "perfect version" of a circuit, let's introduce a new conceptual device: the perfect buffer. As shown in Figure 6.7, this imaginary device measures the voltage at its input without allowing any current to flow into the input (not possible in real life), and provides exactly the same voltage at its output, no matter how much current it has to provide flowing out the output (also not possible in real life). The output current can come from a power supply, so there is no need for the input current to equal the output current. (It is customary not to draw power supply connections in circuit diagrams.) Using this idea of a perfect buffer, the model for complex circuit A can be drawn as shown in Figure 6.8. Again, the perfect buffer is a con-

FIGURE 6.7   An example of a perfect buffer.

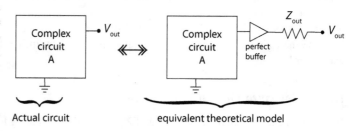

Actual circuit   **equivalent theoretical model**

FIGURE 6.8   Pictorial definition of the output impedance, Z_{out}.

ceptual tool only, and has no actual existence (just as the ideal battery has no actual existence). The perfect buffer is very useful for the construction of theoretical models which describe actual circuits and systems.

Functional Blocks, the Scientific Debugging Process

So far, we've been talking about the outputs of circuits, and have developed a useful model to describe them. We need a similar model to describe the inputs of circuits; we'll develop this model in the next several paragraphs.

To successfully create a complicated measurement set-up, or a complicated electronic circuit, it is absolutely essential to be able to conceptually separate the set-up or circuit into subunits, each of which performs a simple function. For example, as shown in Figure 6.9, perhaps circuit A amplifies the voltage presented at its input (i.e., multiplies the voltage by a constant factor, perhaps 10). Then circuit B uses this amplified voltage to move the electron beam in an electron microscope.

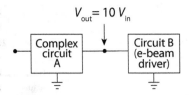

FIGURE 6.9 Two complex circuits

Let's say we have such a circuit, but it doesn't work. Obviously, it's much easier to figure out what's wrong if we can test circuits A and B separately; if only one of them is bad (the usual case), then perhaps it can be replaced, or at the least the job of finding the problem becomes half as big. **This is the most effective way to debug *anything*, whether it be an electronic circuit or a more complicated system: divide it into two parts and test each part. Then divide whichever part is faulty into two parts again and repeat.**

In order to use this debugging technique successfully, it's critical for us to understand how the act of connecting A to B affects the operation of each circuit. Only with this understanding can we test each circuit separately. In most cases, we design the circuits so that their operation doesn't change significantly when we connect them together.

Input Impedance

As we've discussed, the output voltage from A drops if the output has to supply a lot of current. So, to avoid having a significant effect when we connect B to A, we want to design B so that its input doesn't draw a lot of current. To get more quantitative about how much current is too much, it will help to create a model for circuit B. For almost any circuit, one finds that as the voltage applied to the input is increased, the current that goes into the input also increases. (Does this make qualitative sense?) This behavior suggests the model shown in Figure 6.10 to represent the behavior of the input of circuit B. The impedance to ground is called the "input impedance." (It's drawn as a resistor, but it might be something more complicated, such as the parallel combination of a resistor and capacitor.) Again, the perfect buffer and resistor are not really part of circuit B; they are just concepts that we use to build our theoretical model of circuit B. Experience shows that this model works very well for most circuits.

It's clear from this model (and from Ohm's law) that the current flowing into the input terminal increases linearly with the voltage applied to the input terminal. This forms the basic definition of input impedance:

$$Z_{in} \equiv \frac{\Delta \tilde{V}_{in}}{\Delta \tilde{I}_{in}} \tag{6.5}$$

An Example of Complex Input Impedance

If you look at the printing next to a typical oscilloscope input, you'll see that it says "1 MΩ 20 pF" or something similar. This means that the input impedance is really 1 MΩ in parallel with 20 pF of capacitance, as in Figure 6.11.

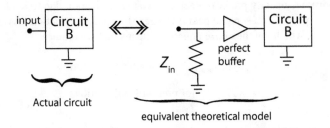

FIGURE 6.10 Graphical definition of the input impedance, Z_{in}.

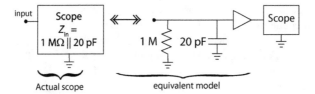

FIGURE 6.11   Model for oscilloscope.

Recall that the impedance of a capacitor is $Z_C = \dfrac{1}{i\omega C}$, where $i = \sqrt{-1}$, $\omega = 2\pi f$ is the angular frequency, and f is the frequency of a sinusoidal signal voltage. You can see that at low frequencies, the impedance of the capacitor becomes very large, so the parallel combination of the input resistance and the input capacitance is dominated by the resistance. However, at high frequencies the impedance of the capacitor becomes very small, so in this regime the input impedance of the scope is dominated by the input capacitance. At intermediate frequencies, you need to take the parallel combination:

$$Z_{in} = Z_{parallel} = \frac{R_{in} Z_C}{R_{in} + Z_C} = \frac{R_{in}/i\omega C_{in}}{R_{in} + 1/i\omega C_{in}} = \frac{R_{in}}{i\omega R_{in} C_{in} + 1}.$$

Combining the Ideas of Input and Output Impedance: Loading Effects

Now that we have our models for the output of A and the input of B, it is easy to compute how much the no-load output voltage of A, $V_{no\ load}$, is lowered when B is connected to it, as shown in Figure 6.12. The output of the perfect buffer in our model for A equals $V_{no\ load}$, since the input of the buffer draws no current and since the output voltage of the buffer always equals the input voltage. The perfect buffer that's part of our model for B draws no current, so it will have no effect on

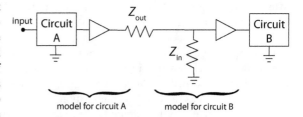

FIGURE 6.12   Model for two circuits connected together, showing effects of output and input impedance.

the circuit fragment between the two buffers. We recognize this fragment, consisting of two resistors, as a voltage divider. We can thus use (6.1) to calculate the voltage at the output of this fragment (which is then applied to the input of B):

$$\tilde{V}_{load} = \tilde{V}_{no\ load} \frac{Z_{in}}{Z_{out} + Z_{in}}. \tag{6.6}$$

So, the output voltage of circuit A is reduced by the fraction $\dfrac{Z_{in}}{Z_{out} + Z_{in}}$ because of the "loading" caused by connecting circuit B.

As discussed above, we usually prefer to keep this voltage reduction due to loading as small as possible. For practical purposes, a reasonable goal is to keep the voltage reduction less than 10% (or 1% for precision applications), which, by the above equation, means that we require

$$Z_{in} \geq 10\, Z_{out} \text{ for most applications, and}$$

$$Z_{in} \geq 100\, Z_{out} \text{ for precision applications.}$$

In most situations, an oscilloscope equipped with a 10X probe has high enough Z_{in} that this condition is met, so you needn't worry about loading when you're using the scope to make measurements. (Of course, it depends on exactly what you're measuring, but most things you'll measure have a sufficiently low Z_{out} to ensure there is negligible loading by the scope.)

How to Measure Input Impedance

Equation (6.6) shows how to determine input impedance experimentally: measure the no-load output voltage from a voltage source with known Z_{out}, then connect it to the device whose input impedance you need to know and measure the voltage again. Rearranging (6.6) gives

$$\tilde{V}_{load} = \tilde{V}_{no\ load}\ \frac{Z_{in}}{Z_{out} + Z_{in}} \Leftrightarrow \tilde{V}_{load}\left(Z_{out} + Z_{in}\right) = \tilde{V}_{no\ load}Z_{in}$$

$$\Leftrightarrow Z_{in} = Z_{out}\frac{\tilde{V}_{load}}{\tilde{V}_{no\ load} - \tilde{V}_{load}} \tag{6.7}$$

For measurements made at DC, this reduces to:

$$R_{in} = R_{out}\frac{V_{load}}{V_{no\ load} - V_{load}}. \tag{6.8}$$

In many cases, Z_{in} can be modeled as a resistor R_{in} in parallel with a capacitance C_{in}. You can show in homework problem 6.5 that, in this case, assuming Z_{out} is purely resistive (i.e., assuming $Z_{out} = R_{out}$) you can rearrange Equation (6.7) to obtain

$$C_{in} = \frac{1}{\omega R_{out}R_{in}}\sqrt{\frac{V_{no\ load}^2}{V_{load}^2}R_{in}^2 - \left(R_{in} + R_{out}\right)^2}, \tag{6.9}$$

where $V_{no\ load}$ and V_{load} are amplitudes (e.g., peak-to-peak amplitudes).

Note that you can measure Z_{in} more accurately if you use a voltage source with Z_{out} comparable with Z_{in}. Usually, this means that you should add a resistor in series with the output of your voltage source, so as to increase the effective output impedance. (Again, for most other purposes, we want a *low* output impedance; this is an exception.) An example is shown in Figure 6.13. We are trying to measure the input impedance of circuit B (which might be a voltmeter). B has a manufacturer's specified value ("spec" value) input impedance of 5 kΩ. (Note that we don't include the "Ω" symbol on circuit diagrams.) We use circuit A, which has a known output impedance of 50 Ω to provide a test voltage. First, we measure $V_{no\ load}$ by connecting A directly to B (top part of figure), and reading the voltage from B. The input impedance of B is 100 times larger than the output impedance of A, so the loading effect is negligible (only about 1%). Next,

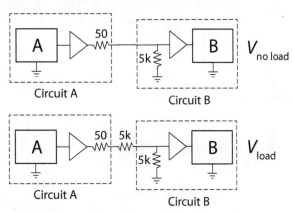

FIGURE 6.13 Measuring input impedance of circuit B.

we effectively change the output impedance of A to $R_{out} = 5050$ Ω by adding a 5 kΩ resistor in series with the output (bottom part of figure). The voltage measured by B under this circumstance is V_{load}. Then we use Equations (6.7), (6.8), or (6.9) to calculate Z_{in} for B.

6.3 Input and Output Impedance: Part 2

How to Calculate Input Impedance by Looking at a Schematic Diagram

First, let's do the very simple example of a voltage divider, as shown in Figure 6.14. Because we want to characterize this circuit by itself, it is customary to assume that whatever the output is connected to draws negligible current, i.e., that it does not appreciably load this circuit. This is equivalent to assuming that the output is left disconnected or "floating." Then it is obvious that the input impedance is simply $Z_{in} = R_1 + R_2$.

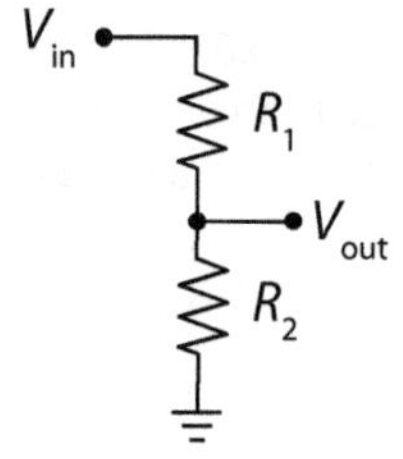

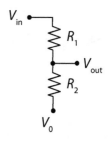

FIGURE 6.14   Voltage divider.

Now let's do a slightly harder example, as shown in Figure 6.15. All we've done is connect the bottom of the voltage divider to a non-zero voltage V_0, rather than to ground. Does this change Z_{in}? Let's see. Recall the definition of input impedance, Equation (6.5):

$$Z_{in} \equiv \frac{\Delta \tilde{V}_{in}}{\Delta \tilde{I}_{in}} = \frac{\Delta V_{in}}{\Delta I_{in}} = \frac{V_A - V_B}{I_A - I_B}$$

where all the voltages are real in this case. We're allowed to choose the two input voltages V_A and V_B for convenience, and then we must find the corresponding two input currents I_A and I_B. Let's choose $V_B = V_0$. Then, clearly $I_B = 0$. When V_A is applied to the input, the resulting input current is $I_A = \dfrac{V_A - V_0}{R_1 + R_2}$, so

FIGURE 6.15   Voltage divider with bottom connected to a non-zero voltage.

$$Z_{in} = \frac{V_A - V_B}{I_A - I_B} = \frac{V_A - V_0}{\dfrac{V_A - V_0}{R_1 + R_2} - 0} = R_1 + R_2$$

So, the input impedance is not changed. Based on this example, I assert that...

> When we consider applying changes in voltage to a point in a circuit, we can compute the resulting change in current by assuming that all fixed voltages in the circuit are at ground. Any fixed voltage "fights" against the change just as much as ground does.

(I have only shown that this works for our one simple example of the voltage divider, but in fact it is always true.)

So, to calculate input impedance of a circuit or system:
Method A:

(1) Assume that the output is left floating.
(2) Pretend that all fixed or well-defined voltages within the circuit (i.e., any voltages provided by a low-output-impedance voltage source, whether DC or AC) are grounds.
(3) Calculate the impedance from the input to ground.

–OR–
Method B:

Use $Z_{in} \equiv \dfrac{\Delta \tilde{V}_{in}}{\Delta \tilde{I}_{in}}$. You can either imagine applying a $\Delta \tilde{V}$ and finding the resulting $\Delta \tilde{I}$, or instead

applying a $\Delta \tilde{I}$ and finding the resulting $\Delta \tilde{V}$. (Usually, it's easier to imagine applying the $\Delta \tilde{V}$.)

How to Calculate Output Impedance by Looking at a Schematic Diagram

Again, we want to characterize the circuit by itself, assuming it's not loading or being loaded by the surrounding circuits. So, it is customary to assume that the input voltage to the circuit is supplied by something with very low output impedance.

Let's do the simple example of a voltage divider, and use the definition of the output impedance, Equation (6.2):

$$Z_{out} \equiv -\frac{\Delta \tilde{V}_{out}}{\Delta \tilde{I}_{out}} = -\frac{V_A - V_B}{I_A - I_B}$$

This time, we'll choose the output currents I_A and I_B, and find the corresponding output voltages. Let's choose $I_B = 0$. This is easy to arrange, just by leaving the output floating (not connected to anything). Then, we can simply apply Equation (6.1) to find the output voltage:

$$V_B = V_{in} \frac{R_2}{R_1 + R_2} \qquad (6.10)$$

Now let the output current be I_A. As shown in Figure 6.16, we can analyze the situation using Kirchoff's junction rule (current into a junction equals current out) and Ohm's law, giving us three equations and three unknowns. We then solve this system of equations to find the output voltage V_A:

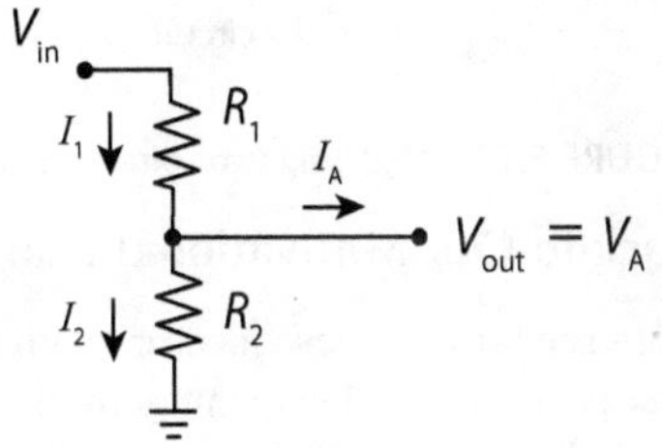

FIGURE 6.16 Currents in a voltage divider.

$$\left.\begin{array}{r} V_{in} - V_A = I_1 R_1 \\ I_1 = I_A + I_2 \end{array}\right\} \Rightarrow \left.\begin{array}{r} V_{in} - V_A = I_A R_1 + I_2 R_1 \\ V_A = I_2 R_2 \end{array}\right\} \Rightarrow V_{in} - V_A = I_A R_1 + V_A \frac{R_1}{R_2}$$

$$\Leftrightarrow V_A = \frac{V_{in} - I_A R_1}{1 + R_1/R_2} = \frac{V_{in} R_2 - I_A R_1 R_2}{R_1 + R_2}$$

So, using this result and Equation (6.10),

$$Z_{out} = -\frac{V_A - V_B}{I_A - I_B} = -\frac{1}{I_A - 0}\left(\frac{V_{in} R_2 - I_A R_1 R_2}{R_1 + R_2} - V_{in} \frac{R_2}{R_1 + R_2}\right) = \frac{R_1 R_2}{R_1 + R_2}$$

We recognize this as the parallel resistance of R_1 and R_2. Note that V_{in} does not appear in this expression; we might have set V_{in} to ground without any effect. Then, our final result would equal the resistance to ground from the output. Although I haven't proved that this result is generally true, it is, and I hope you'll find it reasonable.

So, to calculate output impedance of a circuit or system:
Method A:

(1) Pretend that the input is connected to ground.
(2) Pretend that all fixed or well-defined voltages within the circuit (i.e., any voltages provided by a low-output-impedance voltage source, whether DC or AC) are grounds.

(3) Calculate the impedance from the output to ground.

–OR–
Method B:

Use $Z_{out} \equiv -\dfrac{\Delta \tilde{V}_{out}}{\Delta \tilde{I}_{out}}$. You can either imagine applying a $\Delta \tilde{V}$ and finding the resulting $\Delta \tilde{I}$, or instead applying a $\Delta \tilde{I}$ and finding the resulting $\Delta \tilde{V}$. (Usually, it's easier to imagine applying the $\Delta \tilde{I}$.)

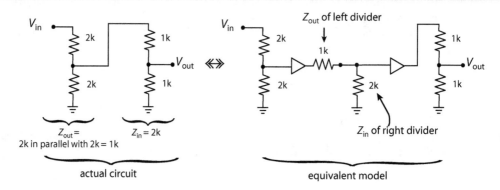

FIGURE 6.17 Modeling two voltage dividers connected together.

Back to Our Motivational Example

Let's see how all these ideas explain the fact that the concatenation of two voltage dividers shown on the first page of this document actually divides the input voltage by 6, rather than the factor of 4 that we might have naively expected. We can draw a model for this circuit, using our ideas of input and output impedance, as shown in Figure 6.17. Although this appears more complicated than the actual circuit, it does exactly the same thing, and has the advantage of modularity: there are three functional units in the model, and they don't load each other at all (because of the perfect buffers). The voltage divider on the left divides by two (i.e., multiplies by 0.5), that in the middle represents the loading, and multiplies the signal by $\dfrac{Z_{in}}{Z_{out}+Z_{in}} = \dfrac{2\,k}{1\,k+2\,k} = 0.67$, and the voltage divider on the right multiplies by 0.5, giving a total multiplication factor of $0.5 \times 0.67 \times 0.5 = 0.17$, corresponding to dividing by 6.

Other Examples, Application to Debugging

Most instruments have the input and output impedance printed on the control panels, near the input or output. For example, the output impedance of your signal generator is given as 50 Ω, and the input impedance of your oscilloscope is given as 1 MΩ in parallel with (typically) 20 pF. Let's assume we're working at very low frequencies, so we can ignore the effect of the input capacitance. If you connect your scope to your signal generator, you will certainly fulfill the condition $Z_{in} \geq 100\, Z_{out}$, so there will be negligible loading at low frequencies.

On the other hand, if you connect the output of the signal generator to a voltage divider with large resistors, and use the scope to monitor the output of the voltage divider, as shown in Figure 6.18, there would

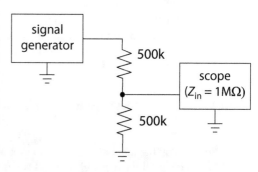

FIGURE 6.18 Using large resistors for a voltage divider can cause problems.

be serious loading problems. As we showed above, the output impedance of the voltage divider is $\frac{R_1 R_2}{R_1 + R_2} = 250\ \text{k}\Omega$. Since the input resistance of the scope is 1 MΩ, the output voltage of the divider would be reduced by

$$\frac{Z_{\text{in}}}{Z_{\text{out}} + Z_{\text{in}}} = \frac{1\ \text{M}\Omega}{0.25\ \text{M}\Omega + 1\ \text{M}\Omega} = 0.8$$

when the scope is connected. To avoid this loading, it would be better to use a 10X probe on the scope. This boosts the scope's input impedance to 10 MΩ, so now the output of the divider is only reduced by

$$\frac{Z_{\text{in}}}{Z_{\text{out}} + Z_{\text{in}}} = \frac{10\ \text{M}\Omega}{0.25\ \text{M}\Omega + 10\ \text{M}\Omega} = 0.98,$$

which is acceptable for most applications.

As a final example, let's look at the combination of two complex circuits we considered before, as shown again in Figure 6.19. Let's say the circuit isn't working. In a threatening, angry voice, your boss says, "You'd better fix this circuit in 2 hours, or else I'll break this egg on your head!" You start by applying a voltage of 1 V to the input, and checking the

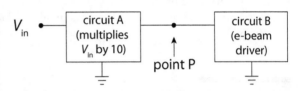

FIGURE 6.19 Two circuits connected.

voltage at point P. If the circuit was working properly, this should be 10 V, since circuit A is supposed to multiply the input voltage by 10. However, instead, you measure 0 V, or at least something close enough to 0 V to be "in the noise." Had you not read this document, you might have said, "Aha! Circuit A is not working!" But now that you know about input and output impedance, you realize that the problem might be with either circuit: circuit A might indeed be dead. Or, something might have happened to circuit B to drastically lower its input impedance (e.g., a short to ground). For example, let's say that circuit A has an output impedance of 50 Ω and is working fine, but a short has caused the input impedance of circuit B to become only 0.1 Ω. Then the voltage at point P would be

$$V_{\text{load}} = V_{\text{no load}} \cdot \frac{Z_{\text{in}}}{Z_{\text{out}} + Z_{\text{in}}} = (10\ \text{V})\frac{0.1\ \Omega}{50\ \Omega + 0.1\ \Omega} = 0.02\ \text{V}.$$

So, to find out which circuit is malfunctioning, you have to disconnect them, and then check the output of A again. If it's still bad, the problem is with A, but if it's now good, the problem is with the input stage of B. (Note: it was clever of you to start by checking the voltage at point P – if it's good, then you know circuit A is working properly, without having to spend the time to disconnect the circuits.)

Input and Output Impedance of Filters

One example of complex output impedance is the low-pass filter, shown in Figure 6.20. This is really just another example of a voltage divider, except that now R_2 is replaced by the impedance of the capacitor. So, the output impedance is the parallel combination of the resistor and the impedance of the capacitor.

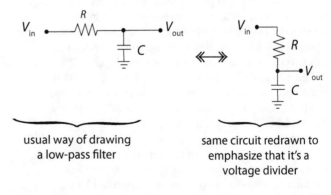

usual way of drawing a low-pass filter

same circuit redrawn to emphasize that it's a voltage divider

FIGURE 6.20 Two ways of drawing a low-pass filter.

To simplify thinking for circuit design, we often use the "worst case" input or output impedance. For the output impedance, the worst case is when the impedance is highest. Since the low-pass filter has an output impedance consisting of a resistor and capacitor in parallel, the impedance is highest at zero frequency (DC), and is then equal to the resistance of the resistor.

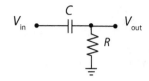

FIGURE 6.21   High-pass filter.

For input impedance, the worst case is when the impedance is lowest. For the low-pass filter, the input impedance would be the series combination of the resistor and capacitor. This has the lowest impedance at high frequencies, when the impedance of the capacitor goes to zero. Then the input impedance is equal to the resistance of the resistor.

You should prove to yourself that the worst case input and output impedances of a high-pass filter (shown in Figure 6.21) are also equal to the resistance of the resistor.

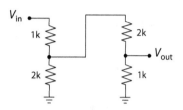

FIGURE 6.22   Circuit for self-test.

Self-test question: use the ideas discussed in this section to show that $\dfrac{V_{out}}{V_{in}} = \dfrac{2}{11}$ k for the circuit shown in Figure 6.22.

6.4 Amplifier Fundamentals

An amplifier multiplies the signal at its input by a constant (at least in the ideal case) factor called the gain. For example, if you apply a square wave with a minimum voltage of −0.1 V and a maximum voltage of +0.3 V to an amplifier of gain 10, the output should be a square wave at the same frequency, oscillating between −1 V and +3 V. Amplifiers are used in virtually all electronic equipment that handles analog voltages, such as the signal coming from a microphone or going to the speaker on a cell phone, or the signal from the light sensing element in a camera. They are also essential in most physics research experiments.

All real amplifiers have limitations. For example, as the frequency of the signal is increased, eventually the gain starts to go down. For most amplifiers, the way in which the gain decreases with increasing frequency is the same as for a low-pass filter. The frequency at which the gain has decreased by a factor of $1/\sqrt{2}$ is called the "bandwidth"; this is analogous to the f_{3dB} for a low-pass filter. All amplifiers add noise to the signal being amplified; you'll explore this more in a later lab. The input impedance of an amplifier would ideally be infinite, but of course it is actually finite, and can usually be modeled as a resistor and capacitor in parallel to ground. The output impedance (ideally zero) is low; sometimes it is set to 50 Ω to match the characteristic impedance of standard coaxial cable, but in other cases it is less than 1 Ω.

Amplifiers can also go into oscillation, which is ordinarily very undesirable. For a commercial amplifier (as opposed to a home-built one), the oscillations are almost always caused by coupling from the output back to the input, via a stray capacitance as shown in Figure 6.23.

There is always some noise present at the output of the amplifier. The noise contains Fourier components at all possible frequencies (up to about 100 times the bandwidth). The stray capacitance from the output of an amplifier back to its input, shown as C_{stray}, together with the input impedance of the amplifier, forms a filter which couples some this noise back to the input. Once at the input, each Fourier component is amplified like any other signal by the amplifier, comes out the front, gets

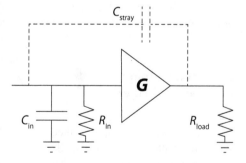

FIGURE 6.23   An ideal amplifier with gain G. The input impedance is modeled as the parallel combination of C_{in} and R_{in}. The output impedance is not shown. R_{load} represents the input impedance of whatever the amplifier is connected to. C_{stray} is an unintentional capacitance resulting from short lengths of unshielded wire at the input and output.

coupled to the input again, and keeps looping around and around, perhaps growing ever larger. For the amplitude of a particular Fourier component to build up, two conditions must be satisfied:

1) The phase shift from the input back around to the input again (via the output and C_{stray}) must be 0°. Otherwise, the new signal coupled from the output would not interfere constructively with the signal from the "previous pass" around the loop.

2) The total gain at the frequency of the Fourier component from the input back around to the input again must be greater than 1.

Since all possible frequencies are present in the noise, if conditions 1 and 2 are met for any frequency, that component will grow in amplitude until it reaches a maximum imposed by some other condition (e.g., the highest voltage the amplifier can output).

Let's look at each of these conditions in more detail.

1) Because the bandwidth effect within the amplifier acts as a low pass filter, it gives a negative phase shift of the output relative to the input, with a shift of 0° at low frequencies, reaching −45° at the bandwidth of the amplifier circuit, and approaching −90° at higher frequencies.

The phase shift from the filter formed by C_{stray}, R_{in}, and C_{in} is more complicated. At low frequencies, the input impedance of the amplifier is dominated by R_{in} (meaning that most of the current that goes through the parallel combination of R_{in} and C_{in} goes through R_{in}). So, C_{stray} and R_{in} form a high-pass filter from the amplifier output to the amplifier input. *If* we could ignore C_{in}, this would produce a phase shift of +90° at frequencies well below $f_{\text{hi}} = \dfrac{1}{2\pi R_{\text{in}} C_{\text{stray}}}$, falling to +45° at f_{hi}, and approaching 0° at higher frequencies.

At the frequency $f_{\text{in}} \equiv \dfrac{1}{2\pi R_{\text{in}} C_{\text{in}}}$, we have $\left| Z_{C\text{in}} \right| = \dfrac{1}{2\pi f C_{\text{in}}} = R_{\text{in}}$, so the magnitudes of the currents going through R_{in} and C_{in} are equal. At frequencies much higher than this, C_{in} dominates the input impedance. In that case, C_{stray} and C_{in} form a voltage divider.

Your turn: show that the output of a voltage divider formed from two capacitors (as shown in Figure 6.24) is in phase with the input, at any frequency.

Combining these two limits, we see that the phase shift from the filter formed by C_{stray}, R_{in}, and C_{in} produces a phase shift of +90° at low frequencies and approaches 0° at high frequencies. Therefore, there is the possibility that this positive phase shift will cancel the negative phase shift from the amplifier, satisfying condition 1.

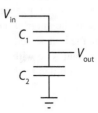

FIGURE 6.24 Capacitive voltage divider.

2) The total gain from the amplifier input to the amplifier output is the gain G of the amplifier, times the effect on amplitude of the low-pass filter associated with the bandwidth of the amplifier (which has f_{3dB} = bandwidth = f_{B}). The filter formed by C_{stray}, R_{in}, and C_{in} reduces the amplitude of the signal at the output by a fraction F as it delivers the signal back to the input. Therefore, the total gain from the input to the output and back around to the input is:

$$\text{total gain} = G\,\frac{1}{\sqrt{1 + f^2 / f_{\text{B}}^2}}\,F$$

There are three ways to keep this below the critical value of total gain = 1 at the frequency (if any) where condition 1 is satisfied. Make G small, make f_{B} small, or make F small. Another way to look at this is to say that high-gain (large G) high-bandwidth (large f_{B}) amplifiers are most prone to oscillation, and if you really need such an amplifier, you must be careful to make

F as small as possible. This means you should reduce R_{in} or increase C_{in} if you can (say by adding a resistor and/or capacitor to ground at the input). However, often this isn't possible, because you need really high input impedance. So, you are left with trying to make C_{stray} as small as possible. This is accomplished by shielding as much of the input and output lines as possible (see Section 6.5 for more on this), keeping unshielded sections as short as possible, and keeping the input and output lines as far apart as you can manage.

6.5 Capacitively Coupled Interference

"Interference" is an undesired signal from outside your apparatus that is picked up by your measurement. The interference can couple into your apparatus by a few different routes. If no precautions are taken, the worst of these is via stray capacitance.

Shown in Figure 6.25 is a schematic of a generic apparatus. C_{stray} and R_{source} form a high-pass filter, allowing some fraction of the voltage from an interference source (such as the AC power line shown here) to appear on the input wire connected to the amplifier. It is then amplified and appears at the output, superposed on the desired signal.

The ratio of input to output voltage for a high-pass filter is

$$\frac{1}{\sqrt{1+f_{hi}^2/f^2}}, \text{ where } f_{hi} = f_{3dB} = \frac{1}{2\pi R_{source}C_{stray}}. \text{ So, to mini-}$$

mize the interference, we want R_{source} and C_{stray} to be as small as possible.

R_{source} is usually fixed by some other experimental consideration. However, if possible, it should be less than 1 MΩ to minimize capacitively coupled interference.

We have much more control over C_{stray}. It can be reduced by shortening the input wire or by moving the amplifier farther from the interference source. However, the most important method to reduce C_{stray} is shielding. If we surround the input wire with a grounded shield, then the stray capacitance couples to the shield, rather than the wire, as shown in Figure 6.26. The wire is capacitively coupled to the shield via C_{shield}. As long as the resistance of the shield to ground is zero, the voltage on the shield stays at zero, so no interference gets through to the input wire. (In reality, the resistance to ground of the shield is small but non-zero, and it's impossible to totally enclose the input wire inside the shield, so it's usually not possible to completely eliminate capacitively coupled interference. However, one can reduce it dramatically with careful shielding, usually to the point where it is negligible.)

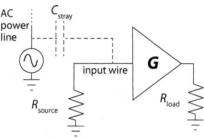

FIGURE 6.25 A generic apparatus. R_{source} is the output impedance of the device connected to the input of the amplifier. The amplifier multiplies the signal at its input by the gain G. R_{load} is the input impedance of the device (for example, a data-taking device) connected to the output of the amplifier. C_{stray} is an unintentional capacitance between the input wire and a source of interference, such as an AC power line.

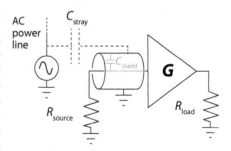

FIGURE 6.26 Reducing interference by shielding.

To arrange shielding, one typically uses coaxial cables for the input wire (with the outer conductor grounded), and/or one can enclose the part of the circuit before the amplifier in a grounded metal box.

To check for residual interference from power lines, connect your scope to the amplifier output, and choose "AC Line" or "Line" as the trigger source. This triggers your scope at the same frequency as the power line, so anything that appears stable on your scope display is interference. (Note that, if your amplifier is operating from AC power, there is probably some internal coupling, so you may not be able to completely eliminate 60 Hz interference.)

Your turn: Why is it less critical to shield the output wire? (Answer below.*)

6.6 Common vs. Ground, Inductively Coupled Interference, and Ground Loops

Common vs. Ground

To gain a deeper understanding of how circuits work, and especially how surprising things can happen when you connect instruments together, we need to distinguish between the terms "common" and "ground."

The "common" is the point in a circuit with respect to which voltages are measured. For our discussion, we will define "ground" to be a point that is connected to "earth ground," i.e., to the dirt outside. (The dirt is electrically conductive if you go down far enough for it to be wet.) In the United States, the third pin on an electrical outlet is connected to ground – there is an actual buried grid of wires that serves to make this connection. Depending on the particular circuit, the common may or may not be tied to ground. For example, for a AAA battery, the common is the negative terminal. The positive terminal is approximately +1.5 V relative to the common, i.e., if you connect a voltmeter across the battery, it reads 1.5 V. You can choose to connect the common to ground; doing so won't affect the reading on the voltmeter or the operation of the battery.

Another example of the difference between common and ground is a typical power supply, as shown in part a of Figure 6.27. It supplies adjustable voltages relative to the banana jack labeled "COM." An example is shown in part b, where the voltmeter on the left measures the voltage difference between the adjustable "+20 V" output and COM. The +20 V output can be adjusted to give voltages in the range 0

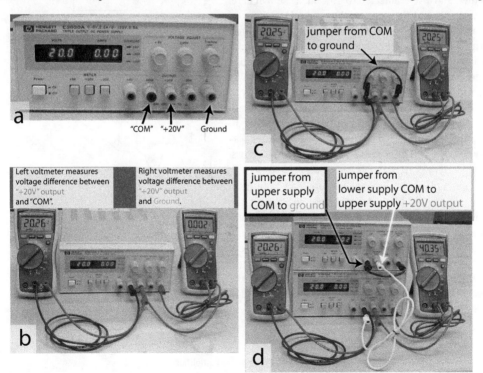

FIGURE 6.27 The difference between COM and Ground.

---

* Two reasons: (1) the output wire is driven by the amplifier, which has a low output impedance, so the weak coupling to interference sources by C_{stray} is much less important. (2) Any interference which does couple to the output wire doesn't get amplified by the factor G.

to 20 V; it is set to 20 V for all of this discussion. The COM is not connected to ground; the power supply works fine without this connection, but the voltage at the +20 V jack is undefined relative to ground, meaning that it might have a significant, perhaps time-varying, additional voltage relative to ground. As shown by the meter on the right, in this case the +20 V output is at 0 V relative to ground, but that's not something you could count on – it's undefined.

If you wish, you can "jumper" COM to ground using a banana cable, as shown in part c; this doesn't affect the voltage difference between the +20 V jack and the "COM" jack. Now the voltage at the "+20 V jack *is* well defined relative to ground – it's at +20 V relative to ground.

You could instead choose to connect the COM to a different voltage, e.g., the +20 V output of a different power supply, as shown in part d. Now, the +20 V jack of the bottom power supply is at +40 V relative to ground, although it's still at +20 V relative to its own COM.

Single-Ended vs. Differential Amplifiers

Now let's apply these ideas to amplifiers. Shown in part a of Figure 6.28 is the model of an amplifier we've been using up to this point. (This type of amplifier is called a "single-ended amplifier," to distinguish it from a "differential amplifier," which we'll discuss shortly.) The coaxial connectors for the input and output are shown explicitly. So far, we've been implicitly assuming that the outer conductors of these connectors are connected together, and also connected to ground, as shown here. Under these conditions, $V_{out} = GV_A$, where V_A is the voltage applied to the inner wire of the input connector, and all voltages are relative to ground.

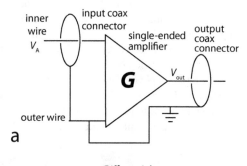

Sometimes, you may want to measure a voltage difference between two points, neither of which is grounded. You'll see in Lab 6B how to do this with a scope. However, if the voltage difference is small enough, you need to amplify it before measuring it. For this case, you need a "differential amplifier," as shown in part b. There are two input coax connectors, and $V_{out} = G(V_A - V_B)$.

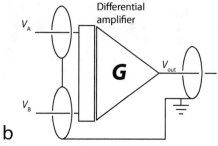

Let's return to conventional single-ended amplifiers. Surprisingly, it is often desirable not to connect the outer conductors of the input and output coax connectors together, nor to connect either of them to ground. (We'll see that omitting these connections can prevent some kinds of interference.) For this circumstance (shown in part c), we refer to the outer conductor of the input connector as the "input common." It is also often called the "signal common."

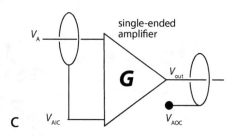

FIGURE   6.28 Single-ended   and   differential amplifiers.

- The voltage at the input common (relative to ground) is V_{AIC}, where the subscript stands for "amplifier input common."

- Similarly, we refer to the outer conductor of the output coax as the output common; the voltage at this point is V_{AOC}, where the subscript stands for "amplifier output common."

- For the rest of this discussion, we will measure all voltages relative to V_{AOC}, i.e., we define $V_{AOC} \equiv 0$.

Roughly speaking, a single-ended amplifier multiplies the difference in voltage between V_A and V_{AIC}. However, this is not a true differential amplifier, so the output voltage (relative to V_{AOC}) has an additive component V_{AIC}, i.e.,

$$V_{out} = G(V_A - V_{AIC}) + V_{AIC} \tag{6.11}$$

(You can see that, if we connect V_{AIC} to V_{AOC}, i.e., set $V_{AIC}=0$, this reduces to the more familiar $V_{out}=GV_A$.)

Inductively Coupled Interference

Once you've carefully shielded your experiment to minimize capacitively coupled interference, which will help enormously, the next most common type of interference you'll need to deal with is inductively coupled interference; a phenomenon called "ground loops" are an important special case.

Inductively coupled interference comes from the emf generated by a changing magnetic field that passes through an important loop in your circuit. The magnetic field usually comes from the power lines. (Since they carry current, they generate magnetic fields.) The emf is given by Faraday's law, $\varepsilon = -\dfrac{d\phi_B}{dt}$, where ϕ_B is the magnetic flux passing through the loop.

Background

We need to discuss an aspect of Faraday's law that is usually glossed over: where exactly does this emf appear in a circuit? We begin by considering an analogous situation that doesn't involve changing magnetic fields: the simplest possible circuit, consisting of an ideal battery and a resistor. We view this circuit in three dimensions, with the z-axis representing the voltage, as shown in Figure 6.29. We define the negative terminal of the battery as the place where $V=0$. Proceeding clockwise, the voltage changes up by the emf of the battery ε as we pass

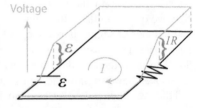

FIGURE 6.29 Voltage changes in perspective.

through the battery, then changes back down by the IR voltage drop across the resistor as we go through the resistor. In order to return back to $V=0$, the voltage drop across the resistor must have the same magnitude as the voltage increase across the battery, i.e., $\varepsilon = IR \Leftrightarrow I = \varepsilon/R$.

Now we imagine dividing the battery in two. For example, if the original battery supplies 10 V, we imagine breaking it into two batteries, each of which supplies 5 V. We also break the resistor into two equal resistors (e.g., we might break a 1000 Ω resistor into two 500 Ω resistors), as shown here. As shown in Figure 6.30, starting at the same place in the circuit (the negative terminal of the left battery), the voltage goes up by $\varepsilon/2$ as we go through the first battery, then down by $\varepsilon/2$ as we go through the first resistor, then up by $\varepsilon/2$ as we go through the second battery, and finally down by $\varepsilon/2$ as we go through the second resistor. There is the same total emf, and the same total resistance, so the current is the same.

Now let's divide the battery into many tiny batteries, with the same total emf, and divide the resistor into many tiny resistors, with the same total resistance, as shown in Figure 6.31. The voltage goes up by a tiny amount each time

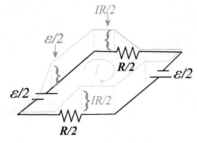

FIGURE 6.30 Battery and resistor each split in two.

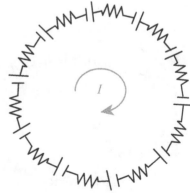

FIGURE 6.31 Battery and resistor each divided into many pieces.

we go through a battery, and down by an equal amount when we go through the resistor next to the battery. The voltage only ever rises by a tiny amount above zero. Again, the total emf and resistance are the same, so the current is the same.

If we divide the battery into an infinite number of infinitesimal batteries, and the resistor into infinitesimal resistors, the voltage remains at 0 as we go around the loop, but the current is still the same, as shown in Figure 6.32. The infinitesimal increases in voltage due to the tiny batteries are continuously canceled by the infinitesimal IR voltage drops across the tiny resistors.

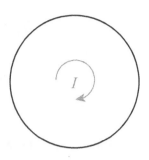

FIGURE 6.32 Battery and resistor divided into infinitesimal pieces.

Now, consider putting a circular loop of uniform wire into a uniform magnetic field that's increasing in time. This creates an emf that drives current in the direction shown in Figure 6.33. The situation is very similar to the previous figure. The emf is spread uniformly around the loop (because of symmetry), and the IR voltage drop is spread uniformly around the loop, so that the voltage is zero at all points, but there's still a current flowing.

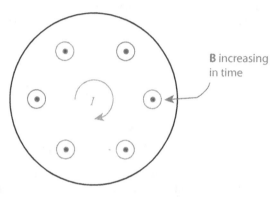

Let's go back to the resistors and batteries. Now we divide up the battery into an infinitesimal number of segments but keep all the resistance in a single resistor. Again, the same current flows because there is the same total emf and resistance. However, now the emf is distributed and the resistance is concentrated. So, the rise in voltage is distributed around the loop, while the IR drop is concentrated, as shown in Figure 6.34.

FIGURE 6.33 Current inducted by changing magnetic field.

If we measure the voltage difference between two points on the loop that are close together, as shown by the V inside a circle, we get nearly zero. However, if we measure the voltage across the resistor, we get a difference of magnitude ε.

This is very similar to what happens if we put a loop of wire in an increasing magnetic field, but with all the resistance concentrated in one place, as shown in Figure 6.35. (We're assuming that the magnetic field produced by the induced current is negligible compared with the field that induces the current; this is usually true.) In most circuits, the loop resistance is in fact concentrated in one place, as shown here. Again, the voltage difference across that resistance has magnitude ε, while the voltage difference between any other two points that are close together is nearly zero.

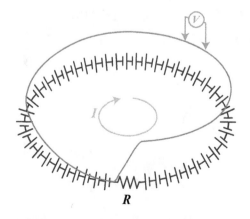

FIGURE 6.34 Battery divided, but resistor not divided.

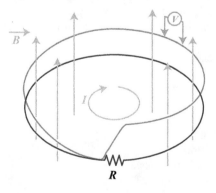

FIGURE 6.35 Resistor not divided, current driven by changing magnetic field.

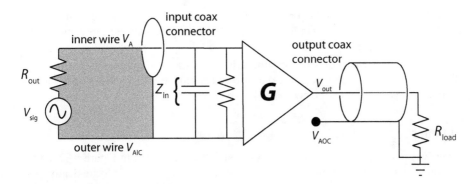

FIGURE 6.36 Inductively coupled interference.

Interference in a Circuit

Now let's apply these ideas to a circuit. In the figure here, we want to use a single-ended amplifier to amplify a voltage V_{sig} from some signal source; this is shown in Figure 6.36 as an AC voltage source, but it could be anything that generates a voltage. The source has output impedance R_{out} and is connected to the input of a single-ended amplifier. As before, V_A is the inner conductor of the input coax connector and V_{AIC} is the outer conductor. For now, we assume no connection between V_{AIC} and V_{AOC}.

The input impedance of the amplifier is shown as a resistor and capacitor in parallel between V_A and V_{AIC}.

The output of the amplifier is connected via a coax cable to a device (e.g., a data taking system) that measures voltage relative to ground; this device is symbolized by R_{load}. As is typical, we'll assume that the outer coax conductor on this R_{load} device is grounded, as shown in the figure.

Now, we introduce a time-changing magnetic field (e.g., produced by the oscillating current I_{AC} in a nearby power line) which is perpendicular to the page. The field through the gray-shaded area creates a time-changing magnetic flux, and so an emf ε_{loop} around the loop. This oscillates at the frequency of the magnetic field, usually the power line frequency. The total emf around the loop is $V_{sig} + \varepsilon_{loop}$. This emf drives a small current around the loop. The input impedance of the amplifier is normally much bigger than the other resistances in the loop (such as R_{out}), so essentially all the IR voltage drop associated with this current occurs across the input impedance, so $V_A - V_{AIC} = V_{sig} + \varepsilon_{loop}$. This voltage difference gets amplified according to Equation (6.11). Because the output contains an amplified version of ε_{loop}, this is inductively coupled interference!

Concept test: If $V_{AIC} = 5$ V relative to V_{AOC} (i.e., $V_{AIC} - V_{AOC} = 5$ V), what is V_{out} in terms of G, V_{sig}, and the ε_{loop} that comes from the time-changing magnetic flux? (Answer below.*)

How to Minimize It

Ordinary shielding with grounded copper or aluminum shields has virtually no effect on magnetic fields – they go right through the shielding. So, you can still get inductively coupled interference. There are four ways to minimize it:

1. Reduce the loop area by using coax cable or cable with the two conductors twisted together. This reduces the magnetic flux, and so the emf. This is by far the most important method of reducing this interference.
2. Move your apparatus away from the source of the magnetic field.

---

* Changing the value of V_{AIC} doesn't change the difference $V_A - V_{AIC}$; only things within the loop change that difference. So, we still have $V_A - V_{AIC} = V_{sig} + \varepsilon_{loop}$. Using Equation (6.11), this gives $V_{out} = G(V_{sig} + \varepsilon_{loop}) + 5$ V.

3. Control where the dominant loop resistance occurs, so that ε_{loop} doesn't get amplified. This is not always possible. (It would not be possible for the example described above, but we'll see an example below where it is possible.)

4. Enclose the sensitive part of your experiment in a material that excludes magnetic fields, such as "mu metal." This type of metal is expensive and has to have thick walls to do its job, so this solution is only used for extremely sensitive experiments.

Ground Loops

For inexpensive amplifiers, V_{AOC} is directly connected to $V_{AIC,}$ as shown in Figure 6.37. (We've used a coax cable on the input side to minimize the critical loop area. Z_{in} is shown schematically as a resistor.) So far, there is no problem; the interference would be minimal.

However, sometimes the device creating V_{sig} must be grounded, e.g., for safety reasons, as shown in Figure 6.38.

This creates a problem because the two ground points are connected together, perhaps by a convoluted path. (For example, if the V_{sig} and R_{load} devices are plugged into different grounded outlets, the path goes from one outlet back to the circuit breaker box, and from there to the other outlet.) This creates a loop with a very large area, shaded in gray in Figure 6.39; this is a "ground loop." Because the area is so large, the magnetic flux from power lines that couples though it can be large, giving a large emf. In this figure, we've labeled the two ground points 1 and 2 because they might well be at different voltages due to the effects of this emf!

This emf accumulates almost entirely along the squiggly line connecting the two ground points, since this is actually much longer than the rest of the wiring. (In the diagram, it looks about the same as the length of the other wiring, but that's just for the sake of fitting the picture on the page.) The emf drives a current. We have no idea where the associated IR voltage drop occurs, because the resistances of the different parts of the loop are unknown. There are small resistances associated with every connection, all in

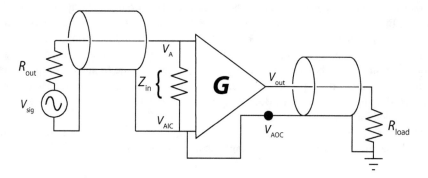

FIGURE 6.37 In an inexpensive amp, the input and output commons are directly connected.

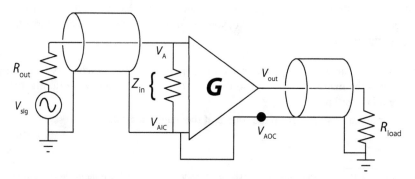

FIGURE 6.38 The source is grounded, perhaps for safety reasons.

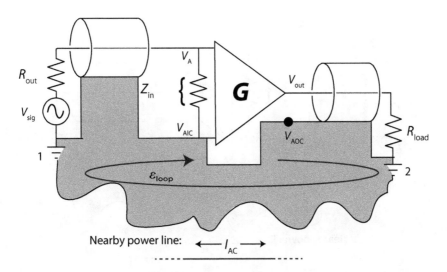

FIGURE 6.39 Because points 1 and 2 are both connected to ground, a loop with larger area is formed, resulting in a large emf.

the range 0 to 1 Ω. It's likely that one of these will be considerably larger than all the others, e.g., because of a little extra corrosion. This dominant loop resistance might easily occur where it will hurt the most.

Any voltage difference that appears within the critical loop (shown with dashed lines in Figure 6.40) between V_{AIC} and V_A will get amplified. So if the dominant resistance of the gray-shaded ground loop is *also* part of the dashed-line critical loop, the ε_{loop} will be amplified. An example is shown in Figure 6.40. (This dominant loop resistance might be a slightly higher than normal resistance in the connection of the outer coax conductor on the amplifier input connector.) A voltage equal in magnitude to the emf for the ground loop, ε_{loop}, appears across this dominant resistance. This voltage appears as part of the voltage difference between the input terminals of the amplifier, i.e., $V_A - V_{AIC} = V_{sig} + \varepsilon_{loop}$. So, ε_{loop} again gets amplified along with the real signal. (However, it's now much larger than in the first example of inductively coupled interference, because of the larger loop area.)

Eliminating effects of ground loops: the manufacturers of amplifiers are aware of this problem. So, in a good amplifier, they intentionally add a small resistance (e.g., 50 Ω) between V_{AIC} and V_{AOC}, as

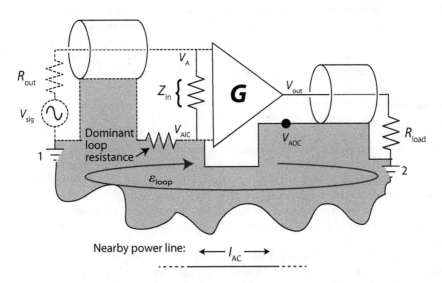

FIGURE 6.40 If there is a dominant resistance in the ground loop that is also part of the critical loop (dashed), the emf is amplified.

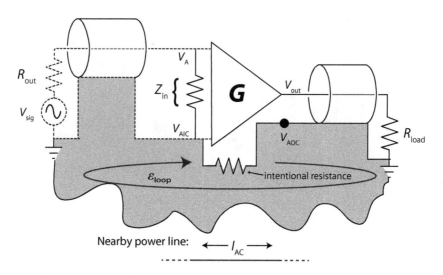

FIGURE 6.41 By intentionally introducing a resistance, the emf is made to appear in a place where it won't be amplified. The resistance in this example is part of the amplifier design.

shown in Figure 6.41. This resistance is big enough to ensure that in most cases it will be the dominant loop resistance. Now, the $\varepsilon_{\text{loop}}$ that appears across this resistance does not appear as a voltage difference between the input terminals, so we have $V_A - V_{\text{AIC}} = V_{\text{sig}}$, as desired. However, now V_{AIC} (relative to V_{AOC}) is $V_{\text{AIC}} = \varepsilon_{\text{loop}}$. Plugging these into Equation (6.11) gives

$$V_{\text{out}} = G\left(V_A - V_{\text{AIC}}\right) + V_{\text{AIC}} = GV_{\text{sig}} + \varepsilon_{\text{loop}}.$$

So, $\varepsilon_{\text{loop}}$ does still appear at the output, but it's not amplified. This is a huge improvement!
If your amplifier is of the type that connects V_{AOC} directly to V_{AIC}, there are still a couple of things you can do to minimize the effect of ground loops:

1. If possible, only ground your circuit in one place. In the example we've been discussing, this would mean not grounding the device that produces V_{sig}.
2. If the device that produces V_{sig} really must be grounded (e.g., due to safety reasons), then, if possible, operate your amplifier on batteries (so that it has no internal connection to ground), and insert a resistor (e.g., 5 Ω) between V_{AOC} and the ground coming from R_{load}, as shown in Figure 6.42. This location is not part of the critical loop between V_{AIC} and V_A, so the voltage drop across this resistor won't be amplified. (In fact, it won't appear at the output at all.)

6.7 Noise

Professional scientists often use the terms "noise" and "interference" interchangeably. However, we will define "noise" as a signal that comes from your sample or apparatus (rather than from outside it) and is not periodic. A typical noise signal is shown by the black line in Figure 6.43. When the noise comes from the sample you're measuring, it often is of scientific interest, and can yield important information, for instance, about the motion of atoms in the sample. We define "noise" as distinct from "interference," which we define as a periodic and always undesired signal, with a source that is usually outside your apparatus, such as power lines.

When the noise comes from the apparatus rather than the sample, it is certainly undesirable. By correct selection and use of apparatus components, you can minimize it, thus changing an impossible experiment to a possible one, or perhaps changing the time needed to complete an experiment from days to minutes.

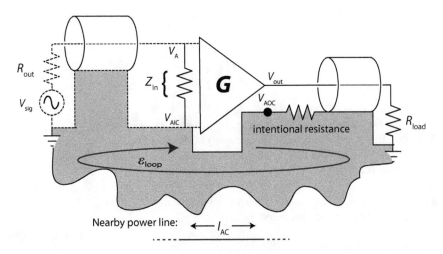

FIGURE 6.42 Even with an amplifier that has the input and output commons directly connected, you can control where the emf appears by adding a small resistor.

Noise Amplitude

To get quantitative, we need to understand some of the statistical properties of noise. We imagine sampling the noise at regular intervals in time; the samples are shown by the gray squares in Figure 6.43. Notice that there are more samples near zero voltage than at larger voltages. To show this distribution of measured values, we divide the voltage axis into small intervals or "bins," and plot how many of the sampled values fall in each bin (the number of "counts"). This is a "histogram." For example, if we divide the voltage axis into 0.5 V bins (e.g., 0 to 0.5 V, 0.5 V to 1 V), the histogram for the 25 samples from Figure 6.43 is shown in the top part of Figure 6.44. It's vaguely shaped like a Gaussian. If we collect a much larger number of samples, then for most sources of voltage noise, the distribution is indeed Gaussian, as shown in the bottom part of Figure 6.44. Noise with this distribution of voltages is called "Gaussian noise."

For a sinusoidal signal, the maximum and minimum values are clear, so we can easily measure the amplitude (from zero to peak), or the peak-to-peak amplitude (from minimum to maximum). However, for a noise signal such as that in Figure 6.43,

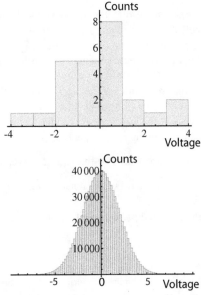

FIGURE 6.43 Black trace: voltage noise as a function of time. Gray squares: samples of the noise, taken at regular time intervals (as indicated by the vertical dashed lines).

FIGURE 6.44 Top: histogram with 0.5 V bins for 25 samples of voltage noise. Bottom: histogram with 0.1 V bins for 10^6 samples.

there is no clear maximum or minimum. To characterize the amplitude, we define the "root mean square" or "rms" amplitude as

$$V_{\mathrm{rms}} = \sqrt{\frac{1}{N}\sum_i V_i^2}.$$ (6.12)

Here, N is the number samples, and the V_i's are the different samples of voltage. Note that V_{rms} is indeed the root (square root) of the mean of the squares of the voltages. This definition is very similar to the definition of the variance, but in this case the average value is not subtracted before squaring.

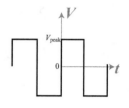

Concept test: Shown in Figure 6.45 is a square wave, oscillating from $-V_{\mathrm{peak}}$ to $+V_{\mathrm{peak}}$. What is V_{rms}? (Answer below.*)

FIGURE 6.45   A square wave.

For a sinusoidal signal, one can show that

$$V_{\mathrm{rms, sinusoid}} = \frac{1}{\sqrt{2}}V_{\mathrm{peak}} \cong 0.707V_{\mathrm{peak}}$$ (6.13)

V_{rms} is only a little smaller than V_{peak}, because the signal spends more time near the peak values than near zero, so that when we take samples at regular time intervals, more of them are near the peak values, as shown in Figure 6.46.

Although one cannot precisely define the maximum and minimum for a noise signal, it's still possible to get a rough idea. First, with your scope monitoring the noise signal of interest, adjust the horizontal scale until the trace looks approximately as shown in the middle part of Figure 6.47, rather than as shown in the top or bottom. (In the top, the s/div setting is too large,

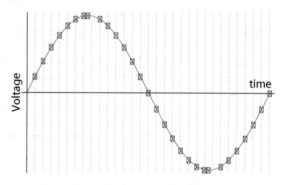

FIGURE 6.46   A sine wave sampled at regular intervals.

so that you can't see the details of the noise. In the bottom, the setting is too small, so that you're not seeing enough swings to positive and negative values to get a good idea of the amplitude.) Next, discard the two or three largest peaks on the screen. The difference between the remaining maximum and minimum can informally be the "peak-to-peak" amplitude.

Concept test: For the square wave, you showed that $V_{\mathrm{rms}} = \frac{1}{1}V_{\mathrm{peak}}$. For the sinusoid, one can show $V_{\mathrm{rms}} = \frac{1}{\sqrt{2}}V_{\mathrm{peak}}$. For the noise waveform shown in Figure 6.43, we can write $V_{\mathrm{rms}} = \frac{1}{A}V_{\mathrm{peak}}$, where V_{peak} is the informal peak-to-peak amplitude. Is A greater than, less than, or equal to $\sqrt{2}$? (Answer below.†)

---

* For all points on the waveform, $V^2 = V_{\mathrm{peak}}^2$. So, it doesn't matter what the time interval between samples is, we'll always have $V_i^2 = V_{\mathrm{peak}}^2$. Plugging this into (6.12) gives $V_{\mathrm{rms}} = V_{\mathrm{peak}}$.

† We see from Figure 6.44 that the noise waveform spends most of the time near $V = 0$, opposite the behavior of the sinusoid. So, V_{rms} will be much smaller relative to V_{peak}, and so $A > \sqrt{2}$. In fact, $A \sim 8$. It's important to bear this large difference between rms and "peak-to-peak" amplitudes in mind when dealing with noise.

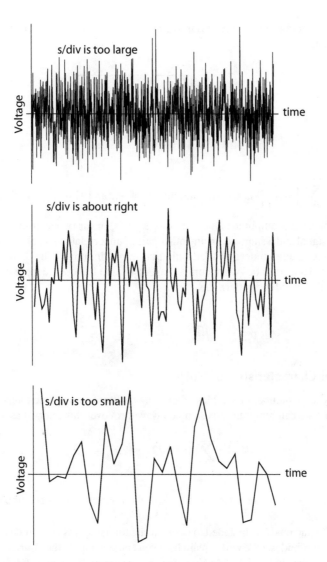

FIGURE 6.47   Adjusting the timescale to estimate the rms value of noise by eye.

Combining Noise Sources

We will see that noise typically arises from a few different sources within the apparatus. So, we need to understand the mathematics of how two different noise signals add together.

Let $V_1(t)$ and $V_2(t)$ be uncorrelated functions with zero average value. ("Uncorrelated" means that the fluctuations in one signal have no relation to the fluctuations in another signal, so that when one signal is positive, the other is equally likely to be positive or negative.) They might be noise signals, sinusoids, or some other signal. To compute the rms amplitude of the combination $V_1 + V_2$, we would take samples at many times t_i, then use

$$V_{\text{rms, tot}} = \sqrt{\frac{1}{N}\sum_i \left(V_1(t_i) + V_2(t_i)\right)^2}$$

$$= \sqrt{\frac{1}{N}\sum_i \left(V_1(t_i)^2 + 2V_1(t_i)V_2(t_i) + V_2(t_i)^2\right)}$$

The middle term, on average, gives zero (assuming no correlation between $V_1(t)$ and $V_2(t)$). Therefore, for a large number of samples,

$$V_{\text{rms, tot}} = \sqrt{\frac{1}{N}\sum_i \left(V_1\left(t_i\right)^2 + V_2\left(t_i\right)^2\right)}$$

$$\Rightarrow V_{\text{rms, tot}} = \sqrt{V_{\text{rms,1}}^2 + V_{\text{rms, 2}}^2}\,, \tag{6.14}$$

where $V_{\text{rms, 1}} = \sqrt{\frac{1}{N}\sum_i \left[V_1\left(t_i\right)\right]^2}$ is the rms amplitude of V_1, and similarly for $V_{\text{rms, 2}}$. Equation (6.14) shows how to combine rms amplitudes for two noise signals, or for two sinusoids of different frequencies, or for a noise signal and a sinusoid, or any other pair of uncorrelated signals with zero average value. This way of combining amplitudes is called "addition in quadrature." (This is the same way that you combine uncorrelated uncertainties when computing error bars.) If there are more than two signals being combined, we simply extend the pattern, i.e.,

$$\Rightarrow V_{\text{rms, tot}} = \sqrt{V_{\text{rms,1}}^2 + V_{\text{rms, 2}}^2 + V_{\text{rms, 3}}^2 + \cdots}. \tag{6.15}$$

Fourier Spectral Characteristics of Noise

Let's collect a very large number of samples of a noise waveform over a time interval T. We know from Fourier analysis that we can represent this sampled waveform over this interval as a sum of sinusoids:

$$V(t) = \sum_n A(f_n)\sqrt{2}\cos\left(2\pi f_n t + \varphi_n\right), \tag{6.16}$$

where $f_n = nf_1$, and $f_1 = \frac{1}{T}$. $A(f_n)$ is the rms amplitude of the Fourier component with frequency f_n, and φ_n is a phase factor. The factor $\sqrt{2}$ is included to compensate for using rms amplitudes rather than the more conventional peak amplitudes. (We will explain how you could compute the Fourier transform of a signal in an experiment in detail in Section 7.3.) Note that the spacing in frequency between the Fourier components is $\Delta f = f_1$, e.g., $f_2 = 2f_1$ and $f_3 = 3f_1$, so the spacing in frequency between these two successive frequency components is $\Delta f = f_1$.

Depending on the source of the noise, the Fourier amplitude $A(f_n)$ can have different functional dependencies on the frequency. One type of noise often used in theoretical models is "one over f noise," for which V^2 is proportional to $1/f$. This means that the Fourier amplitude of the voltage that appears in Equation (6.16) is proportional to $1/\sqrt{f}$, as shown in Figure 6.48. A spike is shown for each f_n, with the height of the spike indicating the value of $A(f_n)$.

The most important type of noise for creating models is "white noise," for which $A(f_n)$ is constant, independent of frequency.

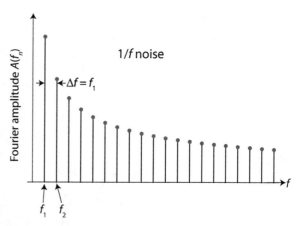

FIGURE 6.48 $1/f$ noise.

An example is shown in Figure 6.49. One important real-world example of white noise is "Johnson noise," the noise created by any object because of the thermal motion of the charge carriers within the object. You can explore Johnson noise in detail in Lab 15A.

Concept test: Why is noise with this spectral characteristic called "white noise"? (Answer below.*)

Consider a white noise source that is connected to the input of a real apparatus. There is always something that limits the highest frequency that can be measured,

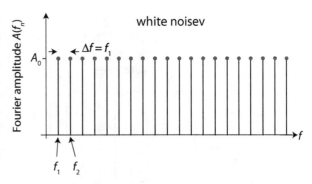

FIGURE 6.49   Fourier spectrum of white noise.

such as the bandwidth of an amplifier. There also is always something that limits the lowest frequency that can be measured; this might be the time interval T over which the data is acquired (which gives a lowest measured Fourier component at frequency $f_1 = \frac{1}{T}$), or it might be a high-pass filter. For now, we will model these effects by a high-pass "brick wall" filter and a low-pass "brick wall" filter; these completely filter out high and low frequencies, while having no effect at all on intermediate frequencies. (Note that this brick wall filtering action is very different from that of the RC filters you're familiar with. For example, a low-pass RC filter has a fairly gradual attenuation as frequency increases. A signal with a frequency 10 times the f_{3dB} is only attenuated† by about a factor of 10.) The result of passing white noise through such a set of filters is shown in Figure 6.50. The range of frequencies that is passed by the filters, B, is called the "Equivalent Noise Bandwidth" (ENBW). The height of the Fourier peaks is A_0, meaning that each Fourier component has rms amplitude A_0.

What is the measured amplitude of this filtered noise signal? The Fourier components are at different frequencies, and are therefore uncorrelated, so we add the amplitudes in quadrature, i.e., using Equation (6.15). The number of components being added is B divided by the spacing Δf between the components, so the total rms amplitude is $V_{rms} = \sqrt{\frac{B}{\Delta f} A_0^2} = \frac{A_0}{\sqrt{\Delta f}} \sqrt{B}$

$$\Rightarrow V_{rms} = v_n \sqrt{B}, \qquad (6.17)$$

where $v_n \equiv \frac{A_0}{\sqrt{\Delta f}}$ is called the "voltage noise density," or the "voltage noise spectral density."

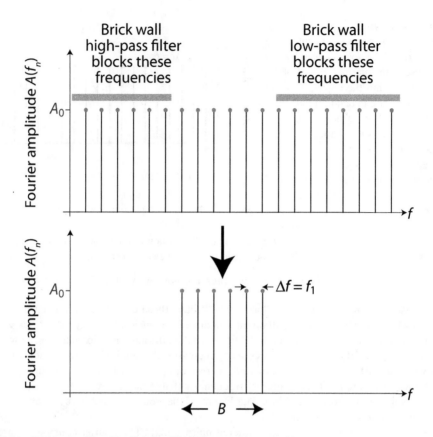

FIGURE 6.50　White noise filtered by "brick wall" high-pass and low-pass filters.

Concept test: What are the units of v_n? (Answer below.*)

What happens if the apparatus has the more usual RC filter characteristics, rather than brick wall filters? We define the transfer function

$$F(f) \equiv \frac{V_{\text{rms, out}}}{V_{\text{rms, in}}},$$

where $V_{\text{rms,in}}$ is the amplitude of the signal before the filters and $V_{\text{rms,out}}$ is the amplitude after the filters. For example, if the system includes a low-pass filter with $f_{\text{3dB}} = f_{\text{lo}}$ and a high-pass filter with $f_{\text{3dB}} = f_{\text{hi}}$, then the transfer function is simply the product of the effects of each filter, i.e., $F(f) = \dfrac{1}{\sqrt{1 + (f/f_{\text{lo}})^2}} \dfrac{1}{\sqrt{1 + (f_{\text{hi}}/f)^2}}$.

Applying this idea to Equation (6.15) gives

$$V_{\text{rms, tot}} = \sqrt{V_{\text{rms,1}}^2 + V_{\text{rms, 2}}^2 + V_{\text{rms, 3}}^2 + \cdots} = \sqrt{\sum_n \left[F(f_n) A(f_n) \right]^2}.$$

For the case of white noise, we have $A(f_n) = A_0$ for all n, so

$$V_{\text{rms, tot}} = A_0 \sqrt{\sum_n \left[F(f_n) \right]^2} = \frac{A_0}{\sqrt{\Delta f}} \sqrt{\sum_n \left[F(f_n) \right]^2 \Delta f}.$$

---

* From Equation (6.17), the units of v_n are $V/\sqrt{Hz}$.

We see that this has the same form as Equation (6.17), with the equivalent noise bandwidth given by $B = \sum_n \left[F(f_n) \right]^2 \Delta f$. In the limit of a long measurement interval, the spacing between frequencies in the sum is very small, and we can approximate the sum with an integral:

$$B = \int_0^\infty F^2(f) \, df = \int_0^\infty \left(\frac{V_{\text{rms,out}}}{V_{\text{rms,in}}} \right)^2 df. \tag{6.18}$$

In homework problem 6.9 you can show that, for a single low-pass filter, the equivalent noise bandwidth is

$$B = \frac{\pi}{2} f_{\text{3dB}}. \tag{6.19}$$

Equivalent noise bandwidth for a single low-pass filter.

Concept test: A white noise source has voltage noise density 50 nV/$\sqrt{\text{Hz}}$. The noise is amplified with an ideal (noiseless) amplifier having a bandwidth of 1 MHz and gain of 10. What is the resulting rms amplitude at the output of the amplifier? (Answer below.*)

Amplifier noise: unlike in the above concept test, all real amplifiers add some noise in the process of amplification. The most common model for amplifier noise is shown in Figure 6.51 by the dashed box. A source with output impedance R_{source} (often called the "source resistance") creates the desired signal V_{sig}. Within the model, there is a source of white voltage noise with voltage noise density v_n and a source of white current noise with current noise density i_n. The current source creates a noise current with rms amplitude $I_{\text{rms}} = i_n \sqrt{B}$. This current flows through the output imped-

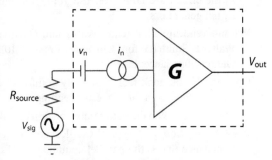

FIGURE 6.51 Noise model for amplifier.

ance R_{source}, creating a voltage $V_{\text{rms}} = I_{\text{rms}} R_{\text{out}} = i_n \sqrt{B} R_{\text{out}}$. This noise is uncorrelated with the noise from the noisy voltage source, so these two noise sources add in quadrature. The output impedance R_{source} adds an additional uncorrelated noise term due to Johnson noise, $V_{\text{rms, Johnson}} = \sqrt{4 k_B T R_{\text{source}} B}$, where k_B is Boltzmann's constant and T is the absolute temperature.

Your turn: what is the total rms noise at the output of the amplifier? (Answer below.†)

As is the case for most models, this does not capture all the behaviors of the real system, but does capture many of them. We could make a more accurate model by replacing the white voltage noise source with a voltage noise source that includes both white noise and $1/f$ noise, and similarly for the current noise source. For a good amplifier, the effect of the $1/f$ noise contribution is negligible above frequencies of about 10 Hz. So, in many cases, the simpler model is good enough.

---

* We use Equations (6.17) and (6.19), with $f_{\text{3dB}} = 1$ MHz. Plugging in the numbers gives 63 μV for the noise at the amplifier input. Multiplying by the gain gives 630 μV.

† We add the three noise sources in quadrature, and multiply by the gain, giving

$$V_{\text{rms, tot}} = G \sqrt{v_n^2 B + i_n^2 R_{\text{source}}^2 B + 4 k_B T R_{\text{source}} B}.$$

6.8 Negative Feedback and Op Amps

This section is available at ExpPhys.com.

6.9 Bode Plots and Oscillations from the Feedback Loop

This section is available at ExpPhys.com.

6.10 Simulation of Analog Circuits

This section is available at ExpPhys.com.

Lab 6A Input and Output Impedance Revisited, Surprising Effects of Capacitance

Learning goals: Come to a deep understanding of input and output impedance, including the capacitive component, and how they affect measurements. Understand that stray capacitances can have significant effects at MHz frequencies, especially when other impedances in the circuit are large. Refresh your familiarity with basic operation of oscilloscopes and signal generators.

Pre-lab reading: First read "Waves and Oscillations: A Prelude to Quantum Mechanics," by Walter F. Smith (Oxford University Press, 2010), Section 1.10, *then* read Section 6.2, "Input and Output Impedance: Part 1."

Optional: If you don't feel reasonably comfortable using an oscilloscope for basic tasks, please visit ExpPhys.com for suggested reading.

Safety: There are no unusual safety concerns for this lab.

Pre-lab question 1: A typical oscilloscope has an input impedance that is the parallel combination of a 1 MΩ resistor and a 20 pF capacitor. What is the magnitude of the capacitor's impedance at 1.9 MHz? Without using a calculator, estimate to within a few percent the magnitude of the impedance of the parallel combination at this frequency.

Pre-lab question 2: Coaxial cable has two conductors: an inner wire of circular cross-section, and an outer conductor in the shape of a hollow cylinder, as shown in Figure 6.52. The space between the two conductors is filled with an insulating plastic.

FIGURE 6.52   A section of coax cable.

   a. Consider two pieces of coaxial cable, both much longer than the diameter of the outer conductor, as shown in part A of Figure 6.53. Piece 2 is twice as long as piece 1. Apply a charge $+Q$ to the inner conductor of each piece and $-Q$ to the outer conductor of each. Let C_1 be the capacitance between the inner and outer conductors of piece 1 and C_2 be the capacitance for piece 2. By thinking about how the charges will distribute on each conductor, the resulting electric field, and the

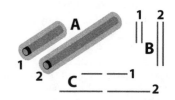

FIGURE 6.53   Three pairs of capacitors.

voltage that is associated with that electric field, make an argument about whether C_1 or C_2 is larger, and by approximately what factor.

    b.   Part B of Figure 6.53 shows two pairs of parallel wire segments. The segments in pair 2 are twice as long as those in pair 1. Again, C_1 is the capacitance between the two segments in pair 1 and C_2 is the capacitance for pair 2. Which is bigger, C_1 or C_2, and by approximately what factor?

    c.   Part C of the figure again shows two pairs of wire segments; this time, the segments in each pair are co-linear. (This makes the capacitance much lower than in part b, but still non-zero.) Again, the segments in pair 2 are twice as long as those in pair 1; the gap length is the same for both pairs. Which is bigger, C_1 or C_2? Is the ratio C_2/C_1 for this part bigger, smaller, or the same as the ratio for part b? Explain briefly. *Hint: in this case, unlike parts a and b, the charge distribution will be very non-uniform along the length of each segment – do you see why? (However, it's still true that the charge distribution is different between 1 and 2.)*

  3. The most common type of coaxial cable in research labs is RG58. The diameter of the inner conductor is 0.85 mm, the inner diameter of the outer conductor is 2.95 mm. The space between the two conductors is filled with polyethylene (dielectric constant: 2.25). Calculate the capacitance per meter between the inner and outer conductors. (It is not difficult to derive the expression for the capacitance from Gauss's Law, but you are also welcome to find the appropriate formula on the internet.)

Introduction

You have already encountered the idea of internal resistance in a previous course. In more sophisticated applications, the internal resistance of a device that creates a signal (e.g., a battery, which creates a DC voltage) is generalized to the "output impedance" (which may be complex), and the internal resistance of a device that measures a signal (e.g., a voltmeter) is generalized to the "input impedance" (which is always complex). The concepts of input and output impedance are absolutely central to the understanding of anything that involves electronics, so you will spend some time in this lab firming up your mastery.

You have probably used an oscilloscope (usually just called a "scope") in a previous course. This is the most important instrument for physics experiments; it acts as your eyes for monitoring electrical signals. Even if you think you're only interested in the DC component of a voltage (i.e., the average voltage), you should check the voltage with a scope – you might find something distressing, surprising, or delightful. In this lab and the next one, you will become more expert in your use of the scope.

Do not use a 10X scope probe in this lab (i.e., connect to your scope using a coax cable, and, where necessary, a coax-to-alligator-clip adapter.) The gender of connectors is determined by the inner conductor, as shown in Figure 6.54

FIGURE 6.54 Four types of coax adapters. For the coax-to-banana, the banana on the "GND" tab side is connected to the outer coax connector, which is often connected to ground.

  1. The manufacturer's specification ("spec") for the output impedance of your signal generator should be printed next to the output connector. (If not, search for the manual for your signal generator on the internet.) Using the method suggested in the reading, measure this output impedance, first at DC and then at about 1.9 MHz. You can create a DC voltage with the "DC offset" control, with the amplitude turned to the minimum. You should assume that the output impedance

is real. As explained in Section 6.5, it's good practice to use coax cabling with the outer conductor of the coax grounded whenever possible. The outer conductor of the coax then acts as a "shield" against interference, giving cleaner results than if you use unshielded wires. Use a coax T, a coax-to-alligator-clip adapter, and a coax male-to-male adapter to attach your loading resistor, which you should fetch yo Your location urself from the assortment your instructor will show you. (With this configuration, you'll have a few inches of unshielded wire in the clip leads, but that's too short to have an effect in this experiment.) Use the same coax-to-alligator-clip adapter to make your measurement of loading and to measure the exact value of your loading resistor. **Why is using the same adapter for both measurements important?**

Compare your results with the spec. **Explain your method and results briefly (with a diagram and a couple of sentences and equations) and calculate the percent discrepancy.** If your discrepancy is more than a few percent for either result, check with your instructor.

2. Similarly, the spec for the input impedance of your scope is printed next to the input connector; note that it is a parallel combination of a resistance and a capacitance. (Again, if necessary, look up the manual.) Using the method suggested in your reading, **measure (to within a few percent) the input resistance and capacitance by making measurements at DC and at 1.9 MHz, using a 1 MΩ resistor connected as shown in Figure 6.55.** As a physicist, you should always want to get the highest signal-to-noise ratio that you easily can. **What value of the AC amplitude maximizes this ratio?**

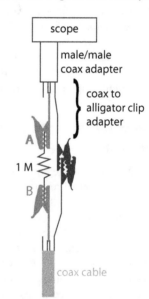

Note that your signal will be somewhat affected when you bring your arm close to the resistor. Make your measurements with your arm farther away. Since the signal has a fair bit of interference, use the cursors on your scope, rather than the automated amplitude measurement. If the interference is at a lower frequency than your signal, press the "Stop" button on your scope to freeze a single frame and then make the measurement. If the interference is at a higher frequency than your signal, place one cursor so it cuts through the middle of the "fuzz" at the top of the sinewave, and the other so it cuts through the middle of the fuzz at the bottom, as shown in Figure 6.56.

FIGURE 6.55   Connecting the resistor for part 2.

Compare your results with the spec values for the input resistance and capacitance, including calculating percent discrepancies. (From now on, whenever you're asked to compare a measured value with spec, you should include the percent discrepancy; I won't bother to remind you of this.) You will likely have a significant discrepancy, with your measured input capacitance significantly smaller than expected. You'll explore this mystery in the following parts of the lab, so don't spend time yet trying to explain it.

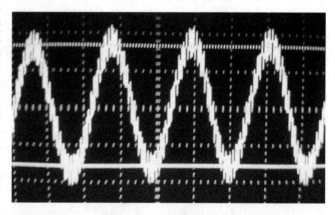

FIGURE 6.56   How to accurately measure the peak-to-peak amplitude when there is significant interference or noise that is higher in frequency than the signal.

3. Repeat the 1.9 MHz measurement of part 2, but this time with the 1 MΩ resistor inside a metal box, as shown in Figure 6.57. (The lid is off the box so you can see what's inside. Note that this is electrically equivalent to the circuit in part 2, just more compact.) **Use your measurement to find a new value for the input capacitance, and compare with spec.** You will likely have a significant discrepancy, again with your measured capacitance below spec, but closer than in part 2. **Devise and describe a model, including a diagram, that qualitatively explains the discrepancies you observed in parts 2 and 3, including why the discrepancy was worse in part 2.** If you're having trouble getting started, see the hint below.*

4. Repeat part 3, but this time using a 100 kΩ resistor. **Come up with a qualitative explanation for why you have measured different values for the input capacitance; illustrate your explanation with a circuit diagram that explicitly shows all relevant resistances and capacitances.** Test your explanation qualitatively by making a measurement with the set-up for part 2, 3, or 4 at a different frequency. **Briefly explain your test.**

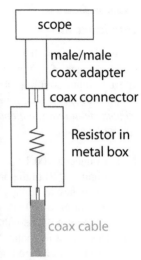

FIGURE 6.57 Connecting the resistor for parts 3 and 4.

Which of the three values for input capacitance (from parts 2, 3, and 4) should be most accurate? Why? Even for the most accurate value, you may find a significant discrepancy from the spec; if so, **explain qualitatively.** (If you're having trouble explaining this discrepancy, see the hint below.†)

For the most accurate method (of parts 2, 3, or 4), should it matter whether the resistor is at the scope end of the cable or the signal generator end? Why or why not? Illustrate your argument with a circuit diagram. Check your answer by testing experimentally at 1.9 MHz, report your result, and revise your thinking if needed.

Lab 6B Intermediate-level Scope Mastery

Learning goals: Become more sophisticated in the use of the oscilloscope, including a deep understanding of 10X probes, differential measurements, and the basics of triggering.

Pre-lab preparation: Read Sections 6.2 and 6.3, and also the material listed at ExpPhys.com for this lab.

Pre-lab question 1: A model for a 10X probe connected to a scope is shown in Figure 6.58. The cable capacitance is 50 pF. To what value should the variable capacitor C_{adj} in the 10X probe be adjusted? (Explain your reasoning.) What will the resulting input capacitance of the 10X probe/scope combination be?

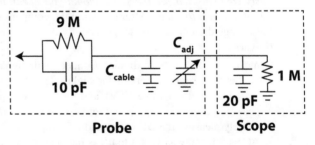

FIGURE 6.58 Model for 10X probe.

---

* Hint: think about the capacitance between the wires connecting to the two ends of the resistor, i.e., to points A and B in Figure 6.55, and note that these wires are longer in the configuration of Figure 6.55 than for Figure 6.57.

† Think about the effect of the male-to-male coax between your resistor and scope.

Introduction

In this lab, you will become reasonably proficient with the scope, including the advantages and disadvantages of 10X probes, how to make differential measurements, the difference between AC and DC input coupling, and the essentials of triggering. These skills are essential for all but the simplest scope measurements. *Hint: you'll be switching back and forth on your scope between the trigger menu and the menus for channels 1 and 2. Be sure you know which menu you're on before you make changes.*

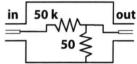

FIGURE 6.59   A 100:1 voltage divider.

1. Connect a 100:1 voltage divider inside a box (as shown in Figure 6.59) to the output of your signal generator. Set the frequency to about 1 kHz. Set the amplitude of the signal generator to 100 mV peak-to-peak, so that the amplitude at the output of the voltage divider should be 1 mV. Connect this signal to channel 1 of your scope using a coax cable. Try the autorange feature – it probably won't work very well for this signal, because it's so small and noisy. Without making any other connections, adjust the scope controls to get the best display you can manage; remember that a good experimentalist always knows what to expect, so use your expectation to set the horizontal and vertical scales appropriately. You should be able to see the sinusoidal nature of the signal. **Briefly describe what you see.** If you're using a conventional signal generator, it has a "sync out" or "TTL" output, which is a square wave at the same frequency as the main output, oscillating between 0 and 5 V. If instead you're using a waveform generator, it may not have this output, but you should be able to configure a second output to give a 5 V peak-to-peak square wave that is in sync with the sine wave output. In either case, connect this second output from your signal generator to the channel 2 input of your scope. Adjust the scope controls to get the best possible display of channels 1 and 2. **Briefly describe what you see. The display should be better than what you saw before. Explain why.** Now, move the cable from channel 2 to the "Ext trig" input of your scope. Adjust the scope controls to obtain the best possible display of channel 1. It should look the same as it did in the previous step. **Explain why.** Call your instructor over to view your result before proceeding. In the next step, you'll use a 10X probe. *Every time you use one, you should make sure it is compensated for the scope you're using it with (as explained in the online materials for this lab); if you know for sure that it's been previously compensated for the same scope, you needn't recompensate it, but if you have any doubt, it's best to check.* After checking the compensation, use a 10X probe to connect from the voltage divider to your scope. **Briefly describe what you see and explain why it looks that way.**

2. Remove the 100:1 voltage divider. Connect your signal generator to your scope using a coax cable. Set your signal generator for slightly less than the highest possible DC offset (e.g., 9 V out of a maximum possible 10 V), and for the lowest possible AC amplitude (or 20 mV peak-to-peak, whichever is larger). *Without ever touching the autoscale button*, use your scope to accurately measure the DC offset and AC amplitude. Note that for accurate measurements of an AC waveform, you want it to fill as much of the screen as possible without going outside the edges. For the most accurate measurement of the DC level, you need the ground and the signal to both be on screen, but as far apart as possible. **Report your method of using the scope for these measurements and your results.**

3. Although 1% tolerance resistors are quite inexpensive, the same is not true for capacitors. In fact, the tolerance on capacitors is often as much as ±20%. So, when precision is important, you should measure the actual value of the capacitors used in your circuit. Capacitors are labeled with their value in pF or µF, so, for example, a 10 nF capacitor will probably be labeled "10000," meaning 10000 pF. Obtain a 10 nF capacitor and a 500 kΩ

resistor and measure their actual values with a DMM. For the capacitance measurement, you need to minimize the lead capacitance, so use a banana-to-coax adapter, a male-to-male coax adapter, and a coax-to-clip-lead adapter to connect the DMM to the capacitor. You will still need to subtract out the capacitance of your leads. Using your breadboard, build a high-pass filter using your resistor and capacitor. **Calculate the f_{3dB} of this filter.** Connect your signal generator to the filter input, and use a coax cable and clip leads to connect the output to your scope. Return the DC offset on your signal generator to 0 and make whatever other adjustments you think are needed to the signal generator controls to make the most accurate and convenient measurement of the f_{3dB} of your filter. *Hint: it is easiest to do this if you use a coax-T to connect the unfiltered output of your signal generator to channel 1 of your scope. You can connect the output of the filter to channel 2. (It's best to stick with this convention for everything you do with your scope for the rest of your life, i.e., display the input signal on channel 1 and the output on channel 2.)* **Compare your measured value for f_{3dB} with expectation, including the percent discrepancy.** (You may have a significant discrepancy.)

4. Connect a 10X probe to channel 2 and adjust it to proper compensation. Using the 10X probe, re-measure the f_{3dB} of your filter. **Report your result. Is the result with the 10X probe or the result from the previous part with the coax cable a more accurate value for the f_{3DB} of your filter? Why? When is it better to use a 10X probe to connect to your scope, and when is it instead better to use a coax cable?**

For the rest of this lab, use only 10X probes to connect to channel 1 and channel 2 of your scope.

5. Obtain a 100 pF capacitor and 10 kΩ resistor, and use them to build a low-pass filter on your breadboard. (Leave your high-pass filter as a separate circuit; you'll use it again soon.) **Calculate the f_{3dB} for this filter. Measure f_{3dB} and compare with your calculated value.** You should find a significant discrepancy. **Explain why, then calculate a revised theoretical value for the f_{3dB} you should measure and compare with your measurement.** *Hint: look at what is printed on your scope probe.*

6. Go back to your high-pass filter. Connect your signal generator to its input. Figure out a way to use your scope to measure the relative phase of the current flowing through the capacitor and the voltage across it, with the frequency set to the f_{3dB} of the filter. *Hint: you need to measure the voltage across the capacitor, but you may not connect either side of the capacitor to ground, since that would disrupt the operation of the circuit.* Make the measurement and **report your method and finding. What did you expect to measure? Why?**

7. **Observing a single, non-repetitive event: Switch bounce.** Most of the time, you use the scope to observe repetitive traces, and the screen updates many times per second. In this exercise, you will practice catching a one-time, non-repeating event.

When you flip a mechanical switch, the contact literally bounces several times over a very short period, causing the output to swing wildly before settling down. This can cause serious problems for digital circuits. For example, if a circuit is supposed to increment a counter each time the switch is flipped, the bouncing might cause several counter increments for each switch flip. (However, it only takes a few electronic components to "debounce"

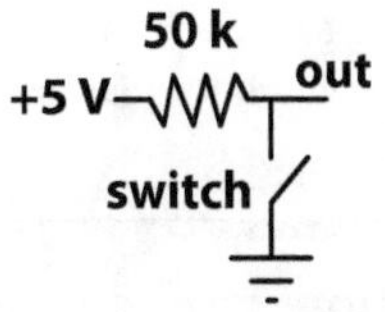

FIGURE 6.60 Circuit to observe switch bounce.

a switch.) Using the switch supplied by your instructor, create the circuit shown in Figure 6.60, and set up your scope to **capture a picture such as that shown in Figure 6.61.** (In this one case, you may use your signal generator to supply the +5 V, since the current is limited by the 1 kΩ resistor. However, in general don't use your signal generator as a power supply; it's not intended to provide substantial current.)

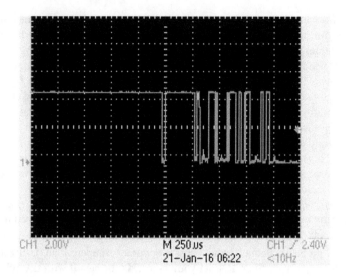

FIGURE 6.61 When a switch is used to change the voltage from 5 V to 0 V, switch bounce causes several swings back and forth over a period of about 1 ms, until the voltage finally settles to the new value.

Lab 6C Introduction to Amplifiers, Capacitively Coupled Interference, and Feedback Oscillations

Learning goals: Understand the purpose of an amplifier and how to use it. Understand the following limitations of amplifiers: bandwidth, maximum/minimum output voltage. Begin the process of explicitly modeling measurement apparatus. Understand how capacitively coupled interference occurs, why it is sensitive to the impedance to ground, how to minimize it with shielding, and why the shield must be grounded. Improve understanding of stray capacitance. Understand how oscillations due to input/output coupling occur, and how to prevent them. Understand how shielding reduces stray capacitances.

Pre-lab reading: Read Sections 6.4 and 6.5.

Pre-lab question: An experimenter observes 5 mVrms of 60 Hz interference on an experiment; the critical wire that is picking up this interference has an impedance to ground of 100 kΩ. What is the stray capacitance to the power lines, which carry 110 Vrms? ("Vrms" means the rms amplitude; for a sinusoid this is the peak-to-peak amplitude divided by $2\sqrt{2}$. We'll discuss this more in the next lab.)

Safety: There are no unusual personal safety concerns. *However*, to protect your amplifier, it is wise to switch the input coupling to ground before connecting or disconnecting anything. (Then switch it to DC.)

Introduction

In many research experiments, the signal is too small to be digitized accurately. Therefore, the experimenter uses an amplifier, a device that multiplies a signal by a constant (e.g., by a factor of 10), to boost the signal before digitizing. Amplifiers are the most common piece of apparatus in physics research. They're also essential for cell phones, televisions, audio players, and automobiles (which use many electronic sensors). For any piece of equipment, including amplifiers, you will only get the best performance if you have a deep understanding of how the equipment works, and what its limits are. This understanding is also essential for troubleshooting. In this lab, you'll gain an intermediate level of understanding for using an amplifier. (In later labs, you'll become a true expert!) You'll also explore capacitively coupled interference; this is the biggest source of interference unless you take steps to minimize it.

1. Set your signal generator to produce a sine wave with about 100 mV amplitude and about 1 kHz frequency. Use a coax T to split the output, connecting one copy of the output to your scope and the other to the A input of your amplifier ("amp"). Connect the output of your amp to your scope; remember that it's good practice to connect the input signal to channel 1 and the output signal to channel 2. (Some amps have two outputs, with different output impedances; if so, use the 50 Ω output.) Set the gain of your amp to 10, make sure all filters are disengaged, and that the input coupling on the amp is set to "AC." (We're only interested in AC signals in this lab, so you should use AC coupling for all the exercises.) Check that the amplifier works the way you expect. Try changing the gain – does the output change in the way you would expect? Try engaging a low pass filter and changing the frequency. Does the filter do what you expect? When you're done experimenting, disengage all filters, go back to gain 10, and make sure the signal generator is set to about 100 mV amplitude. **Measure the bandwidth, defined as the frequency at which the gain drops to $1/\sqrt{2}$ of the nominal value, of your amplifier, and compare with spec.** You should be able to find the spec online or in the manual provided by your instructor. Note that some amplifiers have the peculiar characteristic that, as the frequency is increased, the gain first increases and then decreases. In such a case, the bandwidth may well be significantly larger than the spec. If your amplifier is like this, do not attempt to include the rise in gain as part of the model you will create in the next step. The point of that model is to show the decrease in gain as frequency is increased.

2. The input and output impedances of your amp should be printed next to the corresponding connectors. If not, check online or in the manual provided for these specs. **Draw a diagram that shows a model for your amp that includes these impedances, as well as the decrease in gain at high frequencies that you observed in the previous part.** Your diagram should be based on functional blocks such as the ideal amplifier (show as a triangle with "10X" written inside), perfect buffers, resistors, and capacitors.

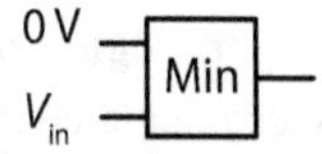

$$0\,\text{V} \boxed{\text{Min}}$$
$$V_{\text{in}}$$

FIGURE 6.62 The output (on the right) equals the minimum of 0 V and V_{in}, i.e., whichever is closer to -∞. A similar block with "Max" written inside would return the maximum of the two inputs.

3. Set the frequency back to 1 kHz. Explore what happens as you increase the function generator's amplitude all the way to the maximum. **Modify your model to include this behavior, using blocks such as those shown in Figure 6.62.**

4. Now, remove altogether the connection to the signal generator and turn it off. In the rest of today's experiments, we're interested in lower frequencies, so engage a 1 kHz low-pass filter on your amplifier. Also, your hands will act as antennas and affect the measurements significantly, so make all measurements with your hands well away from the circuit. Using a male-to-male coax adapter and a coax-to-clip-lead adapter, connect an unshielded length of wire (about 25 cm, but exact length isn't important) to your amplifier's A input. Plug the other end of the unshielded wire into your breadboard, as shown in Figure 6.63, but don't make any other connection to it. Connect the ground wire from your clip lead adapter via another length of wire to a different set of holes on the breadboard, a couple of cm away. (In a later step, you'll use a resistor to connect from your 25 cm wire to this ground.) Adjust the amplifier gain

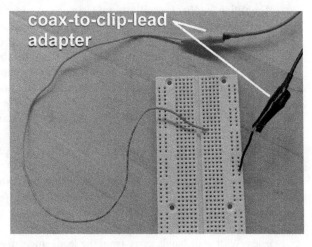

FIGURE 6.63 Arrangement for part 4.

to the largest possible value without overloading. (Do the same for all the following exercises.) Note that it sometimes takes several seconds for your amplifier to recover after it's been overloaded. You should see a roughly sinusoidal signal on your scope. **Briefly explain why it's there and measure its amplitude and frequency.** (*As you should always do from now on, you should report your measurement "referred to the input," i.e., divide the voltage at the output of your amp by the gain.*)

5. Now use your breadboard to connect the input wire (the one connected to the red clip lead) to ground through a 1 MΩ resistor. As you should do every time you significantly change your measurement, readjust the gain of your amp. **Measure** the amplitude of the signal; it should be considerably reduced. **Briefly and qualitatively explain why, including a circuit diagram or two.**

   Predict the amplitude of the signal you will measure when you replace the 1 MΩ resistor with a 100 kΩ resistor. **Show the reasoning for your prediction and check it experimentally.** Next, you will replace the 100 k resistor with a capacitor. **Predict what value of capacitance will produce the same reduction of the signal as the 100 k resistor did.**

   Check your prediction experimentally. (Note that it may take as long as 30 s for the signal to settle down.) You should see a cleaner (more pure) sinusoid than you did with the resistor. **Why is the signal cleaner?**

6. Now connect a coax cable to the input of your amp, instead of the unshielded wire/clip lead assembly, and leave the other end of the coax unconnected. This is just the same configuration as in part 4, except now the inner conductor is surrounded by the outer conductor. The outer conductor is connected to the amplifier's common point when you make the connection to the amplifier. **Measure** the amplitude of the 60 Hz interference. **Briefly explain why this result is different from that in part 4 (i.e., why is it preferable to connect signal sources with coax cables rather than two leads?)**

7. Now use a male-to-male coax adapter to connect one side of a "ground interruption box" to your amp input. (Your instructor will provide an opened version of this box, so you can see what's inside.) This gadget breaks the connections between the outer conductors on the coax cables that connect to either side; the inner conductors of the cables are connected via a wire inside the box. Connect the same coax cable you were using in the previous step to the other side of the ground interruption box. **Measure** the amplitude of the 60 Hz interference. **Explain why this result is different from that in part 6, including a diagram.** *Hint: you can assume that the capacitance between the inner and outer coax cable conductors is much bigger than the capacitance between the outer conductor and the interference source.*

8. **Amplifier feedback oscillations:** Using a coax-to-clip-lead adapter, connect a 10 kΩ resistor between the A input of your amp and ground, i.e., the resistor should connect between the inner and outer conductors of the input coax connector on the amp. Disengage all filters. (Once you've done this on the SR560, you should see that the filter section has the "DC" light on. This is different from the input coupling, which should still be set to AC.) Set your scope to 1 ms per division horizontally and adjust the V/div to get the best display you can manage. You should see, at most, a very small interference signal, and you might see no periodic signal at all. Do you understand why the interference is so small in this case?

   Now, instead of connecting your amp directly to your scope with a coax cable, insert two coax-to-clip lead adapters, as shown in Figure 6.64. (You would never make the connection this way in "real life"; we're doing this to provide a short length of unshielded wire.) Reduce the gain to 500, then increase it one step at a time (e.g., to 1000, then 2000 etc.) until you see a sudden change that's more than you'd expect from the gain increase. You will probably find that the signal is strongly affected by how close you bring your hands to the clip leads. With your hands away from the leads, **measure the frequency (should be at least 10**

kHz) and amplitude of the signal. (Hint: it will probably not be possible to trigger off this signal, so adjust the scope scales as best you can, then press the "Run/Stop" button to freeze the display. You can then make measurements with the cursors.) **Explain why this signal occurs, including a circuit diagram.** (Hint: you may need to review "Amplifier Fundamentals.")

9. (Optional – if time permits.) Experiment with the effect of moving your hand close to the clip leads without touching any part of the circuit. Does the behavior change if you use one hand to touch the metal part of the red output clip lead, and then move your other hand close to the clip leads? (The output voltage of your amp is limited, so this is safe to do.) How can you explain these effects? Hint: you are surrounded by power lines. Some of these are routed through grounded conduits. Therefore, the capacitance of your body to ground and to the power lines is significant.

10. (Optional – if time permits.) You can get oscillations even when you're being careful with shielding, if the source resistance is large. Eliminate the clip leads, so your scope is directly connected (via a coax cable) to your amp output. Connect a grounded 100 kΩ or 1 MΩ resistor in a shielded box (see Figure 6.65) to your amp input, using a male–male coax adapter. Can you still get oscillations if you take the gain high enough?

11. (Optional – if time permits.) In this exercise, you'll explore a common mistake made when breadboarding, one that can be quite frustrating to debug. After this, you'll be able to debug it more easily. Set the amplitude of your signal generator to 1 V, the frequency to 300 kHz, and the amp gain to 1. Using a male-to-male coax adapter and a coax-to-clip-lead adapter, connect an unshielded length of wire (about 25 cm, but exact length isn't important) to your amplifier's A input. Plug the other end of the unshielded wire into your breadboard. Connect the ground wire from your clip lead adapter via another length of wire to a different

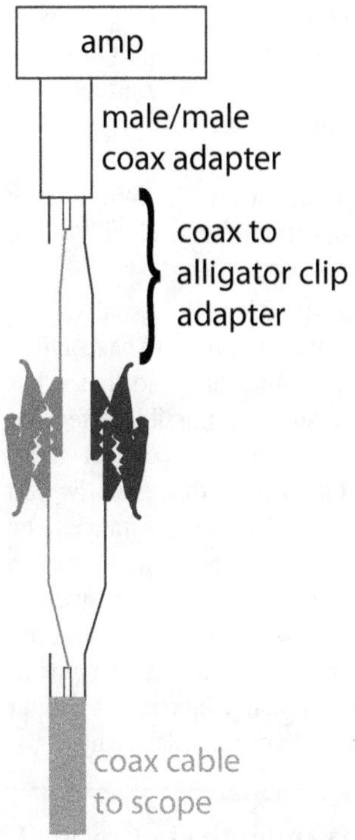

FIGURE 6.64 A silly way to connect to your amp.

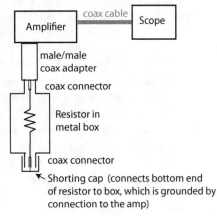

FIGURE 6.65 Connections for Lab 6C part 10.

set of holes on the breadboard, a couple of cm away. Connect your signal generator to the breadboard in a similar fashion, but using shorter lengths of wire. Plug the signal (red) wire from your signal generator into the same row of holes on the breadboard as the red wire from your amp input, so they're connected, as shown in Figure 6.66. Connect the 50 Ω output of your amp to your scope. Check your scope display and verify that everything works the way you expect. Now move the signal wire from the signal generator one set of holes over on the breadboard, so that it's no longer connecting to the amplifier input. **Briefly describe what you see and explain it qualitatively.** Check your explanation by pulling the wire from the signal generator but leaving it next to the wire leading to your amplifier input. **Does the result match your explanation?**

Post-lab question 1: Why is it important for a shield to be grounded?

Post-lab question 2: Usually, you want an amplifier to have a high input impedance, so that it has the smallest possible effect on the circuit you connect it to. This means that usually you want the input capacitance to be as small as possible. However, when you're having trouble with oscillations of the kind explored in this lab, it helps to intentionally increase the input capacitance. Explain why.

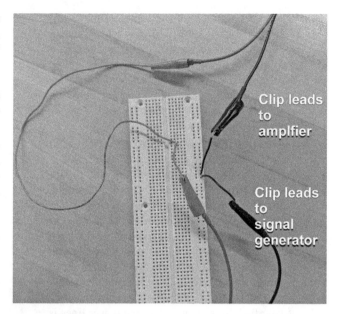

FIGURE 6.66   Arrangement for Lab 6C part 11.

Lab 6D Inductively Coupled Interference and Ground Loops

Learning goals: Consolidate understanding of capacitively coupled interference, especially how it is affected by the resistance to ground. Understand inductively coupled interference and ground loops, including where they come from and how to eliminate them. Improve in ability to synthesize results from several experiments to reach a conclusion. Improve ability to draw correct circuit diagrams from looking at the way instruments are wired together.

Pre-lab reading: Section 6.6.

Pre-lab question 1: A straight wire carries a current of 20 A. What is the magnetic field (in T) it creates at a distance of 1 m from the wire? (You are welcome to look up the relevant equation in a textbook or on the web.)

Pre-lab question 2: A researcher is measuring a small voltage produced by a temperature sensor. The sensor is part of a complex circuit, and the voltage it produces is relative to earth ground. The researcher monitors the sensor using an amp and digitizing circuit, which measure voltages relative to earth ground, but are plugged into a different outlet than the sensor apparatus. Because the sensor circuit and the digitizing circuit are both grounded, there is a ground loop. We'll model the ground wires for these outlets as running parallel for 10 m with a 2 m separation (fairly reasonable assumptions) before they connect together at a circuit breaker box. The area of the ground loop can be approximated as the area between these two wires. A nearby power line runs parallel to the ground wires and produces a magnetic field with an average amplitude of 4.0 μT over the area of the ground loop. The power line has a frequency of 60 Hz, and the current in the line oscillates sinusoidally. What is the amplitude of ε_{loop}?

Pre-lab question 3: Shown in Figure 6.67 is a simplified version of our model for capacitively coupled

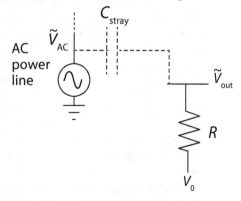

FIGURE 6.67   Model for capacitively coupled interference with the bottom of the resistor at a non-zero but well-defined voltage.

interference. $\tilde{V}_{AC} = V_{AC}\, e^{i\omega t}$ is the complex voltage on the AC power line, and ω is the angular frequency of the power line. (The actual voltage on the power line is $\mathrm{Re}\,\tilde{V}_{AC}$.) R is the output impedance of the device being measured. The voltage $\tilde{V}_{\mathrm{interference}}$ would probably go to the input of an amplifier (not shown). In the version discussed in the pre-lab reading on capacitively coupled interference, V_0 was at ground; we now allow it to be an arbitrary, well-defined constant voltage. Show that the amplitude of interference at the point labeled $\tilde{V}_{\mathrm{interference}}$ is unaffected by the value of V_0.

Note (you needn't show this part): this means that even if V_0 varies in time at a different frequency from the power line (e.g., if V_0 is the voltage created by a function generator), then the amplitude of interference is the same as if V_0 were at ground. This is an important idea: any well-defined voltage, even a time-varying one, plays the same role in reducing the amplitude of interference as ground does.

Introduction: As for most electronics (e.g., computers, televisions), the circuits in your amplifier operate using DC voltages for power. Therefore, your amp includes an AC to DC converter, as shown in Figure 6.68. In what follows, we use this as a source of interference, first using an antenna that is far away from it (at the front of the amp), and then moving the antenna close to it. The converter includes a transformer, diodes, and other elements. The diodes and some of the other elements are "non-linear," meaning that the current doesn't depend linearly on the voltage. Therefore, the current in the

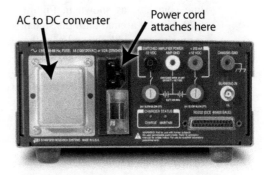

FIGURE 6.68   Rear panel of a typical amplifier.

transformer doesn't depend linearly on the voltage. So, even though the voltage from the power outlet (the main source of electric fields near the converter) is sinusoidal, the current is non-sinusoidal, but still periodic with the same period as the voltage. The current creates a magnetic field. Because this is a dipole field (magnetic field lines always form closed loops), it falls off with distance more rapidly than the electric field.

IMPORTANT: This is a semi-quantitative lab; rough measurements are fine. Don't spend time to measure things precisely.

1. *Capacitively or inductively coupled interference?* Start by making a simple antenna that will be efficient at picking up interference. To do this, connect a few banana cables to get a total length of about 3 m. Coil most of its length into a series of loops, roughly 10 cm in diameter, making sure all the loops are wound in the same direction (e.g., all clockwise). Use a couple of pieces of tape to stabilize this coil. Using a male-to-male coax adapter and a coax-to-banana adapter, connect one end of your banana cable to the "A" input of your amplifier. Use a bit more tape to fix the location of the other end near your amp input, without electrically connecting it to anything. Your apparatus should look roughly as shown in the top part of Figure 6.69, with the loops in front of your amp. **Draw a schematic diagram of this setup; for all such diagrams in this lab, use the symbol shown in the bottom part of Figure 6.69 for your amplifier.**

   In your diagram, use a resistor to ground labeled "R_{load}" to represent your scope. (This represents the input impedance of the scope.)

   1a. Engage a 300 Hz low-pass filter on your amp (since we're focusing on 60 Hz interference) and set the gain to the lowest possible. Set your scope trigger source to "AC Line," set the horizontal scale to 5 ms/div, and **measure the amplitude of the 60 Hz interference.** It's okay if the overload light comes on, as long as the output signal doesn't appear to be "clipped" because it exceeds the maximum output voltage of your amp. (Normally, you would not take measurements with the overload light on, but for today it's fine.) If you do see clipping, then lower the f_{3dB} of the amplifier's low pass filter until it is eliminated. You may need to take it to 30 Hz or lower; that's fine, as long as you keep it the same in the next step.

1b. Place the coil on the top of your amp, near the back where the power line connects, as shown in Figure 6.70. Move it around until the signal is maximized, with the coil near the back of your amp; this may mean that the loop is hanging over a corner of the amp. **Measure the amplitude of the 60 Hz interference and briefly describe the shape of the waveform.** Flip the loop over; **is the shape of the waveform significantly affected?**

1c. Move the coil back to the position in front of your amp (Figure 6.69). Plug the second end of the banana cable (which so far hasn't been plugged into anything) into the ground side of the coax-to-banana adapter. **Draw a new schematic diagram.** Reset the low-pass filter to 300 Hz. Adjust the amplifier gain appropriately. **Measure the amplitude of the periodic signal (referred to the input, of course).**

1d. With the second end of the banana still connected to ground, repeat part b, again moving the coil around until you get the biggest signal. Hint: you should see a difference when you flip the coil, either for this part or for part b. **Are you surprised that there is a signal at the output of your amp, even though you've connected the input to ground (through the coil)? Why or why not?**

1e. Use tape to collapse the loop, minimizing the area of the loop, as shown in Figure 6.71. Repeat parts 1b (with the second banana unconnected) and 1d (with the second banana grounded), **including measurements of amplitude.** In each case, position the collapsed coil so as to maximize the output signal. For this experiment, you should only consider changes of more than a factor of two to be significant; smaller changes may simply be due to the fact that it's impossible to have the coil in the same "place" as it was previously

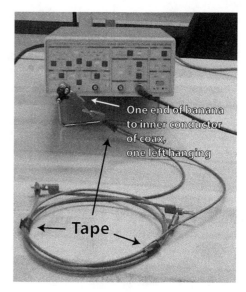

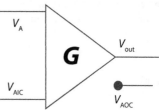

FIGURE 6.69   Top: setup for part 1. Bottom: symbol for amp.

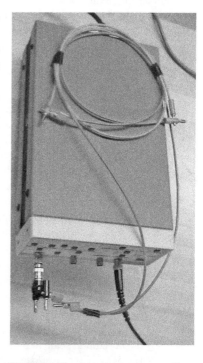

FIGURE 6.70   Arrangement for part 1b.

when it's a different shape. **Qualitatively explain why the signals in parts a–e did or did not differ from each other (e.g., why the signal in part c was different from the signal in part b), and why flipping the coil did or did not make a significant difference in the shape. As part of your explanation, identify in each case whether the main signal was capacitively or inductively coupled interference.** Minor hint: the case of most amplifiers is grounded. You may find it helpful to revisit section 6.5, "Capacitively Coupled Interference."

Remove the two pieces of tape used to collapse the loop, but keep the rest of the tape in place; you'll use this coil in the next part.

2. Ground loops.

    2a. Look at the sample of the "bad ground box" that has the top left off so you can see what's inside. (You should also feel free to explore how things are connected using a multimeter.) Note that there are two types of coax connectors, as shown in Figure 6.72. The inner pins of the two coax connectors are connected together. The outer conductors are connected together via a 20 Ω resistor. By inserting this resistance into a ground loop, we can control where the dominant loop resistance occurs, since even a resistance as small as

FIGURE 6.71   Setup for part 1e.

20 Ω is still much larger than the resistance of the wires and contacts (typically perhaps 1 Ω total). You're also provided with a 100 kΩ resistor inside a box, with each end connected to the inner pin of a coax connector. This represents the signal source that is discussed in the section "Inductively Coupled Interference and Ground Loops"; in this case, $V_{sig} = 0$ and $R_{out} = 100 \text{ k}\Omega$.

    Connect adapters to one end of the resistor as shown in Figure 6.73. Using male–male adapters, connect the other end of the resistor to one end of the bad ground box, and connect the other end of the bad ground box to the input of your amp. **Make a schematic circuit**

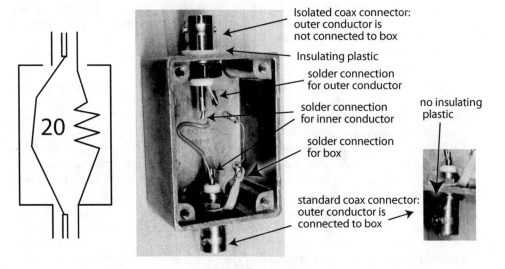

FIGURE 6.72   Sketch and photo of bad ground box; alternate view of bottom coax is shown on the right.

diagram of this completed
apparatus (not a sketch, but
rather a wiring diagram),
using a triangle marked
"*G*" to represent your amp
(don't bother to show the
input or output impedance
explicitly) and a resistor
marked "R_{load}" to represent
your scope. Use the ground
symbol to represent scope
ground; one side of R_{load}
should be connected to
ground. On your diagram,
label the points V_{AIC} (i.e.,
the outer conductor of the
input coax connector of

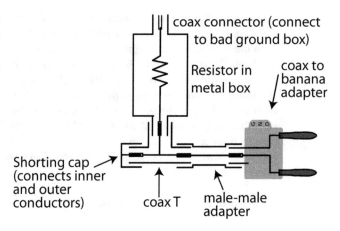

FIGURE 6.73   Sketch of resistor in box with goofy-looking adapters.

your amp), V_A (the inner conductor), V_{out} (the inner conductor of the amp output coax), and
V_{AOC} (the outer conductor). For now, don't show any connection between V_{AIC} and V_{AOC}.
Check your diagram with your instructor before proceeding.

So far, there should be very little 60 Hz interference, if any (less than 10 µV referred to
the input of your amplifier); you should only see some aperiodic noise. If you do see any
significant periodic signal, call your instructor over before proceeding.

2b.   To protect your amp, please switch the input coupling to "GND" now and leave it there
until directed otherwise. As discussed in the pre-lab reading, good amplifiers have a resistor between V_{AIC} and V_{AOC} (i.e., between the outer conductor of the input and the outer
conductor output coax connector). Use an ohmmeter to **measure the value of this resistor
for your amp.** This is there to suppress most effects of ground loops. However, since it
is our object today to observe the effects of ground loops in a simple system, this resistor
gets in the way. We want to "short it out," i.e., to connect a wire in parallel with it. You'll
take advantage of the fact that your amp has two input connectors, A and B; the outer conductors of these are connected together. Similarly, the outer conductors of the two output
connectors are connected together. Use whatever adapters you need, combined with a short
banana cable (not your coil), to connect V_{AIC} to V_{AOC}, i.e., to connect the outer conductor of
the input to the outer conductor of the output, *without* connecting the inner wires together.
(*Do not use a coax cable to connect the output to the input.*) **Modify your schematic to
show the connection you just made.** Have your instructor check your set-up; *once you've
been cleared*, switch the amp input coupling to DC. You should still not see any significant
periodic signal on the scope.

2c.   Your scope has a convenient connection point for ground on the front panel, either a banana
jack or a post you can clip onto (similar to the point used to compensate 10X probes). Use
your coiled banana cable to connect this ground point to the "GND" side of the coax-to-
banana adapter shown in Figure 6.73. (You will need to slide an alligator clip on one end of
your coiled cable to connect to your scope, if it has a connection post for ground rather than
a banana jack.) This simulates being forced to make an extra ground connection even when
you don't want to, e.g., a ground connection that's required for safety reasons. **Modify your
schematic to show the connection you just made.**

Position the coil in roughly the same place you did for Exercise 1d. Tape it securely to
your amp, so that it won't move in the following steps. Until today, you might have expected
this extra connection to lower the level of interference, since you're simply adding another
connection between the shield around your resistor and ground, and you want the shield to
be grounded, right? (Right.)

Measure the amplitude of the 60 Hz signal. Explain why it is higher when the extra ground connection is made, i.e., why it is higher now than it was in step 1d.

2d. Predict *quantitatively* what will happen when you remove the short banana cable you added in part 2b. **Try it and report your results.** If necessary, revise your theoretical explanation. **Explain, including a revised schematic (please redraw it, rather than just modifying your previous schematic) and a comparison of your experimental and theoretical amplitudes.** Hint: remember that your amp includes a resistor between V_{AIC} and V_{AOC}.

2e. Reconnect the short banana cable between V_{AIC} and V_{AOC}. Using any components necessary, reduce the amplitude of the 60 Hz interference to 2/12ths of the amplitude from part 2c *by modifying your circuit*. Rather than guessing, it will be constructive to reason quantitatively with your circuit schematic. You are not allowed to remove the connection between V_{AIC} and V_{AOC}. (If you need a hint, read the next paragraph.) **Draw a schematic of what you did, and explain why it worked.**

This should convince you that, when it is necessary to make a second ground connection, you should consider adding a small resistor (50 Ω would be plenty) between the output common of your amp and the input ground common of the device you connect it to. (This will only work if the output common is not internally connected to ground.)

Post-lab question 1: Section 6.6 claims that the most important way to reduce inductively coupled interference is by using coaxial cable or "twisted pair" cable (in which the two insulated wires are twisted around each other) for the critical loop. What part or parts of this experiment most clearly demonstrated the principal behind this? Explain briefly.

Post-lab question 2: It's good practice to connect a function generator to a circuit using a coax cable with the outer conductor grounded, to prevent capacitively coupled interference. However, in most cases you could connect it with an unshielded wire, and still observe very little if any interference. Why? Hint: in terms of the effect on the amplitude of an AC waveform, a resistor to ground (i.e., a resistor with one end connected to ground) is the same as a resistor to any well-defined voltage.

Lab 6E Amplifier Noise and Introduction to LabVIEW

Learning goals: Understand offsets and how to deal with them. Become more familiar with looking up spec values and understanding equipment manuals. Reinforce understanding of the basic amplifier noise model, including voltage noise density and current noise density. Learn the basics of LabVIEW.

Pre-lab reading: "6.7 Noise."

Introduction: This lab has two parts. In the first, you'll gain a deeper understanding of amplifiers. In the second, you'll begin to learn how to program in LabVIEW. This is a graphical programming language that is widely used in industry and academia. It is wonderfully quick to program in LabVIEW if you're trying to accomplish something that the designers of the language have anticipated, which usually you are. (However, it can be frustrating to try to accomplish something they haven't anticipated!)

Part 1: DC Offsets and Amplifier Noise

1. Every instrument has a "DC offset," i.e., an undesired small value that is added to the desired output. For example, if you're measuring a current with a sensitive meter, the meter might read −0.23 pA when you know there actually is no current flowing. There are two ways to deal with this: (1) adjust your instrument to zero out the offset; or (2) measure the

offset carefully and subtract it from your measurements. In this exercise, you'll use the first approach.

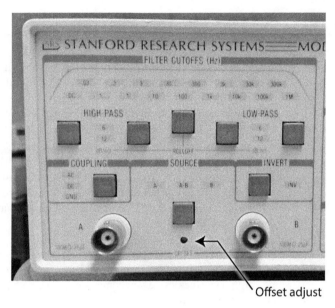

Offset adjust

FIGURE 6.74   Input offset adjust.

Configure your amplifier so that you know it should have 0 V at the output and set the gain to the highest value you can without overloading. Measure the actual output voltage using your scope. Use a small screwdriver to adjust the "Offset adjust" screw on the front panel of your amplifier (shown in Figure 6.74 on a typical amp) until the output is as close to 0 V as you can manage. Note that the required adjustment value may depend on the gain.

The "DC offset" of any instrument changes slowly over time and also depends on temperature. For highly sensitive measurements you need to take these changes into account.

2. In this part, you will measure the voltage noise density of your amplifier. To do this, you'll need a reliable value for the equivalent noise bandwidth (ENBW) of your apparatus. So far, you only know how to calculate the ENBW for a low pass filter. If you did Lab 6C, you know that the bandwidth limit of an amplifier can be approximately modeled by a low pass filter. However, this model is too approximate for a quantitative calculation of ENBW. So, you'll engage an actual low pass filter with an f_{3dB} much lower than your amplifier bandwidth. The ENBW can then be very well approximated as that of the low pass filter. This consideration puts an upper limit on the f_{3dB} of the filter you engage, since it must be much lower than the amplifier bandwidth.

The total noise includes contributions from white noise (what you want to measure) and from $1/f$ noise. If you choose the f_{3dB} of the filter to be too low, your measurement will be dominated by the $1/f$ noise. This consideration puts a lower limit on the f_{3dB} of the filter you engage, since you want it to be large enough that the white noise dominates the $1/f$ noise. In order to nicely satisfy both limits, engage a 30 kHz low-pass filter.

Figure out a way to approximately measure the rms voltage noise density of your amplifier. (Note that your scope is able to measure rms voltages.) Make sure that, in your method, the effects of current noise and the Johnson noise of the source resistance (labeled "R_{source}" in Section 6.7) are negligible. Hint: the input coupling selector for your amplifier works the same way as the selector on your scope; if necessary, review how this works, starting on slide 23 of the scope tutorial on ExpPhys.com (listed under Lab 6B). Make sure your signal is free of any periodic interference (check at a variety of timescales) before you make your measurement. You may need to operate your amplifier on batteries rather than line power to reduce the 60 Hz interference. Experiment with different settings for the s/div on your scope to find one that gives you a fairly stable value for the rms voltage. Check with your instructor before proceeding with the full measurement. **Describe your method and result.** Look up the spec in the manual or online and **compare your measurement with the spec**; for a valid comparison, you must use a gain of at least 10000.

Post-lab question for Part 1: As you've seen in this lab, measuring the voltage noise density is fairly easy. However, the same is not true for current noise density, as you'll explore in this problem.

a) Your amplifier has a spec current noise density of 12 fA/$\sqrt{\text{Hz}}$. In principle, you could measure this by connecting a resistor to ground to the input of your amplifier (i.e., one end of the resistor would be connected to ground, and the other to the amp input, with the resistor inside a shielded box to prevent interference). This resistor acts as the source resistance. You would then measure the noise amplitude at the amp output. Explain why having the value of this resistor be as high as possible will ensure that Johnson noise and amplifier voltage noise have negligible effects on this measurement.

b) Assuming your resistor is large enough that the other noise sources are negligible compared with amplifier current noise, how would you use the measured rms noise amplitude at the amplifier output to calculate the current noise density i_n? (Assume the equivalent noise bandwidth B is known.)

c) In fact, it's difficult to get to the limit described in part b. For any finite R_{source}, the total noise is bigger than what one would expect from just amplifier current noise. However, one could subtract out the theoretical value of the Johnson noise to get the contribution from the current noise, if the Johnson noise doesn't dominate by too much. For what value of resistance would the total noise be twice the noise one would expect from amplifier current noise alone? Assume the spec value for i_n, that the contribution from amp voltage noise is negligible, and that T is 293 K.

d) Say you use the resistance you just calculated. The combination of the resistor to ground and the input capacitance of your amplifier forms an unintentional low-pass filter. If the input capacitance, including the effects of the connectors, is 30 pF total, what is the $f_{3\text{dB}}$ of this filter?

e) Why does the result of the previous part mean that this method for measuring the current noise density for the amplifier won't work? (Hint: re-read the first two paragraphs of part 2 of the lab.)

Part 2: Introduction to LabVIEW

1. Launch LabVIEW, decline the offer to extend the evaluation periods, and click "Create Project." Then choose "Blank VI" to start writing a new program (called a "virtual instrument" or VI).

2. All VIs have a front panel (the user interface) and a block diagram (the guts of the program). Click on the block diagram panel.

3. You will make a program that adds two numbers together. Right click anywhere on the block diagram, then choose Numeric/Add. Position the triangular Add icon where you want it, then left click to release it.

4. If the "Context Help" window isn't showing, activate it from the Help menu.

5. Move your mouse over the Add icon and look at the context help box.
   - LabVIEW was designed to use a left-to-right and top-to-bottom layout. Organize your program in this fashion to the greatest extent possible.

6. Now we're ready to add the user-inputs for the two numbers to be added. Move your mouse over the x-input; your mouse icon changes to a spool of wire. Right click and choose create/control. Note that the name of the control, "x" is highlighted – this is a hint that you should immediately rename it to something more informative. Change it to a name you think is more appropriate by typing the new name. Note that the box under the name says "DBL" at the bottom, meaning that it's a double-precision floating point number, i.e., a real number represented by 64 bits in the computer. It might be a number such as $-3.12456\text{E}-12$. Note further how the input box is connected to the Add triangle by an orange wire; the wire carries the data, and orange indicates double-precision floating point data. In LabVIEW, there are two types of numbers, Double-precision floating point numbers (represented by orange wires on the block diagram) and integers (represented by blue wires).

- The data type being carried by a wire in your program is represented by the appearance of the wire:
    - blue   integer
    - orange   floating point
    - green   boolean (true/false)
    - pink   string (alphanumeric characters)
    - thick line of any previous color: one-dimensional array of one of the above
    - double-line of any previous color: two-dimensional array of one of the above
    - thick patterned purple line is "dynamic data," usually including multiple data elements

7. Go back to the front panel – you should see the input you've just created.

8. Return to the block diagram, and create an input for the y, just as you did for the x. Again, give it an appropriate name.

9. The wire coming from this second input box may be hard to see. This is bad programming practice in LabVIEW – it's important that everything be easily visible. Move your mouse over the input box so that your mouse icon changes to a pointer and drag the box to a better location. Similarly, drag the wire segments so that one can easily see where the wire connects.

   - The hand cursor allows you to adjust settings within your program; the arrow to select items the wire spool to wire up the VI.

   - It is very easy to accidentally introduce tiny wire segments which aren't connected to anything but confuse the LabVIEW compiler. If your program won't compile (as indicated by a broken arrow in the upper left of the screen), always try pressing Ctrl-b to remove the bad wires. (This is also under the Edit main menu.)

   - To delete wires and other program elements, select them using the arrow cursor, then use the delete key (or the cut command.)

10. Just for fun, we will change the y-input from a double-precision floating point to an integer. Right click on the box and choose representation/I32. Note that the box and the wire change to blue, indicating an integer. Snicker at the oafishness of the programmer who forgot to change the icon for integers, so that the icon still displays "1.23," even though you can only type integers into this input.

11. Also note that there's a small orange dot at the input on the Add triangle. This is a "coercion dot," meaning that LabVIEW is automatically converting the data type from integer to floating point to do the addition. Most of the time it's okay to allow LabVIEW to do such automatic conversions, though my personal preference is to show them more explicitly. So, right click in the middle of the blue wire and choose Insert/Numeric Palette/Conversion/To Double Precision Float. Note how the wire changes color, and how the coercion dot is gone.

12. Mouse over the output terminal of the Add triangle (on the point at the right), right click and choose Create/Indicator. As usual, rename it right away to something more memorable. It's important to understand the difference between a "Control" and an "Indicator." A control is a data input from the user to the program, while an indicator is a data display.

13. Go to the front panel. If you mouse over any of the three items there carefully, you can get the mouse icon to change to an arrow, which you can then use to click and drag the item to a new location. Move the three things into a pleasing pattern.

14. Right click on one of the three items on the front panel and choose "find terminal." Do you see what happens?

15. Right click on a different item on the block diagram and choose "find control" or "find indicator." Does the behavior make sense?

16. Save your program, using a suitable filename. You should always save before attempting to run a program. That way, if your program causes the computer to crash (unlikely, but it can happen), you won't lose your effort.

17. On the front panel, type the two numbers you want to add together into the two input boxes. Then, click the arrow in the top left to run your program. Does it work?

18. Now modify your program so that, after adding the two numbers together, you square the result, and display only the squared sum. (Hint: if you remember the "Insert" trick, you can do this in about 15 seconds.) Make sure your program works appropriately.

19. It is good practice in LabVIEW to create sub-VIs as soon as you have an important functional block. As a demonstration of how this is done, in the block diagram, click and drag to create a box that encloses the math parts of your VI, but not the input and output boxes. (Some parts of the wires will also be highlighted.) From the Edit menu choose Create SubVI.

20. Double click on the resulting icon; this brings you to the front panel of the sub-VI you just created. Use the Window menu to go to the block diagram and look at it.

21. Go back to the front panel of your sub-VI, double click the icon in the top right, and use the editor that pops up to change the icon. Go back to the block diagram for your main program and verify that the icon has changed. Save your sub-VI using a suitable name, and close the windows showing the front panel and block diagram of the sub-VI, while keeping the windows for your main program open.

    • Sub-VIs should have all of their inputs on the left, and all of their outputs on the right.

22. So far, your program only executes once and then stops. It would be nicer to have a "live" version. To do this, right click on the block diagram of your main program, choose Express/ Exec Control/While Loop, and then click and drag to enclose your whole program in a While loop. Mouse over the border of the While loop and read the description inside the "Context Help" box. Note how LabVIEW automatically creates a stop button for you. This "STOP" button, in the front panel, will be the button to press when you wish to stop the program.

23. In the front panel, run your program – it should update the squared sum whenever you change the input. To stop execution, click the "STOP" button. Notice the red hexagon (similar to a stop sign) two icons to the right of the run icon. This stop button is essentially a kill switch. Run your program and try stopping it with either this stop icon or the STOP button. What differences do you notice?

24. Right click on the "i" inside the square in the lower left of your while loop and choose Create/ Indicator. Run your program again – can you tell what the "i" represents? (If not, see the next step.)

25. Things may be happening too fast for you to tell what's going on. Stop your program, go to the block diagram, and click the light bulb icon, which is located four icons to the right of the run icon. This slows program execution *way* down, and also shows you the flow of data. Start your program again and watch what happens first on the front panel and then on the block diagram. This "highlight execution" tool is an important debugging aid.

26. On your front panel, right click and choose "Array, Matrix…" and then Array. This inserts an Array indicator. So far, the array only knows that it must hold a list of items, but it doesn't know what type of item. To fix this, insert a Numeric indicator on the front panel, then drag it onto the array.

27. Click and drag down on the lower right corner of the array to show the first several entries in the array. (These will all be "0" for now.)

28. Go back to the block diagram and drag the icon for the array a bit to the right of the while loop, so that it's not inside the while loop.

29. Make a branch from the wire that has the squared sum and connect it to the array. You'll notice that an orange square appears on the border of the while loop where the wire goes through – this is called a "tunnel." You'll also notice a red "X" on the wire segment connecting this box to the array, and you'll see that the arrow you'd normally use to run the program is broken. Mouse over the red X and read the error message.

30. To fix the problem, right click on the tunnel, and change the tunnel mode to "indexing." Now, the wire segment should be a thick orange line, indicating an array of double-precision numbers.

31. Run your program again, and let it go for a few full cycles. (You should still have the "highlight execution" on, so this may take 30 seconds or so.) You won't see any change in behavior until you press the "STOP" button. Does what happens then make sense?

    • LabVIEW includes some "Express" VIs, which are very powerful and easy to use. So, if you can find one matched to your task, using it will usually make your life easier.

32. Close everything you've been working on and open a new Blank VI.

33. From the "Express" part of the insert vi menu, choose Express/Input/Simulate Sig. This is like a signal generator. Accept the defaults in the dialog box. Create a graph indicator for the output (where it says "Sine"). Save and run the program, make sure it works the way you think it should.

34. Now add Express/Signal Analysis/Spectral. Accept the defaults. Connect the output of simulate sig to the Spectral Analysis vi. (The output is also going to the graph.) Create a graph indicator for the FFT output of the Spectrum vi. This shows the Fourier transform of your signal. It's more spread out than you might expect; you'll learn more about this in a later lab. However, the frequency of the peak should make sense. Does it?

35. Every express VI has a dialog box that allows you to change the parameters. Double click on the Simulate sig vi and change a parameter. Re-run your program and make the two graphs respond in the way you expected.

36. Every express VI (and many of the other VIs) has an "error in" and an "error out." These are used to pass error codes through your program, so that if something goes wrong you have a clue of where the problem is. This is especially important when communicating with external devices, which we aren't doing today. Connect the error out of the Simulate Sig to the error in of the Spectrum, and then create an indicator for the error out of the Spectrum. If there were any errors, a message would be displayed in this indicator. Run your program and verify that no error is displayed.

37. The output of most express VIs is in the form of "Dynamic Data," which is a bundle of several data streams including the array of data you usually are interested in, but other pieces of information as well, which depend on the particular VI, but might include things such as the start time and frequency interval. In many cases, you can just wire the Dynamic Data directly to the next VI, as you've done above. However, in some cases, you need to extract the actual array of data. To practice this, insert the "Convert from Dynamic Data" VI, and in the dialog choose "1D array of scalars – automatic." Wire the output of Simulate Signal to the input of Convert from Dynamic Data. Create an indicator for the output of this, and make sure you understand what it's doing.

38. Try the broom button on the right end of the same row of icons that has the highlight execution tool. Pretty neat, eh?

Lab 6F Lock-In Amplifiers

Learning goals: Understand the principles of lock-in amplifiers, their basic operation, and their strengths and weaknesses compared with broadband amplifiers.

Pre-lab reading: "6.7 Noise."

Introduction and Background

In most experiments, it's important to maximize the signal-to-noise ratio (SNR). A high SNR gives higher precision of results, and better ability to distinguish between several possible theoretical models.

So, when you design an experiment, you should make the signal as big as you can. It's worth putting some thought into this: what is it that limits how big you can make the signal?

In many experiments, you apply some stimulus to a sample which then produces the signal you measure. For example, if you want to measure the electrical resistance of a sample, you could force a known current through it (the stimulus) and measure the resulting voltage (the signal). In experiments of this type, if the stimulus gets too big, it can damage the sample, or cause it to start behaving in an undesired way. In our example, if the current is too high, the sample will heat up significantly, changing its resistance. This limit on the stimulus is what limits the signal size in such experiments.

After maximizing the signal, the next step in improving SNR is to minimize the noise. In most experiments, the signal must be amplified in order to be measured accurately. The amplifier inevitably contributes some noise. In the model described in Section 6.7, the total noise at the amplifier output is:

$$V_{\text{noise, out, rms}} = G\sqrt{v_{\text{n}}^2 B + i_{\text{n}}^2 R_{\text{source}}^2 B + 4k_{\text{B}}TR_{\text{source}}B}$$

$$= G\sqrt{B}\sqrt{v_{\text{n}}^2 + i_{\text{n}}^2 R_{\text{source}}^2 + 4k_{\text{B}}TR_{\text{source}}}. \tag{6.20}$$

Pre-lab question 1: The above equation suggests that reducing G should improve SNR. Explain what is wrong with this reasoning.

One important strategy for reducing noise, as suggested by the above equation, is to lower the equivalent

noise bandwidth B, for example, by engaging a low-pass filter, which has $B = \dfrac{\pi}{2} f_{\text{3dB}} = \dfrac{\pi}{2} \dfrac{1}{2\pi RC} = \dfrac{1}{4RC}$.

However, lowering B has a cost: it lengthens the response time of the apparatus, as illustrated in the following problem.

Pre-lab question 2: A researcher has found that the resistance of her sample is sensitive to poison gas in the atmosphere, and so the sample might make a useful sensor. She monitors the resistance by forcing a 1 μA current through the sample. The voltage across it is amplified, then passed through a low pass filter to reduce the noise and improve the SNR. The filter is made with a resistance R and a capacitance C. Say the voltage at the output of the filter under normal atmospheric conditions is V_0. Now, the sample resistance suddenly increases by 25% because poison gas has been introduced in the sample chamber. How long does it take for the voltage at the output of the filter to rise to $1.2V_0$? *Hint: you are probably familiar with the equation for the voltage across the capacitor $V_C = \varepsilon\left(1 - e^{-t/RC}\right)$ for an RC circuit that is being charged by a battery of emf ε, with the capacitor initially uncharged before the switch is closed at $t=0$. If the capacitor is instead initially charged to a voltage V_0, this equation becomes $V_C = V_0 + \left(\varepsilon - V_0\right)\left(1 - e^{-t/RC}\right)$; you can use this as a starting point in your argument.*

Although there is this downside of reducing B, there are many cases where you need to make B very small indeed to get adequate SNR. However, if you make your measurement using a DC stimulus (e.g., the constant 1 μA current in the above example), there is a limit to how low you can go with B. The limit comes from "one over f noise": the noise power in virtually any measurement, whether you're measuring an electrical signal, a biological response, or fluctuations in the stock market, is inversely proportional to frequency! The detailed explanation depends on the particular system being studied, but the general $1/f$ behavior is astonishingly universal.

To account for $1/f$ noise in our model of an amplifier, we must allow v_n and i_n to be frequency dependent. For example, since electrical power is V^2/R, at low frequencies, $v_{\text{n}}^2 \propto 1/f$. At high enough frequencies, the $1/f$ noise becomes negligible, and the noise is instead dominated by other mechanisms that produce white noise, such as Johnson noise, giving a curve as shown in Figure 6.75. (The current noise density behaves similarly.) Thus, using DC or very low frequency voltage signals means fighting the $1/f$ noise of the amplifier.

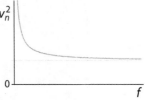

FIGURE 6.75 The square of voltage noise density vs. frequency.

Let's get more quantitative. When v_n and i_n are frequency dependent, we need the generalized version of (6.20):

$$V_{\text{noise, out, rms}}^2 = G^2 \int_0^\infty \left(v_n^2 + i_n^2 R_{\text{source}}^2 + 4k_B T R_{\text{source}} \right) \left(\frac{V_{\text{filter, out, rms}}}{V_{\text{filter, in, rms}}} \right)^2 df, \tag{6.21}$$

where $\dfrac{V_{\text{filter, out, rms}}}{V_{\text{filter, in, rms}}}$ is the ratio of voltage amplitudes at the output and input of whatever filters are present in the system. (This ratio is a function of frequency. There is always some filter present, even if it is only the effective filter caused by the bandwidth of the amplifier.)

Pre-lab question 3: For the example of Pre-lab question 2, assume that the voltage noise density of the amplifier is the dominant source of noise. Assume that B is determined by a low pass filter. Show that, once B has been reduced to the frequency range where $1/f$ noise is dominant, further reductions of B do not improve the SNR. Hint: use Equation (6.21) as a starting point. The integral can't easily be done in closed form, so you'll need to use Mathematica or a similar program to integrate numerically. Choose an appropriate (non-infinite) upper limit for the integral. It suffices to show that lowering the changing the f_{3dB} of the low pass filter from 1 Hz to 0.1 Hz doesn't affect the output noise.

To get around the issue addressed in Pre-lab question 3, scientists often use an alternating rather than a constant stimulus. In other words, they apply a sinusoidal stimulus at frequency f_m and observe the response at the same frequency. For example, we could force a current through a sample, but now use a sinusoidal current (Figure 6.76a) instead of a constant current. If the resistance of the sample varies in time (for example, because the concentration of poison gas changes), as shown in Figure 6.76b, then the voltage (which equals the current times the resistance) is as shown in Figure 6.76c. The "envelope" (shown in black) of the oscillating voltage has a variation in time that's proportional to the resistance. We often say that the rapid sinusoidal oscillation is "modulated" by the slower variation of the resistance. Note that it's important for the period of the current to be short compared with the timescale of the resistance

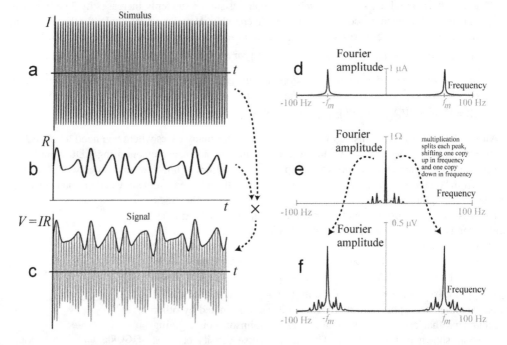

FIGURE 6.76 a–c: Current, resistance, and voltage vs. time. In c, the black line is the "envelope" of the rapid oscillations. d–f: Corresponding Fourier transforms.

variation, so that the envelope accurately represents the variation. (The period shown for the current is chosen for graphical clarity; one would actually use a much shorter period.)

By using a bandpass filter with a very narrow passband (see Figure 6.77) that is centered on the frequency of the AC current, one can still have a low B, and so a low amplifier noise. However, now that the frequencies being passed by the filter are all relatively high, $1/f$ noise is no longer a problem!

Two issues remain: (1) for the least noisy measurement we want as narrow a bandpass as possible. Although one can make B fairly small using combinations of inductors, capacitors, and resistors, it is limited by the resistance the inductor. (2) We need a

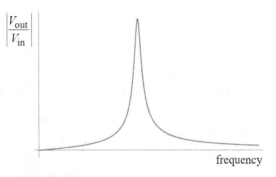

FIGURE 6.77 The voltage versus frequency curve for a narrow bandpass filter.

way to extract or "demodulate" just the envelope of the voltage, which represents the variation we're interested in, rather than the rapidly oscillating voltage itself. A "lock-in amplifier" (or simply "lock-in") solves both issues. It effectively creates a bandpass filter that can be as narrow as desired and can easily be centered on any desired frequency. It also does the job of demodulating the envelope from the rapidly oscillating voltage.

The lock-in is an example of "homodyne" detection. To understand this technique, we need Fourier analysis. Any function of time $y(t)$ that repeats* with a period T can be written as a sum of Fourier components at the angular frequencies $\omega_n = n\frac{2\pi}{T}$:

$$y(t) = \frac{A_0}{2} + \sum_{n=1}^{\infty} A_n \cos(\omega_n t + \varphi_n),$$

where we write the constant as $A_0/2$ because it will make things neater soon.

Since $\cos(\omega_n t + \varphi_n) = \frac{1}{2}\left(e^{i(\omega_n t + \varphi_n)} + e^{-i(\omega_n t + \varphi_n)}\right)$, we can also write

$$y(t) = \frac{A_0}{2} + \sum_{n=1}^{\infty} A_n \frac{1}{2}\left(e^{i(\omega_n t + \varphi_n)} + e^{-i(\omega_n t + \varphi_n)}\right)$$

$$= \frac{1}{2}\left(A_0 + \sum_{n=1}^{\infty} A_n e^{i\omega_n t} e^{i\varphi_n} + \sum_{n=1}^{\infty} A_n e^{-i\omega_n t} e^{-i\varphi_n}\right).$$

To express things more elegantly, we introduce negative frequencies. Since $\omega_n = n\frac{2\pi}{T}$, we have simply that $\omega_{-n} = -n\frac{2\pi}{T} = -\omega_n$. These negative frequencies have no physical meaning; they're merely a mathematical trick for rewriting the Fourier sum. We define

$A_{-n} \equiv A_n$ and $\varphi_{-n} \equiv -\varphi_n$, and use $\omega_{-n} = -\omega_n$

to write

$$y(t) = \frac{1}{2}\left(A_0 + \sum_{n=1}^{\infty} A_n e^{i\omega_n t} e^{i\varphi_n} + \sum_{n=1}^{\infty} A_{-n} e^{i\omega_{-n} t} e^{i\varphi_{-n}}\right)$$

$$= \frac{1}{2}\left(A_0 + \sum_{n=1}^{\infty} A_n e^{i\omega_n t} e^{i\varphi_n} + \sum_{n=-\infty}^{-1} A_n e^{i\omega_n t} e^{i\varphi_n}\right)$$

* The requirement for $y(t)$ to be periodic is not difficult to fulfill; if necessary, we can take the limit $T \to \infty$.

$$\Rightarrow y(t) = \frac{1}{2} \sum_{n=-\infty}^{\infty} A_n e^{i\omega_n t} e^{i\varphi_n} \tag{6.22}$$

where for the last step we defined $\varphi_0 \equiv 0$ and used $\omega_0 = 0\frac{2\pi}{T} = 0$.

To represent a pure sinusoid in this version of Fourier analysis, we write it as a sum of complex exponentials, for example:

$$\cos\omega_m t = \frac{e^{i\omega_m t} + e^{-i\omega_m t}}{2} = \frac{e^{i\omega_m t} + e^{i\omega_{-m} t}}{2}$$

From Equation (6.22), this means that $A_m = A_{-m} = 1$, and all the other A_ns are zero.

Usually, for experiments, we think in terms of frequency f rather than angular frequency ω, and we think of the amplitude as a function of frequency, i.e., we define $A_n \equiv A(f_n)$, where $f_n = \frac{\omega_n}{2\pi}$. The graph of A vs. f is called the "Fourier spectrum." Putting this together with the idea of the previous paragraph, we see that the Fourier spectrum for a single sinusoid of frequency f_m, such as the current, has two spikes, at $+f_m$ and $-f_m$, as shown in Figure 6.76d.

The resistance is more complicated, and so its Fourier transform has many peaks, though all at low frequencies compared with the current, as shown in Figure 6.76e.

Writing the resistance as a Fourier sum, it's easy to see what happens when it is multiplied by a sinusoidal current of amplitude I_0 at frequency $f_m = \frac{\omega_m}{2\pi}$:

$$R(t) = \frac{1}{2} \sum_{n=-\infty}^{\infty} A_n e^{i\omega_n t} e^{i\varphi_n}, \quad I(t) = I_0 \cos\omega_m t = \frac{I_0}{2}\left(e^{i\omega_m t} + e^{-i\omega_m t}\right)$$

$$\Rightarrow V(t) = IR = \frac{I_0}{4} \sum_{n=-\infty}^{\infty} A_n e^{i\varphi_n}\left(e^{i(\omega_n + \omega_m)t} + e^{i(\omega_n - \omega_m)t}\right)$$

This means a peak in the Fourier spectrum of $R(t)$ at frequency $f_n = \frac{\omega_n}{2\pi}$ translates into two peaks in the Fourier spectrum of $V(t)$, one shifted up to frequency $\frac{\omega_n + \omega_m}{2\pi}$ and the other shifted down to frequency $\frac{\omega_n - \omega_m}{2\pi}$. Recall that the period of the current is chosen to be quite short, i.e., the frequency of the current is high compared with the frequencies at which the resistance varies. In other words, $\omega_m \gg \omega_n$. Therefore the pattern of peaks centered on zero frequency in the Fourier spectrum of $R(t)$ is split into two identical patterns, one centered on $+f_m = \frac{\omega_m}{2\pi}$ and one on $-f_m = -\frac{\omega_m}{2\pi}$. This is shown in Figure 6.76f.

To illustrate the operation of the lock-in, we begin with a signal similar to that shown in Figure 6.76c,f, but with a more realistic, rapidly oscillating current; this is shown in Figure 6.78a,e. The schematic diagram for a lock-in amplifier is shown in Figure 6.79. The signal is first amplified. In the process, the amplifier inevitably adds noise, with the result shown in Figure 6.78b,f. In the case shown, the noise level is so high that the signal is barely distinguishable in Figure 6.78b. The peak near zero frequency in Figure 6.78e is due to $1/f$ noise and is shaded in gray for graphical clarity. The amplified signal is then multiplied by a "reference" voltage at frequency f_m; this multiplication is called "mixing." We already know the effect of such multiplication on the Fourier spectrum: each peak is split in two, with one of the resulting peaks shifted up in frequency by f_m and the other shifted

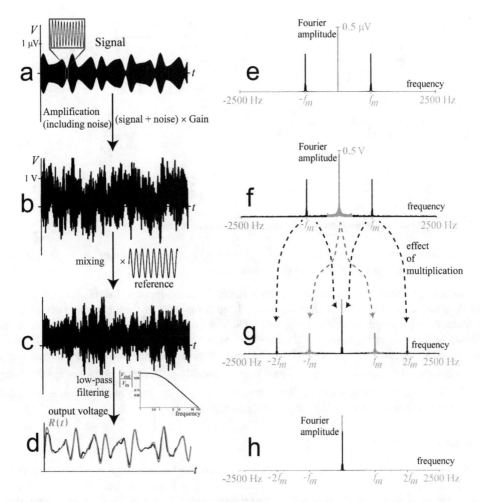

FIGURE 6.78 Operation of a lock-in amplifier. a,e: The signal is formed by forcing a rapidly oscillating current through the sample; the period of the current is much faster than the time scale of resistance variations. (These figures are essentially the same as Figures 6.76c and 6.76f, but the period for the current is more realistic and shorter.) b,f: The signal is amplified, in this case by a factor of a million. Inevitably, the amplifier adds noise. The original signal is so small that the noise almost completely obscures it. In the Fourier transform (f), the peak near zero frequency is due to $1/f$ noise and is shaded in gray. c,g: The signal is mixed by multiplication with a "reference" signal at frequency f_m. Because of the noise, the effect in the time domain (c) is not clear, but in the frequency domain (g), the multiplication splits each peak, shifting one copy to higher frequency and one to lower frequency. This results in one copy of the signal peak pattern near zero frequency, while the $1/f$ noise peak is shifted to $\pm f_m$. d,h: Passing the mixed signal through a low pass filter removes the $1/f$ noise, but retains signal peaks near zero frequency, giving (d) the desired envelope of the signal. The gray trace in d is the $R(t)$ curve that is the desired information, while the black trace is the output of the lock-in.

down by f_m. The results of the multiplication are shown in Figure 6.78c,g. You can see that the signal has been shifted down near zero frequency (and also to $\pm 2f_m$), while the $1/f$ noise is shifted to frequencies near $\pm f_m$. Finally, the mixed signal is sent through a low-pass filter, which blocks out all the $1/f$ noise and most of the white noise, as shown in Figure 6.78d,h. The final result (Figure 6.78d) is a reasonably faithful amplified version of the original signal; it's not possible to completely eliminate the effect of the noise.

Note that the lock-in effectively creates a very narrow bandpass filter centered on f_m, and also amplifies and demodulates the signal. You can think of it as a sophisticated AC voltmeter.

Pre-lab question 4: A lock-in is being used to monitor the resistance of a sample, which is constant in time: $R(t) = R_0$. The applied current is $I = I_0 \cos \omega_m t$. Does the output of the lock-in show (a) a sinusoidal oscillation at angular frequency ω_m and with constant amplitude, (b) a

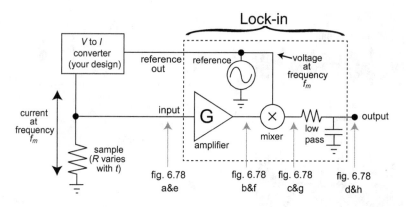

FIGURE 6.79   Schematic diagram of lock-in amplifier, connected to an external circuit.

sinusoidal oscillation at angular frequency ω_m with a modulated amplitude, (c) a constant voltage, or (d) something else? Explain your answer *briefly*.

The lock-in has an additional trick. It can detect the phase of the input signal relative to the reference signal. You can choose to monitor the amplitude of the component of the input that is in phase with the reference, or the amplitude of the component that is out of phase by 90° (either leading or lagging, depending on how you set the knobs), or the amplitude of any phase in between. For example, if the input (of amplitude V_{in}) lags the reference by 45° (for example, because the impedance of the sample includes capacitance), then the amplitude of the in-phase component is $V_{in}/\sqrt{2}$, the amplitude of the 90° component is $V_{in}/\sqrt{2}$, and the amplitude of the 45° component is V_{in}. Some lock-ins allow you to monitor the amplitude of two different phases (e.g., 0° and 90°) simultaneously.

Experimental Procedure

1. Write a LabView program that shows a "live" graph of the voltage that is applied to your A/D, i.e., a graph that updates at regular user-chosen intervals (e.g., once per second), but shows the whole history from the time you started the program. The user-chosen intervals may vary from 0.2 s to 30 s. (Note that, in this approach, you cannot take data nearly as quickly as in the approach you've used before, where you acquire thousands of data points and *then* plot them all at once.) **Include in your report a screenshot of the front panel and also a screenshot of the diagram.**

   To do this, you will need a loop that acquires one data point per iteration and updates the graph each iteration. You will need to pass data from one iteration of the loop to the next iteration. There are two ways to do this, and you'll try both. First, you'll use a "shift register," which is the more versatile option, but results in somewhat messier wiring. Then you'll re-do things using a "feedback node," which is fine for most applications and gives neater wiring.

   To get started, create a While loop. Your program should have a loop iteration time that can be varied from the front panel. There are two ways to do this: (a) right click on the border of your while loop, and choose "replace with timed loop." This is the most powerful method, as it allows you to coordinate the timing of multiple different loops. For the program you're writing today, you only need to wire up the "period" terminal. (b) Add a "timed delay" VI inside your while loop. This is simpler, and fine for many applications. It has two disadvantages: (i) the total time required for each iteration is the sum of the timed delay plus the execution time for the other elements inside the loop. This may result in variation of the iteration time, and certainly means the iteration time is longer than what you enter for the timed delay. However, if the execution time is short compared with the timed delay, this doesn't matter. (ii) You can't coordinate the execution of the loop so that it is synchronized with other loops.

Inside your While loop, add the appropriate VI to acquire a single data point per acquisition; make sure you have the VI set for "Single Point on Demand."

Right click on the right border of your While loop and choose "add shift register." You should see that a square with an upward pointing arrow has appeared on the right border, and a matching square with a downward pointing arrow has appeared on the left border. You use these to pass a variable (a single number, an array, a bundle, etc.) from one loop iteration to the next. Whatever goes into the arrow on the right side during loop iteration i comes out of the arrow on the left side at the beginning of loop iteration $i+1$. You will need to use the "Build array" VI. When you insert this, it has only a single line, but you can drag down on the bottom handle to open more lines. For your graph, use a "Waveform Graph," rather than a "Waveform Chart." Use the "Bundle" VI to create the bundle that goes into the waveform graph, so that the horizontal axis is in seconds, rather than loop iterations.

Make sure your graph resets each time you restart the program. To do this, right click on the left shift register, and choose "create constant." Leave the constant empty.

Add comments and labels where necessary and make your program pretty as well as easy to understand. Make sure that your graph resets each time you start the program. **Save your program to the desktop with the name format "Lab 6F Ralph and Agnes."**

2. Your task in this experiment is to precisely measure the resistance of a 300 Ω resistor. To measure a resistance, one either applies a known voltage and measures the resulting current, or one applies a known current and measures the resulting voltage. We will use the latter approach. We will pretend that the resistor has been cooled close to absolute zero, and that we're measuring its resistance as a function of magnetic field. To avoid heating the resistor, we must limit the current flowing through it to 1 μA. The expected change in the resistance is only 1%. You will simulate this by using a "decade box" resistor, and changing the resistance from the original value of 300 Ω to 303 Ω. How precisely can you measure this change?

We want to contrast the performance of the lock-in on this task with the performance of a broadband amplifier (an amplifier that works with a wide range of frequencies), such as the one you used in Labs 6C, 6D, or 6E. We'll start with the broadband amp, for which you will use a DC current. (For the lock-in, you'll use an AC current.) You will use a commercial voltage supply to provide the current. However, this creates a constant voltage, not a constant current. Let's say you set it to provide a constant 1 V output. What simple component can you add to convert this to a 1 μA current flowing through the 300 Ω resistor, so that the current remains constant to within 0.02% even if the resistance changes by 30%? Check your solution with your instructor before proceeding.

Wire things up in such a way as to permit a good low-noise measurement; ask your instructor for whatever pieces of equipment you think are needed; you may need to do some soldering. Engage a 3 Hz low-pass filter on your amplifier to minimize the effects of noise. Use an oscilloscope in parallel with your computer A/D card, along with your LabView program, to measure the output of your amplifier. **Draw a schematic diagram of your apparatus, including shielding and grounding. Be sure your diagram shows enough that a competent experimentalist could exactly reproduce your work.** (You should indicate the approximate cable lengths.)

3. In an ideal experiment, you would change only one experimental "independent" variable (in this case, the resistance of your decade box), and observe the resulting change in a measured "dependent" value (in this case, the voltage across the decade box). However, in any real experiment, it is difficult to only change one experimental variable. For example, say you observe that the voltage is higher after you increase the resistance. However, maybe the resistance of the decade box depends on temperature, and the temperature of the room is changing over time. Is the change in voltage due to your intentional change in resistance, or instead to the change in room temperature?

Important conclusion: If you think you've observed a change in your measured dependent value caused by the change in your experimental independent variable, you should change the independent variable back, and make sure the measured value also changes back. (If it doesn't, that means that something else is changing at the same time you change your independent variable.) For example, if you're measuring the resistance of a sample as a function of temperature, you should vary the temperature from low to high, and then back from high to low, and make sure that the data points collected when you were increasing the temperature match those collected when you were decreasing the temperature.

As applied to this case, you should measure the voltage across the decade box for about 30 s, then increase the resistance to 303 Ω, collect data for another 30 s, then decrease the resistance back to 300 Ω, and collect data for a final 30 s.

4. **Measure and record the signal-to-noise ratio, as defined above, using your computer graph to estimate both the size of the signal and the size of the noise. Take a screen capture of the voltage vs. time graph and include it in your report.**

5. Now you will repeat this measurement using a lock-in instead of the broadband amp. Instead of the 1 V DC provided by the commercial voltage supply, you will use the reference oscillator that is built into the lock-in to provide a 1 V rms AC voltage at a well-defined frequency. (The amplitude of the oscillator is adjustable, so you'll need to set it to 1 V rms.) Using the same trick as above, you will convert this voltage to a 1 μA rms AC current through the 300 Ω resistor. You will use the lock-in to measure the rms amplitude of the resulting voltage. Set up this circuit and get your instructor to check it before proceeding.

6. Now, spend some time understanding the controls on the lock-in. Examine the various controls, inputs and outputs and try to understand what they do. Proceed only when you feel confident that you understand things fairly well.

7. Set the frequency on the lock-in to 4 kHz.

8. Set the other controls to provide as fair a comparison as possible with the broadband amp experiment. In particular, make sure you've set things to mimic the effects of the 3 Hz low pass filter you used in that experiment.

9. Use the lock-in to measure the voltage across the resistor and **calculate the resistance of your resistor. Compare with expectations.**

10. **Measure and record the signal-to-noise for a 1% resistance change. Compare this ratio with that obtained using the broadband amp and qualitatively explain why one is better than the other. Take a screen capture of the voltage vs. time graph and include it in your report.**

11. Change the "time constant" to 10 s and repeat the experiment. **What is the signal-to-noise ratio now? Compare this quantitatively with expectations. You may find that your result is not consistent with your expectation; if so, qualitatively explain why this happened. What is the downside of using a longer time constant?**

12. **Under what circumstances is it better to use a lock-in than a broadband amplifier? Under what circumstances is it better to use a broadband amplifier?**

Lab 6G Introduction to Op Amps

This section is available at ExpPhys.com.

Lab 6H More on Op Amps

This section is available at ExpPhys.com.

6.11 Homework Problems

This section is available at ExpPhys.com.

7
Fundamentals of Interfacing Experiments with Computers

Walter F. Smith

CONTENTS

We live in a digital world, but we're fairly analog creatures.

– Omar Ahmad (internet entrepreneur)

7.1 Introduction: The Difference between Digital and Analog

In this chapter, we describe the most important issues you need to consider when collecting data on a computer and when performing Fourier analysis in the lab. Some of this may already be obvious to you, but there are some important subtleties that most people find surprising.

What's the fundamental difference between analog electronics and digital electronics? In a typical experiment, a small-amplitude ("low level") signal is filtered and amplified, then converted to digital form, and finally collected on a computer. The voltage on the wire connected to the amplifier input can vary continuously, e.g., from 0 V to +10 mV. Small variations, e.g., from 5.011 mV to 5.023 mV, of this voltage can carry important information about the physical processes being measured. This continuously variable voltage is sometimes called an "analog signal." Say the amplifier has gain 1000. Then, in our example, the output voltage varies continuously from 0 V to +10 V. This voltage is connected to an "analog-to-digital converter" or "A/D" or "ADC," which measures or "samples" the amplified analog signal at regular time intervals (e.g., every millisecond).

Let's consider an "8-bit A/D," such as that used in many digital oscilloscopes. This represents each sample of the analog signal as an eight-digit binary number. We will discuss a simple representation scheme in which the binary number "0000 0000" represents a sample of 0 V, while the binary number "1111 1111" represents a sample of +10 V. The binary digits are written in groups of four to make them easier to read. (In most cases, the actual representation scheme used by an A/D is more complicated, since, for example, it must be able to represent negative voltages.)

The 8-bit A/D has eight output wires. The circuitry in the A/D ensures that the voltages on these wires are always close to one of two values, e.g., 0 V and +5 V. As shown in Figure 7.1, voltages close to 0 V represent a binary "0," and voltages close to +5 V represent a binary "1." For example, in

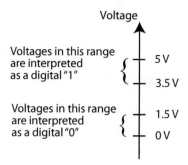

FIGURE 7.1   Interpretation of voltages in binary, for a system with a nominal "high" level of 5 V and a nominal "low" value of 0 V.

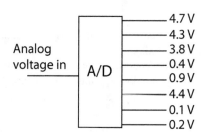

FIGURE 7.2   Analog-to-digital converter.

Figure 7.2 the voltages on the eight output wires represent the binary number (reading top to bottom) "1110 0100."*

The A/D output goes to some other digital circuit, e.g., a chip that stores the number in computer memory. Each input of this chip, and every input of every digital circuit, effectively has a "comparator" that compares the input voltage with a threshold value. As shown in Figure 7.1, for our example system any voltage below 1.5 V is treated as a binary "0," while any voltage above 3.5 V is treated as a binary "1."[†] It's this comparator that distinguishes digital circuitry from analog circuitry. Small variations of the voltage at the input to a digital circuit, e.g., from 4.1 V to 5.0 V, have absolutely no effect on the rest of the circuit, since both these voltages are treated as a binary "1." That insensitivity to small fluctuations means that modest levels of interference have absolutely no effect on digital circuits[‡].

Approaches to Interfacing

The process of connecting a computer to an experiment is called "interfacing." There are four common approaches: (1) using a commercial general purpose analog-to-digital and digital-to-analog interface, typically with USB connection; (2) using a microcontroller; (3) using a specialized instrument (such as a picoammeter) with a digital (usually USB) interface; and (4) doing your own interfacing at the chip level. Our emphasis in this book is on approaches 1 and 2, but much of what you learn here can also be applied to approach 3, and to a lesser extent approach 4.

---

* The A/D shown here has a "parallel output," meaning that the binary representation of the sample is presented at a single instant on multiple output wires. You can also get an A/D with a "serial output." This type has a single output wire, and the binary number is presented as a series of high or low voltages at well-defined increments in time.

† Voltages in the range 1.5 V to 3.5 V might be interpreted either as a binary 0 or as a 1, so the circuit designers ensure that the voltages don't stray into this zone of dreaded uncertainty.

‡ For example, say one of the A/D lines is supposed to be a digital "0," and the actual voltage is put out by the A/D is 0.1 V in the absence of interference. We can tolerate up to 1.4 V of change in this voltage due to interference before the voltage reaches 1.5 V, the threshold above which the next stage might misinterpret the signal as a digital "1."

7.2 Sampling Rate, Resolution, and the Importance of Analog Amplification

The most important considerations when choosing an A/D for your experiment is the time Δt between samples, and the number of binary digits ("bits") of resolution. The sampling rate is defined as*

$f_{sample} = \dfrac{1}{\Delta t}$. A higher sampling rate allows you to more accurately digitize a rapidly varying analog signal; we'll discuss this in detail below.

The number of bits determines the voltage resolution. In our example above, the analog range 0 V to +10 V is mapped onto the binary range 0000 0000 to 1111 1111. There are 2^8 possible binary numbers that we can express with 8 bits, so the smallest change in voltage we can represent is $\dfrac{10\text{ V}}{2^8 - 1} = 39.2\text{ mV}$.

(The −1 in the denominator comes from the fact that the largest number we can represent, 1111 1111, equals $2^8 - 1$.) This is illustrated in Table 7.1. For analog voltages between those listed, the A/D rounds to the nearest value that can be represented. For example, 22 mV would be represented as 0000 0001.

In general,

$$\text{A/D resolution} \equiv \text{minimum change in voltage} = \Delta V = \frac{\text{analog voltage range}}{2^n - 1}. \qquad (7.1)$$

$$n = \text{number of bits}$$

The left part of Figure 7.3 shows the result of digitizing a sine wave with a low-resolution A/D (one with only 3 bits, meaning that there are $2^3 = 8$ possible voltage levels), then re-converting it to analog form (using a "digital-to-analog converter" or "D/A" or "DAC"). The sample rate is much higher than the frequency of the sine wave, so that the time interval Δt between samples is too small to see in this graph.

The right part of Figure 7.3 shows the opposite limit: low sample rate and high resolution. A sample is taken at the times marked by the vertical lines and held until the next sample is taken. Because the resolution is high, the digitized version is precisely correct at the moment each sample is taken.

There are tradeoffs between sampling rate, resolution, and cost. For the highest sampling rates (e.g., what you need in a digital scope), you must accept a limited number of bits of resolution, typically 8 bits (corresponding to $2^8 = 256$ voltage levels) or 10 (corresponding to $2^{10} = 1024$ voltage levels).

It's quite important to amplify the analog signal with an appropriate gain before sending it to the A/D. The goal is to match the maximum and minimum voltages of the analog signal to the input range of the A/D. For

TABLE 7.1

A simple scheme for converting analog voltages to binary

Analog voltage	Binary representation
0	0000 0000
39.2 mV	0000 0001
$2 \times 39.2\text{ mV} = 78.4\text{ mV}$	0000 0010
$3 \times 39.2\text{ mV} = 117.6\text{ mV}$	0000 0011
...	...
$(2^8 - 2) \times 39.2\text{ mV} = 9.9608\text{ V}$	1111 1110
$(2^8 - 1) \times 39.2\text{ mV} = 10\text{ V}$	1111 1111

---

* Note that this is a frequency, measured in Hz, rather than an angular frequency, which would be measured in rad/s.

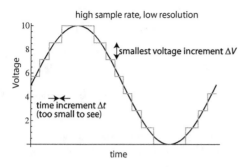
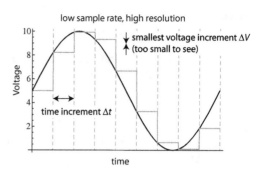

FIGURE 7.3   For both graphs, the thin black trace is the analog input to a 3-bit A/D, while the thick gray trace is the result of re-converting the digital output of the A/D to analog form, using a D/A. Left: very high sample rate, but low resolution (a 3-bit D/A). Because the sample rate is so high, the digitized version changes immediately when the analog input reaches a threshold between two digitized values. Right: low sample rate, but high resolution. At the time each sample is made, the digitized version matches the analog input very closely, but there is a long interval until the next sample.

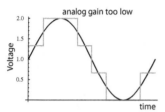
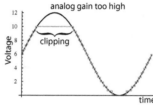
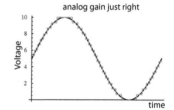

FIGURE 7.4   Results of digitizing using a 4-bit A/D with an input range of 0 V to +10 V. Each graph has a different vertical scale. Note that any A/D you use in the lab will have at least 8 bits, meaning a resolution 16 times better than shown here. However, correct adjustment of the analog gain is still quite important.

example, if the A/D is configured to a range of 0 V to 10 V, and you feed it a signal that varies from 0 V to 2 V, the resulting digitized version will be unnecessarily "steppy," as shown in the left part of Figure 7.4. However, if you use too high a gain, e.g., so that the signal fed to the A/D varies from 0 V to 12 V, then you'll get "clipping," as shown in the middle part of Figure 7.4. The right part of Figure 7.4 shows the optimal case, when the analog signal matches the input range of the A/D. Clipping is a more serious problem than steppiness, because you completely miss what's happening during the clipping periods. So, if you're unsure what range to expect on your analog signal, it's better to use a gain that's low enough that you're certain there will be no clipping.

7.3   The Nyquist Frequency, Aliasing, Windowing, and Experimental Fourier Analysis

When running a new experiment, the first step is to look at the signal in its "raw" form (e.g., voltage vs. time). For most scientists, the next step is to look at the Fourier representation of the signal, i.e., the amplitude (and sometimes phase) as a function of frequency. This can give critical insights about the physical processes being measured and also about sources of interference.

Let's get into the details and see how the process of Fourier analysis is affected by experimental realities. Consider an analog waveform that is observed for a time interval T. We know from Fourier analysis that we can represent the analog waveform over the interval T as an infinite sum of sinusoids:

$$V_{\text{analog}}(t) = \sum_{n=0}^{\infty} A(f_n)\sqrt{2}\cos(2\pi f_n t + \varphi_n), \tag{7.2}$$

Full Fourier series representation of the analog waveform

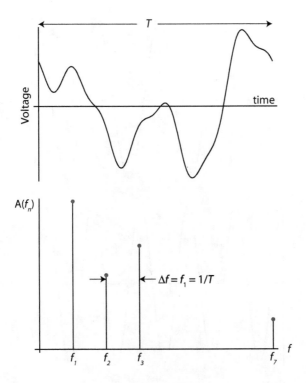

FIGURE 7.5 Top: an analog waveform is measured for a time interval T. Bottom: amplitudes of the four Fourier components that make up the waveform.

where $f_n = nf_1$, and $f_1 = \dfrac{1}{T}$. $A(f_n)$ is the rms amplitude of the Fourier component with frequency f_n, and φ_n is a phase factor. The factor $\sqrt{2}$ is included to convert the rms amplitudes to conventional peak amplitudes. The $n = 0$ term allows non-zero average value. Note that the observation time T (also called the acquisition time) determines the spacing between frequencies in the Fourier series, i.e., the frequency resolution of the Fourier spectrum. An example of the original analog waveform and the corresponding graph of $A(f_n)$ vs. f is shown in Figure 7.5.

However, in an actual experiment, we don't acquire full knowledge about the analog waveform; instead, we only collect samples at time intervals Δt, corresponding to a sampling frequency $f_{\text{sample}} = \dfrac{1}{\Delta t}$.

There is a maximum frequency that these samples can represent: the frequency for which we have two samples per cycle, with one sample at the peak of each period and one at the trough, as shown in the top part of Figure 7.6. Any higher frequency would introduce wiggles between the samples, as shown in the bottom part of Figure 7.6, and the samples cannot represent these extra wiggles. So the highest frequency that can be represented by our set of samples is

$$f_{\text{Nyquist}} = \frac{f_{\text{sample}}}{2}. \qquad (7.3)$$

This is the Nyquist frequency, named after the theoretical engineer and mathematician Harry Nyquist.

A consequence of this highest frequency is that, with our sampled dataset, we can only express part of the full Fourier series shown in Equation (7.2). In other words, there is a maximum value of n:

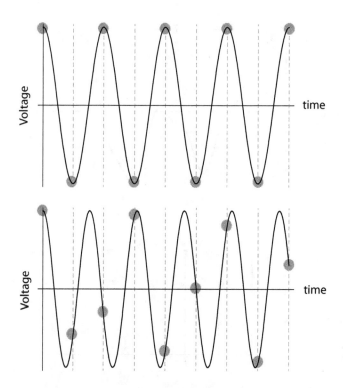

FIGURE 7.6 Top: an analog waveform is sampled at the times shown by the vertical lines. In this case, the frequency of the waveform equals the Nyquist frequency, so that there are two samples per period. Bottom: the waveform frequency is above the Nyquist frequency, so there are extra wiggles between the sampled points.

$$V(t) = \sum_{n=0}^{n_{max}} A(f_n)\sqrt{2}\cos(2\pi f_n t + \varphi_n).$$

It is more intuitive to express the sum in terms of frequencies:

$$V(t) = \sum_{f_n=0}^{f_{Nyquist}} A(f_n)\sqrt{2}\cos(2\pi f_n t + \varphi_n), \text{ where } f_n = \frac{n}{T} \text{ and } f_{Nyquist} = \frac{f_{sample}}{2} \qquad (7.4)$$

Fourier series representation of the waveform that can be reconstructed from the sampled points.

T is the total acquisition time.

Here's a biblical-sounding way to remember the most important points of the above discussion:

The smallest shall determine the largest, and the largest shall determine the smallest.

In other words, the largest time scale of the data acquisition, which is the total acquisition time T, determines the smallest frequency scale of the Fourier spectrum, the spacing between frequencies $f_1 = \Delta f = \frac{1}{T}$. The smallest time scale of the data acquisition, the time Δt between samples determines the largest frequency $f_{Nyquist} = \frac{f_{sample}}{2} = \frac{1}{2\Delta t}$.

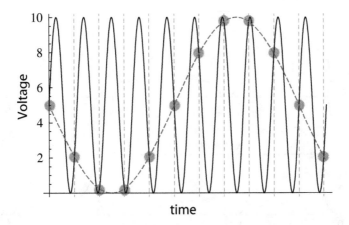

FIGURE 7.7 Thin black curve: analog input to A/D. Samples are taken by the A/D at the times marked by the vertical lines. Gray dashed curve: apparent sinusoid of the sampled points, which is at a much lower frequency than the analog input.

The Fourier amplitudes $A(f_n)$ and phases φ_n can be computed from the samples of the analog waveform using an efficient algorithm called the "Fast Fourier Transform," or FFT. This algorithm is implemented on any instrument you may use that displays a Fourier spectrum, and is available as a subroutine in most high-level programming languages. Most instruments (e.g., oscilloscopes with FFT capabilities or spectrum analyzers) use the representation of Equation (7.4); you may need to select the "Magnitude and Phase" option. However, in some software implementations, the FFT may be returned in a less intuitive representation than Equation (7.4), so that there may be negative and positive frequencies, or (as n increases) the frequencies may start at 0, reach a maximum of f_{Nyquist}, and then go back to zero*.

Aliasing

If the frequency of the analog waveform is higher than the Nyquist frequency, something strange happens. Figure 7.7 shows (solid black curve) ten periods of a sinusoidal voltage applied to the input of an A/D. The sampling frequency is only a little faster than once per period, as shown by the gray dots, so the frequency of the analog waveform is well above the Nyquist frequency.

As shown in Figure 7.8, the FFT for this set of samples has a peak at the frequency of the dashed gray curve in Figure 7.7 that connects the dots, even though that frequency is not present at all in the analog waveform! This is called "aliasing."

Aliasing is primarily a concern if you plan to look at the Fourier spectrum of your data; the samples of the analog waveform as a function of time are still legitimate samples at the moments they were taken. However, if you do need to look at the Fourier spectrum, there are a few things you can do:

- **Use a higher sampling rate,** so that every significant Fourier component present in the analog waveform has a frequency below f_{Nyquist}.

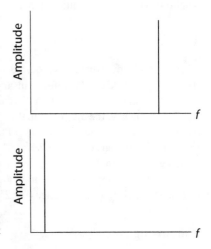

FIGURE 7.8 Top: correct Fourier spectrum for the analog input shown in Figure 7.7. Bottom: Fourier spectrum calculated from the sampled points.

---

* For more details, see "Waves and Oscillations: A Prelude to Quantum Mechanics" by Walter F. Smith (Oxford University Press, 2010), section 8.6, and "Numerical Recipes: The Art of Scientific Computing, 3rd Ed." by William H. Press et al. (Cambridge University Press, 2007), chapter 12.

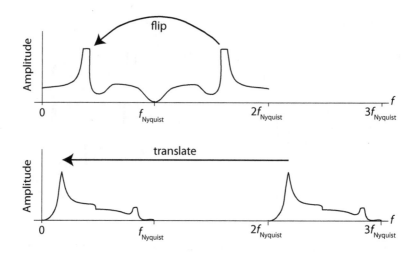

FIGURE 7.9 Top: aliasing "folds" or "flips" Fourier amplitudes in the range $f_{Nyquist}$ to $2f_{Nyquist}$ to the range $f_{Nyquist}$ to 0. Bottom: aliasing translates Fourier amplitudes in the range $2f_{Nyquist}$ to 3 f $f_{Nyquist}$ to the range 0 to $f_{Nyquist}$.

- **Use analog filters on the analog waveform** (before the A/D) to eliminate all Fourier components above $f_{Nyquist}$. Because a standard RC filter has a slow "rolloff" as the frequency is increased (about a factor of 10 attenuation for each factor of 10 increase in frequency above f_{3dB}), you may need to use multiple RC filters in series, though you'll need buffers in between to prevent loading effects. Alternatively, you can use more sophisticated filters, such as active filters or Butterworth filters. These have a faster rolloff than standard RC filters. Note that you have to do the filtering on the analog signal; once it's been digitized, there's no way to tell the difference between the actual frequency and the aliased frequency.

Sometimes neither of the above options is possible, so that aliasing will occur. In this circumstance, it's helpful to know a few additional facts. The aliased frequency is given by

$$f_{alias} = \left| f_{actual} - mf_{sample} \right|, \tag{7.4}$$

where m is whatever integer is required to give the smallest value of f_{alias}. For example, if $f_{sample} = 50\,\text{kHz}$, then $f_{Nyquist} = 25\,\text{kHz}$. Therefore, an analog waveform with $f_{actual} = 26\,\text{kHz}$ exceeds the Nyquist frequency by 1 kHz. Using Equation (7.4) with $m = 1$, we get $f_{alias} = 24\,\text{kHz}$, i.e., instead of 1 kHz above the Nyquist frequency, the aliased signal is 1 kHz below the Nyquist frequency. Extending this, we can see that the frequency range 25 kHz to 50 kHz is "folded back" to the aliased range 25 kHz to 0 kHz, as shown in the top part of Figure 7.9.

Going higher in frequency gives a shift without inversion. For example, if $f_{actual} = 51\,\text{kHz}$ and $f_{sample} = 50\,\text{kHz}$, then (using Equation (7.4) with $m = 1$) $f_{alias} = 1\,\text{kHz}$, so frequencies in the range 50 kHz to 75 kHz are shifted to the range 0 to 25 kHz, as shown in the bottom part of Figure 7.9.

Windowing

Fourier analysis of any kind assumes that the original function is periodic*. For experimental Fourier analysis, the assumed periodicity is the total acquisition time T; this assumption is implicit in Equation (7.3).

---

* Note that for the "Fourier transform" (as opposed to a "Fourier series") the period is infinite. However, taking the Fourier Transform requires knowledge of the original function over the full, infinite extent of time, so it's not relevant for experimental Fourier analysis. The "Fast Fourier Transform" that is used in experiments on a data set taken over a finite time period is really a Fourier series analysis, not a Fourier transform. However, if you try to point this out to someone, they'll likely kick you.

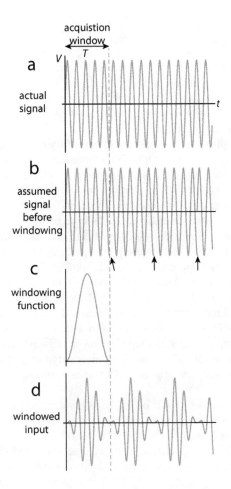

FIGURE 7.10   The windowing process.

For example, suppose the analog input is a sine wave. As usual, we measure it for a time T only, as shown in Figure 7.10a. However, the Fourier analysis process assumes the analog function has a period T, i.e., it assumes that the complete function (if you had measured it over a longer interval) looks as shown in Figure 7.10b. As you can see, this artificially introduces very abrupt changes (marked by the arrows), resulting in an FFT with spurious peaks at high frequencies.

We can avoid this by multiplying the set of samples by a "windowing function," which goes smoothly to zero at $t=0$ and $t=T$; an example is shown in Figure 7.10c. After the multiplication, the sampled data also goes to zero at $t=0$ and $t=T$, eliminating the discontinuity, as shown in Figure 7.10d. Unfortunately, this windowing function changes the waveform, and so makes changes to the Fourier spectrum, such as broadening the peaks and making modest changes to their heights. However, these are usually acceptably small.

In the instrument or software you're using, you can choose between several windowing functions. A good general-purpose choice is the Hanning window, shown in Figure 7.10c, which is just an offset cosine. (It goes from 1 to 0, instead of from 1 to -1.) This gives relatively little peak broadening, and only modest changes in peak heights.

Another good choice is the "flat top" window, so named because it has a Fourier transform with a flat top. This window is designed to give the most accurate peak heights in the Fourier spectrum, at the cost of significant broadening of the peaks. This means that, if there are two peaks close together in frequency, you may not be able to resolve them as separate peaks. You should use this window whenever you plan to use the peak height quantitatively.

7.4 Preview of the Arduino

You are probably familiar with the term "microprocessor," which refers to the central processing unit in a personal computer. This integrated circuit or "chip" requires a great deal of support circuitry to function, including external memory. If one desires to input or output analog voltages, one must add specialty A/D or D/A chips or cards. Even reading or writing digital levels (0 or 5 V) to or from an external device such as a motor controller requires additional hardware.

While microprocessors display amazing speed and versatility for running computer programs, they are poorly suited to simple tasks that require a lot of communication with external devices. For example, a modern washing machine typically is controlled by a computer chip that must interface with water level sensors, switches to turn motors on and off, etc. It is not necessary to have blindingly fast program execution speed, and the program is quite short so not much memory is needed. For such applications, as well as many applications in robotics, one typically uses a "microcontroller" rather than a "microprocessor." The microcontroller always has some built-in memory (additional external memory can be added), and direct software control of several pins that are used for digital input and output (e.g., to turn on a motor). In some cases, the microcontroller also has built-in A/D and D/A functions.

In labs 8B, 8C, and 8D, you can experiment with a microcontroller **board** called the "Arduino." This is a complete working computer with built-in A/D and D/A capabilities and a USB interface that can be used to connect it to a desktop computer; it costs about US$22. The board includes an ATmega328 microcontroller chip, the 16 MHz oscillator that determines the clock frequency for the chip (rather than the 2–3 GHz typical of microprocessors), a chip allowing communication with a bigger computer via USB, a reset switch, some indicator LEDs, and convenient surge-protected connectors for each of the important pins of the microcontroller. (Higher performance versions of the Arduino are available for very affordable prices.)

The microcontroller has 14 pins which can be used for digital input or output. Six of these can be used in "pulse width modulated" mode, meaning that they can easily be put into a state in which they're on for a certain length of time, then off for a certain time. The on/off ratio can easily be varied to 256 different levels, allowing you to vary the average analog output voltage. If desired, you can run this rather steppy output through a low-pass filter, thus creating a true 8-bit digital-to-analog converter. The microcontroller also has six pins for analog input, with 10-bit A/D conversion. There are also several other pins used for more exotic functions such as interrupts. It has 32 KB of non-volatile onboard memory. ("Non-volatile" means that it's like a USB stick – what you write stays there, even if the power goes off.) This may not seem like much compared with the 32 GB of RAM and the thousands of GB of disk memory you might have in a personal computer, but it is plenty for most of the tasks for which one might like to use the Arduino.

The Arduino is an extremely popular microcontroller board, and you can easily find many resources for it on the web, along with a great user community. In addition to the built-in A/D and D/A capability, you can purchase inexpensive "shields" which plug onto the Arduino board and provide higher performance conversion between analog and digital.

You will program the ATmega328 microcontroller by writing code in C on a regular computer, then uploading it to the microcontroller. Once the code has been uploaded, you could disconnect from the computer, and (assuming you provided power), the Arduino would function on its own. In fact, you could unplug the ATmega328 microcontroller chip from the Arduino board, and plug it into some more special-purpose board and it would continue to function.

8

Digital Electronics

Brian Collett

CONTENTS

Any sufficiently advanced technology is indistinguishable from magic.

– Arthur C. Clarke (*Profiles of the Future*, 1973)

8.1 Introduction

Digital electronics deals with signals that take only two possible values, a high value, which we usually write as 1, and a low value, which is usually written 0. We use such two-value signals both to represent the logical values *true* and *false* and to represent the binary digits in the binary representation of a number. This use as binary digits leads us to call such a two-value signal a *bit*, a binary digit.

It is a matter of choice which value represents which truth state. The somewhat more popular choice, *1* = *true*, *0* = *false*, is called *positive logic*, while the opposite choice, *0* = *true*, *1* = *false*, is called *negative logic*. By common convention, we tend to name signals that use positive logic with names expressing their meaning when true. For example, a signal named *Enable* would indicate that something was turned on (enabled) when the signal had the value 1. Signals that use negative logic are given names that either begin with an exclamation mark or are written with a bar over the top. Thus, *!Reset* or $\overline{Reset}$ would indicate a signal which would reset something when the signal had the value 0.

Just as we use addition and multiplication to build complicated formulae out of simpler pieces, so we combine simple logical operations AND, OR, and NOT, to build complicated logical expressions. By themselves, these are extremely simple operations. As we will see, we can use simple electronic circuits, called *gates* or *logic gates*, to perform these simple functions and then combine them into more complicated logic circuits that can e.g., add two multi-bit numbers or store information for later retrieval. We divide logic circuits into two fundamentally different families:

In a *combinatorial logic* circuit, the output of the whole circuit depends only on the current inputs. For example, a circuit to add two binary numbers produces the sum of the numbers for its output and needs no other information.

A *sequential logic* circuit produces an output that depends not only on the current input, but also on the previous state of the circuit; it has a memory. For example, a counting circuit adds one to its output when the input tells it to. That means that the new output depends not only on the input but also on the previous number.

We face two different kinds of challenge in digital logic as in most electronics; we have to be able to figure out what an existing circuit does (*analysis*) and to be able to design a new circuit from an idea of what it should do (*synthesis*). Analysis is fairly straightforward, using *bit following* to turn a circuit diagram into a *truth table*. Synthesis is a multi-step process where we first turn an idea into a truth table, then use formal methods to turn the truth table into a Boolean expression, which we can then turn into a circuit diagram.

8.2 Truth Tables

A truth table shows the output of the circuit for every possible combination of inputs. Every combinatorial circuit has its own truth table and every truth table can be implemented with a combinatorial logic circuit. A logic circuit with n inputs has 2^n possible states. In order to be sure that we have covered all possibilities, we usually arrange the input states in increasing numeric order. If all the numbers from $0 \rightarrow 2^{n-1}$ are present, then you have found all the states.

To find the truth table you must list all the possible input states of the system and decide, by inspecting the desired function of the circuit, what the output should be, since the only possible choices for each input and output are the two binary values 0 and 1. While straightforward this can be time consuming since a circuit with n independent inputs has 2^n possible output states.

When the number of inputs gets large it can become tricky to keep track of all the states and to be sure that you have not missed any. The usual way to keep track of this is to treat the inputs as the bits of a binary number and to put the numbers down in order. If there are no numbers missing, then you know that there are no missing input states.

Once all of the input states have been identified, then you can work back through the table and put in the outputs. If one or more of the input states will not be found in practice, then it will not matter what the output is in those states. You can mark this by putting an X in the output for such a state instead of a 1 or 0.

A simple example may help to make the process clearer.

Example

Consider a multi-way switch for a double flight of stairs. There will be three switches, one on each floor, and any switch can turn the light on or off. Let us call the three switches A, B, and C and the output to the light bulb O. Then we can construct the truth table by following the rule that flipping any one switch (inverting any one input) will make the light change state. If the light is on, then it turns off and vice versa. There are two tricks to producing this table. The first one is that we have to pick a starting state. Let's assume that if all the inputs are 0 then the light is off. (This is quite arbitrary. If we make the other choice, then we will end up with an equally valid but slightly different circuit.) The second trick is making sure that we generate all the possible inputs. We can't go through the states in numerical order this time because of the way our switch flipping rule works. However, we can write the $2^3=8$ numbers down in order and then search through the input states for ones that satisfy the rule. Here are the numbers 000, 001, 010, 011, 100, 101, 110, 111.

We start by assuming that the output, O, is off

A=0, B=0, C=0, O=0

Now we flip one of the switches and change the output state. The very next state in the list differs in only one bit so our next state is

A=0, B=0, C=1, O=1

There are three states that differ from 001 in only one bit. They are 000, 011, and 101. We have already used up 000 so pick 011 and flip the output to get

A=0, B=1, C=1, O=0

This time we can go to 001, 010, and 111. Again, we shall pick the next highest unused state and have.

A=0, B=1, C=0, O=1

That has exhausted the states that have A=0, so it is time to flip A.

A=1, B=1, C=0, O=0

The next set of possible states is 011, 100, and 111. Again, we shall pick the next largest unused stated

A=1, B=0, C=0, O=1

That leaves only two unused states, 101 and 111. Only the first is one flip away, so we have

A=1, B=0, C=1, O=0

and so, the final state must be

A=1, B=1, C=1, O=1

There, that is eight states, the whole lot. Now let us put them in numerical order and write them out as the truth in Table 8.1.

We will reconsider this example below, after we have examined the elementary building blocks or gates.

TABLE 8.1

Three-way light switch

A	B	C	O
0	0	0	0
0	0	1	1
0	1	0	1
0	1	1	0
1	0	0	1
1	0	1	0
1	1	0	0
1	1	1	1

8.3 Gates

8.3.1 Basic Gates

A 1-input gate has two possible states for its single input, producing $2^2=4$ possible 1-input gates. Two of them are really boring, their outputs being always 1 or always 0. We are left with the NOT gate, also called the *inverter*, and its opposite, the buffer. The buffer has its output equal to its input, so the simplest form is just a wire. However, it is sometimes useful to make a copy of the signal with increased power (e.g., to drive a light bulb) and then we use a small amplifier symbol for the buffer, a small triangle as shown in Figure 8.1.

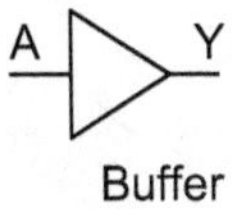

A	Y
0	0
1	1

Buffer

FIGURE 8.1 Amplifier symbol for the buffer.

NOT

The output of a NOT gate is always the opposite of the input. The symbol for this is a small circle that can be combined with any other logic symbol. When used on its own we attach it to the output of a buffer so that the normal symbol is the triangle and circle as shown in Figure 8.2.

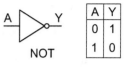

A	Y
0	1
1	0

FIGURE 8.2 Symbol for NOT gate output.

2-Input Gates

Because every 2-input truth table contains four rows, each different gate has a different 4-bit output column. There are $2^4 = 16$ different 4-bit numbers and thus 16 different 2-input gates. Only three of these gates, with their inverses, are interesting enough to warrant their own names. We start with the two logical operations that are familiar from normal life and the English language.

AND

The output of an AND gate is true only if all of its inputs are true. This idea extends beyond two inputs and so Figure 8.3 shows symbols and truth tables for 2-input and 2-input AND gates. The extensions to four and more inputs are obvious.

OR

The linguistic complement of AND is OR. The output of an OR gate is true if ANY of its inputs is true. Here are the symbols and truth tables for 2-input and 3-input OR gates (Figure 8.4).

Note that the AND symbol has a flat back and a semicircular front while the OR symbol is curved on all its edges.

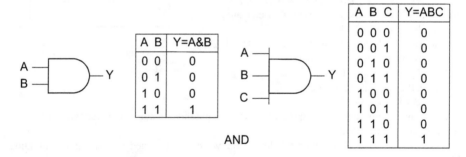

FIGURE 8.3 Symbols and truth tables for 2-input and 3-input AND gates.

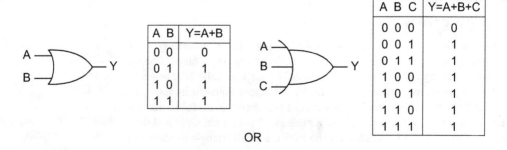

FIGURE 8.4 Symbols and truth tables for 2-input and 3-input OR gates.

XOR

In English we do not distinguish very clearly between the regular OR that we have just described and the less generous version that signifies one choice among many, the exclusive-or or XOR. In this case, the output is true when one and only input is true. This does not transition uniquely to more than two inputs (you have to decide whether the output for the A = 1, B = 1, C = 1 state should be 0 or 1, and there are good reasons for both choices) and you can't buy a 3-input XOR gate so I have only shown the 2-input version in Figure 8.5.

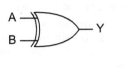

A B	Y=A⊕B
0 0	0
0 1	1
1 0	1
1 1	0

FIGURE 8.5 XOR gate.

NAND, NOR, and XNOR

Each of the gates that we have met so far has a relative produced by adding a not gate to the output. If we negate the AND gate, then we get the NAND gate whose output is false only when A AND B are both true. We can see that the name and symbol of the NAND are derived from the AND gate. NAND is simply NOT AND. The symbol is made up from the AND symbol followed by an inverter in Figure 8.6.

Just as NAND is really NOT AND, so NOR is really NOT OR. As with AND and OR, we can as easily extend NAND and NOR to three or more inputs, though I have not shown them.

It is worth noting that NOR is in some sense the opposite of AND. Each has truth table (Figure 8.7) with a single 1 and three 0s in the output column but, where AND is true if all inputs are true, NOR is true only if all inputs are false. We shall see this again in a little while as a De Morgan theorem.

Just like AND and OR, XOR has its negative equivalent, the unpronounceable XNOR (it sounds almost like snore!). This interesting name stands for *exclusive NOR* but it really means *not exclusive OR* as you can see in the truth table and symbol (Figure 8.8).

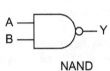

A B	Y=Ā&B̄
0 0	1
0 1	1
1 0	1
1 1	0

FIGURE 8.6 NAND gate.

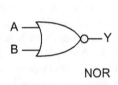

A B	Y=Ā+B̄
0 0	1
0 1	0
1 0	0
1 1	0

FIGURE 8.7 NOR gate.

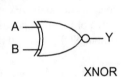

A B	Y=Ā⊕B̄
0 0	1
0 1	0
1 0	0
1 1	1

FIGURE 8.8 XNOR gate.

8.3.2 Multi-Gate Circuits

Out of these seven basic gates we can build everything else. Let us look at how we combine gates to form new functions and how we find the truth tables for the new functions. Figure 8.9 shows a complex gate made of several simpler gates. You can see how we wire the input and output terminals of the individual symbols together to get more complicated logic functions, just as we wire individual resistors, capacitors, and diodes together to get complex circuits. The final circuit has two inputs, A and B, and a single output, O.

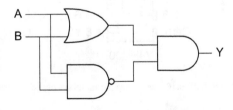

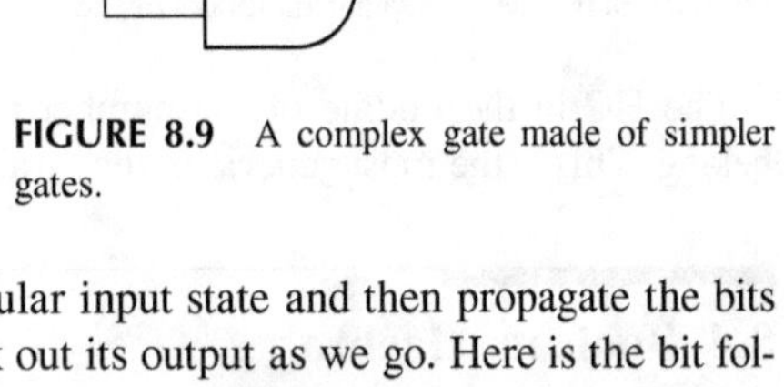

FIGURE 8.9 A complex gate made of simpler gates.

We can find out what this circuit does using *bit following* to build the truth table step-by-step. We set up a particular input state and then propagate the bits through the gates, using the truth table of each gate to work out its output as we go. Here is the bit following process applied to this circuit. We will go through the input states in numerical order so that it is easier to build the table. The first step is to set A = 0 and B = 0.

We propagate these bits along the wires to the inputs to the first layer of gates. The upper OR gate gives us $0+0=0$ while the lower NAND gate gives us $!(0\&0)=!0=1$. Thus, the inputs to the last AND gate are 0 and 1, giving us an output of 0 and the first line of our truth table (Table 8.2).

The next input state is $A=0$, $B=1$. Together these make the output of the OR gate a 1 and also make the NAND gate output a 1. That gives the final AND gate an input of $1\&1=1$, so that the output of the circuit is true, and we have the second line in our truth table.

TABLE 8.2

Outputs for every possible input

A	B	Y
0	0	0
0	1	1
1	0	1
1	1	0

The next state does exactly the same thing since the circuit is completely symmetric in A and B.

Finally, when both inputs are 1 the NAND gate outputs a 0, forcing the last line of the table to read $A=1$, $B=1$, $Y=0$.

We have now found the output for every possible input and built Table 8.2, which we at once recognize as the truth table for an XOR gate. So this circuit is one of many ways to make an XOR gate out of simpler NAND, NOR, and NOT gates.

8.3.3 CMOS Logic Gates

Individual logic gates can be constructed from MOS FET transistors. Manufacturers produce a variety of such gates in integrated circuit (IC) form, with anything from a few gates to billions of gates in a single package. The most useful family of logic ICs is the 74 series, so-called because they all have names containing numbers that begin 74.

These are available in a variety of speeds and other characteristics but they all share common packages and pin numberings. They all come in *dual-inline (DIP)* packages, with two rows of pins either side of a plastic body that has a marker at one end. With the marked end at the top the pins numbers start at 1 on the top left pin and run counterclockwise round the body as seen in the Figure 8.10.

All of these chips require power and ground to be connected to the top right and bottom left pins and it is also good practice to connect a 0.01 µF capacitor from power to ground, as close to the IC as possible.

Figure 8.10 shows three of the most popular members of the 74 series, the 7400 quad 2-input NAND gate, the 7402 quad 2-input NOR gate, and the 7404 hex NOT gate or hex inverter.

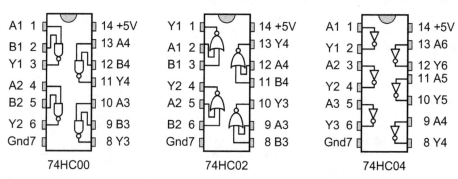

FIGURE 8.10 Most popular members of the 74 series.

The HC in the middle of the number tells us the specific speed/power family to which these ICs belong. This is the most generic of the families and suitable for all but the fastest work.

8.4 Boolean Algebra

Bit following is fine for simple circuits but for more complicated ones we need to learn a little about the *Boolean algebra* developed in the 19th century by George Boole. We manipulate symbols and operators that stand for binary values and binary gates in a way that looks a lot like ordinary algebra. This algebra will allow us to analyze circuits without bit following in most cases and will lead us to methods for finding a circuit to implement any truth table.

8.4.1 Variables

Just as in ordinary algebra, we use names to represent binary values. The difference between algebraic variables and binary variables is that a binary variable can take on only the values 1 and 0. A, B, Closed, and Ready are all examples of possible binary variables.

8.4.2 Operators

The operators of Boolean algebra are the simple gates that we have already met, AND, OR, and NOT. Unfortunately, there are several notations in common use. I will use one common one, but you should be prepared to recognize any of them.

NOT

The expression NOT A can be found written in at least the following ways in various different books.
NOT A, A*, A', *A, 'A, /A, -A, !A, $\overline{A}$
I prefer the last two because they are easy to write.

OR

Here are some of the ways that the inclusive OR can be written:
A OR B, A | B, A∪B, A + B
I prefer the last symbol because it reminds us that OR operation in Boolean algebra behaves a lot like the addition operator in regular algebra.

AND

Here are some of the ways to write AND:
A AND B, A & B, A∩B, A B, A·B, A×B
Again, I prefer the last two forms because the AND operation plays the role of multiplication in Boolean algebra (0 AND anything = 0).

XOR

XOR is maybe a little less common and so has fewer representations. Here are some usual forms for the XOR operation; again, the preferred one is last:
A XOR B, A^B, A~B, A⊕B

8.4.3 Expressions

Now we have the tools to write Boolean expressions. Let us begin by translating our 3-gate version of the XOR gate into an expression. The left two operators generate the outputs A + B and !(A×B). These form the inputs to the final AND gate so the final expression is Y=(A + B)·!(A×B), which we know must be equal to A⊕B.

8.4.4 Algebraic Relations

We can obviously extend this to circuits that are as complicated as we want, though the resulting expressions may get *very* complicated. Boolean algebra allows us to manipulate expressions into simpler forms using a set of rules, some of which are similar to those for ordinary algebra.

Associativity

The binary operations of Boolean algebra obey an associative rule just like that of ordinary algebra.

$A + (B + C) = (A + B) + C$

$A \oplus (B \oplus C) = (A \oplus B) \oplus C$

$A \cdot (B \cdot C) = (A \cdot B) \cdot C$

This means that you can write strings of ANDs or ORs without parentheses because it doesn't matter what in order you do the operations. Thus, we can write

$A + (B + C) = A + B + C$

and, as already mentioned, we can buy AND and OR gates that have three or more inputs in addition to the more common 2-input ones.

Commutativity

The binary Boolean operators also obey a commutative law so that the order of terms in an expression does not matter.

$A + B = B + A$

$A \oplus B = B \oplus A$

$A \cdot B = B \cdot A$

Distributivity

Finally, the Boolean operations obey a distributive law in just the way that the arithmetic operations of multiplication and addition do. The only difference is that Boolean algebra has two different forms of addition so it has two distributive laws.

$A \cdot (B + C) = (A \cdot B) + (A \cdot C)$

and

$A \cdot (B \oplus C) = (A \cdot B) \oplus (A \cdot C)$

The Boolean operators obey a precedence rule in the same way that arithmetic operators do. AND has higher precedence than OR so that we can write an expression such as $(A \cdot B) + (A \cdot C)$ without parentheses as $A \cdot B + A \cdot C$ and the precedence tells us that we have to do the AND operations first and apply the OR to output of the AND operations. It is usually best not to rely too heavily on precedence. It is better to include a few unnecessary parentheses than to risk confusion.

De Morgan's Theorems

Boolean algebra has two rules that have no counterpart in ordinary algebra. They are called De Morgan's theorems after their discoverer. These allow you to replace an AND term by an OR term and vice versa. Here are the two theorems.

$$A \times B = \overline{\overline{A} + \overline{B}}, \text{ which can also be written } \overline{(A \times B)} = \overline{A} + \overline{B}$$

and

TABLE 8.3

Boolean facts

$A + B = B + A$	$(A \cdot B) \cdot C = A \cdot (B \cdot C)$	$A \cdot !B + !A \cdot B = A \oplus B$	$A \cdot B + !A \cdot !B = !(A \oplus B)$
$A \oplus B = B \oplus A$	$A \cdot (B + C) = A \cdot B + A \cdot C$	$A + 0 = A$	$A + 1 = 1$
$A \cdot B = B \cdot A$	$A \cdot (B \oplus C) = (A \cdot B) \oplus (A \cdot C)$	$A \oplus 0 = A$	$A \oplus 1 = !A$
$(A + B) + C = A + (B + C)$	$(A \oplus B) \oplus C = A \oplus (B \oplus C)$	$A \cdot 0 = 0$	$A \cdot 1 = A$
$A \times B = \overline{\overline{A} + \overline{B}}$	$\overline{(A \times B)} = \overline{A} + \overline{B}$	$A + A = A$	$A + !A = A \oplus !A = 1$
$A + B = \overline{\overline{A} \times \overline{B}}$	$\overline{(A + B)} = \overline{A} \times \overline{B}$	$A \oplus A = A \cdot A = 0$	$A \cdot !A = 0$

$A + B = \overline{A} \times \overline{B}$, which can also be written $\overline{(A + B)} = \overline{A} \times \overline{B}$

These are a lot more powerful than the simple rules of algebra that we have previously seen. They are quite easy to prove by simply writing out the truth tables for the two sides and showing that they are the same. Since a Boolean expression is completely described by its truth table, *any two expressions that have the same truth table are equal to each other.*

Some Useful Equivalents

Table 8.3 is a table of useful Boolean equations. Most of these are ones that we have already met. The rest are straightforward facts that are very useful for simplifying expressions.

8.5 Logic Design

We now have most of the tools that we need to pursue a design from a description of the desired function to a complete circuit diagram.

Simple circuits can often be handled by recognition as we did with the multi-gate example above. More complex circuits require more systematic tools. There are two methods that are guaranteed to convert any truth table into a valid logic equation that can be simplified as needed.

8.5.1 Sum-of-Products

The sum-of-products method constructs an expression from a truth table one line at a time. In fact, one non-zero line at a time. The general form of the description that this method produces is a set of equations of the form

Out = expr1 + expr2 + expr3 + ...

where each expr is true for exactly one combination of input bits. This is realized with AND gates to compute the expressions and a final OR gate to combine them. The OR gate ensures that there is a result of 1 in the output only for inputs that make one of the AND expressions true.

We can express the whole method with this algorithm.

Sum-of-Products Algorithm

```
Treat each output column separately. There is one equation for each column.
Write the name of the output column followed by the = sign.
For each row in the truth table:
    If the output is 0 do nothing.
    If the output is 1 then for each input variable:
        If there is a 1 in the current row, then write down the name of the
        variable.
        If there is a 0 in the current row, then write down
        NOT the name of the variable.
    Join all the signal names with ANDs; for clarity, write parentheses around
    the combination.
If there is more than one AND expression, join all the expressions with ORs.
```

The sum-of-products method is guaranteed to work but is most efficient for truth tables that have many 0s in the output and fewer 1s. If there are more 1s than 0s then the next method is better. If there are equal numbers of 1s and 0s then use whichever one you prefer; they are of equal complexity.

Example

Let's look again at our three-floor light. There is only one column so we will get one equation. We start by writing Out =, then move on to look down the entries in the Out column.

The first row has a 0 in the output column, so we don't do anything with it.

The second row has a 1 in the output column, so we have to use this row. There are zeros in the A column and in the B column so these two inputs must be negated. The C entry is a 1 so it appears un-negated and the expression for this row is $!A \cdot !B \cdot C$

The third row also has a 1 in the output column and so contributes an expression to the answer. A is again negated but this time the 1 in the B column tells us that B is not. C now has a zero and so appears negated. The complete expression is $!A \cdot B \cdot !C$.

The fourth row has a 0 in the output and can be ignored.

The fifth row has a 1 and contributes $A \cdot !B \cdot !C$.

The sixth and seventh rows are both ignored.

The eighth row has a 1 and contributes $A \cdot B \cdot C$.

The complete expression that describes this truth table is then

Out = $!A \cdot !B \cdot C + !A \cdot B \cdot !C + A \cdot !B \cdot !C + A \cdot B \cdot C$

or, adding parentheses for safety (which I strongly recommend),

Out = $(!A \cdot !B \cdot C) + (!A \cdot B \cdot !C) + (A \cdot !B \cdot !C) + (A \cdot B \cdot C)$

8.5.2 Product-of-Sums

Just as the sum-of-products makes use of the OR gate's property of having only one 0 entry in its output column, so the product-of-sums makes use of the AND gate's property of having only one 1 entry.

Product-of-Sums Algorithm

```
Treat each output column separately. There is one equation for each column.
Write the name of the output column followed by the = sign.
For each row in the truth table
     If the output is 1 do nothing.
     If the output is 0 then for each input variable:
          If there is a 1 in the current row, then write down NOT the name of
          the variable.
          If there is a 0 in the current row, then write down the name of the
          variable.
     Join all the signal names with ORs and write parentheses around the
     combination.
If there is more than one OR expression, join all the expressions with ANDs.
```

This method works down the truth table producing an expression for every row that has a 0 in the output column. Its final result looks like this:

Out = expr1 $\times$ expr2 $\times$ expr3 $\times \ldots$

Here is our example worked the new way:

Example

Again, we have only one column so that we start with Out =

The first row is a zero so use it. The entry under A is a 0 so we write down A. The entries under both B and C are also 0s so we also write B and C The whole expression becomes $(A + B + C)$.

The second and third rows are both 1s and are ignored.

The fourth row is a 0 and again input A is 0 so that we write A, but B and C are both one so that we end up with $(A + !B + !C)$.

The fifth row is 1 and is ignored.

The sixth row is a 0 and produces (!A + B + !C).
The seventh row is a 0 and produces (!A + !B + C).
The eighth row is a 1 and is ignored.
The complete expression is Out = (A + B + C) × (A + !B + !C). × (!A + B + !C). × (!A + !B + C)

This time the parentheses are *required* to make sure that we get the order of the operations right because AND has a higher priority than OR.

8.6 Common Logic Functions

While any logic device can be built up from its truth table there are some common functions that can serve as useful building blocks as well as examples of the kinds of circuits that we can construct from simple gates.

8.6.1 Coders/Decoders

A coder or decoder translates one binary code into another. The terms are pretty much interchangeable, but we typically use the term coder for a circuit that has more inputs than outputs and the term decoder for a circuit that has more outputs than inputs.

One such code is the standard binary counting sequence in which the bit-patterns represent numbers in base-2, illustrated in Table 8.4 on the right. Another such code is a one-of-N code, in which only one of the N possible outputs is true at any one time. You might use a one-of-N code if you had an output device made of a set of lights, each with a number written on it, and then lit one light at a time to tell you which state the system was in.

TABLE 8.4

Binary code

Binary				Dec
0	0	0	0	0
0	0	0	1	1
0	0	1	0	2
0	0	1	1	3
0	1	0	0	4
0	1	0	1	5
0	1	1	0	6
0	1	1	1	7
1	0	0	0	8
1	0	0	1	9
1	0	1	0	10
1	0	1	1	11
1	1	0	0	12

Example

A 1-of-N decoder is ideal for controlling the floor display of an elevator. Unless you are in a very old elevator or a quite new one with an LED display, the floor display consists of a panel of numbered lights. Since the elevator is only on one floor at a time, only one of the lights is lit at any moment. Internally, the elevator controller uses a binary code to keep track of which floor the elevator is on and then uses a 1-of-N decoder to produce a set of signals, only one of which is turned on at a time. These signals then control the lights in the floor display.

2-Line to 4-Line Decoder

A 2-line to 4-line decoder converts between a 2-bit input and a 1-of-4 line output where three of the outputs take one truth value while the fourth, selected, value takes the other. The two inputs, A and B, form a 2-bit number, whose value selects which of the four output bits will be made active. The truth table (Table 8.5) shows the positive output logic version where the inactive outputs are 0 and the active output is a 1.

TABLE 8.5

Truth table

B	A	Y3	Y2	Y1	Y0
0	0	0	0	0	1
0	1	0	0	1	0
1	0	0	1	0	0
1	1	1	0	0	0

The output bits are numbered according to the value of the associated input pattern. For example, the bit-pattern 10 that represents the number two turns on the Y2 output.

There are commercially available decoders like this including the 74138 3-line to 8-line decoder, the 74139 dual 2-line to 4-line decoder, and even the 74154 4-line to 16-line decoder. Curiously, these have negative logic outputs, where all the outputs but one are 1 and the selected output is 0.

Encoders

There are not very many examples of encoders because there are very few many-wire codes to turn into few-wire codes. The best known is called a priority encoder. It takes a set of individual input bits each of which has a value, or *priority*, assigned to it and it outputs a binary number corresponding to the signal with the highest priority.

Let's look at the simplest non-trivial example, the 4-line to 2-line encoder. As a 4-input circuit, it must have 16 distinct input states but with only two output bits it can have only four output states. Thus, an output of 10 (binary representation of 2) means that input 2 is high but input 3 is not. The highest priority active input is input 2.

The full truth table for this would be very large but we can reduce it if we rewrite the table using 'x' entries to indicate "don't care" states then we see the problem is much smaller. With only four output states we really only need four lines in the truth table, as shown in Table 8.6.

Note that the output state 00 is ambiguous. It could mean that input I0 is active and no other or that no input is active. This is not usually a problem. These devices are normally used in situations where the output is only meaningful when at least one of the inputs is active.

Since the x entries mean that we don't care what values those inputs take, we completely ignore them when building our equations. We only consider those inputs that have specified values.

Thus, the equations become
$O0 = (!I3 \cdot !I2 \cdot I1) + I3$ and $O1 = (!I3 \cdot I2) + I3$,
which are very simple.

TABLE 8.6

Priority encoder

I3	I2	I1	I0	O1	O0
0	0	0	0	0	0
0	0	0	1	0	0
0	0	1	x	0	1
0	1	x	x	1	0
1	x	x	x	1	1

One use for priority encoders is found in analog-to-digital converters, which take an analog voltage, a voltage that can take any value, and output a binary number that is proportional to the input voltage. For example, such a converter might output the number 2 for all voltages between 2 V and 3 V.

Consider the 2-Bit Analog-to-Digital Converter circuit in Figure 8.11. This takes a voltage on the line marked *V*in and outputs a 2-bit binary number approximation to the voltage on outputs O1, O0. The funny triangular things with two inputs are comparators that compare the voltages on their inputs and output 1 if the voltage at the positive input is greater than the voltage at the negative input, else 0. For example, if we set *V*in to 2.43 volts then the bottom three comparators will output a 1 because 2.43 > 0, 2.43 > 1, and 2.43 > 2 but the third will output a 0 since 2.43 < 3. We have seen above that if we put input 0111 into our 4-line to 2-line priority encoder then its output will be O1 = 1 and O0 = 0, the binary code for the number 2. You can try some other voltages and see that this circuit converts any voltage between 0 V and 4 V into a one-digit (2-bit) approximation to the voltage!

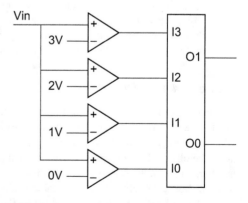

FIGURE 8.11   2-bit analog-digital converter.

8.7 Arithmetic Logic

This class of circuits performs arithmetic and logic functions on whole sets of bits at once. They treat the bits as the binary digits of a number in base-2. This section assumes familiarity with binary arithmetic.

The most basic arithmetic operation is the addition of two numbers and it starts with the addition of two single bits and the circuit called the half-adder.

8.7.1 Half-Adder

Since $1+1=2$, and the binary representation of 2 is 10, the half-adder will have two inputs, A and B, must produce two output bits, a sum and carry that we will call C and O (for output). It is easy to construct the truth table in Table 8.7.

We immediately recognize that $C=A\&B$ and $O=A\oplus B$ so that we get the circuit of Figure 8.12.

We cannot add two multi-bit numbers using only half-adders because we need to account for the carry from one digit to the next. For that we need the full-adder.

8.7.2 The Full-Adder

Our goal is a circuit to add two multi-bit numbers. Let's look at the process of adding two 2-bit numbers and we shall see what tools are needed. For example, the Figure 8.13 shows the process of adding $3+1$ to get 4.

We see two things. First, the result of adding two 2-bit numbers is a 3-bit number, not a 2-bit number. Second, because we have to deal with the "carries" from one column to the next, the fundamental operation that takes place in each column is not the addition of two bits but of three, two inputs and a carry from the previous column. That means that our Full-Adder circuit needs to have three inputs, A, B, and Ci (the carry in from the previous stage) and two outputs, the sum, O, and the carry out, Co. Trying all combinations, we get the truth table as shown in Table 8.8.

We recognize that last column from the light switch problem. It is
$O=Ci \oplus A \oplus B$.

TABLE 8.7

Half-adder

A	B	C	O
0	0	0	0
0	1	0	1
1	0	0	1
1	1	1	0

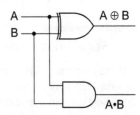

FIGURE 8.12 Half-adder.

```
  1   1
  0₁  1
-------
1   0   0
```

FIGURE 8.13 Adding $3+1$ to get 4.

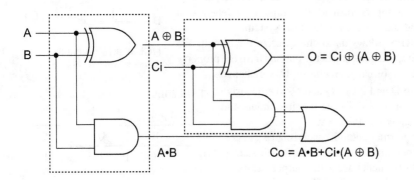

FIGURE 8.14 Full-adder circuit.

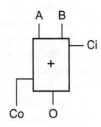

FIGURE 8.15 Full-adder symbol.

The carry output, Co, is a bit more complex. If we apply the sum-of-products method and simplify the resulting expression we get

$C = A \cdot B + Ci \cdot (A \oplus B)$.

Now we see that $A \oplus B$ is a common sub-expression, so we won't need to make that twice. It is also the main part of a half-adder and we usually implement the full-adder with two half-adders, as shown in Figure 8.14.

It is often convenient to have a shorthand symbol for a common building block, so we get the symbol shown in Figure 8.15.

We can now design circuits to add numbers of any width, using one full-adder per bit.

TABLE 8.8

Full-adder

Ci	A	B	Co	O
0	0	0	0	0
0	0	1	0	1
0	1	0	0	1
0	1	1	1	0
1	0	0	0	1
1	0	1	1	0
1	1	0	1	0
1	1	1	1	1

8.8 Sequential Logic

All the combinatorial circuits that we have seen so far share the property that the output pattern of bits depends only on the input pattern. You cannot make a computer from combinatorial circuits alone because a combinatorial circuit cannot modify its behavior based on previous events. A computer has to have some way of remembering previous inputs and modifying its current behavior based on things that happened earlier. So we need to add to our armory of logic circuits a memory circuit. Ideally, we want to be able to present a single bit to such a circuit and to say "remember this bit." The circuit should then hold that piece of information at its output until we tell it to change. Once we have such a circuit, called a *latch*, we can collect latches together and build memories that can hold more than one bit. This section describes how we can combine our logic gates to build such a latch and then how to combine latches with some combinatorial logic to produce circuits that can go through a sequence of states all by themselves. We call such circuits *sequential* logic.

8.8.1 The Flip-Flop

The secret to building a memory circuit lies in making one simple change to the way we have assembled logic circuits. So far, circuits have had clearly different sets of inputs and outputs. Now we are going to mix things up by using the output of a circuit as one of its inputs. This is called *feedback*, because we feed the output back to the input as shown in Figure 8.16. It is called a *flip-flop* or a *bi-stable*.

We cannot write a truth table for this system because the output is no longer a function only of the inputs. Instead, the output now depends on both the current inputs and the previous state of the system.

Some tedious bit following allows us to characterize the circuit. We find that for three of the possible input states the outputs have the property that they are always inverses of each. Accordingly, we change the names of the outputs to Q and !Q, or Q and $\overline{Q}$, to emphasize that relationship and we can then show the possible inputs and outputs in the *state table* shown in Table 8.9.

The top three rows are normal enough, each input state corresponds to a unique output state. However, the fourth state is new. We find that when both inputs are 1 the output can be in either the state Q=1,!Q=0 or the state Q=0,!Q=1, it remembers its previous state! That is, the actual output in this *memory* state depends on the previous state. If we start with inputs = 10 so that

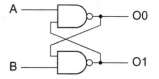

FIGURE 8.16 A flip-flop.

TABLE 8.9

!S-!R flip-flop

!S	!R	Q	!Q
0	0	1	1
0	1	1	0
1	0	0	1
1	1	Q	!Q

the outputs are Q=0, !Q=1 and go to input=11 then the system will stay with Q=0, !Q=1. Similarly, if go from the input=01 state to input=11 we end up with Q=1, !Q=0. However, if we go from the rather odd input=00 state to the input=11 state then we have NO IDEA what the final state will be! Because we cannot remember this state it is often called the *unstable* state.

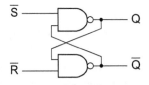

FIGURE 8.17 $\bar{S} - \bar{R}$ flip-flop.

With this state table in mind, we re-name the inputs as shown in Figure 8.17. A 0 value on the upper input forces Q=1. We say that this *sets* Q and so we name the input !S, "S" for set and the not sign because 0 is the active state (this is a negative logic signal). Similarly, a zero on the lower input forces !Q to 1 and thus makes Q=0. We say that it *resets* Q and name the signal !R.

We call this system a *bi-stable latch* because it has two stable output states (10 and 01) and it holds (latches) that state when the inputs return to 11. It is also called the !S-!R flip-flop or, if typography is with us, the $\bar{S} - \bar{R}$ flip-flop. There is an alternate version, built from two NOR gates, whose inputs are active in the 1 state instead of the 0 state, that is called the R-S flip-flop.

8.8.2 Switch De-Bouncing with the $\bar{S} - \bar{R}$ Flip-Flop

The $\bar{S} - \bar{R}$ flip-flip is very useful for curing an annoying problem with mechanical switches. Switches usually consist of various springy bits of wire moved around by levers and the springiness leads to the problem of *contact bounce*. When a switch is moved from one position to the other, the connection is not made or broken smoothly. Instead, the bits of metal bounce around for a short time (usually milliseconds) before settling into the new state. So, if we connect up a simple single-pole single-throw switch and a resistor, it may produce the logic signal shown in Figure 8.18.

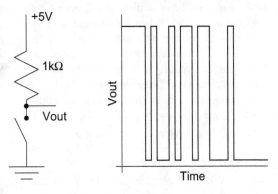

FIGURE 8.18 Switch bounce.

When the switch is closed to bring the output from +5 V down to 0 V, the contacts will bounce against each other for a few milliseconds. The output will not be a clean 1→0 transition but a messy thing like the one shown. It doesn't even take on only the values 0 and +5 because stray capacitance in the circuit forms an RC time constant with the 1 k resistor. You may get all sorts of bumps and spikes until the metal stops bouncing and the output settles down to a clean 0.

There are many situations where a noisy signal like this would be a major problem. Think, for example, of using this as the input to a circuit that counted pulses. You would have no idea how many counts you would get for a single flip of the switch. When a signal must make clean transitions, a noisy signal like this must be de-bounced. The usual way to do this is with a single-pole double-throw switch and an $\bar{S} - \bar{R}$ flip-flop, as in Figure 8.19.

When the switch is thrown, the contacts bounce as usual, but the bouncing takes place only between the moving element and the contact to which it has moved. The moving contact does not bounce between the two fixed contacts because they are far too far apart.

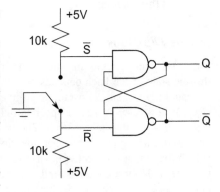

FIGURE 8.19 De-bouncer.

Let us follow what happens as we move the switch. At first, the switch is holding the $\overline{S}$ input of the flip-flop low, so the Q output is high. When we move the switch, the $\overline{S}$ input goes high as soon as the contact is broken. Nothing happens to the output as the inputs are now in the memory state. Very soon, the moving contact hits the lower fixed contact and brings the $\overline{R}$ input low. This forces the flip-flop to change state and Q goes to 0. Now the contact bounces, allowing the $\overline{R}$ input to go high again. A high input on $\overline{R}$ takes the flip-flop back to the memory state and again nothing alters; the output stays at Q=0. The contact may bounce for a while, taking the $\overline{R}$ input between 0 and 1, but not altering the Q output. The output will not change until the switch is moved all the way back to the up state. Thus, each throw of the switch alters the output exactly once, as we want.

The Transparent Latch

Now our bi-stable is not quite the memory circuit that we seek. It does have a simple sort of memory, but we would like more control over when it remembers and when it learns. If we add two more NAND gates and a NOT gate, then we have a real memory circuit as shown in Figure 8.20.

Table 8.10 is the state table for this circuit. When the clock input is high, the Q output follows the data input. When the clock is low, the latch remains in the state that it was in at the instant when the clock went low. Because the data flows straight through the latch when the clock is high, this circuit is called a transparent latch.

It is easy to understand how this circuit works. When the clock is low, both of the input NAND gates have high outputs so the $\overline{S}$-$\overline{R}$ flip-flop is in its memory state. The Data line is logically disconnected from the circuit and cannot affect the outputs. When the clock is high, the $\overline{R}$ input is equal to Data, while the $\overline{S}$ input is equal to Data. Thus, when Data is high, it forces S low, and sets the Q output. When Data is low, it forces R low and resets the Q output.

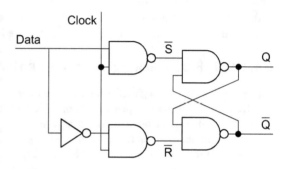

FIGURE 8.20   Transparent latch.

TABLE 8.10

Latch-state table

Clock	Data	Q	!Q
0	0	Q	!Q
0	1	Q	!Q
1	0	0	1
1	1	1	0

D-Type Flip-Flops

There are many circuits in which it is important that the state of the latch depends on the value of its input only at a single instant determined by the clock. In this case, we need a special kind of latch called a *D-type* flip-flop or latch. The transparent latch is not a D-type latch; when the clock input is high the output follows the input. Let us look at the state table for a common kind of D-type latch (Table 8.11). As you can see, we have had to enlarge our notation again. The notations Q_n and Q_{n+1} refer to the states of the Q output before and after the next active clock edge while the little upward arrow

TABLE 8.11

D-type flip-flop state table

Clk	D	Q_{n+1}	!Q_{n+1}
0	0	Q_n	!Q_n
0	1	Q_n	!Q_n
1	0	Q_n	!Q_n
1	1	Q_n	!Q_n
↑	0	0	1
↑	1	1	0

in the clock column shows us that this is a *positive edge triggered* flip-flop, that is, the output changes only following a rising edge on the clock input and stays fixed the rest of the time.

When a transition does occur, the effect is to force the Q to the same state as the D input. In this case, D stands for data since it brings in the data that the flip-flop is to remember. While this one was activated by a rising edge, D-type flip-flops are available in both positive and negative edge triggered forms.

D-Type Flip-Flop Symbols

All D-type flip-flops share the same basic symbol (Figure 8.21), the rectangle with the Q output coming from the top half and the $\bar{Q}$ from the bottom half. The D input is placed opposite the Q input to emphasize its role as the source for the new Q value. The edge-driven clock is indicated by the little wedge on its input. The clock goes right into the center of the symbol to show that it affects both Q and $\bar{Q}$ equally. By itself, the wedge shows that the output changes only a rising edge. A falling edge triggered latch adds a little circle to indicate the inversion. As shown in the figure, D-type latches are also available with set and/or reset inputs that bypass the clock (though the *set* is usually named *preset* in these cases).

8.8.3 Simple Counters

The simple memory circuits that I have described allow us to build whole new classes of logic circuits. One such new class is the *counter*; a circuit that responds to a clock input by stepping through a set of output states in numeric order. Because an n-bit binary number can have 2^n different states, it takes n bits to count up to 2^n. Thus, to count from 0–3 will take two bits and thus two flip-flops.

1-Bit Counter or Divide-by-2

The simplest counter has only one bit. It counts 0, 1, 0, 1, etc. We refer to this as a divide-by-2 since for every two rising edges that happen on the clock there is only one rising edge on the output of the divide-by-2. We can make a divide-by-2 by connecting the !Q output of a D-type flip-flop to its own D input. This leads to a fundamental instability. If the Q has the value 1 then D is set to 0, so that after the next clock edge Q will become 0. But that will make D go to 0 and so Q will reverse on the next clock edge. The circuit and several cycles of the input and output are shown in Figure 8.22. We see that the output changes state every time the clock goes high so that the output frequency is half the input frequency.

2-Bit Counter

We make the simplest 2-bit counter by cascading two of these divide-by-2 counters to make the divide-by-4 circuit shown in Figure 8.23. The outputs just decrease in frequency by a factor of two each time, but I have annotated the output states to show that they go through the four possible states in arithmetic order,

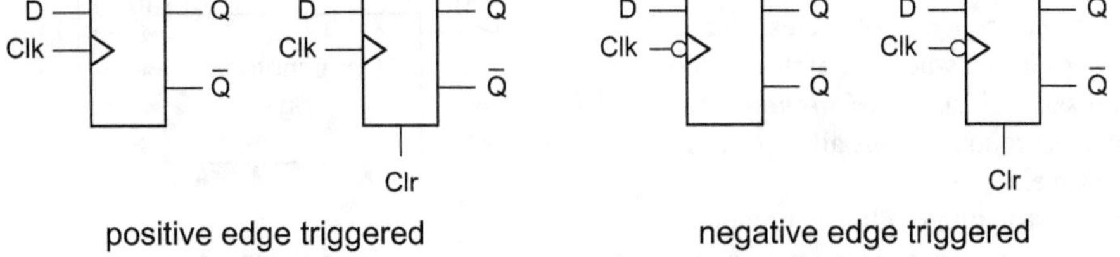

FIGURE 8.21 Edge triggered flip-flop symbols.

albeit counting down. They count down because we used positive edge triggered flip-flops. We can count up by using negative edged triggering.

The top line of the diagram shows the clock input. We create that so we can make it whatever we like. Here I have made it a series of identical pulses forming a 1:1 square wave. I assume that the flip-flops start in the O0=O1=0 state and watch the outputs through five incoming pulses. Note that time proceeds from left to right, as you would see it on an oscilloscope.

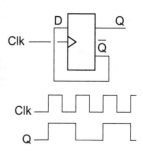

FIGURE 8.22   Divide-by-2.

As expected, the output cycles through the four 2-bit numbers in the standard counting sequence. We can see how this happens. Each time the clock has a falling edge, O0 changes state because its D input is tied to the !Q output. That makes O0 a square wave at one half the frequency of the input. Since O0 is the clock input for O1, O1 changes state every time O0 falls and the O1 signal is again a square wave at half the frequency of its clock input, O0. Because of the way the resulting square waves line up, the two output bits march through the four possible states in numeric order.

This counting scheme can be extended to count up to any power of 2 that you like. Using three latches you can build a counter with eight states that can count from 0 to 7. With four latches you have 16 states and can count from 0 to 15 and so on. This kind of counter, where the output of one stage is used as the clock input for the next stage, is called a *ripple counter* because of the way information ripples along from one stage to the next. The problem with this kind of counter is that, since the clock for one stage is the output from the previous stage, it takes time for changes to propagate

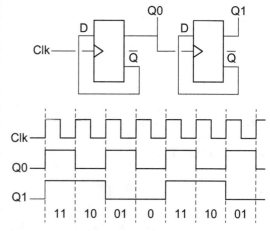

FIGURE 8.23   Divide-by-4 ripple counter.

gate through the whole counter and during the brief interval that the change takes there are unwanted, incorrect *glitch* states present on the outputs. The next section will show how to eliminate these glitch states and will produce a method of designing more general sequential circuits than simple counters.

8.9 Synchronous Logic

The secret to removing the glitches inherent in a ripple counter is to clock every flip-flop with the same signal. That way all of the outputs will change as nearly simultaneously as the hardware will allow. The problem is that, at the moment, they will all do the same thing. We can solve this problem by adding a combinatorial logic to steer the flip-flops into the right sequence of states. We call such a logic circuit, where all of the flip-flops share the same clock, a *synchronous* circuit because synchronous means all happening at the same time.

The basic structure of all synchronous circuits is the same. There is a *latch*, a set of flip-flops driven from a single clock, to store the current state of the system and collection

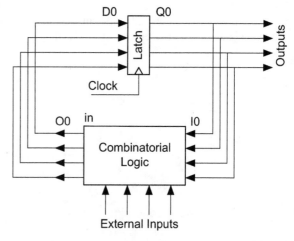

FIGURE 8.24   General synchronous circuit.

of combinatorial logic that makes sure that the system goes through its states in the correct order (see Figure 8.24). Because the combinatorial logic decides which state the system will go to next, it is often called *steering logic*.

The outputs of the latch are connected to the inputs of the combinatorial steering logic, which may also have external inputs. The combinatorial circuit uses the values of all its inputs to compute a set of outputs that are connected to the latch inputs. This feedback makes talking about the system a little difficult. I have tried to clarify things by labeling the different points in the circuit. I call the latch outputs Q, and the inputs D, while I call the combinatorial circuit's inputs I (for internal inputs) or E (for external ones) and its outputs O. In reality, the Qs and the Is are the same signals as are the Os and the Ds but we need to keep them logically separate.

The latch stores the instantaneous *state* of the system as a binary number on the individual Q outputs. When a clock pulse arrives at the latches, the outputs all change at the same time and the system enters a new state. Which state it enters is completely determined by the values present on the inputs to the latch just before the clock pulse. Those inputs come from the combinatorial logic and their values are determined by the current state (the set of Qs) and the external inputs. Each different combinatorial circuit produces a different sequence of states and so produces a different synchronous system.

Synchronous systems are far more general than the simple counters that we have seen so far. In addition to various kinds of counter we can build synchronous circuits to control many kinds of sequential task. Some familiar examples include controllers for traffic lights and elevators. A traffic light controller must not only make sure that the lights work their way through the correct color sequence but may also take into account inputs from pedestrian buttons and from sensors buried in the road that can tell the system when cars are waiting at the light. The ultimate example of a synchronous system is a digital computer. Here the external inputs include the instructions stored in memory and so the memory instructions control the operations of the computer.

8.9.1 Describing Synchronous Systems

We describe combinatorial circuits with truth tables and sequential circuits with state tables. Synchronous systems are special cases of sequential systems and so can be described by state tables. However, the state tables can get very large and hard to understand. We often need a more convenient way of describing large synchronous systems. One common method is the *State Diagram*.

A state diagram is a graphical picture of the operation of a synchronous system that is somewhat similar to the flow charts often used to describe computer programs. In this, we draw each of the states of the system in a little bubble and then we draw arrows between states showing the transitions that are possible. We label the bubbles with the names of the states and we label the arrows with the conditions that cause the transitions to be taken.

It is normal practice to label each transition with the names of those external inputs that are true (1) and leave off those inputs that are false (0). Any transition that happens when all external inputs are false is left unlabeled. Obviously, if there are no external inputs then the transitions are left unlabeled.

Examples

Figure 8.25 shows the state diagram for a synchronous divide-by-3 counter.

The arrows make it clear that the transitions are strictly one-way. The circuit always goes from state 00 to state 01 and never the reverse. Since there are only three bubbles the system automatically produces a divide-by-3 effect.

A more elaborate synchronous system would have external inputs that control its behavior. We can design such a more advanced 2-bit counter. We will start with a full 4-state counter so it will need two bits in its latch. We shall add an external input Up that controls the direction in which the counter counts and an input, Reset, that forces

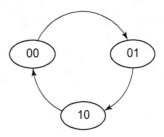

FIGURE 8.25 State diagram.

the counter to the 00 state. Figure 8.26 shows the state diagram for this more advanced counter.

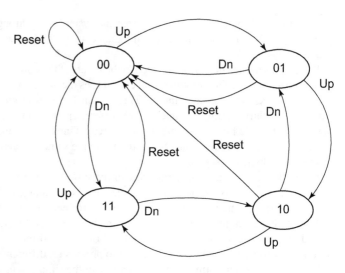

When the external inputs are both 0, the counter cycles round anti-clockwise, counting down. When Up is set the counter cycles round clockwise, counting up. When Reset is 1 the counter is forced into the 00 state. What is not clear from the diagram is what happens when both Reset and up are set. In that case, Reset should win and so the lines labeled Up should be labeled Up•!Reset but I left that off to make the diagram clearer.

FIGURE 8.26   Up-down div-4 state diagram.

Once you have created a state diagram, it is fairly simple to design the combinatorial logic circuit that will implement a given synchronous machine.

8.9.2 Designing Synchronous Circuits with D-Type Flip-Flops

Designing a synchronous circuit really means designing the combinatorial steering logic that controls the sequence of state. We design a synchronous system based on D-type flip-flops in a series of stages.

1. Decide how many flip-flops make up the system. Since a set of n binary bits can represent any of 2^n unique states: a system that has anywhere between $2^{m-1} + 1$ and 2^m states requires m flip-flops.
2. Draw the state diagram for the system.
3. Compute the values of the D inputs for each output state.
4. Write a truth table to generate the D values from the outputs.
5. Convert the truth table into a combinatorial circuit using standard methods.

TABLE 8.12

Initial truth table for a synchronous design

Q2	Q1	Q0	D2	D1	D0
I2	I1	I0	O2	O1	O0

I usually start step 4 by creating an empty table with headings like Table 8.12.

This emphasizes the fact that each Q output is connected to the corresponding I input and each O output is connected to the corresponding D input.

Next, I fill in the Q values with the sequence through which I want the system to run. If there are any external inputs, then I will need to write out a different sequence of Qs for each different set of external inputs.

Finally, I can complete the table by filling in the D/O side of the table. This is made very simple by the way in which a D-type flip-flop operates. When a clock pulse arrives, the flip-flop copies the state of its D input and that becomes the new Q. Thus, the D/O side of the table is filled in with the Q patterns in a different order. An example should clarify this.

Example

Divide-by-3 counter using D-type flip-flops.

We start by deciding how many flip-flops we need to implement the system. Since it has to have three unique states, we need two flip-flops (capable of handling up to four unique states).

We already have the state diagram on the previous page.

If the circuit is to go through that sequence of states, then when the counter is in the state 0,0, the next state must be 0,1, and so the D inputs must also be 0,1. Similarly, if we are in state 0,1 then the D inputs must be 1,0, the outputs for the next state.

When we put all this together, we get Table 8.13.

We fill in the truth table using the top set of column headings, Q and D. Now we design the combinatorial circuit using the bottom set of headings, I and O. In terms of these headings we can write down logic expressions for the two output columns using our usual methods. In this case we find that

$O1 = I0$ and either $O0 = !(I1 + I0)$ or $O0 = !(I1 \oplus I0)$,

where we have a choice of expressions for O0 because we have not specified what will happen if the counter is in state 1,1. Either of these expressions will give us a counter that operates correctly, assuming that the counter starts off in a legal state.

TABLE 8.13

Div-3 logic table

| Q1 | Q0 | D1 | D0 |
I1	I0	O1	O0
0	0	0	1
0	1	1	0
1	0	0	0

8.9.3 Excluded States in Synchronous Logic

The D-type divide-by-3 counter of the previous example illustrates a common problem with synchronous systems. In normal operation the counter should never be the state 1,1. We call this state an *excluded* state since it is excluded from the state diagram. In practice we cannot be sure that the state will never occur and that can lead to problems.

Let's look at what happens if the counter somehow gets into state 1,1.

If we choose the simpler expression, $O0 = !(I1 + I0)$, then when the output state is 1,1 the inputs will be $I1 = 1$, $I0 = 0$. That means that the next state will be 10 and the counter is back into the normal cycle.

If we choose the more complex expression, $O0 = !(I1 \oplus I0)$, then when the output state is 1,1 the inputs will be $I1 = 1$, $I0 = 1$. In that case, the next state will also be 1,1. So, if the counter ever gets into this illegal state, then it is stuck there and will never emerge.

So the first version is much safer; it has a self-recovery mechanism. Therefore, we will choose the first version and we can now design the circuit shown in Figure 8.27.

I have drawn this circuit to emphasize how the circuit fits into the general pattern of a synchronous device. I have enclosed the latch section and combinatorial section in dotted boxes to make them stand out. You would not normally draw the circuit this way but would lay it out however was convenient.

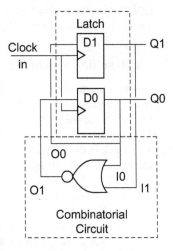

FIGURE 8.27 Synchronous divide-by-3 circuit.

8.9.4 External Inputs

All of the synchronous systems that we have seen so far operate independently of the outside world, apart from their clock signal. In practice we often want the behavior of the system to depend on events in the real world and so must supply extra inputs, external inputs, to the combinatorial logic piece of the synchronous system.

TABLE 8.14

Logic table for 2-bit up/down counter

External		Q1	Q0	D1	D0	External		Q1	Q0	D1	D0	External		Q1	Q0	D1	D0
Reset	Up	I1	I0	O1	O0	Reset	Up	I1	I0	O1	O0	Reset	Up	I1	I0	O1	O0
0	0	0	0	1	1	0	0	0	1	0	0	0	1	1	0	1	1
0	0	1	1	1	0	0	1	0	0	0	1	0	1	1	1	0	0
0	0	1	0	0	1	0	1	0	1	1	0	1	X	X	X	0	0

Let us look at a more complex example to see how to include the effects of external inputs. We have already seen the state diagram for a 2-bit up/down counter with synchronous reset (2-bit Up-Down counter with Synchronous Reset). Let us construct the truth table for this system. It will need two external input columns, two Q/I columns, and two D/O columns. The simple case Up=0, Reset=0, looks a lot like the divide-by-3 with one extra state (Table 8.14, left and center). When Up=1 and Reset=0, the counter functions as a normal 2-bit up counter (Table 8.14, center and right). Finally, if the Reset=1 then the logic must steer to the 0,0 state regardless of the other inputs (Table 8.14, right bottom).

Now we are ready to write the equations. Since there are many more 0s than 1s in the output columns, I shall use sum-of-products to find the equations.

O0=(!Reset•!Up•!I1•I!0)+(!Reset•!Up•I1•!I0)+(!Reset•Up•!I1•!I0)+(!Reset•Up•I1•!I0)

O1=(!Reset•!Up•!I1•!I0)+(!Reset•!Up•I1•I0)+(!Reset•Up•I!1•I0)+(!Reset•Up•I1•!I0)

With a little Boolean algebra, we can simplify these equations somewhat to find

O0=!Reset•!I0 and O1=!Reset•!Up⊕(I1⊕I0)

These give us the circuit of Figure 8.28.

8.9.5 Resetting Synchronous Circuits

As we have seen, there is a problem with synchronous logic systems which have states that+are not normally encountered when the system is operating. Such states are called *forbidden states* or *excluded states* and the complete set of such states is called the *forbidden zone*. If the system gets into one of these states, then one of two things can happen. The logic can push it back into a normal state or the system can get stuck in the forbidden zone. It often costs extra logic to make sure that the system gets itself out of the forbidden zone if an accident takes it there. The system designer has to decide whether the extra logic is worth it or not. Often the answer is not.

If the designer decides not to provide the extra logic to recover from the forbidden zone automatically, then she must make some provision for getting the system back into a standard state. This is usually done by providing a reset line that forces the system into a good starting state, often the all 0 state.

All flip-flops start up in random states and so you could be certain that your system would occasionally start up in a forbidden state and then you would be stuck. Indeed, the classic symptom of an unintended forbidden state is a circuit that works most of the time but sometimes won't work when it is first turned on. It can usually be made to work if you just turn it off and back on again, but it is obviously better to design it correctly instead.

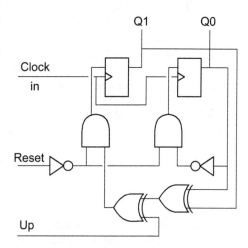

FIGURE 8.28 Synchronous 2-bit up-down counter.

8.10 Introduction to Verilog

This section is available at ExpPhys.com

Lab 8A Digital Logic

Learning goals:
- Design simple combinatorial logic systems using the sum-of-products method and De Morgan's theorem.
- Build familiarity with gated latches, ripple counters, and modulo counters.
- Design simple sequential systems with D-type latches and the sum-of-products method.

Apparatus: Electronics designer, signal generator, dual-channel oscilloscope and probes, 74HC00, 74HC02, 74HC04, 74HC11, 74HC74, de-bounced pushbutton.

Pre-lab reading: Read Sections 8.1–8.9. Make sure that you are familiar with both de Morgan's equations and the sum-of-products design technique.

8A.1 Combinatorial Logic

Familiarization

We'll start with a simple exercise to give you a feeling for working with the chips and to see the truth tables in action. Use the row of switches along the bottom of your breadboard to generate input signals and connect outputs to the row of LEDs in the top right.

Task 1: Not Gate

First, fill in what you **expect** to be the truth table of a **NOT** gate, also called an **inverter,** in the *Predicted* table (Table 8.15).

Now take a 74HC04 hex inverter (six NOT gates) and **connect up** *one* of the gates with its input coming from a switch and its output going to one of the LEDs on the breadboard. You must also wire the top right corner pin, pin 14, to 5 V and the bottom left corner pin, 7, to ground. This pattern holds for almost all digital logic chips (Figure 8.29).

By testing both values of the input and noting the outputs, **fill in** the *Measured* table (Table 8.16).

Does this indeed implement the NOT function? **Explain.**

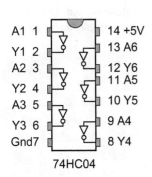

A1 1
Y1 2
A2 3
Y2 4
A3 5
Y3 6
Gnd7

14 +5V
13 A6
12 Y6
11 A5
10 Y5
9 A4
8 Y4

74HC04

FIGURE 8.29 Pinout for 74HC04 hex inverter+.

TABLE 8.15

Predicted

In	Out
0	
1	

TABLE 8.16

Measured

In	Out
0	
1	

Task 2: NAND Gate

Now let's do the same thing for a NAND gate. Again, start by **filling in** the values that you expect into the *Predicted* truth table (Table 8.17).

Take a 74HC00 quad 2-input NAND gate (Figure 8.30) and **connect up** one of the gates. Again, the inputs come from switches and the output goes to one of the LED indicators.

Note: Don't forget to wire up power and ground!

By trying all possible inputs and noting the outputs, **fill in** the *Measured* truth table (Table 8.18).

Does this indeed implement the NAND function? Again, **explain**.

TABLE 8.17

Predicted

A	B	$\overline{A\&B}$
0	0	
0	1	
1	0	
1	1	

TABLE 8.18

Measured

A	B	$\overline{A\&B}$
0	0	
0	1	
1	0	
1	1	

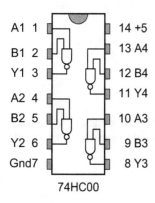

A1 1   14 +5
B1 2   13 A4
Y1 3   12 B4
A2 4   11 Y4
B2 5   10 A3
Y2 6   9 B3
Gnd7   8 Y3

74HC00

FIGURE 8.30 Pinout for 74HC00 quad NAND.

2–4 Line Decoder

In the digital realm, we encode everything as patterns of bits. The most common interpretation is as binary numbers. With two binary bits we have four possible states and can encode the integers 0–3. A 2–4 line decoder takes a binary number, on two input lines, and decodes it into a one-of-four output. Thus, it has two inputs and four outputs. Table 8.19 is the truth table that we want.

Note that the outputs are inverted from what you might expect. This is partly for historical reasons and partly to make the assignment more interesting.

Pre-lab question 1: Use the **product-of-sums** technique from the chapter to find a set of logic equations that implement this truth table. You will need to create four separate equations each with two inputs, A and B.

Pre-lab question 2: Use De Morgan's theorems to **design** (manipulate equations) and **draw** a circuit with two inputs and four outputs, using only NAND, and NOT gates to implement those equations.

TABLE 8.19

Truth table for 2-4 line decoder

Inputs		Outputs			
B	A	Y3	Y2	Y1	Y0
0	0	1	1	1	0
0	1	1	1	0	1
1	0	1	0	1	1
1	1	0	1	1	1

Task 3

Build and **test** the circuit. **Describe** its operation and any problems you experience.

8A.2 Sequential Logic

Combinatorial logic always does the same thing. Adding memory allows circuits to behave one way sometimes and a different way at other times. This is the basis of **sequential logic**.

The Bi-Stable Latch or S-R Flip/Flop

The basis of all sequential logic is the bi-stable latch, which you will build from two of the NAND gates in a 7400. Since its inputs are active-low, it is most properly called an $\bar{S}$-$\bar{R}$ Flip/Flop or Latch, since a zero on the $\bar{S}$ pin will *set* the Q output to 1 while a zero on the $\bar{R}$ input will *reset* the Q to 0.

Pre-lab question 3: Fill in both the NAND truth table (Table 8.20) and the truth/state table for a simple bi-stable latch (Table 8.21).

Note that the state names in Table 8.21 are just for use in the lab exercise.

TABLE 8.20

Truth table for NAND

A	B	A&B
0	0	
0	1	
1	0	
1	1	

TABLE 8.21

Truth table for bi-stable latch

A	B	Q	$\bar{Q}$	State Name
0	0			Unstable
0	1			Q1
1	0			Q0
1	1			Memory

Task 4

Cross-connect two of the NAND gates from a 74HC00 into a bi-stable latch as shown in Figure 8.31.

Verify that, when both inputs are at logic 1, the circuit can be in one of two stable states and that it *remembers* which of the two inputs was last at logic 0. You can also check the non-stable input state 0,0 and show that its output is NOT remembered when you return the switches to 1,1. This $\bar{S}$-$\bar{R}$ Flip/Flop demonstrates the incredible power of the simple idea of feedback – of taking the output from the circuit and recycling it as an input.

In Table 8.22, **write** a sequence of inputs and outputs to demonstrate the memory behavior and label each state.

Explain how your sequence of events demonstrates the memory behavior of the latch and make clear which of inputs A and B plays the role of $\bar{S}$ and which the role of $\bar{R}$.

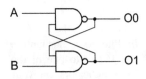

FIGURE 8.31 Bi-stable latch.

The D-Type Latch

Memory becomes more powerful when we can control what it remembers. This leads to the D-type edge-triggered latch.

TABLE 8.22

A sequence of applied inputs and observed outputs to demonstrate memory behavior of the bi-stable latch

In1	In2	Out1	Out2	State name

Pre-lab question 4: Fill in the truth table (Table 8.23) for a 74HC74 D-type latch (ignoring the preset and clear inputs, which are assumed to be tied high).

Pre-lab question 5: Explain the meaning of the subscripts n and n+1 on the output entries.

TABLE 8.23

Truth table for D-type latch (also known as a flip-flop)

Clk	D	Q_{n+1}	$\bar{Q}_{n+1}$
0	x		
1	x		
↑	0		
↑	1		

Counting

The new behavior of the bi-stable latch came from adding feedback to a pair of gates. If we once again add feedback, then we get another useful new behavior.

Task 5

Wire up one half of a 74HC74 D-type latch as shown in Figures 8.32 and 8.33. **Remember** that we must still connect the power supply pin, pin 14, to 5 V and must ground pin 7! The pins marked Hi can be connected straight to +5 V.

Connect the *Clock In* line to a pushbutton and *V Out* to one of the LED indicators on your designer. Push the button several times.

Describe and **explain** what is happening.

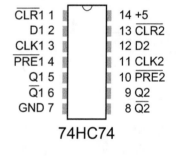

FIGURE 8.32 Divide by 2.

FIGURE 8.33 Pinout for 74HC74 dual flip-flop.

Task 6

Next wire up the other half of your 74HC74 in the same way to get the 2-bit counter of Figure 8.34. This time drive the *Clock* with a 1 kHz, 0–5 V square wave from the Sync or TTL output of your signal generator and connect outputs *V Out1* and *V Out0* to the inputs of your oscilloscope.

Draw the output and annotate your diagram to show that the two outputs form the two bits of a binary counting sequence.

Does the counter count up or down? Why?

How does the frequency of output 0 compare with the clock frequency? **Why?**

What about output 1?

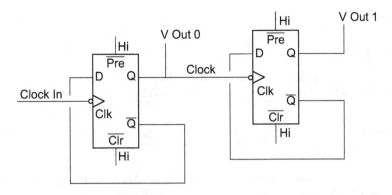

FIGURE 8.34 2-bit binary counter .

8A.3 Synchronous Sequential Machines

If you increase the clock frequency to 1 MHz and look carefully at the moment that O0 transitions from $0 \rightarrow 1$ then you will see that there is a short gap before O1 makes the transition from $1 \rightarrow 0$. This glitch arises because the output of one flip-flop is the input to the next and it takes a few nanoseconds for the flip-flop to respond to the change in its input. We can build counters that are better as well as more flexible by driving all the flip-flops with the same clock and using additional gates to control the way that the bit-pattern changes.

Task 7

Redraw the outputs from your 2-bit counter when it is driven from a 1 MHz clock to illustrate the transition glitch described above.

Note: You will have to experiment carefully with the horizontal scale and with the triggering to see the glitch. Consult your instructor if you are having difficulty finding it.

Task 8

Design (i.e., write sum-of-products equations for) a synchronous divide-by-5 machine using three D-type flip-flops. It has five distinct 3-bit states 000, 001, 010, 011, 100, respectively, which then repeat (Table 8.24).

> **Use a De Morgan theorem** to convert the equation for bit 0 to use a single 2-input NOR gate.
>
> **Build and test** your counter. You should need two 74HC74 ICs, one 74HC02 quad 2-input NOR gate, and one 74HC11 triple 3-input AND gate (see Figure 8.35).

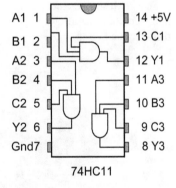

FIGURE 8.35 Pinout for 74HC11 triple 3-input AND.

TABLE 8.24

Divide-by-5

Q2	Q1	Q0	D2	D1	D0
0	0	0			
0	0	1			
0	1	0			
0	1	1			
1	0	0			

Draw the output as you would see it on a 4-channel oscilloscope looking at the clock and the three output bits. Note you will have to do this two bits at a time. Be sure to trigger your oscilloscope on the *slowest* output signal!

Label your figure to show the five distinct states.

Discuss why your synchronous counter counts UP, while the earlier ripple counter counted down.

How would you make your divide-by-5 counter count down?

Lab 8B Controlling the World with Arduino

Learning goals: Write simple digital input and output programs in the Arduino ecosystem. Interact with the Arduino program through the serial terminal.

Apparatus: Arduino Uno board, PC with Arduino environment installed, LEDs and 300 ohm resistors, small solderless breadboard, and jumper wires.

Pre-lab reading:

- Read Section 7.4 "Preview of the Arduino."

- Read through the Arduino tutorial by Limor Fried, also known as Lady Ada, at www.ladyada. net/learn/arduino. The first six lessons of this introduce many of the basic ideas of programming and show you the Arduino way of doing things.

- Explore the Arduino website at www.arduino.cc. In particular, you should work through the *Getting Started* section found under the *Resources* menu and should spend some time familiarizing yourself with the *Reference* section of the *Resources*. Any time you are not sure what something does in an Arduino program you can find more information in the *Reference* section.

Introduction: Your Arduino Uno is an inexpensive (~$20) board that has a complete micro-controller (an Atmel ATMEGA328) on it. This is the long black chip just above the POWER and ANALOG IN labels in Figure 8.36.

The small square chip, just above the shiny oval can in the upper left quadrant of the board is a USB interface chip that allows us to put programs into the micro-controller and then to communicate with them from a PC.

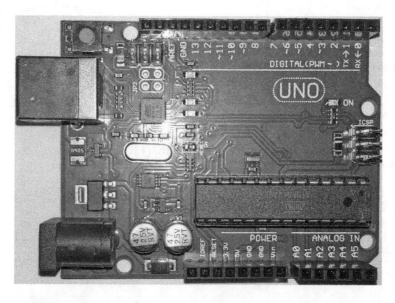

FIGURE 8.36 Arduino Uno board.

The Arduino Uno is one piece of a complete ecosystem of products under the Arduino name. In addition to the actual computer (called Uno for the rest of this lab to distinguish it from the software) there is a complete programming system designed for this hardware and also called Arduino. There is an icon for it on your desktop and when you first start the program it will look something like Figure 8.37.

The white area is a fairly standard text editor in which you will compose your program. A new program (called a **Sketch** in Arduino parlance) has the simple structure shown here. It consists of two separate pieces, two **functions**, one called *setup* that will be executed just once, when your program first starts, and one called *loop* that will be called repeatedly thereafter. So the basic rules for the Arduino to execute a sketch are

1) Run the code in setup

2) Run the code in loop

3) Go back to 2

Note that the code runs in what we call an **infinite loop**. That is, it keeps repeating the same task time and again until the power is turned off.

FIGURE 8.37   Arduino.

Getting Started

If you have not already done so, **start Arduino**. You should get a window very like the one in Figure 8.37.

Now **plug** the USB cable attached to your Arduino into the computer. If all is well then the green LED marked ON should light up in the upper part of the board. If not, then please call your instructor.

A Simple View of Our Micro-controller

There is a lot going on inside the ATMEGA328 on your Uno, but for now we will take a very simple view. A micro-controller is very snazzy device for playing with digital and analog voltages or *signals*. Such signals appear to the world as pins on the chip connected to wires (the row of sockets that stick up from the surface of the Uno) that can either carry information from the outside world into the controller (**inputs**) or carry information out to the world (**outputs**). Because they carry information into and out of the chip they are called **Input/Output** or **I/O** pins.

Switches have only two states, *on* and *off* so they naturally represent binary digits and we call such signals *digital*. Voltages can take a continuous range of values. They represent smoothly varying quantities that are analogous to values in the real world, so we call such signals *analog*.

The pins have names that are printed on the Uno board beside the pins. The top row of pins in Figure 8.36 are labeled DIGITAL and there are 14 pins numbered 0–13. The bottom row of pins is split into two sections. The left-hand section contains pins that convey power into and out of Uno and they are labeled POWER. The righthand section has six pins labeled A0–A5 that convey analog voltages into the board. This section is labeled ANALOG.

If we connect one of the digital I/O pins to a device in the outside world then we can turn the device on and off from our program. For example, there is a yellow LED (labeled L in Figure 8.36) that is connected to digital pin 13 and we can turn the LED on and off by changing the state of the pin.

Controlling the Pins

In the Arduino world, we have two simple commands that allow us to control the state of each pin and one to test the state. The first is called pinMode and it looks like this:

```
pinMode(pinNo, mode);
```

The pinNo and mode are called **parameters** or **arguments**. They are numbers that tell the command which pin to work on and what to do with it. We will look at them in little bit.
The second command is called digitalWrite and looks like this:

```
digitalWrite(pinNo, state);
```

If the pin is set up as an output, then this command allows you to turn the switch on and off.
The final command is called digitalRead and we shall learn more about it in a little while.

pinNo

Since there are 14 pins in the connectors the pinNo should be a number between 0 and 13. If you put something else there then the command will just do nothing! If you put in a valid number then this command will determine whether that pin, the one numbered pinNo, is to be used as an input or an output.

mode

This can take one of the values INPUT or OUTPUT. If you make the pin an INPUT, then you can test the voltage on the pin and learn whether it is close to ground (state 0) or close to 5 V (state 1). Any value that is sent to the pin with `digitalWrite` is completely ignored. If you make the pin an OUTPUT, then you can control the voltage on the pin using the `digitalWrite` command.

State

The pins each have two output states. If you set the state to HIGH, then the pin will be at 5 V and if you set it LOW then the pin will be at ground.

Your First Program

Here then is our first program (Listing 8.1, below).
Note: I have added line numbers to facilitate explanation These are NOT part of the program, but they make it easier to talk about. You can make them appear in the Arduino window by checking *Display Line Numbers* in the Arduino preferences.

LISTING 8.1: OUR FIRST PROGRAM

```
1 /*
2 * Blink
3 * Basic Arduino example. Turns the on-board LED on pin 13 on for one second,
```

```
 4 * then off for one second, repeatedly. Uses the LED_BUILTIN defined by Arduino.
 5 * Brian Collett
 6 */
 7
 8 //the setup routine runs once when you press reset:
 9 void setup() {
10   //initialize the digital pin as an output.
11   pinMode(LED_BUILTIN, OUTPUT);
12 }
13
14 //the loop routine runs over and over again forever:
15 void loop() {
16   digitalWrite(LED_BUILTIN, HIGH);  //turn the LED on (HIGH is the voltage level)
17   delay(1000);            //wait for a second
18   digitalWrite(LED_BUILTIN, LOW);  //turn the LED off by making the voltage LOW
19   delay(1000);            //wait for a second
20 }
```

You should now recognize many of the lines. Let's look at the rest.

Lines 1–6

These lines are one style of **comment**. A comment is text that is *completely ignored* by the computer. It is just there to make the program more readable. This kind of comment, a **multi-line comment**, runs from the opening /* all the way up to and including the closing */.

You may put anything you like into a comment. Good practice *requires* you to write a comment at the top of *every* program that you write. This comment briefly describes what the whole program does and includes the names of the people who wrote the program. Comments in the body of the program should say why each section of code is there, *what* it does – the code itself should tell you *how* it does it.

Lines 7 and 13

These lines are blank lines that are put in to make the program a little more readable. You are allowed to insert blank lines pretty much anywhere. It is good practice to use them to separate different sections of code from each other, as here.

Lines 8, 10, and 14 and the Ends of Lines 16–19

These lines are a different style of comment, often called a **C++-style comment** for historical reasons. A C++-style comment starts with the character pair // and includes all the characters up to the end of the line. Like any comment, these lines are completely ignored.

You can see that there are two ways to use these comments, either as lines by themselves or as the ends of active lines. Such comments can add English explanations to make the code clearer.

Lines 11, 16, and 18

These lines are the ones that we were expecting. Line 11 tells the computer that we want to use pin LED_BUILTIN as an output. Lines 16 and 18 change the state of the pin, bringing it first to +5 V and then to ground to turn the LED on and off. The name LED_BUILTIN is provided by the Arduino system and is much more human friendly than the bare number 13, which is its value. This is because the Uno board comes with an LED and driving circuitry connected to digital pin 13 (labeled '13' on the board itself).

Lines 17 and 19

These lines introduce us to another Arduino command, `delay`.

$$delay(numberOfMilliseconds)$$

Any digital computer runs on a timescale that is astonishingly faster than our timescale. If we want a program to run on a human timescale then we normally have to slow it down to our speed. This is such a common operation that we have a built-in command to accomplish it; delay simply pauses the execution of the program for some length of time. The amount of time is given as a number of milliseconds so that the calls of `delay(1000)` cause the computer to do *nothing* for 1000 mS = 1 second.

Note: Without the delays this program will turn the LED on and off so fast that you will not see it flash. You will explore this a little later in this lab.

Making it Work

Edit the text in your Arduino window to match the code in Listing 8.1. Note: We usually use tabs to insert more than one whitespace character. Lines 10 and 16–19 each begin with a single tab character.

In the toolbar just above the edit window there is a set of buttons. Let's look at them from left to right (Figure 8.38).

FIGURE 8.38   Arduino toolbar.

Verify

The big check mark tells Arduino to look through the text of your sketch and look for any errors. If it finds errors then it will print error messages in the black message pane below the white edit pane. We don't often need to do this separately.

Upload

The right-pointing arrow is the most useful button. It tells Arduino to do a whole set of things:

1) Save the program and give it a name if it does not already have one. Arduino will suggest a name using the current date, for example `sketch _ jan23c.ino`. You probably want to change it to something more useful. For example, I might call this first program `Blink1.ino`. This is the same as the Save button, the one that looks like a down arrow.

   Note: You *must* leave the extension as `.ino` or Arduino will not be able to find the program later.

2) Check the program for errors and turn it into an executable form if there are none. This is the same as the Verify button.

3) If the Verify worked then transfer the program down the USB cable into the Uno and start it running.

New

The picture of a sheet of paper will save the current sketch and start a brand new one with the usual empty Setup and Loop functions.

Open

The up arrow will save the current sketch and bring up a dialog box to allow you to open a previous sketch. Remember that it will only open .ino files.

Save

As described above, the down arrow will save the current sketch, allowing you to choose a new name if you have not done so yet.

Terminal

The magnifying glass at the right hand of the toolbar will open a separate terminal window that you will use to communicate between the PC and Uno later in the lab.

Let's Do It!

Compile and run your first Arduino program by pushing the Upload button. If it does not result in a nice blinking red, then consult your instructor.

Note: You will be asked for a new name for the program. Please choose something meaningful!

More Blinkys!

Once your first program works, it is time for you to do something on your own. At the moment, the LED blinks at a single fixed rate. In the next part of the lab you will alter the program so that the LED blinks in a different way.

Start a new program, either by pressing the New button or by going to the File menu, selecting Save as... and choosing a new name. Get your new program to look the same as your first one if it does not already do so.

Task 1

You first job is to make the LED flash in a more complex pattern. I would like the new program to Blink in a pattern of two long ONs followed by one short ON of about half the duration. The OFF times should all be the same, about as long as the short ON except that after the last short ON I want you to put in a significantly longer off. We are after something like this

OONN off OONN off ON o--------f OONN off OONN off ON o--------f OONN off OONN off ON o--------f

Start by modifying the header comment to reflect the new purpose and then modify the code. NOTE that copy and paste followed by a little judicious editing is a great thing.

Describe and explain what you did to make this happen and attach the working code to the lab.

Task 2

Wire up a second LED to your Uno. Pick one of the other digital output pins (0–12) and connect the LED and a 300 Ω resistor between the pin and ground as shown in Figure 8.39. Now modify your program so that it flashes the built-in LED and your new LED alternately, with a 1 Hz overall rate.

Again, describe and explain what you did to make this happen and attach the working code to the lab.

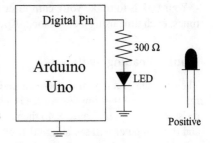

FIGURE 8.39 Driving an LED with Arduino.

Pre-lab Task

Explain why we added a 300 Ω resistor between the pin and the LED and **justify** the value that we chose.

If we want to drive several different LEDs, e.g., the seven LED segments of a numeric display, which all share a common positive terminal, **why** must we use a separate resistor for each LED and cannot just put a resistor in the common positive line?

Task 3

Now let's add some new programming ideas! First, we have the idea of a variable, a value with a name, much like the 'x' of mathematics. We can tell Arduino that we want to use a name to store a number with a line of code called a **declaration**, which looks like this.

```
int count = 0;
```

This tells Arduino to set aside a storage location big enough to hold a number, to give it the name "count," and to make its initial value 0.

Now that we have a variable, we can change its value and we can test its value and thus make our computer count. We change the value of a variable using an **assignment** statement, like this

```
count = count + 1;
```

Viewed mathematically, this is a piece of total nonsense – the value cannot be equal to a larger value. It is slightly unfortunate that computer languages were invented at a time when there were only the usual characters on a typewriter to choose from. It would be far easier for the beginner to understand if it were written

```
count ← count + 1;
```

and read "count becomes count plus one," which is what is meant. So, the meaning of an assignment is something like "evaluate the expression on the right of the equals sign to produce a number and then replace the value of the variable whose name is on the left with that value." This makes clear that a reference to count on the right side of the equals sign will use the *old* value. The value will not change until the very end of the statement.

We need one more piece of code, called a while loop. It looks like this

```
while (count < 5) {
    //Do something
    count = count + 1;
}
```

This will perform the lines of code that lie between the braces (the curly brackets) so long as the test in the first line succeeds. In this case, it will happen so long as count is less than 5. In order to make this useful we must alter the value of count inside the loop, as the third line does.

Your task is to **make your computer flash** the built-in LED three times and then flash your LED six times, each time using a count loop. **Append** your commented code to the lab.

Input – Reading Switches

The opposite of a digital output is a digital input. This takes the form of a binary truth value, a true or false value, and the simplest source is a switch connected to a resistor. If we wire up a switch and a 10 k resistor as shown in Figure 8.40 then, when the switch is open the resistor will pull the input up to 5 V and the computer will see a digital 1, when the switch is closed the voltage will fall to 0 V and the computer will read a digital 0. We can tell the computer that we want to use a pin as input with a line such as

```
pinMode(PF_0, INPUT);
```

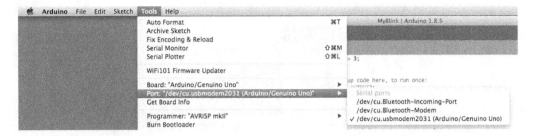

FIGURE 8.40 Selecting the Arduino serial port.

and then obtain its value with the function

$$digitalRead(pinNo);$$

Unlike commands such as `pinMode` and `digitalWrite` that we met above, this "returns a value." That is, when the computer performs this operation it ends up with a number representing the state of the switch using the code that 0 (logical false) means that the switch is closed while any other value (logical true) means that the switch is *not* closed. We can save this piece if information for later by saving it in variable using a simple assignment statement such as

$$int\ switchUp = digitalRead(pinNo);$$

or, far more usefully, we can *test* the value using an if statement such as

```
if (digitalRead(PF_0) == 0) {
    digitalWrite(PF_1, LOW);
} else {
    digitalWrite(PF_1, HIGH);
}
```

Here the first line first reads the value of the input pin and then takes one of two different actions depending on the result of the test. In this case, the red LED will be lit if the button reads HIGH (non-zero, not pressed) and turned off if the button read LOW (zero, pressed).

We can use this in a little program that will allow us to control the red LED with a switch connected to digital pin 2.

```
/* Light LED when Pin 2 goes HIGH */
int Switch=2;    //Now we can use the name Switch for pin 2
void setup()
{
    pinMode(LED_BUILTIN, OUTPUT);
    pinMode(Switch, INPUT_PULLUP);  //The PULLUP adds a resistor making it
                                    //easy to add a switch
}
void loop()
{
 //put your main code here, to run repeatedly:
 if (digitalRead(Switch) == 0) {
    digitalWrite(LED_BUILTIN, LOW);
 } else {
    digitalWrite(LED_BUILTIN, HIGH);
 }
}
```

Task 4

Extend the program to work with two pushbuttons, one controlling the built-in LED and one controlling your external LED. **Include your commented code with the lab writeup.**

Arduino Examples and analogWrite

The Arduino community has provided a large number of example programs to demonstrate various features of your Uno and of the Arduino ecosystem. You can access these by going to the File menu in Arduino and selecting Examples. You will get a long menu of examples broken up into different categories. I suggest you spend a few minutes looking at some of the ones that do not require external hardware, these are the ones in the top section with the numbered sub-sections.

Task 5: List the Ones You Have Explored and Say What Each Does

Task 6

Under the Basics section there is an example called **Fade**. It requires an LED and resistor connected to digital pin 3. Move your LED/resistor pair to pin 3 and try the example. **Describe** in your own words **what** it does.

Now connect your oscilloscope in parallel with the LED/resistor and observe the waveform on the oscilloscope. **Describe what** you see and **explain how** it changes the brightness. For this you will probably want to look up any new functions on the Arduino web site at www.arduino.cc/reference/en/.

Serial Programming

You have now seen how to get single bits of data into and out of the computer with pinMode, digitalWrite, and digitalRead. Another way to communicate with your programs is using a keyboard and display, in much the way that you communicate with the PC. The Uno does not have a keyboard or a display but there is a way to connect it so that it can "borrow" the keyboard and display of the PC. We do this with a **serial** communication link that the Arduino system provides for us. The communication channel shares the USB connection that powers Uno and allows us to program the micro-controller. When you plug the Uno into the computer, the serial connection is automagically created and given an internal name. On a Windows computer this will be something like "COM14" and on a Mac it will be something like "/dev/cu.usbmodem2031". You can find the name by selecting Port from the Tools menu as seen in Figure 8.40.

Then you can access the port from the Arduino software by pressing the magnifying glass button at the right end of the toolbar. It will bring up a terminal window that looks like Figure 8.41.

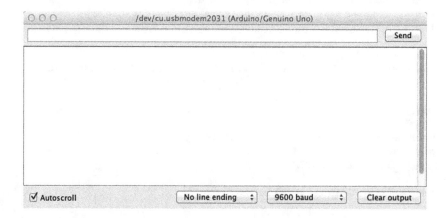

FIGURE 8.41 Serial terminal.

Any characters that your Uno sends to its serial port will be placed in the large window and you can type characters to be sent to Uno in the small window at the top. They will be sent when you press *Send*. In the interests of sanity, you should probably set the line ending to *Carriage return* using the leftmost button in the bottom toolbar. This will make each box of characters that you type be sent to Uno as a separate line.

Our First Words

Inside an Arduino program, the serial port that allows us to talk to our PC is represented by an object called Serial. This object can execute several useful commands. The first one that we need is called begin. It tells the micro-controller to turn on the serial port (which actually uses digital pins 0 and 1, labeled RX and TX in Figure 8.36) and tells it what speed the PC is expecting. We send the command to the Serial object by joining them with a period, like this

```
Serial.begin(9600);
```

This line of magic code *must* be put in the setup routine of any program that wants to talk to the PC. We refer to this as *opening the serial port*.

Once the serial port has been opened you can send data out of it. You can look at the Arduino reference materials to see the full range of output options, but the most common ones are `print` and `println`. Each of these sends one item to the PC in text form. The difference is that `println` tells the PC to go to a new line after it has displayed the item. Both `print` and `println` can print a variety of different values, integers, characters, and strings being the most common. For example, the following code

```
int val1 = 52;
Serial.print("There are ");
Serial.print(val1);
Serial.println("weeks in a year.");
```

will print

There are 52 weeks in a year.

on a line by itself in the serial terminal. It is a very good idea to *always* print something immediately after you call Serial.begin, that way you know if it worked or not.

Pre-lab Assignment: Printing in the Mind

Look at the program in listing 8.2 on the right. Using what you have just learned, **predict** what will appear in the serial window when the program is run.

LISTING 8.2: SIMPLE SERIAL PROGRAM

```
int Count = 0;

void setup() {
    Serial.begin(9600);
}

void loop() {
    Serial.print("Count = ");
    Serial.println(Count);
    Count = Count + 1;
}
```

Task 7: Printing in the World

Make a new Arduino program with the code from listing 8.2, run it, and **compare** the result with your prediction. **Explain** any differences in your writeup.

Getting Input

We have learned to send data from Uno to the PC. Now we want to go the other way. There is a `Serial` command called `read()` that will read one character from the PC. The problem is that there may not be anything for it to read. In that case it will just return 0, which is boring. `read()` is most useful when mixed with another command, `available()`. A call to `Serial.available()` will return the number of characters that have been typed on the PC and not yet read. If there are no characters waiting then it will return 0, which is the same thing as returning `false`. So we can sit waiting for characters and sending them back to the PC with code like Listing 8.3.

Note: You enter text into the box at the top of the serial window and then press the SEND button to dispatch it to Uno.

LISTING 8.3: SERIAL INPUT

```
int Ch=0;//Store a char here

void setup() {
    Serial.begin(9600);
    Serial.setTimeout(-1);
}

void loop() {
    if (Serial.available()) {
        Ch=Serial.read();
        Serial.write(Ch);
        Serial.println(Ch);
    }
}
```

Task 7: Run the code in listing 8.3 and **describe** and **explain** the output. You may need to refer to the Arduino website and you may need to consult with your instructor.

What does setTimeout do?

The Serial input routines in Arduino are impatient. `Serial.read` always returns immediately, with or without a character. Fancier input routines, such as those to read numbers, give you only a very short time (about 1 second) to type something in before they give up on you and return anyway. This is not the behavior that we usually want. If we ask to read a number, then we want the computer to wait until we are good and ready and not just give up on us. `Serial.setTimeout()` allows us to change how long Serial waits for us. The value is in milliseconds. A value of −1 will make all serial reads wait several days before giving up on us. I recommend that it become a habit to include it in *every* program that does input.

Getting Numeric Input

We have seen how to read characters with the `read()` command but probably the most common things that we want to read are numbers. Serial has some support for this, though it is not as convenient as you might like. The `parseInt()` function tries to interpret the incoming text as an integer and returns the value of the number that it finds. This will read characters until you type something that is not a digit and will return the value of the resulting number. Unlike read (), parseInt () has Serial.available() built into it so that it will wait (so long as you set the timeout to a very long time) until you type.

So you can read a number with code like this

```
int inputValue = Serial.parseInt();
```

This line collects valid digits typed into PuttyTel and assembles them into a number, which it puts into the variable inputValue once you enter a character that is not a valid digit. I normally end numbers by pressing the "return" key.

Task 8: A (Mindlessly) Simple Calculator Program

Use two such code segments to write a truly trivial calculator. It should read two numbers from the PC keyboard (serial terminal again) and then print out their sum. If you knew some programming before last week, then make it do its job prettily, with prompts for the input and some explanation of the output.

Include a suitably **commented** listing of your program as part of your writeup.

Controlling Uno from the Terminal

One of the most common uses of a micro-controller is as an interface between the PC and hardware in the real world. This usually requires that the PC be able to tell the micro-controller what to do, and possibly ask it to send data back. The serial port is the simplest way to achieve this. Listing 8.4 is a simple program that is an extension of the Fade program that allows you to control the brightness of the LED that you connected to pin 3 by sending single-letter commands from the terminal.

Note that we are very careful to call analogWrite only when there is a new character. You must not call analogWrite more than about 500 times per second or strange things start to happen.

Note also the trick of splitting a single line string over two text lines by enclosing each piece of the string in double quotes. The compiler automatically joins adjacent quoted strings into a single string.

Run this program and **verify** that you can control the brightness of your LED as expected.

LISTING 8.4: LED CONTROL FROM SERIAL PORT

```
int LED=3;    //A common trick to make the program more readable
int Brightness=20;  //how bright the LED is, initial val is dim

void setup() {
    //make LED an output and set initial brightness.
    pinMode(LED, OUTPUT);
    analogWrite(LED, Brightness);
    //open the serial port and write a helpful message.
    Serial.begin(9600);
    Serial.setTimeout(-1);
    Serial.println("LED brightness controller.");
    Serial.println("Press r to dim the LED"
        "and R to increase the brightness.");
}
void loop() {
    //Only do anything if we have input waiting
        if (Serial.available()) {
            int ch=Serial.read();
            if (ch == 'r') {//Note characters need single quotes!
                Brightness=Brightness - 5;
```

```
            analogWrite(LED, Brightness);
        } else if (ch == 'R') {
            Brightness=Brightness+5;
            analogWrite(LED, Brightness);
        }
    } else {
        Serial.println("Unrecognized command!");
        Serial.write(7);  //Ascii code 7 is the BELL character
    }
}
```

Task 9: Putting It All Together

Finally, invent your own new system. Design an original combination of switch and/or keyboard inputs and LED outputs, give them a behavior, and demonstrate your system to your instructor.

Lab 8C Interfacing an Experiment with Arduino

Learning goals: Interface a microcomputer with analog signals in the real world. Control a temperature using both a simple bang-bang thermostat algorithm and with a realistic servo algorithm.

Apparatus: Arduino Uno, host PC, 100 k B = 3950 thermistor, 20 k fixed resistor, small heating element, ULN2803, solderless breadboard, 5 V 2A power supply, heat resistant tape.

Pre-lab reading:

- Read the Wikipedia article on thermistors at
  - en.wikipedia.org/wiki/Thermistor
- and the Wikipedia article on bang-bang control at
  - en.wikipedia.org/wiki/Bang–bang_control
- then skim the Wikipedia article on Proportional-Integral-Derivative control at
  - en.wikipedia.org/wiki/PID_controller
  - Pay particular attention to the Control Loop example section.

Interacting with the World

In the previous lab, we looked at single-bit interactions with the real world, reading switches, and turning LEDs on and off. However, the real world is full of quantities that are far more subtle. For example, an LED has many states between off and on, states from barely visible to full brightness, and a real quantity such as temperature takes on a continuous range of values. Such quantities require real numbers to represent them perfectly, but we make do with integers and a consequent loss in accuracy.

Controlling the World

In the first Arduino lab, we met the `analogWrite` method that gives us a way to output an integer through a rather simple digital-analog converter (DAC) that uses a method called pulse-width modulation (PWM). This generates a pulse at a fixed frequency of 490 Hz with an on-duration that varies between 0% and 100% of the period producing an average output voltage of

$$Vout = DACvalue \times 5 \text{ V}/256 = DACvalue \times 0.0195 \text{ V}$$

The actual output is a pulse that is either 0 V or 5 V and we have to add something to do the averaging in order to convert the output to a controllable value. With the LED we used, our brains to do the averaging. The LED was either on or off but once the rate of pulsing gets high enough (about 100 Hz for a source like the LED) the brain cannot respond fast enough and sees simply the average value. The same sort of thing happens if we drive a motor from a PWM signal or if we apply it to a resistor to generate heat (see below). If we need to put the output into something that is fast enough to respond to the individual pulses, such as another electrical circuit, then we have to do the average ourselves. This is usually done with a simple low-pass filter. For example, a low-pass filter with a cut-off frequency of 10 Hz will produce an output voltage that has only 65 mV of ripple (about 4 DAC units) while still allowing the voltage to change at rates up to 10 Hz.

Measuring the World

Your Arduino Uno has a built-in multi-channel analog-digital converter that is accessible through the six ANALOG IN pins near the POWER connector on the Uno.

You met the analog-digital converter (ADC) in the first lab in this sequence. It converts an analog voltage into a digital number that is proportional to the input voltage. The ADC in the Uno is a 10-bit system that converts voltages into a number between 0 and 1023 using the formula

$$ADC value = \frac{1024 \times Vin}{Varef}$$

where Vin is the voltage on the analog input pin and $Varef$ is the reference voltage on the AREF pin in the upper left of the Uno board in Figure 8.36. This is normally left unconnected, in which case, the system uses the 5 V power as the reference so that the usual formula is

$$ADC value = \frac{Vin}{0.00488 \ V} \quad \text{or} \quad Vin = ADC value \times 0.00488 \ V$$

Measuring Temperature

There are several ways to convert temperature into an electrical signal. The most accurate methods for temperatures near ambient use either thermocouples, in which two dissimilar metals generate a tiny voltage that is proportional to the temperature of the junction, or fine platinum wires whose resistance increases slightly with temperature. Both of these produce small signals that need amplification before they can be fed to a computer. Less accurate, but much more convenient, is the **thermistor**, a small resistor with a resistance that decreases strongly with temperature rise. It can be made to give a signal that is large enough to read with the computer. The drawback is that the variation is quite non-linear so that converting between resistance and temperature is a little complicated, but that is what computers are for!

From Temperature to Resistance

Over a range of at least 200°C, the resistance and the actual temperature of a thermistor are related by the Steinhart-Hart equation which can be written

$$\frac{1}{T} = \frac{1}{T_0} + \frac{1}{B} \log \left(\frac{R}{R_0} \right)$$

where T is actual temperature in Kelvin, T0 = 298.15 K is a reference temperature of 25°C, R the actual value of the resistance, and R0 the value of the resistance at 25°C. B is a constant provided by the manufacturer of the thermistor and reflects the materials from which the thermistor is constructed. In this lab we will be using a thermistor that has R0 = 100 k at 25°C and a B = 3950.

For example, at 100°C, our thermistor will have a value given by

$$\frac{1}{373.15} = \frac{1}{298.15} + \frac{1}{3950} \log\left(\frac{R}{100}\right) \text{ or } R = 100 \times e^{\left(\frac{3950}{373.15} - \frac{3950}{298.15}\right)} = 7 \text{ k}$$

From Resistance to Voltage

There are several ways to convert a resistance into a voltage. The simplest is to put the sensor in series with a fixed resistance forming a voltage divider. Other methods include running a constant current through the sensor and measuring the voltage (this is what a multi-meter does when measuring resistance) and making the sensor one leg of a so-called bridge circuit. We will use the simplest method, resulting in the circuit of Figure 8.42.

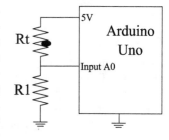

FIGURE 8.42   Temp input.

I have put the thermistor, Rt, the one with the black dot on it, as the top leg. This way, the output voltage will increase as the temperature increases. The fixed resistor, R1, should be chosen to have a value roughly in the middle of the range of values that you expect in operation. In our case, we expect values in the range 100 k down to about 7 k so a value of 20 k is a good choice.

Task 1

Build the circuit of Figure 8.42. **Write** and **demonstrate** a program to print the voltage on pin A0 once per second. **Append** your commented program to the writeup.

You will have found that the temperature value is not perfectly steady. The signal suffers from small random fluctuations called **noise**. We can diminish the effect of the noise by averaging the signal. **Alter your code** to measure the temperature ten times per second but to add the ten samples together then divide by ten to get the average. **Print** this average once per second and **compare** the results with your first program.

Task 2

Look up the log() function in the Arduino documentation and **modify** your program to print the temperature as well as the voltage. **Append** your commented program to the writeup.

Explain the behavior of the system when you hold the thermistor between your fingers and when you blow on it. **Comment** on its accuracy.

Controlling Temperature

Now that we can measure temperature, we want to move on to control. For this we need a transducer that can modify the temperature. In this case, we will use a small heating element. This is basically a wire-wound resistor designed to withstand high temperatures without damage. When we run a current through the resistor it will start to warm up by Joule heating as it dissipates power $P = VI = I^2R$.

In order to get the resistor hot, we will need to dissipate significant power in the heater and thus will need to run significant current, far more than the 20 mA maximum that the Uno can deliver to any pin.

Switching High Currents

Controlling significant currents from micro-controllers is a common enough function that there are a large number of single-chip solutions available. These generally contain high-current transistors to

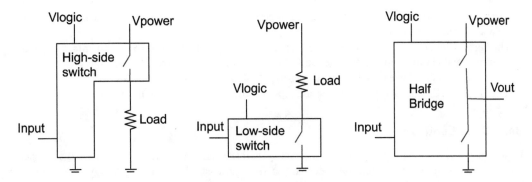

FIGURE 8.43 Switch geometries.

do the actual switching and then level-adjusting circuits that allow the 0–5 V output signal of an Uno, or the 0–3.3 V output signal of an Arduino Due to turn the switch on and off. Such drivers come in various different configurations and voltage/current ratings, as shown in Figure 8.43. Low side switches are the simplest but require the load to float above ground. High-side switches allow the load to be connected to ground but are more complex internally and typically require two power sources. The fanciest switches are the half-bridges that combine a low-side and high-side switch into one unit. Two of these form a full-bridge which can drive current through a resistive load in both directions while needing only a single power supply.

The chip we will use is a long-time industry favorite, the ULN2803 (Figure 8.44). This has eight identical low-side switches, each of which can carry up to 0.5A of current. This comes in an 18-pin package and is unusual in having no power supply pin. Sixteen of the pins carry signals into and out of the chip, one is the shared ground, and the last is connected to protection diodes that can be used when driving inductive loads such as motors. We want to switch currents up to at least 1A so I have chosen to parallel four of the outputs. This splits the load current across four separate switches, reducing the heating of any one switch.

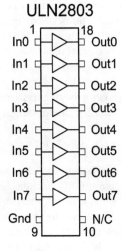

FIGURE 8.44 ULN2803.

Task 3

Add the extra components to your circuit to get the circuit of Figure 8.45, on the right. Tape the thermistor to the side of the heater so that it makes good thermal contact. Note that you must ground pin 9 of the ULN2803, even though there is no way to show this on the figure.

Note: While it is currently unimportant which digital output pin we use on the Arduino, it will become important later. The Arduino Uno supports the `analogWrite` function on pins 3, 5, 6, 9, 10, and 11 only. We will need that feature later so it is good to use pin 9 now.

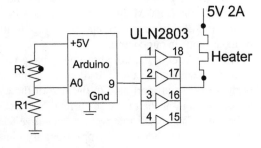

FIGURE 8.45 Thermostat.

Write and run a program that will turn the heater on for 30 seconds and then off for 30 seconds, printing out the average temperature every 1 second, then repeat.

Printing numbers is nice, but it would be helpful to be able to visualize them. The Arduino environment provides the *Serial Plotter* tool as an alternative to the *Serial Terminal*. Like the terminal, the plotter is accessed from the *Tools* menu in the Arduino application. If your program simply prints one temperature value per line then it is already set up to use the plotter.

Use the *Serial Plotter* function of to **plot** the temperature as a function of time for at least one complete cycle and **explain** the shape.

The *Serial Plotter* can plot more than one line on the same graph. It will take several values, separated by commas and ending with a new line, and plot each on a separate line. Because the graph has only vertical scale you may have to scale your values to get them to all plot readably. **Modify** your program to plot both the ADC value and the temperature. **Append** the new program to the lab writeup.

Bang-Bang Control

Now that we can both measure temperature and generate heat, we are ready to control the temperature. The simplest control system has been in use since at least 1830 when Andrew Ure patented the bi-metallic thermostat. The algorithm is simple, if the temperature is too low, turn the heater on. If the temperature is too high, turn it off. Such a system, that is either fully on or fully off is called a Bang-Bang Control system, because it bangs back and forth between two states.

We think of any control system as a device to minimize the difference between a desired value and the current value. We call that difference the **error** of the system. The control system adjusts the value of its output, the **response** of the system, with the goal of minimizing the error. The bang-bang controller is simply the result of making the response into a binary output. Control systems were first studied by James Clerk Maxwell (of Maxwell's Equations fame) in the late 19th century as a way to improve speed control for steam engines. The field developed over the next half century to meet the needs of improving industry and for military uses, such as guiding torpedoes.

Task 4

Implement the bang-bang control algorithm to regulate the temperature of your heater/thermistor system at 60°C, about as hot as you can safely touch. You should make the set point (the 60°C value) into a variable both for clarity of programming and to allow the value to be changed later. It will suffice to print the average temperature and to turn the heater on or off about once per second. **Run** the program and **plot** the temperature as a function of time. **Discuss** the **shape** of your plot and the accuracy/constancy of the final temperature.

Proportional Control

While a house furnace can only be turned on or off, and should not be switched too often, our heater coil can dissipate power at any of a wide range of rates. The power dissipated is $I \times R$, so varying the current will vary the power. While our switch allows only full current or no current to flow, we can drive the switch from a PWM signal to generate a wide range of different heating powers. The instantaneous current is still either on or off, but the average heating will depend on the PWM value. This allows us to be more subtle in our control.

The next level of control above bang-bang is called **proportional control**. Instead of having the response be a binary value, we allow the response to take on a range of values and we make the size of the response proportional to the size of the error, hence the name. We must also make sure that we get the sign of the response correct. We need an *increase* in temperature to result in a *decrease* in heating. That means that we need *negative* feedback. The system has an extra degree of freedom because we can choose the proportionality constant, which we call the **gain** of the system. The larger the gain of the system, the less error it will tolerate. However, if we make the gain too large, then the system will become unstable, and instead of regulating, it will tend to create wild swings of temperature.

The power switch is already wired to a pin that supports PWM so you just need to switch from turning the pin on and off with digitalWrite to setting a value with `analogWrite`. Here is a suggested piece of code to incorporate into your program. It assumes that the set point is called `TSet` and that the most recent temperature reading is in variable `CurrentTemp`.

```
float error = Tset - CurrentTemp;
int response = int(Gain * error);
if (response > 255) response = 255;
if (response < 1) response = 0;
analogWrite(9,response);
```

So, if the set point is above the current temperature, we get a positive error and so will apply current to the heater. The further the temperature is from the set point, the more current we will supply. Once the temperature rises above the set point, we will get a negative response, which we interpret as an instruction to turn the heater off. I suggest an initial gain of about 50.

Task 5

Convert your thermostat program to use proportional control instead of bang-bang control. Again, you should compute the average temperature and update the response about once per second. Try at least two values of the gain. In each case, **plot** the temperature as a function of time and **discuss** the quality of the regulation.

Task 6

Add to your program the ability to modify the set point temperature using commands from the serial port. You may either use single letters to increase or decrease the set point or allow the user to enter a number to set the new temperature. Note that you cannot do this while using the *Serial Plotter* so you will have to rely on the list of numbers to see the variation in temperature.

Demonstrate your control over the set point and **append** the code to the writeup.

Proportional-Integral Control

Proportional control suffers from a simple problem. It cannot regulate the temperature *at* the set point. In order for there to be a response, there must be an error. The solution is to add a term that depends on the integral of the error. When the system is far from equilibrium this term will be large and will change rapidly, driving the system back to where it belongs. When we reach equilibrium, the term will stop changing but will still have a value large enough to provide the response required. Since the response depends on both the current error and the integrated error, we call this **proportional-integral** or **PI** control.

We can add an integral term to our controller quite simply. We need to add a global variable, I shall call it IntError, to build up the integral. We put this at the top of our program, before setup() and loop().

```
float IntError = 0.0;
```

Then, we alter the controller code in loop() to become

```
float error = Tset - CurrentTemp;
IntError += error;        //Note can increase or decrease!
int response = int(PGain * error + IGain*IntError);
if (response > 255) response = 255;
if (response < 1) response = 0;
analogWrite(9,response);
```

Note that we now have two gain variables, PGain and IGain. These should also be globals, and reasonable initial values are 20.0 and 0.1 respectively.

Task 7

Add the integral term to your controller and experiment with gain values. **Compare** the accuracy of your PI controller with the accuracy of the simpler controllers using graphs of the temperature as a function of time.

Lab 8D Arduino Motor Control

Learning goals: Understand basic principle of high current switching with computers. Control brushed DC motors and stepper motors from an Arduino.

Apparatus: Arduino, solderless breadboard, small DC motor, 5 V bipolar stepper motor, small power FET, 1N5181 diode, 0.01 μF 50 V capacitor, L293D, DRV8825 stepper-driver board, 5 V and 12 V or greater 1–2A motor power supplies.

Pre-lab reading: Read the tutorial on Stepper Motors at learn.adafruit.com/all-about-stepper-motors.

Moving Onward

One common laboratory task for a micro-controller is moving things around. For example, optical spectrometers typically rotate gratings (or, more rarely, prisms) to allow different wavelengths of light to pass through. In this lab, we will explore several different kinds of motor and learn to control them from our Arduino.

Brushed DC Motor

This is the simplest kind of motor, the kind that you learned about in first year E&M and that you find in everything from the vibration buzzers in cell phones to battery powered drills and saws. A DC voltage drives current through coils that are located in a strong magnetic field and the resulting $I\mathbf{L}\times\mathbf{B}$ force rotates the motor shaft. The more voltage you apply to the motor, the more current you push through the coils and the faster it will turn.

No useful motor will run from the 20 mA current available from an output from our Arduino, so we will need some kind of interface between the computer and the motor. If we are content to drive the motor in a single direction, then we can use a single **transistor** to turn the motor on and off.

The Transistor

A transistor, whose name is a conflation of "**trans**fer res**istor**," is the component that does almost all the real work in all of our digital and analog circuits. It is a three-terminal device (i.e., it has three leads) and a voltage applied between two of the leads controls the current that flows between a different pair of leads. The symbol (Figure 8.46) is designed to illustrate this. The **drain** and **source** leads are connected by a heavy vertical bar (the **channel**) showing that current can flow between them. The **gate** is shown isolated from the channel. No DC current flows into or out of the gate but the voltage between the gate and the source controls the current flowing between the drain and the source.

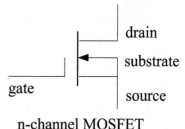

FIGURE 8.46 nFET.

When the voltage between gate and source (the **gate-source** voltage, V_{GS}) is less than a minimum called the **threshold** voltage NO current can flow from drain to source. We say that the transistor is **off**.

When the gate-source voltage is well above the threshold, the channel behaves as a low value resistor (typically <1 ohm) and a large current can flow from drain to source. We say that the transistor is **on**.

In the intermediate range, the drain-source resistance falls as the gate voltage rises and we are in a region where we can make amplifiers, like the op-amps that you met in an earlier lab.

The Transistor as Switch

An nFET with a low threshold voltage makes an extremely good switch. With its source connected to ground, the FET can be turned on or off by a digital signal from the Arduino, as shown in Figure 8.47.

Since no DC current flows in the gate lead, we can control a large motor current with no current draw from our Arduino, exactly what we want for our motor. There are two remaining issues.

First, DC motors are electrically extremely noisy. As the motor spins the current through the coil is interrupted and reversed hundreds of times per second and the rapidly changing currents radiate electrical noise strongly. We deal with this by using a separate power supply for the motor from the one powering the Arduino and by putting a capacitor across the leads of the motor.

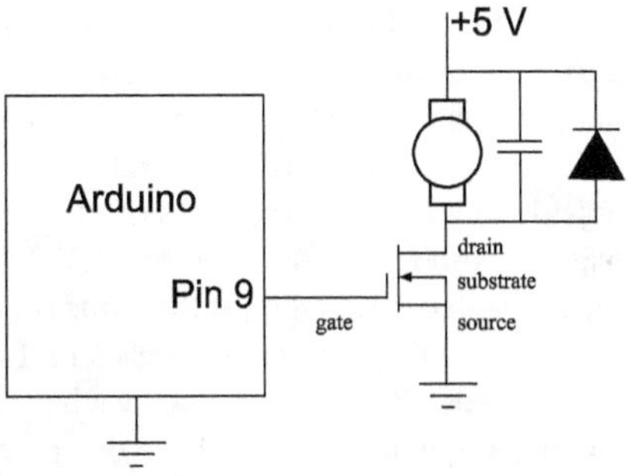

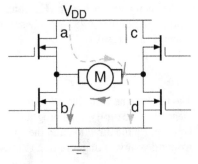

FIGURE 8.47 The new symbol is the motor.

Second, because there is energy stored in the magnetic field of the motor and in the kinetic energy of its rotation it can be hard to stop a motor. When we tell the transistor to turn off the motor resists the change and generates a high voltage that can damage the FET. We deal with this by putting a diode in parallel with the motor. Under normal circumstances the diode is turned off but if the motor tries to send the voltage on the FET above the power supply voltage then the diode turns on and shorts the current to the power supply. This is called a **free-wheeling diode** because it disconnects the FET from the motor in a similar way to the free-wheel gear on a bicycle that allows the pedals to drive the wheels but does not allow the wheels to drive the pedals.

Task 1

Build the circuit of Figure 8.47 and demonstrate that you can turn your motor on and off under control of the Arduino. Use a 1N5181 Schottky diode and a 0.1 μF capacitor. Make the motor run for 2 seconds and stop for 2 seconds then repeat. Append your program to the lab.

Speed Control

With the LED we saw that the integrating effect of our eyes allows us to use pulse-width modulation to vary the brightness of the LED over a continuous range. The inertia of the DC motor provides the same sort of integration so we can again use pulse-width modulation to control the motor speed. Friction means that there is some minimum power that we can apply to the motor and still have it turn but we can still vary the speed from off to full speed.

Task 2

Alter your motor control program to vary the speed of the motor as it runs. Try to have it speed up from rest to full speed over a 2 second period, slow to rest again over a further 2 second period, then rest for 1 second before repeating. You will have to experiment to find the lowest mark-space ratio that will actually move the motor. Append your program to the lab.

Reverse?

Alas no. Not with our simple switch. Reversing a motor requires a significantly more complicated switch than our simple transistor. At minimum, we need a four-transistor **H-bridge** switch to provide direction control. The principle is illustrated in Figure 8.48.

If we turn on transistors a and d but turn off b and c then current will follow the short-dashed path from V_{DD} to ground. The current will flow from left-to-right through the motor and it will turn in the clockwise direction.

FIGURE 8.48 H-bridge.

If we turn off a and d and then turn on b and c, current will follow the long-dashed path. In this case, current still flows from V_{DD} to ground, but now it passes through the motor from right-to-left and the motor turns in the counterclockwise direction.

If we turn a and c off and b and d on or vice versa, then we place a short across the motor and it stops quickly. If we turn all the FETs off, then the motor coasts to a stop and we had probably better have free-wheeling diodes across all of the FETs (as we did with in the first part) to protect them. We must NEVER turn a and b or c and d on at the same time. If we do then all the current that the power supply can generate will flow through the two series FETs, almost certainly melting and destroying them.

At the cost of four transistors, we have built a circuit that drives current in either direction through the motor even though we only have one polarity of power supply. The problem is that it is now hard to turn on the upper two transistors, a and c. Their sources are *not* at ground and so the output voltage from the Arduino is unlikely to be great enough to turn them on. Fortunately, there are a lot of complete H-bridge systems available in single chips with all of the gate drive circuits built-in. Most such ICs also have circuitry to make it impossible to turn on the FETs in one of the forbidden modes.

The L293

One such common bridge chip is the L293D (Figure 8.49) made by Texas Instruments. This contains two full H-bridge circuits, with free-wheeling diodes built-in. Each bridge can switch up to 1A at up to 36 V with logic inputs as low as 0–3.3 V, though it will need significant heat-sinking for high power loads.

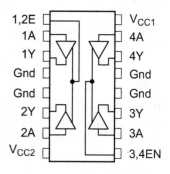

The four ground pins near the center of the chip not only carry the full motor current but also conduct heat out of the chip. For moderate loads they may provide enough heat-sinking for the whole chip. They work best when soldered into a printed circuit board.

There are two separate power supplies. V_{CC1} supplies the logic part of the chip and is usually connected to the 5 V power from the computer. V_{CC2} is the power supply for the motor. It can be as large as 36 V and must be able to supply the current for whatever loads we are switching.

FIGURE 8.49 L293D.

Task 3

Wire up your motor and L293 to the Arduino as shown in Figure 8.50. There are now three wires going from the Arduino to the motor controller, Enable (En), A1, and A2. For the moment, Enable should be set HIGH and the state of the motor controlled from the other two inputs. Experiment with the four possible settings of A1 and A2 and **write a table** showing the behavior of the motor for each case. **Explain**, with the aid of figures like Figure 8.48, how these four inputs states produce these results.

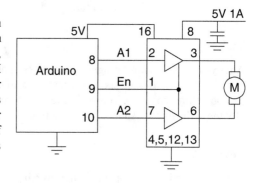

FIGURE 8.50 Driving a DC motor with an L293D.

Task 4

The Enable line turns off all four switches in the bridge that it controls. We can use this line to implement PWM. **Modify** your program to do PWM (analogWrite) on pin 9 and **demonstrate** that you can now control the direction and the speed of your motor using commands sent over the serial connection. **Append** your working program to the lab writeup.

Stepper Motors

The brushed DC motor is the simplest of motors, and one of the most efficient. If you want to make something move rapidly without too much control then a brushed DC motor is good choice. Stepper motors are far less efficient – you get significantly less torque for the same current – but in exchange they offer precise positioning. A stepper motor has a fixed number of stable positions (commonly 200), when current is flowing, and it can be shifted from one position to the next. This makes a stepper perfect when you want to position something. For example, stepper motors provide the precise motions that make 3D printers possible. The cost for this positioning accuracy is some complexity in drive. In addition to the precision, a stepper motor has a longer working lifetime than a bushed DC motor because there is no commutator to wear out; the coils are fixed and the magnets are on the rotor.

A stepper motor has two coils that need to be turned on and off in sequence and in both directions. This means that we need two H-bridges to drive a single stepper motor so that a single L293D makes a good simple stepper-driver, so long as the stepper motor is well matched to the power supply as there is no easy way to control the coil current.

Task 5

Replace the DC motor with a stepper motor and wire it into the circuit of Figure 8.51. There are several sequences of inputs that will drive the motor. The simplest turns one coil on at a time following the sequence of Table 8.25 below and putting a short delay between each step. Write a program to generate such a sequence and verify that you can drive the stepper motor in either direction. **Experiment** with the length of the delay to find the maximum speed at which your motor will turn. **Give** the maximum speed in rotations per second and **describe** how you measured that speed.

Note: You need to make sure that you connect the four wires going to the stepper as shown, one coil to pins 3 and 6 and the other to pins 11 and 14. You can check the resistance with a multi-meter to make sure you get it right.

Table 8.25 shows the normal stepping sequence. If you work through the rows from top to bottom the motor will turn in one direction. If you work through the rows from bottom to top then it will turn in the opposite direction.

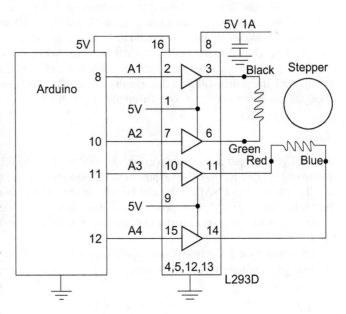

FIGURE 8.51 Driving a stepper with an L293D.

TABLE 8.25

Normal stepper sequence

Step	Coil A	Coil B
1	+	+
2	+	−
3	−	−
4	−	+
1	+	+

The entries in the Coil columns show the direction of current flow in each coil. You need to translate this into values of A1, A2, B1, and B3 to send to the motor.

A Better Way

A simple dual H-bridge like the L293D is a very versatile component but it is not the most convenient way to drive a stepper motor. First, there is no simple way to control the current in the motor so the power supply must be carefully matched to the motor. Second, a change in stepping pattern requires reprogramming the computer. The popularity of 3D printers and small stepper-driven machine tools such as laser engravers has resulted in a flood of easy-to-use stepper drivers appearing on the market. These interface to the computer through two wires, **Step** and **Direction**, and incorporate current limiting for the motor as well as a choice of stepping pattern. In addition to the standard pattern these offer patterns that further divide each step into microsteps, as many as 32 microsteps per fullstep being common.

Figure 8.52 shows the layout of the popular DRV8825 stepper-driver board, which is pin compatible with several other such driver boards. The actual driver chip is hidden under the silvery heat-sink visible in the bottom half of the board. With the heat-sink, the chip can drive more than 1A per coil from motor supply voltages as high as 48 V. If you add a cooling fan then the current can reach about twice that. Like the L293D, the DRV8825 operates with two separate power domains. The logic signals and the ground on the lower right lie in one domain that operates on 0–3.3 V or 2–5 V. The outputs (A1, A2, B1, B2, sorry; the two chips use the same names for different functions) and the top right Vmotor and Gnd pins form the second domain that can use up to 48 V. As shown, the pin names are actually printed on the bottom of the chip, but that is not very much use when wiring the thing!

The little silvery circular component just below the T of "Top View" is a variable resistor that sets the maximum current that will flow through the motor. The chip uses PWM to limit this current independently of the power supply voltage. This makes it much more flexible.

Task 6

Rebuild your stepper motor system using your DRV8825 driver following the circuit of Figure 8.53. We have to tie the RESET and SLEEP pins high to allow the chip to operate and we are going to ignore the FAULT output. The ENABLE pin can be left unconnected because it has a pull-down resistor on the board. Use the same 5 V motor power supply that you used in Task 5.

Rewrite your program to verify that you can spin the motor in both directions. Is the top speed still the same?

Now **replace** the 5 V motor power supply with a 12 V or higher supply and remeasure the top speed. **Discuss** your findings in your lab writeup.

Task 7

We have so far ignored the pins M0–M2. These pins control the stepping sequence used by the motor. **Tape** a small flag onto the output shaft of your stepper to monitor its position accurately. **Demonstrate** that it takes 200 steps to produce one 360° rotation of the shaft.

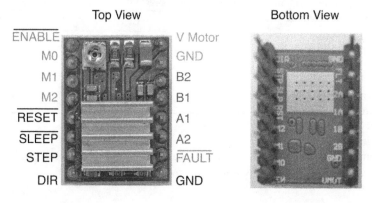

FIGURE 8.52   DRV8825 stepper-driver.

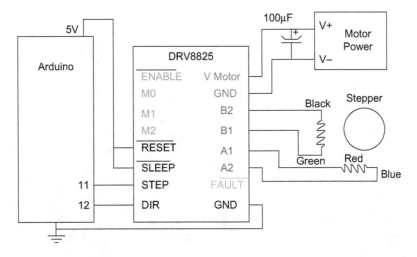

FIGURE 8.53   Driving stepper w/DRV8825.

Now **connect** pin M0 to the 5 V logic supply from the Arduino. **How many steps** does it now take to complete one 360° rotation?

Explore the other possible settings for the three M inputs, which have pull-down resistors built-in so that they are at logic 0 unless pulled high externally. **Include** a table describing your findings in your writeup.

Lab 8E Field Programmable Gate Arrays (FPGAs)

This lab is available on ExpPhys.com

Brian Collett is the Winslow Professor of Modern Science in the Physics department at Hamilton College. He traces his interest in electronics back to the radios that he took to pieces as a small child. Gradually, he learned to build instead of destroying; earning a PhD in physics along the way, and now creates apparatus and computer programs both for teaching and for neutron physics research. He has taught electronics for the past 40 years and his creations for research range from part of a 2D X-ray camera for biological research, through sensors for underwater robots and a midi-telephone for music installation, to multi-channel computer-controlled power supplies for nuclear physics research. When not engrossed in physics, he is likely to be involved in classical music, cooking, or trying to keep up with a 200-year-old farmhouse (Figure 8.54).

FIGURE 8.54 Brian Collett as Bunthorne in Gilbert and Sullivan's "Patience."

Using Field Reprogrammable Gate Arrays (FPGAs)

9

Data Acquisition and Experiment Control with Python

Paul Freeman and Jami Shepherd*

CONTENTS

I think as automation gets even more and more prevalent, we're going to need to learn how to code. Everybody does.

            – Reshma Saujani (founder of *Girls Who Code*)

Besides black art, there is only automation and mechanization.

            – Federico Garcia Lorca (Spanish poet)

In this chapter, you will learn how software is used to control data acquisition in the laboratory and how you can automate experimental tasks. To support this, the chapter includes an overview of programming best practices and hands-on Python exercises. You will need to refresh your programming skills

---

* Previously published as Jami L Johnson.

and implement Python programs in a variety of ways to complete the laboratory tasks. The chapter uses waveform generation and analysis of sound waves as an example problem. Overall, this will give you a first look at how to automate all aspects of an experiment, including experimental planning, experimental setup, data acquisition, and data analysis.

Learning Goals

- Learn how software tools are used in experimental research
- Develop Python skills for laboratory automation and basic signal processing
- Learn how to systematically design an automated experiment
- Critically assess hardware specifications
- Complete a fully automated audio experiment: generation, detection, and analysis
- Advanced: integrate new hardware into an automation framework

9.1 Overview

For many experiments, data acquisition is a simple process: drop a ball and record the time it takes to fall; apply heat to a solution and record the change in temperature; measure the height of a growing plant. In these cases, data acquisition is a straightforward part of the scientific process.

While many experiments produce simple data, many others produce results which are orders of magnitude more complex: track the motion of the stars in the night sky; measure bacteria populations in a petri dish; monitor diseased tree populations in a large forest. Experiments such as these can be performed more easily, quickly, and reliably by harnessing technology. In many cases, automating the scientific process with computers can enable experiments that were previously impossible.

This chapter explores the use of software for conducting experiments and automating laboratories. We will use a simple example to demonstrate several levels of automation while teaching skills that transfer to elaborate, real-world, laboratory setups.

9.1.1 Automation Technologies

Computer software (along with computer-controlled hardware) is commonly used for automating experiments. The level of automation can range from minimal (e.g., turning equipment on and off) to fully autonomous (e.g., start/stop, perform experiment, collect results, provide analysis). For highly specialized experiments, like the Large Hadron Collider, all the software and hardware needed to acquire data is custom developed as part of the experiment. But not all technology is so specialized. For a majority of experiments, a number of technological solutions are available and customizable to assist in streamlining the data acquisition process.

Programming Languages

Computer programming languages, such as C, Java, and Python, have long been used to automate repetitive calculations and simple software tasks. Over time, computer-controlled hardware emerged. Now, many instruments that were once operated manually in laboratories can be controlled via one (or several) of the popular computer programming languages.

LabVIEW

Although technically a programming language itself, one of the most common commercial data acquisition platforms is LabVIEW. LabVIEW was developed by National Instruments, released in 1986, and has had many major updates since. LabVIEW is a graphical programming language which creates abstractions of hardware devices as VIs (virtual instruments). Each instrument and programming function is then represented to the user as a block, each with inputs and outputs. Users construct experiments by connecting outputs from one block to inputs of others. The visual nature of LabVIEW makes it more intuitive than other programming languages. It is also well supported by hardware manufacturers. Together, these attributes have led to the popularity of LabVIEW within the scientific community.

Dash

An important challenge of data acquisition is creating a user interface (UI) which neatly presents all the controls of an experiment to the user. LabVIEW has excelled because it has succeeded in making this process intuitive to users.

Dash is an effort to bring modern experimental controls into the open source world. It is built on top of Python and using the Dash libraries users can *lift** a Python application into a web application. Inputs and outputs to the Python code are then presented to the user as familiar interfaces, such as dropdown menus and graphs.

> *Regardless of whether you use Dash, I think that Python-based approaches to data acquisition will continue to gain market share over GUIs such as LabView, because Python is free, increasingly popular in the sciences, and easy to learn.*

– Jack Parmer[†]

The emergence of technologies like Dash indicate that the days of data scientists staring at text on terminal screens are numbered. Being able to initiate, control, and review experiments from mobile devices is now a minimum standard for data acquisition and data automation technologies.

PLACE

PLACE is short for Python Laboratory Automation, Control and Experimentation, and is an effort to make an open source automation tool that can be used by people without advanced programming experience (Figure 9.1).

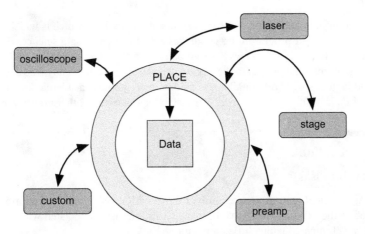

FIGURE 9.1 PLACE controls the instruments and coordinates access to the resulting experimental data. The timing and execution of commands for each instrument is performed by PLACE and instruments must provide PLACE with any result they want to have written into the dataset. This helps ensure consistent and reliable data for each experiment.

---

* *Lifting* is an operation where you map inputs and outputs to another system.
† www.quora.com/Are-there-alternatives-to-LabView

PLACE emphasizes data acquisition through the use of a shared, NumPy-based, data record array, into which all instruments and modules record data. The result is a single (but possibly very large) array of data containing all the information for a given experiment. The PLACE framework controls each instrument's ability to access the data record array.

A *PLACE instrument* is any piece of hardware automated by PLACE. Each instrument requires its own PLACE plugins, but these are easily created from existing Python scripts. For example, if you have a Python-controlled temperature sensor, you can easily adapt this code into a PLACE plugin and introduce automated temperature sensing into any of your future experiments.

Along with automation support, PLACE provides a graphical dashboard through which you can initiate, monitor, and retrieve all your experiments and data. The dashboard is also web-based, meaning you can track your experiments remotely on your laptop, tablet, or mobile devices.

The experiment in this chapter includes an advanced section pertaining to PLACE. In this section, you are given the opportunity to adapt Python scripts into a PLACE module and use it to perform an automated experiment.

9.1.2 What This Chapter Is Really About

As you work through this chapter, you will likely encounter tasks that are new and/or unfamiliar. Some of these technologies and concepts are challenging if your exposure to them in the past is limited. If you have never written a computer program, some of the tasks in this chapter may be difficult. However, the topics presented in this chapter are intended to provide valuable exposure to the types of data acquisition technology frequently found in laboratories across the globe, and we aim to make the chapter accessible to anyone willing to put in the effort. Unlike other chapters in this book, this chapter focuses on how software is used to enhance the scientific process. In doing so, it is unavoidable that many computer science topics must be discussed. Many of these topics cannot be covered fully within the limited text of this one book.

For students entirely new to Python, Section 9.3 will provide introductory material and exercises that will be useful for getting up-to-speed with the required skills for the automation experiments. However, some topics may require outside reading. Wherever possible, suggested resources have been provided. Nonetheless, interested readers should endeavor to fill in gaps in their knowledge with the substantial resources available online.

Why Python?

The primary programming language covered in this chapter is Python, which is widely used in industry and universities. Many other programming languages are also used in data science. Examples include: R, LabView, MATLAB®, and Java. So why is Python used in this chapter? Along with being one of the most popular programming languages in the world,* Python is also completely free and easy for anyone to download and begin using. Additionally, it is an extremely popular academic language. Any knowledge acquired while learning Python is readily transferable to learning other programming languages.

9.2 Safety Precautions

9.2.1 Automation Risks

Turning over experimental control to automation carries with it risks that are important to understand and consider before the experiment begins. In many cases, the risks associated with automation can be minimized or even eliminated with a little preplanning.

The specific experiments presented in this chapter do not present a large safety risk. The LabJack hardware which you'll use to interface your experiment to a computer is unlikely to be dangerous in the event of a failure or if left unattended. Nonetheless, standard electrical safety precautions should be considered when using LabJack voltage outputs.

---

* www.tiobe.com/tiobe-index/

Including Safety Precautions Within Programs

When you are writing programs to control laboratory hardware, you should read the specification of the hardware to determine the valid input/output ranges for various values. In most cases, the hardware will be smart enough to prevent you from setting invalid values, but it is better not to test this. Instead, program the limits into your software so that users cannot accidentally start a program with illegal values.

How to Stop a Running Experiment?

In some cases, it may be necessary to stop a running experiment. Most hardware and software have abort methods. Hardware devices can often be stopped by simply turning them off. Software is usually stopped by performing a keyboard command or closing the window running the application.

For the LabJack, you can stop the device by unplugging the power cable. Usually, the USB cable will be supplying the power to the LabJack, but this may not be the case in all laboratory setups.

To stop the execution of software, it is often necessary to hold down the Ctrl key and press "C," which will send a kill signal to the running program. In some cases, programs will ignore this, so closing the window of the executing program will also typically work.

9.3 Python: An Introduction and Primer

This lab will require installation of several software packages and drivers. To help ensure they are up to date, installation instructions and tips for installation can be found in the section for this chapter at the book's website, ExpPhys.com.

9.3.1 Programming Best Practices

Writing computer programs correctly, the first time, is a challenge – even for programming experts. This chapter is not going to teach you to be a programmer. This chapter will, hopefully, open a window into automation software and how it is used within scientific communities and provide hands-on experience to get you started.

Here are a few tips and tricks for developing better code.

Comments

Comments are sections of a program which are ignored during execution. It is a bit like writing notes in the margin of a book – it allows the programmer to make notes within the code to help clarify the intent.

Each language has its own way of denoting a comment. In Python, a # is used to begin a comment. All characters after the # symbol are ignored.

Comments are used by programmers for a variety of reasons, and programmers do not often agree on a single "correct" way to use comments. However, almost all agree on the following guidelines:

Comments should clarify.

Generally speaking, a comment should add information that cannot be communicated in the code itself.

```
if x > 10: # 10 is the maximum possible
    raise ValueError("x value out of range")
```

This comment clearly explains why 10 is being used. This is a good use of comments. It would also be possible to document this without needing to include a comment, by defining a well-named constant:

```
maximum_possible_x = 10
if x > maximum_possible_x:
    raise ValueError("x value out of range")
```

Comments should not duplicate the code.
It is not useful to have a comment that restates what your code is doing.

```
# calculate x
x = 10 * 10
# print x
print(x)
```

As you write code, you should use comments in a way that makes sense to you. When working on a team, reaching an agreement on how comments will be used becomes an important part of the collaboration process.

Variables and Immutability

When you record values into a laboratory notebook, you might write "x = 144" at one point and then later write something like "a = 2 * pi * x" somewhere else. This makes sense. However, it would be confusing if you had written "x = 12" somewhere else on the page because now it is not immediately clear which "x" we need to use to compute "a." And yet, this is a common habit for beginner programmers.

In computer science, the notion of *variables* seems to have brought with it the idea that variables can (and should) be changed. However, after decades of confusion, many programming languages are embracing the idea that, in most cases, the value of a variable should not be changed.

As you write your code, try to avoid overwriting existing variables with new values. Consider this example:

```
v = 7
r = 0.8
d = 4
t = 10
x = v**2 / r
print (x)
if x < 100:
    x = d * t
    print(x)
r = x * 2
```

What is the value of r at the end of this code? What is the value of x? The answer to these questions is not immediately obvious. Instead, we could have written the same code like this:

```
v = 7
r = 0.8
d = 4
t = 10
x = v**2 / r
print (x)
if x < 100:
    x_prime = d * t
    print(x_prime)
    r_prime = x_prime * 2
else:
    r_prime = x * 2
```

There are two important changes in this version of the code. The first major change is that none of the variable values are ever reassigned. They are assigned once and then left alone. This makes it easy to reason about them when you come back to edit the code. *What is the value of* r? This question is easy now. It's still 0.8. It hasn't changed!

The code also clarifies that it computes a new value, r_prime, and the value is the result of one of the two outcomes of the if statement. Previously, the reuse of the variable x obscured two different outcomes.

Values in programming that cannot be changed are called *immutable*. Python does not support immutability in this way,* meaning you can change any value you like. Many other programming languages have started including immutable values by default, with some functional programming languages using immutable values exclusively.

So, while Python still allows you to write code as you like, you may find that thinking of variables as immutable leads to code that is easier to read and understand.

Functions

Functions serve at least two purposes:

1. Write code once and use it many times.
2. Break up code and put a label on it.

Skilled programmers know when to use functions to improve their code. Writing functions forces you to think about what your code is doing, so it can be sorted into logical blocks. As you learn to code, practice putting sections of your code into functions.

Test Often

If you write 100 lines of code, run it, and get an error, then one of those 100 lines of code (at least) has a bug in it. That's a lot of code to look through. But if you test after writing only five lines of code, then if you get an error the search will be much easier.

As soon as you write a line of code and are unsure of exactly what it does, you should be thinking about how soon you can test it. You might need to write a few lines of code to finish up the current section, but you should be looking for that opportunity as soon as it presents itself.

Code Style

Each programming language tends to have its own style. Sometimes style is hotly debated, and sometimes it is very well defined. In any case, how should you decide what style to use?

Most languages will publish official style guides. Python uses a document called PEP8 to encourage consistent style. There are various tools to help enforce the PEP8 guidelines onto your code, but even just reading through it quickly is a good start. Despite the code style you might prefer, if you work on someone else's code, it's generally polite to conform to their style.

9.3.2 Self-Guided Python Tutorial

Even if you are already familiar with Python, please follow the procedures below for installing the particular Python environment we assume.

Installation and Module 01A

1. Go to ExpPhys.com; under the section "Installing Required Software," follow the instructions for how to install Python on your computer.

If you're already familiar with Python programming, you can skip ahead to the warm-up experiment in Section 9.4.

---

* Technically, Python *does* support immutability, but this is mostly hidden from the user.

2. Go to ExpPhys.com; under this section and step, follow the instructions for how to download the first of a series of Python tutorials.

3. Work through this notebook. Even if you have previous experience in programming, you should skim through the reading, but do all the exercises. If you don't have previous experience, you should carefully read and work through all the exercises. Remember that we're using Python 3, so ignore everything relating to Python 2, which will soon be obsolete.

4. Note that for all the exercises except the first one, you need to insert a new cell for Python code. Use the menu near the top of the browser page to do this (e.g., Insert/Insert Cell Below). Also, at one point you have to figure out on your own the formatting command needed to print an integer.

If you've done all the exercises, you should be able to answer the following questions easily:

Exercise 1 Concept test: *Are you more than half a billion seconds old?**

Exercise 2 Concept test: *What's the fourth digit after the decimal point when the fine structure constant is raised to the fifth power, with the result expressed in scientific notation?*†

Exercise 3 Concept test: *Somewhere in your code, you had to type-cast the user input. Why?*‡

Exercise 4 Concept test: *Write code that prints each element in sequence of a list of any length.*§ *(Make sure your code doesn't rely on knowing the length of the list.)*

Exercise 5 Concept test: *If you pronounce phonetically the word you get from the letters you're supposed to select out of "supercalifragilistic," does it sound more like "pee shall tee," "suffragette," or "Ural glee?"*¶

Exercise 6 Concept test: *Modify your code so it picks out the letters divisible by 6. What animal do you get for "supercalifragilistic?"***

You've accomplished something significant! Do something to celebrate!

Module 01B

1. Go to ExpPhys.com; under this section and step, follow the instructions there for how to download module 01B.

2. Start working through the module. About a quarter of the way through, you'll encounter the line:

```
num = len(mylist)   # gets length of input list
```

The text following the # is a comment. That text is ignored by Python. It makes the code easier for humans to understand. Adding comments helps you to clarify your code, and also makes it much easier for you to understand, either for yourself when you come back to it after several days, or for someone else who's never met you. We discussed code comments in a prior section.

3. Work through the module up to Exercise 8. Remember for this exercise that you need to have previously loaded the numpy package (e.g., by executing "import numpy as np"), and then you need to refer to it when you execute simple math functions, e.g., use np.sqrt rather

---

* Assuming you're at least 16 years old, then yes.
† 3
‡ Because the user input is returned as a string, so you have to use the int function to type-cast it to an integer before you add 10 to it. (You can do the user input, the math, and the type-casting all in one line, if you like. Do you see how?)
§ The code below assumes the list is in the variable MyList:
```
for i in range(len(MyList)):
    print(MyList[i])
```
¶ I got "pcialt," which could be pronounced "p shall tee."
** Cat.

than just `sqrt`. Also, we encourage you to use σ in your code, since it makes it slightly more readable than "sigma." To get the σ, type `\sigma` followed by the tab key.

Exercise 8 answer: 0.176…

4. In the "Making Vectors and Matrices, 1-D and 2-D Arrays" section, the module authors use the term "tuple." In Python, a tuple is similar to a list, except that it cannot be changed, and uses parentheses instead of square brackets. It is typical to use tuples (rather than arrays) as inputs to Python functions (such as `np.zeros`) when the values won't change as a result of executing the function.

Concept test for breakpoint 5: (Do the breakpoint exercise first.) Predict what will happen when you execute `print(arr[-1,-1])`, then try it.

Concept test for Exercise 9:

Create a function that takes as inputs an amplitude and a frequency, and returns:

1) an array of time values spaced apart by 1 msec, ranging from 0 up to 20 msec, and

2) an array of values equal to $A \sin \omega t$, where A is the amplitude, t is the array of time values, $\omega = 2\pi f$, and f is the frequency.

Make sure you do this as a function, not just as a piece of code. As you should do for every function you ever write, include a docstring that describes what it does, then lists and describes the inputs (A and f), then lists and describes what is returned (the time and sin arrays). Test the docstring using the help function. Also, include at least one comment in your function. Then, write a program that calls your function with $A = 1.3, f = 12$, and prints a complete sentence, reporting the time and corresponding sin value for the 12th entries of the two arrays (the time array and the sin array); arrange to have the time printed to three digits after the decimal point, and the sine value to four digits after the decimal point.[*]

5. Work through Exercise 9, then skip ahead to "Plotting with Matplotlib."

6. Work through the module until you get to "Note that, as you'd expect, the exponential function is linear in a semilog plot." After that line, skip ahead to Exercise 11.

7. Do Exercise 11; define functions for `f1` and `f2`. As part of your code, create an array called n that contains the integer values 0 through 10.

Concept test for Exercise 11: using the same array n, make the same plot, but dividing the values of n by 10, so that the x-axis ranges from 0 to 1, and the graph shows plots of x and x^2 (ranging, of course, from 0 to 1). Do not create a new array of x values. Answer below.[†]

---

[*] Answer: At t = 0.011, sine = 0.9588. (Remember that the 12th element of an array has array index 11.)

[†]
```
def f1(n):
    '''
    identity function
    inputs: n (any number)
    returns: n
    '''
    return n
def f2(n):
    '''
    square function
    inputs: n (any number)
        returns: n squared
    '''
    return n**2
n = np.arange(0,11) #create the array of integers
#Note that we need both x and y arrays in the call to the plot function.
plot(n/10,f1(n/10),'r-',label="n")
plot(n/10,f2(n/10),'b-',label="n squared")
xlabel("n")
ylabel("f(n)")
legend(loc=2)
```

8. Stop after completing Exercise 11.

Module 01C

1. Go to ExpPhys.com; under this section and step, follow the instructions there for how to download module 01C.
2. Read through the module only as far as the piece of pseudocode.

Additional Content

Go to ExpPhys.com, and work through the "additional content" for this section.

9.3.3 Working with Python Files

If you are going to start with Jupyter notebooks for your Python code, you can skip this section for now.

Tools like Jupyter notebooks make it very easy to work with Python code, as a lot of the file management aspects of writing code are hidden away in the background. However, there are some programming tasks that cannot be accomplished with a Jupyter notebook, such as using the PLACE data acquisition framework. For access to the full power of Python, you will use the following steps:

1. Write a Python file (a text file with a .py extension) using your favorite code editor.
2. Provide your Python file to the Python runtime.
3. Output prints to the display or is saved into files.

These three steps (or very similar ones) are common across a huge range of programming languages and date back for decades. A strong understanding of this process is extremely useful if you want to make programming a part of your life. Therefore, this section will quickly summarize how to perform the above steps. The instructions assume you are using Windows, but tips for Mac and Linux user are included when possible.

Writing a Python File

You will need a code editor in order to create a Python file correctly. The editors that come with Windows are not ideal for editing code, so they are not recommended. Fortunately, there are many free options available online. If you don't have your own preference, install Notepad++.*

Open your code editor. You will be able to write code into the editor as you did with the cells in a Jupyter notebook. However, in this case, you will not be able to run the code in the editor and immediately see the output. Jupyter Notebook uses an *interactive Python shell* interface, which allows you to run code on-the-fly. The standard Python runtime does not work this way.

When you have written your code, you need to save it somewhere. You will need to know the location so you can run your program in the next step. On Windows,† you can save this into your Documents folder, or make a new folder (e.g., called "Code") somewhere. It doesn't really matter where you save it, as long as you know where it is. And remember, it isn't enough to know the folder it is in, but you need to know the entire *path* to your file. So, for example, if your username is `coderX1A`, and you saved a file named `sample_code.py` into the Windows `Documents` folder, the path to your file is usually something like:

```
C:\Users\coderX1A\Documents\sample_code.py
```

---

* While Notepad++ is not available for Mac or Linux, users will find many code editors available online. Ask your friends to show you their favorite editor!
† Mac and Linux users: save your code anywhere, just make sure to remember the location!

You can usually find this information in the file properties or in the information provided by the file explorer utility. Figure 9.2 shows the typical information you can expect to find in the file properties window.

Run a Python File

You should have already set up Anaconda in a previous step. So, you will simply need to launch Anaconda Prompt,* which is a program that will let you run Python programs. When you launch it, you will see a mostly black window with some text in it, similar to Figure 9.3. This is your command prompt, which is a place where you can manually launch programs and applications. If your username is `coderX1A`, as above, the text in this window might look something like this:

```
(base) C:\Users\coderX1A>
```

This text is telling you the path (or location) you are currently viewing. You change the current directory by using the `cd` command. So, continuing with the example, to get to the directory that has our program in it, you would change to that directory by typing:

```
cd C:\Users\coderX1A\Documents
```

or, since the directory is inside the directory you are currently in, you could also just type:

```
cd Documents
```

(See additional hints on ExpPhys.com for how to navigate directories and do other common tasks in the Anaconda Prompt window.)

Now you are ready to run your program. Just type `python.exe`,† followed by the name of your program, like this:

```
python.exe sample_code.py
```

Python Output

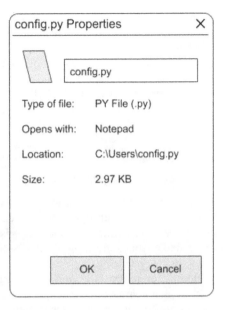

FIGURE 9.2 In Windows, you can often find the location of a file in the file properties.

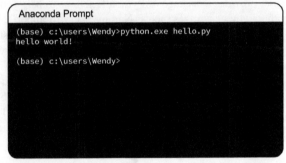

FIGURE 9.3 The Anaconda Prompt provides a terminal environment for running Python code, similar to what is available on Mac and Linux systems.

When you run your code, any of your output that is not written into a file will print into the Anaconda Prompt window. Any files you create will be saved into the location specified in your code. After the program completes, you will be able to run it again, or run another program.

It is very normal to develop your program using the following steps:

1. Edit code in your text editor.
2. Save your code.
3. Run your code in the Anaconda Prompt.
4. Examine the output.
5. Repeat steps 1–4 until the output is what you want.

---

* Mac and Linux users can use Terminal to run Python code, and it will be very similar to the Anaconda Prompt
† Only Windows users need to add `.exe` to the end. Mac and Linux user will just type `python`. In fact, Windows users will usually find they can leave off the `.exe` and it will still work.

Tools like Jupyter Notebook make it much easier to write small Python programs because they take care of all the steps shown above. Tools that handle multiple development steps are called *integrated development environments* (or IDEs). However, there are still many Python programs which cannot be created within a Jupyter Notebook, and you will be required to know how to deal with Python programs in the real world. You should practice writing Python programs in your editor and running them in Anaconda Prompt until you feel comfortable with this development cycle. Professional software developers frequently use IDEs to speed up their workflow but will also use more traditional methods when needed.

9.4 Warm-up Experiment

After ensuring you are comfortable with Python in the previous sections, we are ready to start automating an experiment with Python. This warm-up experiment serves two purposes. First, it will help to ensure that your hardware and software is installed correctly and working as intended. This is often referred to as your *software toolchain*. Second, it should get you thinking about how to solve problems using computer code.

From the exercises above, you should be comfortable creating an array containing a sinusoidal signal, calculating a Fourier transform, and plotting 1D signals in Python. The primary packages we suggest for this are NumPy or SciPy for scientific computing and matplotlib for plotting. These packages are provided as default with the Anaconda distribution, and a wealth of tutorials are available online.

You will be using a LabJack interface unit to connect your experiment to a computer. The manufacturer provides a package of Python functions to help you use it. Using this package is analogous to the other Python packages such as Numpy. You will import the LJM package (also known as a "library") and send parameters/variables to functions which complete specific tasks. The functions in the LJM library facilitate sending and receiving data to and from the LabJack. This is accomplished using *Modbus TCP* protocol. This allows values to be read and/or written from various Modbus *registers* of the LabJack. Each register is specified by a name. The LabJack website has a useful Modbus Map,[*] that details the registers available on each device. For example, STREAM_SCANRATE_HZ is a register on T4/T7 devices used to set the number of times per second that data will be streamed, i.e., it sets the sampling rate. We will use these registers in the warm-up experiment, and in greater detail in Section 9.5.

The learning goals of the warm-up experiment are:

- Validate LabJack driver installation
- Validate Python installation
- Test LabJack operation
- Read data from the LabJack
- Print data to the screen using Python
- Plot data using Python
- Save data using Python

As a simple experiment, the temperature sensor built into the LabJack can be used as a data source. Alternatively, please feel free to use one of the analog voltage inputs on the LabJack if you feel comfortable doing so, connecting it to a signal generator or other signal source.

9.4.1 Materials

- LabJack (T4 or T7)
- Windows or Linux computer with a USB port and the following software

---

[*] The link to the Modbus Map is provided on ExpPhys.com.

- Jupyter notebook or Anaconda Python (running Python 3.4+)
- LabJack driver
- LabJack Python libraries*

See instructions at ExpPhys.com for installing the driver and libraries.

9.4.2 Complete Warm-Up Experiment

For this warm-up experiment, you will write a program that reads the temperature (from the sensor on the LabJack) once every second, over a period of 1 minute, plots the temperature as a function of time, and saves the data to a file.† You should decide whether it is better to plot the measurements on the screen when they are taken or plot them all at the end. *What are the advantages/disadvantages of plotting the data as the measurements are taken? What are the advantages/disadvantages of only plotting the data at the end?* (Some answers below.‡)

Your program should follow the good programming practices discussed in this chapter and also safely stop the program and close connections to your devices. Also, it is a good idea to include tests that make sure you are not violating the constraints of your system (e.g., voltage limitations, sampling rate).

The first step to automating an experiment is making a plan. One way to do this is to first write an outline in *pseudocode*. Using pseudocode means writing an outline of the control flow and variables your program will need, without worrying about the specific functions and syntax used in the final program. This outline should detail every step in the automation process.

Since you may be new to coding, we provide the pseudocode below. However, *before you look below, make your best effort right now at writing the pseudocode yourself.*§

Next, you need to convert the pseudocode into actual Python. Below, we provide a few pieces of advice and snippets of code to get you started. Many of the code examples provided can be copied directly into your program.

Create a File

The first task is to create a new Python file. You can create a Jupyter Notebook or use a regular Python file in your favorite code editor.

Imports

In the tips below, we give each import command followed by details of how to use the functions in the imported package. However, in your actual program, all the imports should be at the beginning of the program, as outlined in the pseudocode.

---

* The LabJack Python library comes with example scripts that are very helpful to get started working with the LabJack.
† There is a Jupyter notebook for this section provided on ExpPhys.com.
‡ Plotting each data point as it is taken gives better feedback for the experimenter, allowing them to abort or change the experiment if things aren't working. Waiting to the end and plotting all the data at once allows the data points to be taken at the maximum possible speed.
§ Imports:

```
        Laback Python Package
        MatPlotLib
        Numpy
        Python timing functions (needed to get the 1 second interval between data points)
        Csv
Open connection to LabJack
Loop from n = 1 to 60
        Read the temperature
        Plot the new data point
        Wait until 1 s has elapsed
Write the data file
Close the LabJack connection
```

Communication with the LabJack

After installing the Labjack Python package on your computer, you can import it to your Python program:

```
from labjack import ljm
```

This package contains methods and functions that access data from the LabJack. These commands are implemented in the same way that you implemented the numpy package in the previous exercises, except rather than typing `numpy.` or `np.` in front of the command, you will type `ljm`, as demonstrated below.

Open a Connection to the LabJack

Connect to the LabJack using the `openS()` function, and save the connection ID as a variable (called "handle") to be used for future communication with the Labjack. The connection ID is saved as an integer, as are all LabJack addresses.

```
# open the first found labjack
handle = ljm.openS("ANY", "ANY", "ANY")
# type(handle) is int
```

Read the Current Temperature

The eReadName function reads a single value from the Labjack. The first input is the connection ID to the LabJack, and the second is the name of the variable to be read.

```
result = ljm.eReadName(handle, "TEMPERATURE_AIR_K")
```

The name of the variable comes from the LabJack Modbus map. In this instance, the Modbus register is TEMPERATURE_AIR_K. When this register is taken as a parameter by eReadName, the temperature is read. Later in this chapter you'll see examples in which registers are written to, rather than read from.

Storing Data

It would be a good idea to decide on a **data structure** into which all your measurements can be stored. Lists are well supported in Python and would be a good option, but arrays are preferred when working with the NumPy library. *What is the difference between an array and a list?* There are other options, too. *What other data structures might work for this data?* The data structure you choose is simply a design decision and is not inherently right or wrong.

Imports for Timing

Here we import the necessary packages for timing the acquisition of data. The time package in Python provides various time-related functions.

```
import time
```

Pause the Program for 1 Second

The `sleep` function pauses Python for a set amount of time, in seconds.

```
time.sleep(1)
```

Imports for Plotting and Saving Data

Import the packages and functions required for data visualization and manipulation.

```
import matplotlib.pyplot as plt
import numpy as np
```

Plot a List of Numbers

Below is an example of plotting a list of data. You can start with this framework to plot the temperature data.

```
squares = [1, 4, 9, 16, 25]
plt.plot(squares)
plt.show()
```

Save a List of Numbers

There are many options for saving data with Python. Below is one example of saving the data as a text file.

```
cubes = [1, 8, 27, 64, 125]
np.savetxt('cubes.txt', cubes)
```

Close Connection to the Labjack

When we are finished recording data, we need to close the connection to the Labjack using the connection ID to our device.

```
ljm.close(handle)
```

Wrap Up

When you have completed all these tasks, you should have created a basic toolchain, whereby you can acquire data from a hardware device, store the data in a data structure, display the data, and save the data.

As you move forward into more elaborate scripted experiments and automation, you will benefit from the solutions you found while completing these simple tasks. Taking time to investigate, implement, and test a hardware/software toolchain is an important part of software development, and a vital part of software automation.

9.5 Experiment

In this section, we will build upon the skills you learned in Section 9.4, but with "real" experimental hardware. With this exercise, we provide a Jupyter notebook to guide your programming process (see ExpPhys.com). Recall from the exercises in Section 9.3, that each cell in a Jupyter notebook can be formatted as a *code cell* to write Python code, a *markdown cell* in which text can be formatted, or *raw cells* supporting various output formats, including LaTeX and HTML. Markdown and code cells should be sufficient for this exercise but using LaTeX for equations always adds a nice touch. A traditional Python .py file is equally as effective for completing this experiment, and which environment you choose is a matter of preference.

The goal of this lab is to fully automate a waveform analysis experiment. You will write a program capable of generating, detecting, analyzing, and saving a temporal audio waveform. You will apply the good programming practices from previous sections in this chapter to create an automation script using Python language. Further, you will demonstrate that you can use your automated experimental system to understand several physical characteristics of waves.

In addition to the primary task, an advanced experimental activity is included. During this advanced activity, you will learn how to fully integrate your automated experiment into the PLACE framework.

9.5.1 Materials

- LabJack T7 (or T4)
- Windows, Linux, or Mac* computer with a USB port and the following software
  - Anaconda Python (running Python 3.4+) or Jupyter Notebook[†]
  - LabJack driver
  - LabJack Python libraries
- Source: speaker or transducer, e.g., PRO-SIGNAL ABT-410-RC – Transducer
- Receiver: microphone + amplifier, e.g., SparkFun Electret Microphone Breakout BOB-12758

9.5.2 Hardware Limitations

Please read Section 7.2 before going further.

Every experiment requires consideration of hardware constraints in order to account for safety hazards, ensure accurate data acquisition, and avoid damaging equipment. *Before you begin programming, write down the limitations unique to your system, including*:

Data Acquisition (LabJack)

- *Output (DAC0)*: What is the voltage range of your output digital output pin? Can you tune the amplitude resolution? What output/stream rates (i.e., how many data points per second) are supported? How do these output rates affect the frequencies of analog sine waves you can generate?
- *Input (AIN0)*: What are the maximum and minimum voltages the analog inputs can handle on the LabJack? What is the voltage resolution? Can you tune the voltage resolution? What is the maximum data rate this input can record? How does this impact the frequencies you can reliably record?

Instruments

Study the specification sheets for each of your instruments.

- *Source.* What are the specifications and limitations of your source? For example, what are the maximum and minimum voltages allowed? What frequency bandwidth is supported? Is there a minimum voltage threshold required to generate an audible signal?
- *Receiver.* What are the specifications and limitations of your receiver? What is the frequency bandwidth? What supply voltage must be provided? What are the maximum/minimum signal voltages expected? What voltages can be output from the microphone and is that within the range of the input voltage the LabJack supports?

Communication (USB)

- In this experiment, we are using USB communication to send/receive data from the computer/Python. What rates are supported? Is this limiting to this experiment?

Throughout the lab, it is important to stay within these constraints to accurately generate and record your signal and avoid damaging components.

---

* The advanced portion of this lab uses the PLACE software, which is not yet fully supported on Mac systems. It would be best to choose Windows or Linux if you intend to continue into that section.

[†] Scripts created in Jupyter Notebook will be more difficult to adapt into PLACE automation scripts in Section 9.6.

9.5.3 Experimental Setup

The analog output (DAC0) and ground should be connected to the (+) and (-) terminals of your source (e.g., speaker/transducer). You will program the DAC port to send a waveform to your source.

The analog input (AIN0) should be connected to the output of your microphone. The VS terminal of the LabJack supplies a constant 5 V, which can be used to power the microphone.

For the components listed in the Materials section, it looks something like Figure 9.4.

9.5.4 Understanding LabJack Streaming

Before you start programming, it's important to understand how the LabJack handles the tasks of sending and receiving signals. To do so, the LabJack implements *streaming*, which allows sending and receiving signals in parallel, accurately sampled in time, and at high transmission rates. We provide a general overview here, and more detailed instructions in Section 9.5.5 "Plan the Software Workflow". There is additionally a wealth of information on the LabJack website, and we provide direct links in the web resources.

To generate a waveform for this experiment, you must define a sinusoidal signal made up of discrete values equally spaced in time and send this signal to your source via an analog output port on the LabJack. At the same time, you should record the waveform detected by your microphone receiver via an analog input port. The LabJack achieves this using Stream Mode, which records data at a specified scan/sampling rate using the Modbus TCP protocol.

LabJack uses a ring buffer to hold data to be streamed using the Stream-Out mode. The length of the signal (in number of samples) must be within the buffer memory limitations of the LabJack. The maximum buffer size for the LabJack T7 is 2^{15} samples. For your program, define a maximum number of samples for your signal that is less than or equal

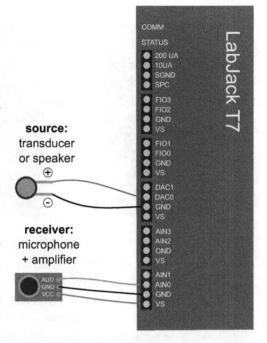

FIGURE 9.4 Example experimental setup.

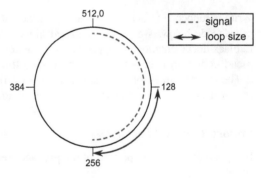

FIGURE 9.5 Example usage of the ring buffer for Labjack streaming. Note, the buffer size is indicated in number of 16-bit samples (2 bytes), therefore the 512-sample buffer corresponds to 1024 bytes.

to this value. The LabJack will continuously stream the portion of the signal contained within the last N samples defined by the loop size. For example, let's say you have defined a buffer size of 512 samples, but your signal is only 256 samples in length (Figure 9.5). Your samples will fill the first half of the buffer, and the rest of the buffer will be empty. If your loop size is set to 128 samples, the first 128 samples of your signal will be output once, and the LabJack will continuously output the last 128 samples of your signal in a loop. In this case the first 128 samples are *only output once*. However, your loop size and signal size can be the same. For example, in this situation setting the loop size to 256 will output the entire signal repeatedly. The empty 256 samples in your buffer will be ignored.

9.5.5 Plan the Software Workflow

In this experiment, we aim to achieve the following goals, as pictured in Figure 9.6.

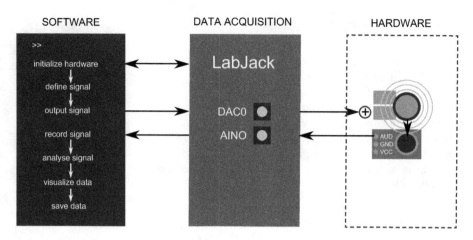

FIGURE 9.6 Workflow for software, data acquisition, and hardware.

- Create a user-defined sinusoidal audio waveform
- Output waveform via an analog output (e.g., DAC0)
- Record signal on an analog input (e.g., AIN0)
- Plot the numerical and experimental signals
- Analyze the signal by computing the Fourier transform
- Plot the numerical and experimental results
- Save the results to a CSV file

Write pseudocode to achieve the goals outlined above by filling in the missing details in the example pseudocode below. We provide detailed hints below each heading, but we recommend studying the LabJack documentation to supplement your understanding. We recommend reading through the sparse pseudocode fully once before going back and filling in the gaps.

The web resources for this chapter contain a variety of useful links to LabJack resources. You should review these resources and use them, as needed, to complete the experiment.

Import Required Packages

It is good practice to import all of the packages required for a given program at the beginning of your script.

- Matplotlib for plotting
- NumPy for array manipulation
- LabJack for communication with DAQ
- Import necessary math constants or functions
- CSV package for saving data

Define Signal

You must define the data you want to send the LabJack, sample by sample. Define a sinusoidal signal with predefined frequency, amplitude, and phase. By defining your signal requirements first, you are better equipped to set up the LabJack streaming parameters. Be sure to incorporate the limitations of your source and receiver you outlined previously in Section 9.5.2 "Hardware Limitations."

- Create a sinusoidal waveform with a user-defined number of data points per period, phase, amplitude, and frequency

- Test that the signal is within the amplitude limits of the source
- Test that the signal is within the achievable sampling rate of the LabJack
- Ensure the Nyquist criterion is enforced (data rate must be at least twice the frequency of the sinusoid)
- Note that the length (in number of samples) of your signal must fit in your ring buffer.

Open Connection to LabJack

This is the same in the warm-up experiment.

Prepare Output Streaming

You must define the output parameters to tell the LabJack which signal you want to generate, from which port, and under what conditions.

- Define the address of the output port. The following method is required.

  `ljm.nameToAddress(name)`

  For example, one of the lines in your program might be:

  `out_address = ljm.nameToAddress("STREAM_OUT0")[0]`

  This function takes as an input a Modbus register name (in this example, "STREAM_OUT0," which is the register used to stream data to DAC0) and returns its address and type in a tuple. When you use this method, you will follow it with a [0], because this selects the address part of the tuple. The address has type integer. Addresses are always integers in Labjack. To find the names of Modbus registers, do an internet search for "Labjack Modbus map," then use the pull-down menu to search the area of interest (e.g., "STREAM").

- Prepare the output port. You will need to:
  a. turn the port DAC0 off (in case it was on already)
  b. set the target port to DAC0
  c. define the buffer size*
  d. declare the loop size for streaming, i.e., the number of data points that will be written out in a repetitive loop.
  e. turn the port on

  Each of these tasks can be achieved using the following Python method, but must be called independently for each register.

  `ljm.eWriteName(handle, name, value)`
  Writes one value to the Modbus register specified by name. The Jupyter notebook for this lab guides you through the necessary function calls.

- Write the signal to DAC0 as an array. Now, you are ready to fill your buffer with the signal you defined previously. You can send the samples that define your signal to the buffer for this goal using the following method.

  `ljm.eWriteNameArray(handle, name, numValues, aValues)`
  Writes consecutive values in the array aValues to the Modbus register specified by name and return an ErrorAddress. numValues is the number of consecutive points in the signal, aValues. The Jupyter notebook for this lab guides you through this function call.

---

* Note that the length of your signal must also be less than or equal to the buffer size.

Prepare Input Streaming

Analogous to the output streaming, you must prepare the analog input port AIN0 to record data.*

- Define the address of the input port using the `ljm.nameToAddress` method
- Choose whether to use single-ended or differential readings[†] using `ljm.eWriteName`
- Define the amplitude range you want to record with using `ljm.eWriteName`

Start Streaming

After defining the input/output parameters, we are ready to start streaming both the input and output! Only two commands are required here.

- Start streaming using the following method

  Initializes a stream object and starts streaming.
  ```
  ljm.eStreamStart(handle, scansPerRead, numAddresses, aScanList, scanRate)
  ```

  - `scansPerRead` determines how many scans will be produced with each call to `eStreamStart`
  - `aScanList` is an array of addresses to stream (the addresses of your input/output ports)
  - `numAddresses` is the size of `aScanList` (in this case, `numAddresses = 2`)
  - `scanRate` determines the number of scans per second

- Read the data using the following method

  Returns data from the stream buffer to variable ret. The data is located in ret[0].
  ```
  ret = ljm.eStreamRead(handle)
  ```

  Note that `eStreamRead` returns a list of data equal to `scansPerRead * numAddresses`, but only the first `scansPerRead` multiplied by the number of *input* channels are valid. In this case, only the first half of the data (you are using one input channel and one output channel) are valid.

Stop Streaming

In order to stop streaming, you should use the following method:
```
ljm.eStreamStop(handle)
```
Stops LJM from taking any more data from the specified device.

You could place this command at the end of your script, so that it is executed after the code above is executed once. To make your code more practical, you construct your script to allow the streaming to run[‡] until a keyboard interrupt is performed,[§] at which point the command to stop streaming is executed and your script will move on to subsequent analysis, visualization, and saving.

Disconnect the LabJack Connection

When you are done talking to the LabJack, it is good practice to disconnect, as in the warm-up experiment.

---

* Again, take into account the hardware constraints you jotted down in Section 9.5.2.
† See ExpPhys.com for links to more information
‡ This could involve a while loop to continuously call "eStreamRead."
§ Check out Python's try-except statement and KeyboardInterrupt exceptions.

Analyze Data

Once you have recorded your data, perform an analysis to visualize the frequency spectrum of your recorded signal. You may consider plotting the spectrum of both your output and input signal to compare. This can provide insights into additional hardware limitations or considerations.

- Compute Fast Fourier Transform (FFT) of the numerical and experimental signal. We recommend the `numpy.fft.fft()` function to compute the Fourier Transform and combined with `numpy.fft.fftfreq()` to provide the list of the frequencies used for the FFT. For more information about using these functions, utilize Python's built in `help()` command. For example, `help(numpy.fft.fft)` will provide the documentation for computing a 1D Fourier transform.

Visualize Data

Plot your data and spectrum! Write down the steps required to visualize both your numerical and experimental data. Note that it may be beneficial to scale or normalize the experimental and/or numerical signals for comparison.

Save Data

You can choose to write the data and spectra in separate files, or as unique columns in the same file. We recommend using the CSV package, the writer object, and writerow method. The `help()` function provides insights on these functions, and a wealth of example code is available online.

After saving, open the file and see if this data is as expected. You can also try opening the data with alternative software (e.g., Microsoft Excel, Google Sheets, or LibreOffice Calc) and plotting the data to compare and validate your results

- Save signals and spectra to CSV file

9.5.6 Create Automation Script

Once the experiment is fully understood, you are ready to develop a Python automation script to achieve the desired experimental outcome.

Use the pseudocode above to guide your programming. Your program can be executed using the command line/terminal, a Python integrated development environment (such as Spyder or the Jupyter notebook), or the Custom Scripts module in PLACE. There is an example Jupyter notebook on ExpPhys.com that provides coding snippets to get you started. The subsections with no code are for you to complete. ExpPhys.com also provides valuable links to the LabJack's documentation.

9.5.7 Performing Useful Science with Your Experimental System

Now that you have our basic automation script for your experiment in Python, it is straightforward to perform new and meaningful experiments with minimal additional programming. Use your automation script to complete the following exercises, which only require changing your input signal.

Simple Harmonic Motion

This exercise should have been completed as you created the script above. For completeness, ensure that you can complete this exercise with your script.

Create a sinusoidal signal $s(t)$ with a user-defined frequency, amplitude, and phase in Python. Send the signal to your experimental hardware via your automation script. Verify the frequency of the resulting signal with the FFT.

Resonance Measurement

Solids have a natural resonant frequency, at which the material most readily vibrates. This includes the material in your source (often a piezoelectric crystal). Vary the frequency of your waveform and observe the resulting FFT to validate the resonance frequency of your source. Explain.

Superposition of Waves

A. Create two sinusoidal signals $s_1(t)$ and $s_2(t)$, each with a unique user-defined amplitude, frequency, and phase. Superimpose the two signals $s(t) = s_1(t) + s_2(t)$. Generate and record the waveform with your audio system using your automation script. Plot the numerical and experimental waveforms, as well as their Fourier transforms, and describe your results.

B. What is the relationship between the frequencies of the two sound waves and the beat frequency? Are there any hardware considerations to take into account in order to observe the beat frequency?

C. With a non-zero amplitude for each wave $s_1(t)$ and $s_2(t)$, at what times is the amplitude of resulting waveform $s(t)$ zero? Validate this result numerically and experimentally.

Advanced Problem: Square Waves

A square wave with a fundamental frequency f_0 is made up of the superposition of f_0 with an infinite number of odd harmonics. Mathematically, this equates to:

$$V_{\text{square}} = \text{Amplitude} \sum_{k=1}^{\infty} \frac{\sin\left[2\pi(2k-1)f_0 t\right]}{2k-1}$$

where $V_{\text{square}}(t)$ is the voltage. Create a square wave and send the signal to your experimental setup and record the temporal signal and frequency spectra. Discuss your results, including the effects of the hardware limitations.

9.6 Advanced Lab: Leverage the PLACE Framework

Now that you have constructed a working Python script, you will extend its capabilities by transforming it into a PLACE module capable of integrating into the PLACE framework. The PLACE framework provides an experimental ecosystem where modules share a user interface, metadata, output data files, and plotting. It also allows modules to work together to easily perform any number of advanced experiments.

At the core of PLACE is what's become known as the config-update-cleanup cycle. For each module, you will need to specify what occurs during each of these steps. The *config* phase is where you prepare for the experiment. The *update* phase is where the experiment is run and where one or more iterations of specific tasks are performed. The *cleanup* phase is the final phase, where any final steps are taken to wrap up the experiment.

The Python script you developed in the first part of the experiment will be your starting point. In this section, you will start moving parts of that script around, slowly transforming it into a PLACE module. When you are done, you should be able to operate your module within the PLACE web application, just as if it was one of the other pre-installed PLACE modules.

PLACE disclaimer: PLACE has been developed by the authors of this chapter (and many others!) at the University of Auckland over the past 5 years. The primary goal of PLACE is to provide a solution to laboratory automation that is completely free, open source, easy to use, and flexible. While significant progress is being made all the time, it is also still in development. Feel free to notify us (via the PLACE GitHub page) if you encounter any problems.

If you earlier skipped Section 9.3.3 "Working with Python Files," please read it now; for this section, you will not be able to use Jupyter notebooks, but you can cut and paste code that you developed there into a code editor.

Primary tasks:
- Download/setup the PLACE source code
- Outline the PLACE module
- Move configuration values into PLACE config
- Move user adjustable values into a PLACE user interface
- Modify output data to use PLACE data file
- Modify plotting to use PLACE plotting
- Create a user interface

Before you begin:
- Prerequisites (from previous sections)
  - Python
  - Anaconda
  - Text editor (suitable for editing code)
- Uninstall any existing versions of PLACE (if applicable)
- Install GitHub Desktop*

Setting-Up PLACE Development

Congratulations, you're about to become a software developer! Unlike most applications, which just install themselves, you will be contributing your own code to an existing software project. This means you must get the *source code*[†] from the project's GitHub *repository*,[‡] add your own code, and then build the project yourself.

Using the GitHub Desktop, *clone*[§] the PLACE source code from:

```
https://github.com/palab/place
```

Also, because PLACE is in active development, there is a special *branch*[¶] of the project intended for this lab. This way, you won't find out that all the code has changed since this book was published. To access it, you simply need to *checkout*[**] the branch named "labjack." You can checkout the branch within GitHub desktop, or you can do it from the Anaconda Prompt.

Next, using Anaconda Prompt, navigate to the directory where you cloned the PLACE source code. In this directory you will find all the code needed to run PLACE, as well as the code for some of the default hardware modules included with PLACE. Don't worry too much about all the files in this directory. You won't need to do much with them.

When you've reached the directory, if you didn't checkout the branch using GitHub Desktop, you can checkout the branch now by running:

```
git checkout labjack
```

---

* Only available for Windows (Linux user can use Git without needing GitHub Desktop).
† Source code is a term used to describe code used to build a larger application.
‡ A GitHub repository is the location of the "master copy" of the code.
§ Cloning is like copying or downloading, except that Git remembers the origin of the code. If you need help cloning code to your computer, consult the entry for this section on ExpPhys.com.
¶ Git allows you to maintain different versions of your code. Each of these is called a branch and is named because it often splits off an existing branch. Two distinct branches can continue being developed independently.
** Checking out a branch will change all the code in the directory to match the branch. It can even create or delete files as needed. If you have changed files in the project directory, you may not be able to checkout a different branch.

Next, you need to invoke conda[*] to install the other software PLACE needs to run:

```
conda install -c freemapa/label/labjack place --only-deps
```

The argument *only-deps* is short for *only dependencies*. This means Conda will install all the software required by PLACE but will not install PLACE itself. As with GitHub, the label "labjack" will ensure that you get the right packages for this lab.

After the dependencies are finished, you will manually install the custom version of PLACE you downloaded by running this Python command:

```
python setup.py develop
```

This will install the files in the PLACE directory so you can run the PLACE application in *development* mode. While you are in development mode, if you make changes to the files in this directory, you will immediately see the changes the next time you run PLACE.

You can now start PLACE anytime by running the command:

```
python -m placeweb
```

This will launch a web server which will facilitate running PLACE experiments. When the server has started, it will print a local web address onto the Anaconda prompt window. If you put this address into your desktop or mobile web browser, you should see the PLACE user interface. You can now begin the process of moving control of your Python script from the previous section into the PLACE framework.

It should be noted that if you need to remove the PLACE installation (and leave development mode) later, just go back to the directory and run:

```
python setup.py develop --uninstall
```

Outline the PLACE Module

If everything has gone correctly, you have PLACE installed and you are ready to develop your own module. It's time to put on that Python programming hat again.

As mentioned earlier, PLACE uses a *config-update-cleanup* cycle. All plugins written for PLACE must adhere to the schedule of this cycle. To ensure this, Python code written for PLACE is required to be a *subclass* of a PLACE plugin type. There are several types of plugins, but this experiment will make use of the PLACE *Instrument* class.

By being a subclass of Instrument, your code will be required to implement three Python functions, named `config`, `update`, and `cleanup`. PLACE will automatically call these functions; you just need to write the code describing what should happen during each of these events. After completing the first section of the lab, you likely already have most of the code you need. You will just need to arrange it in a way that PLACE can make use of it.

A special file has been included for this lab. It is located at:

```
place\plugins\labjack_t_series\labjack_t_series.py
```

If you don't see this file, you probably haven't correctly checked out the labjack branch from GitHub, so go through the steps above to double check that you have the correct branch.

If you have never created a Python subclass (or *any* class) before, you will find this template useful in starting your PLACE plugin. As you complete the next sections of this lab, write your code into this file.

---

[*] Anaconda is a software package that helps work with the tool conda, in the same way that GitHub is a software package that helps work with the tool git.

PLACE Configuration Values

To construct a plugin that will work with PLACE, you will need to adjust many aspects of your Python script. This will allow you to take advantage of the features already written into the PLACE framework. The first thing you will ask the framework to manage for you are your *global configuration values*. Global configuration values are those values that tend not to change but may not be the same for everyone.

Consider this simple example:

```
name = 'Roger'
print('Your name is', name)
```

If the person running this is named Roger, then everything will work out. But if their name is not Roger, they will need to change the value of the variable. In a distributed software application, you typically can't go into the code and change the value of something, so your module would forever use Roger as the value of the variable.

PLACE uses a *configuration file* to get around this issue. A configuration file is accessible to the user but is read by the program as it is running. Typically, PLACE stores this file in the user's home directory as .place.cfg. Values that need to be changed are stored in this file and the program will read them as it runs.

If you wanted to store the name in the configuration file, you would update the above example to look something like this:

```
from place.config import PlaceConfig
name = PlaceConfig().get_config_value(
        'test_instrument', 'name', 'Roger'
)
print('Your name is', name)
```

This code performs three tasks:

1. import the PLACE configuration file
2. get the value, specifying:
    a. plugin name = 'test'
    b. key = 'name'
    c. default value = 'Roger'
3. use the value

The plugin name should be the same for all the values you want to store. This way, all your configuration values will be grouped together in the configuration file. Using "LabJack" (or something similar) is recommended.

The final value, the default value, is optional. If it is not specified, the module will print an error message telling the user they must add this value to the configuration file before PLACE will run. By including it, PLACE will use the name Roger until the user changes it to their own name, but it will never give the user an error. You should provide default values whenever possible.

When working with the LabJack for these experiments, you will find many values which could be stored in the configuration file. For example, there are a couple different stream outputs available on the LabJack. In your Python script, you would have chosen a stream (ex. STREAM_OUT0) and used it throughout your code. This works great until you encounter a situation where you need to use another stream. Perhaps you already wired up another experiment to STREAM_OUT0 and you don't want to change it. Maybe STREAM_OUT0 is acting weird and you want to test your experiment on a different one. If you put the stream into the PLACE configuration file, you (or another user) can easily change it to one of the other options, without needing to edit the Python code.

The Python code looks like this:

```
output_name* = PlaceConfig().get_config_value(
        'LabJack, 'output_name', 'STREAM_OUT0'
)
```

Every time you run a PLACE experiment, the above code will look for a value under the key 'output _ name' under the 'LabJack' heading in the .place.cfg file on your computer. It will assign that value to the output_name variable. If the value cannot be found in the the .place.cfg file, PLACE will save the default value, "STREAM_OUT0," into the configuration file and also load this value into the variable.

Now go through your own script and modify the code to store configuration values in the configuration file.

PLACE Experimental Variables

There is a second type of configuration value used by PLACE, referred to as an *experimental variable*. These are configuration values which are frequently changed to modify the experiment. These are also the ones you will later control with the PLACE user interface.

Perhaps your code needs to wait for a bit. If it does, you may have written a line of code that looks something like this:

```
wait_time = 10.0
```

If you want the user to be able to change this value, you would either need to follow the instructions in the previous section (to make it a global configuration value) or you could make it an experimental variable by changing it to this:

```
wait_time = self._config['wait_time']
```

All the values in self._config† will be sent into your program from the user interface, and this is where you will pick them up.

This LabJack experiment requires several values that you will want to be able to change easily. Since you are working with waveforms, one obvious value over which to have control is the frequency. This value could be stored in the PLACE configuration file, but that wouldn't make sense because the user would need to update this file every time they wanted a new frequency. It is much easier to let them change it within the PLACE user interface.

As you saw above, loading values sent in from the user interface is simple, so we can write the code to load a frequency like this:

```
frequency‡ = self._config['frequency']
```

The same thing could be done for values like *amplitude* or the *number of harmonics*. As you work with the experiment more and more, you will learn the values you want to control. You can always come back and make updates later.

Now go through your code and update your variables as needed. It is important to remember that global configuration values are derived from the PLACE configuration file and experimental variables are derived from the user. If you put too many values into the configuration file, the user will be frustrated

---

* If you need to access your configuration values in other PLACE methods (config, update, cleanup), you can save them into self, the class object (ex. self.output_name).

† The underscore ('_') at the beginning is required by PLACE.

‡ As with configuration values, if you need access to these values in the other PLACE functions, make sure to save them into the self class object.

when they always have to edit the file. Conversely, if you put too many values into the user interface, the settings may become cluttered and confusing. Finding the right balance is important.

Remember that any value which you have not updated to be either a PLACE global configuration value or a PLACE experimental variable will not be able to be changed by anyone using your module later, so read through all of your code a second time to make sure you've accounted for all the values that might need to change.

PLACE Data

Data generated by PLACE modules is gathered together into a single file and delivered to the user via the PLACE web application. To work nicely within the PLACE framework, your module must save data in a specific format. The format is very permissive, provided a few guidelines are followed:

- Your data must be put into a NumPy record array
- Your module must save the same data "shape" at each update

Essentially, what this means is that if you save ten numbers on the first PLACE update, PLACE will expect ten numbers from each subsequent update. This requirement improves memory management for the data array and allows large amounts of data to be handled without issue.

The PLACE documentation* has an extensive tutorial on saving data, which you are encouraged to read. However, the following code (taken from the tutorial) shows how you might save four temperature readings from a fictional probe during each PLACE update.

```
temp1 = read_from_probe(1)
temp2 = read_from_probe(2)
temp3 = read_from_probe(3)
temp4 = read_from_probe(4)
data = [temp1, temp2, temp3, temp4]

heading = self.__class__.__name__ + '-temperature' # †
dtype = np.dtype([(heading, np.float64, 4)])
record = np.array([(data,)], dtype=dtype)

return record
```

In this experiment, you are probably reading data from the LabJack using the `eStreamRead(self.handle)` method. Once you have read the data you want from the stream, you would save it into the PLACE data file by returning it at the end of the update function.

```
# ... < other code > ... #
dtype = np.dtype([(
    'LabJack-data',   # heading for your data ‡
    np.float64,       # type of your data §
    LOOP_SIZE * 2     # number of data points ¶
)])
return np.array([(
    data,             # the actual data
)], dtype=dtype)
```

---

* The PLACE documentation is available within the application itself. Start the PLACE server and navigate to the PLACE web address in your browser. Then click on "Documentation" at the top.

† `self.__class__.__name__` is a special Python variable that holds the name of the Python class. In this case, it will be the name of the PLACE instrument you are using.

‡ The heading will distinguish your data from data recorded by other PLACE instruments.

§ float64 is a 64-bit floating point type. Many other types can be used depending on the values you need to store.

¶ Remember: you must return the same number of points on every update or PLACE could reject your data.

Update your code to save data into the PLACE data file. If some of the process is unclear, refer to the PLACE documentation within the web application for assistance.

PLACE Plotting

Being a web application, PLACE is not able to fully support plotting using Matplotlib. In some cases, Matplotlib plots will display correctly, but their use is not encouraged. Instead, PLACE provides support for basic plotting using a different library. In cases where more elaborate plots are needed, users are encouraged to replot using another application after the PLACE experiment completes.

Plotting should be performed during the update phase. The most recent plots will be displayed in the web application for the user. Basic plots can be displayed very easily:

```
self.plotter.view1("figure title", [1, 2, 3, 4])
```

Slightly more complicated plots can be created with a little more effort:

```
line1 = self.plotter.line(
    [1,   2,   3,    4],
    label='x'
)
line2 = self.plotter.line(
    [1,   4,   9,   16],
    label='x squared'
)
line3 = self.plotter.line(
    [1,   8,   27,   64],
    label='x cubed'
)
line4 = self.plotter.dash(
    [1,  16,  81,  256],
    color="pink",
    label='x "quarted"'
)
self.plotter.view(
    "figure title",
    [line1, line2, line3, line4]
)
```

Using one of the plotting methods above is preferred, but if you already have code to create a Matplotlib figure, you can still use it! PLACE will render the figure as an image and send it to the web application:

```
self.plotter.png("figure title", figure) #*
```

Go through your code and replace any of your Matplotlib plots with version that are compatible with PLACE.

PLACE User Interface

User interaction with PLACE is through a web-based graphical user interface. As with most aspects of PLACE, the internal workings of the framework are designed to handle the majority of this user interface. However, there are still a few aspects which must be manually customized. You will define these custom parameters in an Elm file. As with the Python section, an Elm template file has been created for LabJack at the following location:

---

* If you are using pyplot, the figure is hidden away from you. You can access the figure by using the gcf() function.

elm\plugins\LabJack.elm

Elm is a very different programming language from Python, and this lab does not expect you to learn Elm programming. Instead, the template file walks you through what to put into this file step-by-step, and examples are included for each section. It is important that you take your time and get the syntax right, as Elm errors may be difficult to decipher with limited experience with the language.

During most of the Elm file construction steps, you will be configuring something for each variable you want the user to be able to change via the web interface. One step will have you configure the *type* of each value. Another step will have you configure the default value. There are several steps like this. It is important that you try to list the variables in the same order for each step, so PLACE can correctly construct the user interface for you.

When you have filled in the template file, you will need to *build* it. Elm code cannot be directly executed by your computer, so we need to *transpile*[*] it into something else. Elm programs build into JavaScript, which can then be executed on an internet browser. If you have not installed Elm, you will need to do that before continuing.

Note: PLACE currently requires Elm 0.18. If you use a new version of Elm, it is unlikely you will be able to complete this experiment.

To install Elm, you can often just use the command:

```
npm install -g elm@0.18
```

If you do not have npm, you will need to install Node.js, but this is very well supported by all major operating systems. Install Node.js,[†] restart the Anaconda Prompt, and then repeat the steps above. If this fails, you may need to find an installation guide online that matches your system.

Once Elm is installed, you will be able to build the UI components for your PLACE plugin. Go into the directory of LabJack.elm, and use the following command to build your file:

```
elm-make LabJack.elm --output LabJack.js
```

Running this will either give you an error or make the JavaScript file. If you get an error, you will need to try to correct the error and rebuild. When you no longer get an error, copy the JavaScript output file and replace the one in the placeweb directory:

placeweb\static\placeweb\plugins\LabJack.js

Testing PLACE Plugin

At this point, you should be able to run everything and see if it's working the way you want. Before you do this, ensure that you have completed the following tasks:

- PLACE source code downloaded
- 'labjack' branch checked out from Git
- PLACE dependencies installed
- PLACE installed in development mode
- labjack_t_series.py updated with your Python code
- LabJack.elm template file completed
- LabJack.elm built into LabJack.js and copied into placeweb directory
- PLACE server started

---

[*] In programming, when you transpile your code, you are producing code in another programming language, that performs the same task as the original code.

[†] Choose "Add to Path" if prompted during installation.

With the PLACE server running,* you can navigate to the web address listed (default: http://127.0.0.1:8000) and see the PLACE webapp. PLACE loads into the "History" view and shows any prior experiments you have completed. Click on "New Experiment" button.

From the "Plugins" menu, click on "LabJack" to add the LabJack module to the user interface screen (ignore if it has already been added). Click the tab labeled "LabJack" and click the check box to open the user interface options for LabJack. You should see the options you selected when you created the LabJack user interface.

If everything up to this point has worked, you should be able to configure the options and run an experiment. Select the desired options from the user interface. At the top of the user interface, select the number of updates you'd like to run. If you haven't written your code to modify its configurations across many updates, you can probably leave this set to 1.

Click "Start" to begin the experiment. While you are testing your plugin, you should monitor the output from the PLACE server. If something goes wrong, the server will often print out the error message.

If your initial test passes, congratulations! If not, you will have to try to determine what went wrong. This can be one of the most challenging aspects of programming, so patience and persistence will be rewarded.

After you get one configuration running, try many others to make sure it is working the way you expect. And now that it's automated, running all these tests will be a lot easier. Once you have a working PLACE plugin, it is easy to integrate it with other modules. You can also write more elaborate plugins for LabJack, which adjust the experiment based on the PLACE update number. There are nearly endless possibilities!

9.7 Homework Problems

Problem 1: The city has hired you to do a traffic survey. You have been given permission to put GPS on every bicycle in the city for one month. Your job is to use the GPS readings to produce a dataset. One city official asks you what information will be available in the dataset you produce. What do you tell them? *Hint: you can (and should) think up several answers but remember that one valuable or creative answer might be better than several trivial answers.*

Problem 2: Write an automation script to play "Twinkle Twinkle Little Star."

Problem 3: You have been given a robot to help out in your day-to-day tasks. It isn't programmed to do anything right now, but that's okay, because it can be programmed with pseudocode. Its dexterity should allow it to do anything you can do. Write a pseudocode program to automate a repetitive task in your life. *Hint: your program should draw from concepts such as variables, loops, conditional statements, and functions.*

Problem 4: Complete Problem 2 using PLACE, where the notes of the song (e.g., ccggaag...) and the duration of each note in seconds (e.g., 1111112...) are inputs. Create a PLACE module such that one note is played by the LabJack during each update phase.

---

* Reminder: you can start the PLACE server by running `python -m placeweb` at the Anaconda Prompt.

FIGURE 9.7 Paul Freeman in Queenstown, New Zealand.

Paul Freeman is currently a software developer for Sylo, developing distributed applications and decentralized technologies. He also volunteers support to the University of Auckland in several research areas, including laboratory automation and genetic sequencing, and teaches a Python workshop to physics and math students each semester. He thinks everyone should learn to code and enjoys making it fun and easy for others. Paul is also an artist, baker, writer, and runner who enjoys food festivals and the theater (Figure 9.7).

FIGURE 9.8 Jami Shepherd at Milford Sound, New Zealand.

Jami Shepherd is a research fellow for the Dodd-Walls Centre for Photonic and Quantum Technologies at the University of Auckland. Her research focuses on imaging and sensing with light and sound waves, with a focus on medical imaging applications. Outside of the lab, she enjoys playing soccer, hiking, and enjoying a good coffee with friends (Figure 9.8).

10

Basic Optics Techniques and Hardware

Walter F. Smith

CONTENTS

10.1 Laser Safety

Lasers are widely used in current physics research because of the narrow range of wavelengths emitted by the laser, the high intensity, the high degree of collimation, and the long coherence length.*

Depending on the power level, the hazard associated with a laser can be extreme or negligible. The first thing you should find out when using a laser is its class. Class 1, 1M, 2, and 2M lasers (classes I, II, and IIa in the old system) are safe to use without protective eyewear, as long as you're not viewing the beam through optical instruments (e.g., a microscope or magnifying glass); your blink response is quick enough to protect your eyes. You should never intentionally look into the beam. Class 3R lasers (class IIIa in the old system) can technically be used without protective eyewear, so long as the wavelength is visible and you're not viewing the beam through optical instruments.† These lasers are "potentially hazardous under some direct and specular reflection viewing conditions if the eye is appropriately focused and stable, but the probability of an actual injury is small." Many consider it good practice to use protective eyewear with class 3R lasers, especially when creating the setup; check with your instructor. For class 3B and class 4 lasers (classes IIIb and IV in the old system), you must always use protective eyewear, as well as protective barriers and curtains to prevent accident exposure to others. For class 4 and the high end of class 3B lasers, you must also use skin protection. Make sure your protective eyewear is designed for the wavelength of laser you're using.

Concept test (answer below‡): If you're using a blue laser, should your protective eyewear be blue or orange?

In addition, you should adhere to the following standard practices. Accidental exposure to the laser is most likely when you are creating your setup. So, if you are using anything above a class 2 or 2M laser, you should do the setup using reduced beam power and/or with a filter directly in front of the laser. You should block or turn off the laser before adding a new optical element. When at all possible, keep the optical components close enough together that it's not possible for someone to put their head between them. (This is usually advantageous for optical reasons as well as for safety.) Arrange your setup so that the beam is in a horizontal plane, well below eye level. Remove shiny jewelry from hands and wrists, to avoid accidental reflections. If you need to bend down (e.g., to pick up something you dropped), first turn off the laser or block it at the source.

When wearing eye protection, you're usually unable to see the beam at all, since that wavelength is completely filtered. So, to do the alignment, you should shine the beam onto a viewing card, which you can purchase from most optics supply companies (e.g., ThorLabs); make sure you have one that's appropriate for the wavelength you're using. For blue lasers, the fluorescence shifts the wavelength emitted from the card to red, so that you can see it through your eyewear. The cards are less effective for red lasers, since fluorescence can only shift the emitted light to longer wavelengths. However, you can usually still see the spot on the card. If you don't have the special fluorescent card, a piece of white paper is often adequate; the paper is treated with fluorescent chemicals to make it appear "whiter than white." In some cases, fluorescent index cards can work well.

Another method is to set up a low power laser that is collinear with the higher power laser, and at a different wavelength, so that you can easily see it while wearing the eye protection suited for the higher power laser. It's important to be as scrupulous as possible about arranging the collinearity, and even

---

* The coherence length can be thought of as the length of a photon. It is important in applications involving interference. For example, to observe interference in a setup where the laser is split into two beams which are then brought back together, the path lengths must be equal to within a coherence length. For gas lasers (e.g., HeNe lasers), the coherence length is typically tens of cm. For diode lasers (e.g., those used in a laser pointer), the coherence length is less than a mm. Both of these can be dramatically extended by frequency stabilization techniques, so that coherence lengths of tens or thousands of meters are possible. The coherence length for non-laser sources is a fraction of a mm. As you can show in Problem 10.1 (available on ExpPhys.com), the coherence length x is related to the range of wavelengths emitted by the laser by $\Delta\lambda \cong \lambda^2/\Delta x$, and to the range of frequencies Δf emitted by $\Delta f \cong c/\Delta x$.

† ANSI Z136.1-2014 *American National Standard for Safe Use of Lasers.*

‡ Blue glasses transmit blue light, so would be totally ineffective in protecting against a blue laser. Orange glasses absorb blue light.

when you've aligned things optimally with the low power laser, some fine tuning will be needed with the high power laser.

Before turning on a laser, you should check with your instructor about whether you need to complete additional laser safety training.

10.2 Lasers

You are most likely to encounter two types of lasers in physics research: gas lasers and diode lasers. The decision between types is usually based on the desired wavelength. Gas lasers (e.g., HeNe) inherently have good coherence length (tens of cm), a very narrow band of emitted wavelengths (with a "linewidth" $\Delta\lambda$ less than 10 pm), and excellent collimation. They are available in polarized and "randomly polarized" versions. Unfortunately, the randomly polarized version is not "unpolarized" (meaning a superposition of all possible polarizations), but instead has a polarization that often varies over timescales of tens of seconds as the laser warms up. Full warm-up requires an hour or more, and even them the polarization may be affected by changes in room temperature. Therefore, I recommend you purchase only polarized lasers, which are usually about the same price.

Diode lasers are available in standard and frequency-stabilized versions. The standard versions are much less expensive than gas lasers. The linewidth is a few nm, with correspondingly short coherence length (a few tenths of a mm at best). The central wavelength can vary by a few nm between nominally identical units and depends somewhat on room temperature (about 0.2 nm/°C). When possible, purchase a "laser module," which includes optics to produce a fairly well collimated beam; however, often the beam profile is elliptical rather than circular.

The best frequency-stabilized diode lasers ("external cavity diode lasers") are much more expensive than low-end gas lasers, but have linewidths of less than 1 fm and can be tuned over a small range (typically about 0.1 nm).

Concept test (answer below*): What is the coherence length (essentially the length of a photon) for a laser with a 1 fm linewidth and a wavelength of 633 nm?

10.3 Optical Hardware

Optical Tables and Breadboards

Optics setups are usually constructed either on an optical "breadboard" (a metal plate with an array of tapped (i.e., threaded) holes), or an optical table (a table with an array of tapped holes in the top surface). Optical tables and breadboards provide three functions: (1) the precision array of tapped holes allows convenient mounting of optical elements. (2) Tables and breadboards have a high ratio of stiffness to mass, so that if there are vibrations, everything moves together as much as possible, resulting in little relative motion. (3) Tables and breadboards provide some isolation from vibrations in the floor.

The choice between a table and the various types of breadboards involves a tradeoff between cost and the other desirable aspects. The least expensive option is a breadboard made from solid aluminum. This has the lowest stiffness-to-mass ratio and is also the most strongly affected by thermal expansion. However, it is adequate for all but the most demanding experiments; more costly options are usually only needed when data must be taken over periods of days, so that thermal expansion becomes critical, or when a laser beam must remain focused on the same spot over several hours to a tolerance of less than 100 μm, or for experiments that are extremely sensitive to vibrations. (For all the experiments described in this book, a solid aluminum breadboard is fine.) Aluminum breadboards have the advantage that they can be easily moved. You can easily get some isolation of the breadboard from building vibrations by putting an array of "Sorbothane" disks between the breadboard and the table it is placed on; this

---

* $\Delta\lambda \cong \lambda^2/\Delta x \Leftrightarrow \Delta x \cong \lambda^2/\Delta\lambda = 400$ m.

elastomer is engineered for use in vibration isolation applications.* It's essential to put the breadboard on a very sturdy table.

The next step up is breadboards made from steel with an internal honeycomb construction, which enhances stiffness. These have significantly less thermal expansion. The thicker the breadboard, the higher the stiffness-to-mass ratio, but the heavier (and so harder to move) it is. The high-end choice is optical tables, which can include pneumatic legs for the best possible vibration isolation.

In all cases, to minimize the effects of vibrations and thermal expansion, it's best to make optics setups as compact and close to the breadboard/table surface as possible. For example, if two parts of an interferometer are located only 10 cm away from each other, the distance between them will be changed by thermal expansion by half as much as if they were separated by 20 cm, and they will vibrate in unison at least twice as well. However, it's important not to drive yourself crazy by making things so close together or so low down near the table surface that it causes headaches with installing additional components or making adjustments. For the vertical positioning, it's worthwhile to think ahead and consider which component in your setup will require the largest vertical distance from the table-top to the center of that component (e.g., a large sample mount, or something that must be mounted on a stage that rotates in the horizontal plane), then set everything else to that height.

When purchasing the table or breadboard, you must make the fateful decision between imperial (inch-based) threads on the holes and spacings between the holes, or metric (cm-based). It's best to stick with one system or the other as much as you can.

Posts, Postholders, and Pedestals

The most common method of mounting optics is the post/postholder system shown in Figure 10.1. This allows convenient adjust of height. The posts and postholders are each available in a wide range of heights. Each optic element is held in a mount, which is attached to the top of the post using a small diameter (8–32 or M4) screw. In some cases, this is a "setscrew," i.e., a screw with no head, but instead with a hexagonal socket for an Allen wrench on one end. In other cases, the mount is attached with a "capscrew," i.e., a screw with an Allen head.

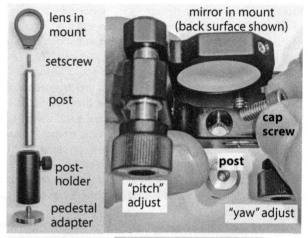

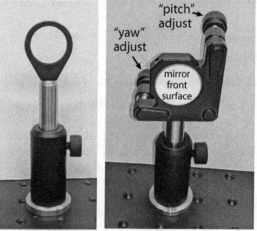

FIGURE 10.1 Post and postholder system. Top left: some optics attach to the post via a setscrew. Bottom left: assembled version. Top right: other optics attach via a cap (Allen head) screw. The mirror mount allows adjustment of the forward/backward tilt ("pitch") and the left/right azimuthal angle ("yaw"). Bottom right: assembled version.

---

* I recommend 1 inch diameter, 0.5 inch thick disks, with a durometer rating of 30. These can be obtained with stick-on adhesive from IsolateIt.com. You can determine how many disks to use using the desired weight of about 1 kg per disk. However, avoid positioning them more than 40 cm apart, since this will lead to "drumhead" behavior. If necessary, you can cut them in half to avoid this.

FIGURE 10.2 Top images: screwing a postholder into the optical table. Bottom row: After disassembly, if the screw was put with the Allen head up (toward the post holder), it can be easily removed whether it winds up in the table or in the postholder.

The postholder has a large diameter (1/4–20 or M6 thread) threaded hole at the bottom. Often, this is used to secure the postholder directly to the tabletop, using a setscrew, as shown in Figure 10.2. Screw the setscrew partway into the postholder, then screw this assembly into the tabletop. When you are ready to disassemble things, you unscrew the assembly from the tabletop; it's essentially random whether the setscrew will wind up partly screwed into the tabletop or partly screwed into the postholder. Note that the setscrew has a hexagonal (Allen) socket on one end only. This should go on the end that screws into the postholder, not on the end that screws into the table. That way, when you unscrew the postholder from the tabletop, if the screw remains partly screwed into the tabletop, you can use an Allen key to remove it. If instead it remains partly screwed into the postholder, you can reach with an Allen key through the top of the postholder to remove the screw.

Screwing the postholder directly into the tabletop makes it easy to align optics in straight, perpendicular, or 45° lines. However, in many cases, one needs more flexibility with positioning in the horizontal plane. Of the various ways to attach something to a tabletop, the clamping fork system shown in Figure 10.3 gives the most flexibility and is the fastest. It's best to purchase the fork with a "captive screw," since this saves time hunting for the screw.

In some cases, such as when a large object must be supported or when working in a lab that doesn't use the clamping fork system, it's better to use a base with a larger footprint. In this case, first attach the

FIGURE 10.3   Clamping fork system, including captive screw and base adapter for postholder.

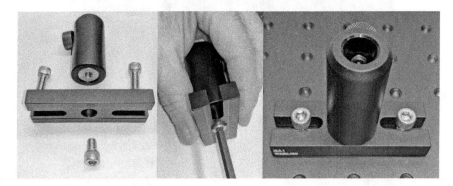

FIGURE 10.4   Attaching a postholder via a base.

base to the postholder by putting a screw through the recessed area on the base, then screw the base to the tabletop, as shown in Figure 10.4.

An alternative to the post and postholder system is to use a pedestal, as shown in Figure 10.5. These provide more rigidity, and so are preferred for applications that are very sensitive to vibrations such as interferometers. However, the height adjustment is limited. One can purchase various heights of pedestals, and usually the critical elements are all mounted on the same height. However, some additional adjustability is available via spacer rings, which can be purchased in a few different thicknesses.

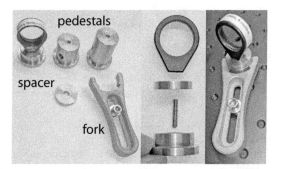

FIGURE 10.5   Left: pedestal system. Middle: assembly including spacer ring; note long setscrew. Right: fully assembled with spacer ring.

10.4 Optical Elements

A recurring theme when dealing with optical elements such as lenses and filters is the need for labeling. In some cases, an expensive optical element can become almost useless if it is not labeled. For example, if you find an unlabeled round piece of something that looks like transparent glass, it could be a quarter or half waveplate (you'll learn more about those later in this chapter), but determining which wavelength it's designed for requires multiple tests. So, the mount for the optical element should be clearly labeled, including the wavelength or range of wavelengths the element is designed for. If the element is removed from the mount, it should immediately be placed into a labeled container or a labeled small cloth bag. (It is helpful to keep the containers for the optics elements you're using in a box which is kept near your optics table or breadboard.) When labeling a mounted optic, it's best to use a label printer; handwritten labels rub off too quickly. (It's adequate to use handwritten labels for storage bags or containers, since these are handled less.) It requires significant discipline to adhere to these practices, but it pays off hugely.

Lenses

The most important parameter for a lens is the focal length, i.e., the distance beyond the lens at which an incident parallel beam (such as a laser) is brought to a focus. The next most important parameter is the range of wavelengths the lens is intended for; this is determined by the material the lens is made from (e.g., fused silica lenses are needed for ultraviolet), and the anti-reflective coating.

However, you may need to deal with unlabeled lenses. If you need an accurate value for the focal length f, you can accurately measure the distance s from a source to the lens, and the distance i from the lens to the image, and use the "thin lens equation":

$$\frac{1}{f} = \frac{1}{s} + \frac{1}{i} \tag{10.1}$$

However, often an estimate for f is all that's needed. You can roughly determine the focal length by finding the distance from the lens at which an image of the overhead lights is formed.

Concept Test (answer below*): Does this give an overestimate or an underestimate of the focal length?

The "power" of a lens is defined as the inverse of the focal length, and has units of diopters, where one diopter equals an inverse meter. If one lens is positioned just after another (so that the two are "in series"), the total power is approximately equal to the sum of the individual powers.

To bring a parallel incident beam to the minimum diameter focus, the lens should have a hyperboloid shape.[†] One can purchase such "aspheric" lenses, but they are expensive, especially for large diameters. It's less costly to make lenses with a spherical surface, so almost all lenses have this shape. However, it has the drawback that rays parallel to the optic axis but far from it are brought to a focus at a different point than rays close to the optic axis; this is called "spherical aberration."

Lenses can of course be used to form images; for such applications, double convex lenses (curved on both sides) are often best. However, the main use of lenses in this book is for controlling the diameter of laser beams. For such applications, it is best to use plano-convex lenses. To minimize spherical aberration, the lens should be oriented so that parallel rays enter or exit the lens on the curved side. (See, for example, Figure 10.11.)

One can purchase pre-mounted lenses, which have the advantages of being pre-labeled, easier to handle, and easier to combine with other optical elements by screwing them together. However, they're about 40% more expensive than unmounted lenses. It is not difficult to put an unmounted lens into a mount. The mount has a thin lip on one side. Place the lens into the mount, so that it rests against the lip. The lens is held in place with a retaining ring, which should be tightened with a "spanner" (a special wrench), as shown in Figure 10.6. Avoid using improvised tools (such as tweezers) to tighten the retaining ring, since this leads to frustration and occasionally a scratched lens.

---

* Since the overhead lights are not actually infinitely far away, the light from them is somewhat divergent at the position of the lens, so this measurement is an overestimate of the focal length.

† *Optics, Third Edition* by E. Hecht (Addison-Wesley, 1998)

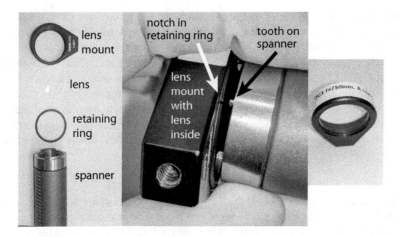

FIGURE 10.6 How to mount a lens. Left: parts. Center: engaging the spanner with the retaining ring. Right: fully assembled.

FIGURE 10.7 Flip mirror.

Mirrors

Most mirrors have a reflective surface made from aluminum (about 90% reflectance for visible wavelengths) or silver (about 97% reflectance). Aluminum mirrors withstand high humidity and accidental fingerprints better than silver mirrors. The price is about the same. In most cases, the mirror is coated with a thin transparent film, such as SiO_2, to limit tarnishing and reduce the effect of scratching. Mirrors are installed in mounts that allow fine adjustment of the angle around a vertical and a horizontal axis, as shown in Figure 10.1c,d.

Occasionally, it's useful to use a "flip mirror," which can be folded down to allow the beam to pass over it, or folded up to direct the beam in a different direction, as shown in Figure 10.7. Unfortunately, the angle adjustment screws for such mirrors are not very precise, so you should arrange for the primary path to be the one that's used when the mirror is flipped down. An alternative that is bulkier, but more precise, is to combine a conventional mirror mount with a flip adapter.

Neutral Density Filters

Neutral density filters reduce the power of light. For example, this allows the experimenter to explore a range of powers that varies over a few orders of magnitude, a broader range than is usually possible using the electrical power adjustment on the laser (if there is one).

They are rated by the "optical density," or OD, where

$$\boxed{\frac{\text{Power in}}{\text{Power out}} = 10^{\text{OD}}}$$

(10.2)

Concept test (answer below*): An OD 1 filter and an OD 2 filter are placed back to back. What is the ratio of the output power to the input power for this assembly?

The power reduction is achieved either by reflecting or absorbing part of the light. The absorptive type of filter is preferred in most cases.† For these, the ratio of input to output power varies by about 10% over the range of optical wavelengths. They can be purchased with or without anti-reflective coating,

The mounting system for filters is the same as for lenses. Filters can be purchased mounted or unmounted, with the mounted version costing significantly more. However, one often wishes to combine filters by screwing them together, which is much easier with mounted filters.

Beamsplitters

It is often necessary to split a laser beam, most frequently to create two paths for an interferometer or for sensing the laser power. One can purchase plate beamsplitters, in which one side of a glass plate is coated with a partially reflecting layer, or cube beamsplitters (see Figure 10.8).

Both types absorb little of the incident power, but conventional cube beamsplitters absorb up to 15%. Therefore, for high power applications, "laser line" cube beamsplitters (which absorb very little at a specified wavelength) or plate beamsplitters are preferred. Plate beamsplitters are more difficult to mount, and are more easily damaged, so the cube type is usually preferred.

Both types are available in "non-polarizing" and "polarizing" varieties. For polarizing beamsplitter cubes, the transmitted beam is horizontally polarized to a very high degree, while the reflected beam is primarily vertically polarized, but may contain up to 5% horizontally polarized light. The transmitted beam of the non-polarizing type has only a modest effect on the polarization of the beam; for cube beamsplitters, the ratio between vertically and horizontally polarized light is altered by at most 7%. However, the polarization of the reflected beam can be substantially changed. The intensity ratio of reflected horizontal and vertical polarizations is very similar to that in the incident beam, but the phase between them can be changed. For example, incident light that is polarized at 45° relative to the horizontal will have a transmitted beam that is also 45°, but the reflected beam may well be elliptically polarized.

---

* The OD 1 filter reduces the power by a factor of $10^1 = 10$, and the OD 2 filter reduces it by another factor of 100, for a total reduction of a factor of 1000. Note that the OD of the combination is simply the sum of the individual ODs.

† Reflective neutral density filters are more difficult to use, in part because the reflective surface is unprotected, so that they are easily scratched by being touched. Also, the reflected beam carries much of the power, leading to annoyance and also safety concerns. (One safe way to deal with the reflected beam is to direct it into a "beam stop," an object that can safely absorb all the laser power, by orienting the filter at an angle slightly away from perpendicular to the beam.) However, these filters are more uniform in the ratio by which the power is reduced as wavelength is varied; the ratio is constant to within a few percent over the range of optical wavelengths. They are somewhat better for high power applications, since absorbing the power in the filter can lead to damage or thermal expansion (and associated beam distortion) of the filter. However, the advantage over absorptive filters is only about a factor of two. Ordinary reflective neutral density filters should not be stacked immediately adjacent to each other, since reflections amongst the various surfaces can lead to undesired interference effects and also to surprising changes in the OD of the stacked assembly. However, for no increase in price one can purchase "wedged" filters, in which the reflective surface is slightly angled relative to the mount, eliminating these issues.

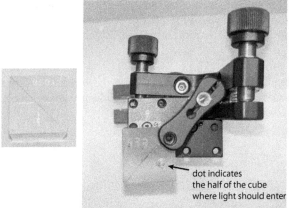
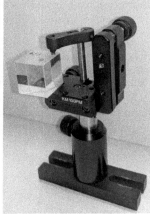

dot indicates
the half of the cube
where light should enter

FIGURE 10.8 Cube beamsplitters. Left: some are clearly marked with the intended beam path. Center and right: for others, the half of the cube on which the beam should enter is shown by a dot. The cube is held in a kinematic prism mount.

(See Section 11.2 for the explanation of elliptical polarization.) The effect on the polarization of the reflected beam depends on the type of beamsplitter, so it is best to test experimentally.

Because of subtleties with anti-reflection coatings, the incident beam should come in on the side marked with a dot or with an incoming arrow, as shown in Figure 10.8. Beamsplitter cubes are usually mounted on prism mounts, which provide angle adjustments similar to those in a mirror mount. To hold the cube onto the mount, loosen the clamp on the clamping arm, lower the arm until the nylon-tipped setscrew just touches the cube, then tighten the clamp. Finally, gently tighten the nylon-tipped setscrew by a half-turn or so; overtightening can bend the support post for the clamping arm, and cause stress in the cube which impairs its performance.

Polarizers and Waveplates

Control of polarization is essential for many physics experiments and applications. Polarization is controlled with beamsplitters (see above), polarizers, and waveplates.

The ideal polarizer only transmits light that is polarized along the transmission axis. Real polarizers are quantified with the "extinction ratio." This is defined, assuming a polarized input beam, as the ratio of maximum transmission power (transmission axis aligned with input polarization) to minimum transmission power (transmission axis perpendicular to input beam). For most applications, film polarizers (the least expensive kind) are adequate; their extinction ratio is at least 100 and is more than 500 over most of the visible range.

As with lenses, film polarizers are mounted using retaining rings. However, excess stress can impair the performance of the polarizer. So, they should be mounted using a "stress-reducing retaining ring," which incorporates an elastomer O-ring. The retaining ring should be tightened only until it is slightly snug – enough to keep the polarizer from moving, but no more.

Waveplates are used to convert between linear and circular polarization, and to change the angle of polarization. As with polarizers, they have a unique axis that is perpendicular to the laser direction, and which determines their action; for waveplates, this is called the "fast axis." See Section 11.2 for more details. Waveplates are more sensitive to stress than polarizers, so I recommend purchasing the premounted version, even though it is a little more costly.

The transmission axis (for polarizers) and the fast axis (for waveplates) is marked by the manufacturer, often by a line on the edge of the optic. However, the marking for a particular polarizer or waveplate can be off by as much as 5°. Therefore, when mounting the optic, it's important to determine the actual transmission or fast axis experimentally. For polarizers, you can roughly check the transmission axis by observing light reflected off the floor, e.g., from the window (see Figure 10.9). This light is partly polarized parallel to the floor, so that you should see the least reflected light when the polarization axis

FIGURE 10.9 Roughly checking polarization axis. Left: the light from the overhead lights down the hall is reflected off the floor. The reflected light is mostly horizontally polarized. The transmission axis of the polarizer is horizontal, so the light gets through. Right: the transmission axis is vertical, so the light is blocked.

is vertical. To find the polarization axis more precisely, mount the polarizer on a breadboard or optical table, and use a polarized laser or another polarizer with known transmission axis.

To find the true fast axis of a quarter wave plate (qwp), start with horizontally polarized light, and pass it through the qwp with the marked fast axis at 45° above the horizontal. Put a linear polarizer after the qwp and observe the strength of the transmitted beam as the linear polarizer is rotated. If the fast axis is actually at 45°, the beam emerging from the quarter wave plate will be circularly polarized, so that the strength of the beam transmitted through the linear polarizer is independent of the polarizer angle. If it isn't, make small adjustments to the angle of the qwp until the beam strength doesn't vary as the linear polarizer is rotated. The true fast axis of the qwp is now at 45°.

To find the true fast axis for a half wave plate (hwp), start with horizontally polarized light, and pass it through the hwp with the marked fast axis at 45° above the horizontal. The light emerging from the hwp should be linearly polarized, with the polarization reflected about the fast axis. So, if the fast axis is truly at 45°, the light emerging from the hwp should be polarized vertically. Use a linear polarizer after the hwp to measure the actual polarization angle. The true fast axis is at half this angle.

Polarizers and waveplates should be installed in rotary mounts, so the angle of the transmission axis or fast axis can be easily varied. It's worth the small extra expense to purchase rotary mounts with adjustable zero, so that the transmission or fast axis can be easily and accurately aligned with the scale on the mount.*

For most rotary mounts, the angle is read at the top, but usually one wants to measure the angle of the transmission or fast axis relative to the horizontal (see Figure 10.10a). Therefore, you should mount the optic with the transmission or fast axis aligned with the 90 degree mark of the rotary mount. Then, when you read "0" degrees (at the top of the rotary mount), the axis is horizontal. **IMPORTANT: although this is the smart way to mount optics, many labs use the convention of aligning the transmission or fast axis with the 0 degree mark on the rotary mount. If you're using an optic that was mounted by someone else, make sure you know what convention they used. If you mount the optic yourself,**

---

* It is possible, with patience, to align the axis of the optic with the numerical scale on a rotary mount without adjustable zero. First, make a tiny mark (about 1–2 mm long) with an extra fine point Sharpie on the front surface of the optic at the position of the axis of the optic. Set the optic into the mount, with the Sharpie mark about 25° counterclockwise from the desired orientation and tighten the retaining ring with the spanner. The Sharpie mark will likely still be too far counterclockwise. Loosen the retaining ring slightly while exerting as little downward pressure as possible with the spanner, then retighten while exerting more downward pressure. You should find that the Sharpie mark has advanced a little clockwise. Repeat until it's at the desired position.

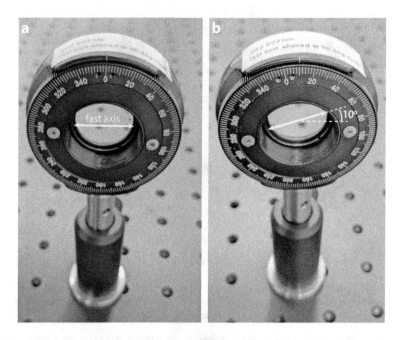

FIGURE 10.10   (a) Rotary mount marked at top, but angle measured relative to horizontal. (b) optic rotated by 10°.

mark the mount clearly, e.g., "fast axis aligned with 90 deg mark." Also, you should orient the mount so that the markings are on the "downstream" side of your setup, i.e., so that when you're look-ing at the markings you are looking back up toward the laser. This way, when you rotate the optic 10° counterclockwise, the reading on the mount (read off the top of the mount) will be "10" degrees, and the transmission or fast axis will be at 10° relative to the horizontal (see Figure 10.10b).

10.5 Beam Expanders

It is common to need to expand the beam to a larger but still collimated beam (one having parallel rays). There are two ways of doing this, as shown in Figure 10.11. The version with the diverging lens is preferred, because it's more compact and avoids concentrating the beam to a small volume. (For a high power beam, this can heat the air to the extent of causing fluctuations in the beam path.) However, for low power lasers such as those used in this chapter and Chapter 11, one often uses the version with converging lenses, since there are usually more focal lengths available in a typical lab.

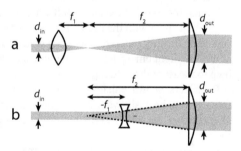

FIGURE 10.11   (a) Keplerian beam expander. (b) Galilean beam expander; note that $f_1 < 0$.

Concept test (answer below*): What is d_{out}/d_{in} for the version in Figure 10.11a, in terms of the focal lengths? How about for the version of Figure 10.11b?

10.6 Alignment

As you'll learn in Lab 10A.2 "Precision optical alignments," good alignment of a laser requires two mir-ror bounces. There are three common ways to arrange this, as shown in Figure 10.12; the choice between them depends on where you need to fit the rest of your optics.

---

* Both parts: $d_{out}/d_{in} = |f_2/f_1|$.

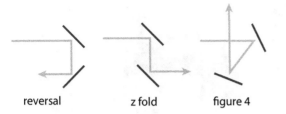

reversal   z fold   figure 4

FIGURE 10.12 Three methods for achieving two mirror bounces.

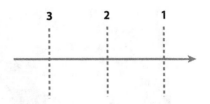

FIGURE 10.13 The beam will pass through the left two elements before hitting the rightmost one. Install the farthest downstream element (number 1) first, then number 2, then number 3.

It is usually best to start with a beam that is aligned with the grid holes (as instructed in Lab 10A.2 "Precision optical alignments"), and then position each new element along this beam, without moving the beam itself. If there will be several optical elements in a row, start with the one that is farthest "downstream," i.e., farthest from the laser, as shown in Figure 10.13. That way, you have a clean beam for alignment, one that is unaffected by upstream optics.

When you introduce a new optical element, you should center it on the beam. If the element is a 1-inch circular optic, targets such as the one shown in Figure 10.14 make this easy.

Usually, you want to align an optical element exactly perpendicular to the beam. One way to do this is by observing the small part of the beam that is back reflected from the element. Align the element so that the back reflected beam goes straight back in the direction of the incident beam. Sometimes, it is helpful to have the incident beam come through a card with a hole in it, so that the back reflection can be seen more easily, as shown in Figure 10.15.

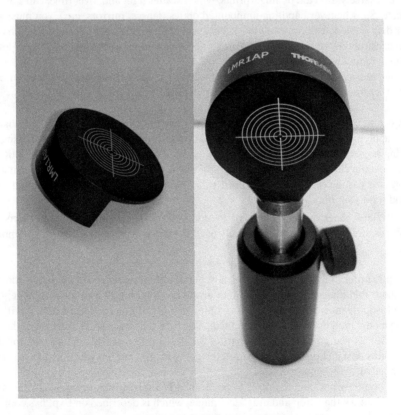

FIGURE 10.14 A target can be hung on a one-inch circular optic for alignment.

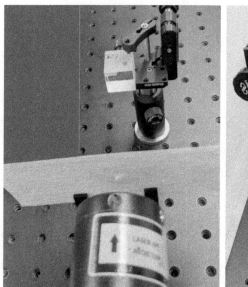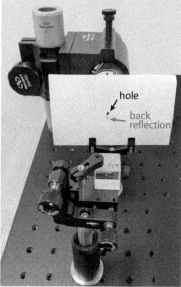

FIGURE 10.15   Back reflection method of alignment, including card with hole.

10.7 Protection, Storage, and Cleaning

All optics are adversely affected by fingerprints, which scatter light and, over time, can actually etch the optical surface. So, whenever handling unmounted optics (lenses, mirrors, etc.), you must wear gloves. Some prefer disposable latex or nitrile gloves, because they provide better dexterity, while others prefer cloth gloves because they're reusable. You can handle mounted optics without gloves, as long as you're very careful not to touch the optical surfaces.

Despite your best efforts at avoiding fingerprints, it's inevitable that you will need to clean optics. There are two cardinal rules, which must **never** be violated:

1. Use a can of compressed gas or a rubber bulb to blow off the optic, removing all visible dust, before applying other cleaning methods. Otherwise, the dust will act as an abrasive in later cleaning stages, scratching the optic. Do not use your own breath to blow off the optic, as there is always some chance of accidentally getting tiny droplets of spit on it. (Ick!) If using a can of compressed gas, be sure to hold it within 15° of vertical; if you tip it further, liquid will come out and stain your optic.

2. Use only lens paper, a clean cloth, or a cotton swab for cleaning. Other products, such as KimWipes, are too scratchy. A previously used cloth may have abrasive particles embedded in it.

To clean an optic (after blowing with air), the most commonly recommended method is the "tissue with forceps" method: fold a piece of lens tissue into a wad, grab with a pair of locking forceps, and apply a few drops of isopropanol. Shake any excess isopropanol off; if you have too much, it will get into the threads of the mount, and draw dirt out from them. Wipe the tissue across the surface of the optic, rotating the lens tissue as you go so that a new part of the tissue is constantly coming into contact with the optic; this prevents dust pickup up by one part of the tissue from scratching the optic. Discard and replace the tissue as needed.

If the above doesn't work, and the optic to be cleaned is a mirror that protrudes from its mount (the usual case), you can try the "drop and drag" technique, which is described on ExpPhys.com.

If the above techniques are not effective, you have two options:

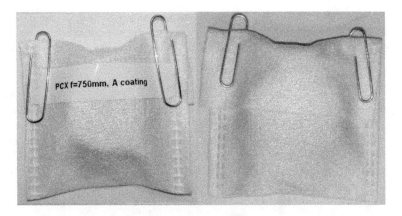

FIGURE 10.16 Cloth bag for storing optics elements, kept closed with paper clips.

(1) Remove the optic from its mount, and clean with the full immersion or "drop and drag" techniques; see the link on ExpPhys.com for a full description of these techniques, which are very gentle.

(2) In research, one must balance the need for getting things done quickly (which truly is quite important) with the need to protect optics for decades of useful life. It takes a good deal of time to remove the optic from its mount, clean it using one of the techniques above, remount it, and re-align it. Therefore, **if the optic is not expensive** (e.g., a standard mirror with a protective coating or a standard lens for optical wavelengths), you can clean it without removing it from the mount by breathing on it with an open mouth ("Haaaaah") to coat it with a thin layer of moisture from condensation, then wiping it with a *clean* cloth (not your shirt!) or lens tissue, just using your gloved finger (rather than forceps) to move the cloth or tissue, similar to the way you'd clean eyeglasses. *Make sure you have observed cardinal rule 1 (blowing it off) before you wipe it. If the optic is anything other than a basic inexpensive lens or mirror, you should spend the time to remove it from the mount and clean it in one of the more gentle ways.*

You should store optics elements in their original containers, or in labeled cloth bags, to prevent scratching. Small cloth bags are available for low cost from optics suppliers. Keep them closed with paper clips, as shown in Figure 10.16.

10.8 Organization

More than in any other area of experimental physics, it is essential to keep your optics well organized. I'll be more than usually dictatorial here, to efficiently impart lessons I've learned from bitter experience. Each optical element or piece of hardware is a significant expense and put together the costs add up more quickly than you can imagine. Without significant attention to organization, you will waste a great deal of time and money.

Labeling

Unlike electronic components, optics elements are usually not labeled after you take them out of the box. However, each one may be suited only to a limited wavelength range, and there is no way to tell this by looking. There is no way to tell the difference between a quarter wave plate and a half wave plate just by looking. Some more exotic optic elements look just like a quarter wave plate but have a completely different function. Although you can re-measure the focal length of a lens, you waste time doing so. So, you must label every optic element as soon as it is put in a mount, as shown for example in Figure 10.10. The label must include the wavelengths for which the optic is designed; in the case of lenses, you can use a shorthand for the type of anti-reflective coating, as long as it will be understood clearly by others working in your lab. If you remove the optic from the mount, you must immediately place it in a labeled box or bag. You must use a label maker to create the labels; handwritten labels rub off quickly.

Storage

You should store optics elements and hardware in a storage cabinet with multiple small drawers. If you're the one setting this up, spend some time figuring out an organization scheme that makes sense to you, or use the one recommended on ExpPhys.com. Have a cardboard box underneath your optics bench, and as soon as you remove an optic element from its small box or bag, put the box or bag in the cardboard box. That way, when it's time to disassemble your setup, you have all the containers ready. Don't delay in returning the optics parts to the proper storage cabinets when you disassemble the setup.

Tools Organization

Mount a magnetic strip on or near your optics bench. Keep the most common Allen wrenches stuck to it.

Lab 10A The Quantum Eraser,* Simple Version[†]

> **Learning goals:** Become familiar with research-grade mounting systems for optics and with beam splitters.
>
> **Pre-lab reading:** 10.1–10.8, plus the coverage of polarization of light in your intro physics book
>
> **Safety notes:** You will be using a class 2 laser. Therefore, you do not need to use safety glasses or other unusual precautions. Do not intentionally look into the beam or point it at people. You should adhere to the other standard laser safety practices described in Section 10.1.

10A.1 Introduction

A central theme in quantum mechanics is the interplay between the behavior of a physical system and experimentalists' knowledge about the state of the system. The idea that the reality of a physical system depends on the information about the system that can be measured with the apparatus takes some getting used to, but one cannot understand quantum experiments without accepting this principle. The canonical example of this phenomenon is the two-slit experiment, in which the amplitudes for photons passing through one slit or the other interfere and result in wavelike interference patterns – but only if it is not possible *even in principle* to determine which slit a photon went through with the apparatus used.

Classical Polarization and Interference

As you've reviewed in your intro physics text, for light passing through a polarizer, only the component of the electric field parallel to the transmission axis gets through, i.e., $E_{\text{transmitted}} = E_{\text{incident}} \cos\theta$, where θ is the angle between the incident polarization and the transmission axis. Since the intensity is proportional to E^2, we see that Intensity $\propto \cos^2\theta$; this is Malus's Law.

In the classical view, the interference pattern observed on a screen is due to constructive or destructive interference between the various electric (or magnetic) field waves traveling via all possible paths from the source to the point on the screen.

---

* Parts of this lab were written by Jacob Seely and David Colletta, and are used with their permission.
[†] You can do a more sophisticated version, using one photon at a time, in Lab 20C. That lab gives much deeper insights into the quantum mechanics.

Quantum Polarization and Interference

In the quantum view, to understand the interaction of a photon with a polarizer, we resolve the incident wavefunction into a new one expressed in the basis of the axes parallel and perpendicular to the transmission axis. Putting this anthropomorphically, the photon is required to "choose" whether it has polarization along the transmission axis (and so gets transmitted) or instead has polarization perpendicular to the transmission axis (and so gets blocked). The probability amplitude Ψ of "choosing" to be parallel to the transmission axis is proportional to $\cos\theta$, and the fraction of photons that gets through is proportional to $|\Psi|^2 \propto \cos^2\theta$, in agreement with Malus's Law.

In the quantum view, the interference pattern observed on a screen is due to constructive or destructive interference between the various parts of the wavefunction Ψ traveling via all possible paths from the source to the point on the screen. If there is a way to determine the path that the photon traveled, then the multiple paths are collapsed to a single path, and there is no interference.

10A.2 Precision Optical Alignments

Walking the Beam

In many optical experiments, you need to get the beam to go precisely through the centers of several optics elements. The procedure for doing so requires two mirrors, and by the end of this exercise you'll see why.

If you are planning to do 10A.3 "Mach-Zender Interferometer and the Quantum Eraser," you should use a polarized laser* for this exercise, with the polarization set at 45° relative to the horizontal. Use a linear polarizer to check the polarization angle.

Create the setup shown in Figure 10.17, with mirror 1 mounted on a post and postholder and mirror 2 mounted on a short pedestal. You are provided with two cards with small holes in them, also mounted on short pedestals. Position them so the beam reflected off mirror 2 passes through both holes, with the cards separated by about a half meter along the beam direction; clamp everything in place.

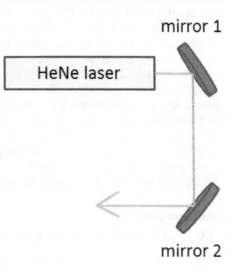

FIGURE 10.17 Initial setup.

Change the settings of both angle control knobs on both of the mirrors, so that the beam is nowhere close to going through the holes. Now, use the combination of the four adjustments (two knobs on two mirrors) to get the beam going back through both holes; you're only allowed to adjust the mirrors – don't move the cards. Figure out and record the systematic procedure for making the adjustments. Why do we need two mirrors to accomplish the task of getting the beam to go through both holes?

Before you go on, it's important that you have a full understanding of the above questions, so please check the "answers" on ExpPhys.com.

Aligning a Laser with the Grid of Holes

Many experiments require aligning the beam exactly (within a very small tolerance) parallel to the grid of holes on the optical table or breadboard.

Your goal is to adjust the two mirrors so that the reflection from mirror 2 is precisely parallel (horizontally) to the rows of screw holes in the table, and also precisely parallel in the vertical direction to

---

* If a polarized laser is not available, you can use an unpolarized laser with a polarizer set to 45° in front of it. However, as the laser warms up, the distribution of polarization modes changes, so the intensity of the beam emerging from the polarizer varies over periods of several seconds up to several minutes. For this lab, that's okay, as long as you know to expect it.

the table surface. The technique you will learn can be applied to a wide variety of optical setups.

Mount an iris mounted to a baseplate in such a way that the height and horizontal position can be continuously adjusted. Near the mirror (i.e., mirror 2), put two screws in adjacent table holes along the beam path, and screw them all the way in. Put two more screws as far down the table as possible. Roughly adjust the tilt and (if needed) positions of both mirrors by hand so that the beam appears parallel to the table and to the rows of screw holes. Now place the base plate used to support the iris with its edge flat against the closest set of screws and adjust the position of the iris so the beam goes through the iris. DO NOT ADJUST THE MIRROR! (This is "step 2" in Figure 10.18. Now move the iris to the farther set of screws and use the fine angle adjustments on the mirror to send the beam through the iris. DO NOT ADJUST THE IRIS! (This is "step 3.")

Now bring the iris back to the first set of screws and repeat this process a few times until the beam goes through the iris at both positions

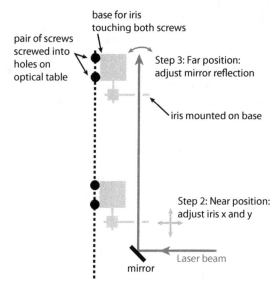

FIGURE 10.18 Method for aligning beam with grid of holes. Image courtesy of Prof. Enrique Galvez, who developed this method.

without any adjustment. If the beam hits the edge of the mirror as a result of this alignment, you must adjust the placement of this mirror and redo the alignment.

Your beam should now be parallel to the screw holes (horizontally) and to the breadboard surface (vertically). That's an important accomplishment!

10A.3 Mach-Zender Interferometer and the Quantum Eraser*

Your goal is to create and align the version of a Mach-Zender interferometer shown in Figure 10.19.

This requires very precise alignment of all the components, so if you can get it to work you should be proud of yourself! We assume this is the first time you've created such a precise setup, so we'll walk you through it. The components you are adding should all be on short pedestals, held in place with "forks."

Insert Polarizing Beam Splitter Cube and Align the Beam with the Table

The polarizing beam splitter (PBS) splits the beam with one of the exit beams polarized horizontally and the other vertically. If the incident beam is polarized at 45°, then each of the exit beams has equal power. Ensure the cube is placed in its holder so that light is not blocked by the holder and so the incident beam hits the cube on the side with the black dot on it. The reflected beam should roughly be parallel to a row of holes. Using the iris and the same procedure as above, align the beam reflected out of the cube with the grid of holes.

Q1: Why isn't the alignment of the transmitted beam affected by your adjustments to the PBS cube's tilt as you align the reflected beam?

Insert Mirrors 3 and 4, and Align the Beams

Place one mirror in the path of the beam that is reflected from the PBS cube and a second mirror in the path of the beam that is transmitted. Tilt them so that they are both reflecting the laser toward the spot

---

* This is adapted from a lab by Prof. Mark Beck of Reed College.

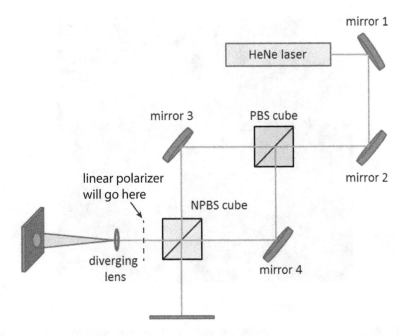

FIGURE 10.19 Mach-Zender interferometer for quantum eraser experiment.

where you will place the non-polarizing beam splitter (NPBS) cube. Using the iris technique, align the beams that are reflected from these mirrors with the grid of holes. Erect a screen in the path of each beam just off the edge of the optical table. You will use these screens to align the beams from the two paths, and eventually to observe the interference pattern.

Insert NPBS cube and Align the Beams with the Table

This is the trickiest part of the alignment, since it requires that both beams hit the inside of the cube at the same place. First, roughly position and secure the cube so that the faces of the cube are perpendicular to the beams coming from mirrors 3 and 4 and the beams appear to meet in the same place inside the cube. It doesn't matter which side the black dot is on, because the setup is symmetrical – beams are coming from both sides. If you have correctly aligned the beams with the table, you should only need to adjust the horizontal position of the beams so that they overlap inside the cube by changing the angles of the mirrors, as they should be at the same vertical level already.

The goal is to have both pairs of beams coming out of the cube overlap so that you only see one spot on each of the screens. More than likely, you will see two spots on the screen at this stage of the alignment. This is because the beams that are transmitted through the cube are unaffected by the tilt of the cube, but the reflected beams are. Adjust the fine tilt on the NPBS cube mount until the spots are as close together on the screens as you can get them. You may find that the spots no longer meet at the same point inside the cube. In this case, the cube must be repositioned slightly and the tilt alignment must be repeated. After a few iterations, both conditions for alignment (spots meet at the same place inside the cube, and spots on the screens overlap) should be satisfied.

When the alignment seems good enough, put a diverging lens in the path of either beam after the cube. On the screen placed after the diverging lens, you should see a large circular spot of laser light, without any strong interference fringes.

Adding the Final Polarizer

Now, add a polarizer at 45° at the spot labeled "linear polarizer will go here" in Figure 10.19. (Don't screw it in place.) On the screen placed after the diverging lens, you should now see a "bull's eye"

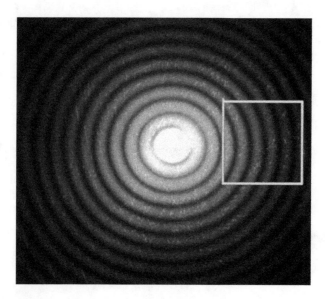

FIGURE 10.20   Bull's eye interference pattern.

interference pattern, with alternating bright and dark interference fringes, as shown in Figure 10.20. (In fact, you will likely only see part of it, as suggested by the small square in the figure.) This occurs because of angle-dependent path-length differences between the beam that bounces off mirror 3 and the beam that bounces off mirror 4, in a manner very similar to the angle-dependent path-length differences that occur in a two-slit interference experiment. In this case, there is circular symmetry around the laser beam axis, so instead of getting a pattern of vertical interference lines (as in an experiment with two vertical slits), you get the "bull's eye" pattern. If you do not see the pattern or part of it, try making *small* adjustments to the NPBS angles. If you still don't see fringes, ask your instructor for help.

Make small adjustments to the NPBS angles until you're seeing part of the pattern that is a little off center, as suggested by the square in Figure 10.20. (This makes it easier to tell whether or not strong fringes are present.)

Understanding Interference, and the "Quantum Eraser"

Now remove the polarizer you just added. **Q2: What do you observe on the screen?**

To understand what happened, we have to consider the effect of the polarizing beam splitter cube.

Q3: Using the linear polarizer, determine the polarization of both the transmitted and reflected beam after the PBS cube. Given the results of this exercise, and thinking in the classical picture, why do we see no interference pattern when the polarizer is removed? Why do we see interference when the polarizer is present?

Q4: Explain your results using the quantum mechanical picture. *Hint: this experiment is very similar to a two-slit interference experiment, with the beam that bounces off mirror 3 playing the role of the light emerging from one of the slits, and the beam that bounces off mirror 4 playing the role of the light that emerges from the other slit. As you know, for the two-slit experiment, if you do any measurement that allows you to determine which slit the light went through, the interference pattern is destroyed. So here, if there is even the possibility of doing such a measurement, the interference pattern will be destroyed.*

Q5: Why is this experiment called the quantum eraser? *Hint: what is being erased when you add the polarizer at 45°? Think in terms of whether a measurement would allow you to determine which path a particular photon had followed.*

Q6: Why was it important to use short pedestals when building this setup, rather than posts and post holders?

11

Laser Beams, Polarization, and Interference

Justin Peatross and Michael Ware

CONTENTS

Better laser, better science, better life.

– Ruxin Li

We can easily forgive a child who is afraid of the dark; the real tragedy of life is when men are afraid of the light.

– Plato

11.1 Introduction

Lasers are ubiquitous in modern society. They transmit the vast bulk of all information on the internet through fiber optics, are commonly used for medical tasks from dentistry to surgery, play crucial roles in everyday technologies like printing and manufacturing, and are common in almost all disciplines of scientific inquiry. Getting a basic understanding of how to manipulate and control them is an essential task for most experimental disciplines.

Learning Goals

This chapter provides theoretical context and laboratory exercises that are of practical relevance to experimental laser physicists. We first examine polarization devices, namely polarizers and wave plates, which are commonly used to manipulate and characterize light polarization. We only consider light with a definite polarization; we do not treat unpolarized or partially polarized light. In the second section of this chapter, we examine the field structure of a common laser beam, including its focal characteristics and wavefront curvature. These laboratory exercises are designed to illustrate concepts and develop skills used in laser experiments.

Additional Reading

A more in-depth treatment of the concepts explored in this chapter can be found in chapters 3, 6, and 10 of *Physics of Light and Optics* by Peatross and Ware, available at optics.byu.edu.

Pre-Lab Questions

Q1 We frequently encounter natural sources of polarized light in everyday life, for example, reflections of sunlight from the surface of water or the scattered blue light coming from the sky. Observe these sources of light through a polarizer and rotate it to different angles. Explain how these sources of light become polarized.

Q2 What is circularly polarized light and how does it differ from linearly polarized light? Why would reflecting circularly polarized light from a mirror strongly distort the polarization state (i.e., cause it to become elliptical)?

Q3 Some people have a misconception that a laser beam stays the same size as it propagates indefinitely, or that it has a uniform intensity within a circular cross-section, suddenly dropping to zero outside of that circle. Why are these unrealistic physical characteristics?

Q4 Verify that Equation (11.19) is a solution to Equation (11.18).

11.2 Polarization

Consider a single-frequency electric-field plane wave, propagating in the z-direction in free space. (A magnetic field wave necessarily accompanies the electric field.) Maxwell's equations require the electric field to point perpendicular to $\hat{z}$, and so the field may be written as

$$E(z,t) = \left(xE_x + yE_y\right)e^{i(kz-\omega t)} \tag{11.1}$$

The real part of this expression corresponds to the physical field. Writing the field in this complex format allows for convenient tracking of phase.* In addition to amplitude and phase of the field, the vector $\hat{x}E_x + \hat{y}E_y$ describes the *polarization* of the light field. If only E_x is nonzero, then we would say the light is polarized along $\hat{x}$ (horizontal), and if only E_y is nonzero, then we would say that the light is polarized along $\hat{y}$ (vertical). When both E_x and E_y are nonzero, the two components can also have different phases, which makes for interesting polarization states, as will be discussed later.

We now consider two types of devices that can manipulate polarization: a *polarizer* and a *wave plate*. A *linear* polarizer, commonly just called a polarizer, absorbs the vector component of the electric field that is perpendicular to the *transmission axis* so that the transmitted light is linearly polarized, as shown in Figure 11.1. A *wave plate* is made from a birefringent crystal wherein the index of refraction that light experiences depends on the orientation of its polarization. Wave plates have the appearance of thin windows and do not absorb light. Rather, a wave plate introduces a relative phase delay between field components oriented along the *slow axis* and the *fast axis*, as shown in Figure 11.2. These two axes

---

* Note that the complex field components may be written in the form $E_x = |E_x|e^{i\varphi_x}$ and $E_y = |E_y|e^{i\varphi_y}$, where φ_x and φ_y are their phases. We may then write $E(z,t) = x|E_x|e^{i(kz-\omega t+\varphi_x)} + y|E_y|e^{i(kz-\omega t+\varphi_y)}$. Using Euler's formula $e^{i\theta} = \cos\theta + i\sin\theta$, the physically relevant expression for the field (i.e., the real part) is seen to be $\hat{x}|E_x|\cos(kz-\omega t+\varphi_x) + \hat{y}|E_y|\cos(kz-\omega t+\varphi_y)$.

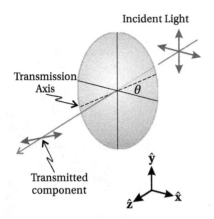

FIGURE 11.1 Light transmitting through a polarizer oriented with transmission axis at angle θ from x-axis.

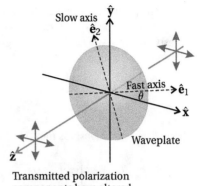

FIGURE 11.2 Wave plate interacting with a plane wave.

are associated with higher and lower refractive indices, respectively.

To describe how a polarizer or a wave plate affects light, we need to write the electric field components in terms of a vector basis that is natural to the orientation of the optical element. Let the direction of the transmission axis of a polarizer be specified by the unit vector $\hat{e}_1$ and the absorption axis of the polarizer be specified by $\hat{e}_2$. Let these orthogonal unit vectors lie in the x-y plane with $\hat{e}_1$ oriented at an angle θ from the x-axis, as shown in Figure 11.2. Similarly, if we are considering a wave plate, we can let $\hat{e}_1$ and $\hat{e}_2$ align with the fast and slow axes of a wave plate, respectively. By inspection of the geometry in Figure 11.3, $\hat{x}$ and $\hat{y}$ are connected to $\hat{e}_1$ and $\hat{e}_2$ via

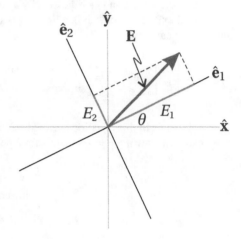

FIGURE 11.3 Electric field components written in the $\hat{e}_1, \hat{e}_2$ basis.

$$\hat{x} = \hat{e}_1 \cos\theta - \hat{e}_2 \sin\theta$$

$$\hat{y} = \hat{e}_1 \sin\theta + \hat{e}_2 \cos\theta \tag{11.2}$$

Substitution of (11.2) into (11.1) yields for the electric field

$$E(z,t) = (\hat{e}_1 E_1 + \hat{e}_2 E_2) e^{i(kz - \omega t)}, \tag{11.3}$$

where

$$E_1 \equiv E_x \cos\theta + E_y \sin\theta$$

$$E_2 \equiv -E_x \sin\theta + E_y \cos\theta. \tag{11.4}$$

We can introduce the effect of a polarization device, either a polarizer or a wave plate, by injecting a factor ξ acting on the $\hat{e}_2$ component. After traversing the device, the field is written

$$E_{\text{after}}(z,t) = (\hat{e}_1 E_1 + \hat{e}_2 \xi E_2) e^{i(kz - \omega t)} \tag{11.5}$$

In the case of the polarizer, we use $\xi = 0$, while for a wave plate we use $\xi = e^{i\varphi}$, which introduces a relative phase shift φ between the two polarization components. We now have the field after the polarizer, but in the $\hat{e}_1$, $\hat{e}_2$ basis. To write the fields in the original $\hat{x}$, $\hat{y}$ basis we invert (11.2) to obtain

$$\hat{e}_1 = \hat{x}\cos\theta + \hat{y}\sin\theta$$

$$\hat{e}_2 = -\hat{x}\sin\theta + \hat{y}\cos\theta \tag{11.6}$$

Substitution of these relationships into (11.5) together with the definitions (11.4) for E_1 and E_2 yields

$$\begin{aligned}
E_{\text{after}}(z,t) = &\hat{x}\left[E_x\left(\cos^2\theta + \xi\sin^2\theta\right) + E_y\left(\sin\theta\cos\theta - \xi\sin\theta\cos\theta\right)\right]e^{i(kz-\omega t)} \\
&+ \hat{y}\left[E_x\left(\sin\theta\cos\theta - \xi\sin\theta\cos\theta\right) + E_y\left(\sin^2\theta + \xi\cos^2\theta\right)\right]e^{i(kz-\omega t)}
\end{aligned} \tag{11.7}$$

Notice that if $\xi = 1$ (i.e., no modification by a device), then (11.7) reduces back to (11.1).

The field exiting the device is a linear mixture of E_x and E_y, which can be represented using matrix algebra. We can rewrite (11.7) as

$$\begin{bmatrix} E_x \\ E_y \end{bmatrix}_{\text{after}} e^{i(kz-\omega t)} = \begin{bmatrix} \cos^2\theta + \xi\sin^2\theta & \sin\theta\cos\theta - \xi\sin\theta\cos\theta \\ \sin\theta\cos\theta - \xi\sin\theta\cos\theta & \sin^2\theta + \xi\cos^2\theta \end{bmatrix} \begin{bmatrix} E_x \\ E_y \end{bmatrix} e^{i(kz-\omega t)} \tag{11.8}$$

The 2×2 matrix in this equation is called a *Jones matrix*.

We can write down the Jones matrix for a polarizer by inserting $\xi = 0$:

$$\begin{bmatrix} \cos^2\theta & \sin\theta\cos\theta \\ \sin\theta\cos\theta & \sin^2\theta \end{bmatrix} \tag{11.9}$$

Polarizer with transmission axis at angle θ

The crystal for a *quarter wave plate* is cut to a thickness such that light polarized along the slow axis is delayed by a quarter cycle ($\pi/2$ radians) relative to light polarized along the fast axis. We therefore introduce the phase factor $\xi = e^{i\pi/2} = i$, which corresponds to a quarter of a full 2π phase cycle. The Jones matrix for that device becomes

$$\begin{bmatrix} \cos^2\theta + i\sin^2\theta & (1-i)\sin\theta\cos\theta \\ (1-i)\sin\theta\cos\theta & \sin^2\theta + i\cos^2\theta \end{bmatrix} \tag{11.10}$$

Quarter wave plate with fast axis at angle θ

A *half wave plate* delays the slow axis by a half cycle. To get the matrix for a half-wave plate, we introduce the phase factor $\xi = e^{i\pi} = -1$, and the Jones matrix becomes:

$$\begin{bmatrix} \cos^2\theta - \sin^2\theta & 2\sin\theta\cos\theta \\ 2\sin\theta\cos\theta & \sin^2\theta - \cos^2\theta \end{bmatrix} = \begin{bmatrix} \cos 2\theta & \sin 2\theta \\ \sin 2\theta & -\cos 2\theta \end{bmatrix} \tag{11.11}$$

Half wave plate with fast axis at angle θ

These matrices operate on *Jones vectors*. The Jones vector for light polarized along the x-dimension is $\begin{bmatrix} 1 \\ 0 \end{bmatrix}$, where we have factored out the vector the overall field strength E_0 in order to just concentrate on the bare essence of the polarized state. The Jones vector for light polarized along the y-dimension is $\begin{bmatrix} 0 \\ 1 \end{bmatrix}$, and for light linearly polarized along some angle α, the Jones vector is $\begin{bmatrix} \cos\alpha \\ \sin\alpha \end{bmatrix}$. The Jones vectors for several polarization states are given in Table 11.1.

Example: Horizontally polarized light traverses a quarter-wave plate with fast axis at $\theta = 45°$. What is the polarization of the beam that exits the device?

Solution: At $\theta = 45°$, the Jones matrix for the quarter-wave plate (11.10) reduces to

$$\frac{e^{i\pi/4}}{\sqrt{2}} \begin{bmatrix} 1 & -i \\ -i & 1 \end{bmatrix} \tag{11.12}$$

Quarter wave plate with fast axis at $\theta = 45°$

The overall phase factor $e^{i\pi/4}$ in front is unimportant unless the beam of light is to interfere with another beam. When the quarter wave plate operates on the horizontally polarized light, we get

$$\frac{1}{\sqrt{2}} \begin{bmatrix} 1 & -i \\ -i & 1 \end{bmatrix} \begin{bmatrix} 1 \\ 0 \end{bmatrix} = \frac{1}{\sqrt{2}} \begin{bmatrix} 1 \\ -i \end{bmatrix} \tag{11.13}$$

The resulting Jones vector is shorthand notation for the following field:

$$\mathbf{E}(z,t) = \mathrm{Re}\left[E_0 \frac{1}{\sqrt{2}} e^{i(kz-\omega t)} \hat{\mathbf{x}} - E_0 \frac{i}{\sqrt{2}} e^{i(kz-\omega t)} \hat{\mathbf{y}} \right] \tag{11.14}$$

$$= \frac{E_0}{\sqrt{2}}\left[\cos(kz - \omega t)\hat{\mathbf{x}} + \sin(kz - \omega t)\hat{\mathbf{y}} \right]$$

Right circular polarization

The field in the y-dimension leads the field in the x-dimension by a quarter cycle. That is, the behavior seen in the y-dimension happens in the x-dimension a quarter cycle later. The field never goes to zero simultaneously in both dimensions. The strength of the electric field is constant, and it rotates in a circular pattern in the x-y dimensions, similar to that seen in Figure 11.4. This type of field is called *circularly polarized*, spiraling in the *right-handed* sense in this case. We use the convention of calling right-circular polarization the state that spirals in the same sense as a common right-hand screw. The opposite convention is used by some other authors.

In the above example, a quarter-wave plate (properly oriented) can change linearly polarized light into circularly polarized light. A quarter wave plate can perform the reverse operation as well. When traversing a quarter wave plate with fast axis oriented at θ, right-circularly polarized light produces linear polarization oriented along $\theta + 45°$ and

TABLE 11.1

Jones vectors for several common polarization states

Linearly polarized along x	$\begin{bmatrix} 1 \\ 0 \end{bmatrix}$
Linearly polarized along y	$\begin{bmatrix} 0 \\ 1 \end{bmatrix}$
Linearly polarized at angle α (measured from the x-axis)	$\begin{bmatrix} \cos\alpha \\ \sin\alpha \end{bmatrix}$
Right circularly polarized	$\frac{1}{\sqrt{2}}\begin{bmatrix} 1 \\ -i \end{bmatrix}$
Left circularly polarized	$\frac{1}{\sqrt{2}}\begin{bmatrix} 1 \\ i \end{bmatrix}$

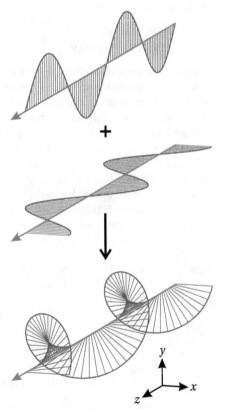

FIGURE 11.4 The combination of two orthogonally polarized plane waves that are out of phase results in elliptically polarized light. Here we have *left-handed* circularly polarized light, spiraling in the opposite sense specified by (11.14).

left-circularly polarized light produces linear polarization oriented along $\theta-45°$. One can prove this for arbitrary θ or check it at a few specific angles such as $\theta=0$, 45°, 90°.

As seen in the next example, a half wave plate is useful for rotating the polarization angle of linearly polarized light by varying amounts while preserving the linear polarization state.

> **Example:** Calculate the effect of a half wave plate at an arbitrary θ on horizontally polarized light.
> **Solution:** Multiplying by the half-wave matrix (11.11), we obtain
>
> $$\begin{bmatrix} \cos 2\theta & \sin 2\theta \\ \sin 2\theta & -\cos 2\theta \end{bmatrix} \begin{bmatrix} 1 \\ 0 \end{bmatrix} = \begin{bmatrix} \cos 2\theta \\ \sin 2\theta \end{bmatrix} \tag{11.15}$$
>
> The resulting Jones vector describes linearly polarized light at an angle $\alpha=2\theta$ from the x-axis (see Table 11.1).

Lab 11A   Polarization and Jones Vectors

As you perform these lab exercises, align the beam carefully so that it is reasonably centered on each component, and notice back-reflections to ensure that each optical element is aligned perpendicular to the beam.

11A.1   Optical Activity

Many organic molecules have a definite *handedness* or *chirality*, meaning that the image of the molecule in a mirror is distinct from the original molecule. Life on Earth uses *right-handed* sugar molecules and *left-handed* amino acids. Because of the right-handedness of the sugar molecules in corn syrup, right- and left-circularly polarized light experience slightly different refractive indices. This phenomenon is called *optical activity*. As a result, right circularly polarized light undergoes a phase delay φ relative to left circularly polarized light as it travels through the syrup.

Determine how much right-hand circularly polarized light (with $\lambda=633$ nm) is delayed (or advanced if φ is negative) with respect to left-handed circularly polarized light as it goes through approximately 3 cm of Karo syrup (the neck of the bottle).

Hint: as shown in Figure 11.5, you will want to send linearly polarized light through the syrup and use a polarizer afterward to see how the linearly polarized light reorients to a new angle. Note that you can write linearly polarized light as equal amounts of left- and right-handed circularly polarized

light: $\begin{bmatrix} 1 \\ 0 \end{bmatrix} = \frac{1}{\sqrt{2}} \left(\frac{1}{\sqrt{2}} \begin{bmatrix} 1 \\ i \end{bmatrix} + \frac{1}{\sqrt{2}} \begin{bmatrix} 1 \\ -i \end{bmatrix} \right)$. Introduce a relative phase delay φ to the right circularly polarized

light by writing

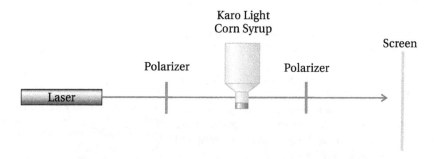

FIGURE 11.5   Lab schematic for 11A.1 "Optical Activity."

$$\frac{1}{2}\begin{bmatrix} 1 \\ i \end{bmatrix} + \frac{e^{i\varphi}}{2}\begin{bmatrix} 1 \\ -i \end{bmatrix}$$

and show that this can be written as

$$e^{i\delta}\begin{bmatrix} \cos(\varphi/2) \\ \sin(\varphi/2) \end{bmatrix}$$

where the overall phase δ is unimportant. Compare this with the Jones vector for light linearly polarized along angle α (see Table 11.1).

11A.2 Quarter Wave Plates

Create linearly polarized light oriented along $\alpha = 30°$ by placing a polarizer in the path of a laser beam ($\lambda = 633$ nm).

(a) For each of steps 1–3 below, calculate the Jones vector of the transmitted light and identify the polarization state associated with the Jones vector at each step.

(b) Measure the fraction of the original intensity that transmits through each step using a power meter and compare these measurements with the predicted values. Be sure to set the angles relative to the horizontal axis, consistent with the angle convention used in Figure 11.6. Hint: intensity is proportional to $|A|^2 + |B|^2$ for the Jones vector $\begin{bmatrix} A \\ B \end{bmatrix}$.

(1) Place a quarter wave plate next (after the initial polarizer), with fast axis at $\theta_1 = 60°$.

(2) Next, place a polarizer with its transmission axis at $\theta_2 = 90°$.

(3) Finally, place a second quarter wave plate with fast axis at $\theta_3 = 30°$; the arrangement should now be as shown in Figure 11.6.

11A.3 Circular Polarizer

Construct a right-circular polarizer by reorienting the three elements in the previous exercise: set the first quarter wave plate with fast axis at $\theta_1 = 45°$, the polarizer at $\theta = 90°$, and the final quarter wave plate with fast axis at $\theta_3 = -45°$.

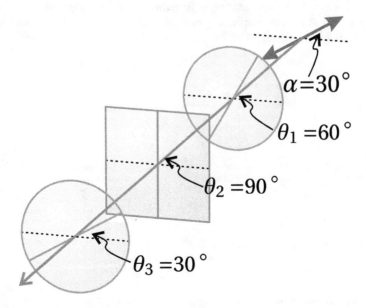

FIGURE 11.6 Arrangement for 11 A.2 "Quarter Wave Plates."

(a) Calculate the Jones matrix for this system. (Answer below.*)

(b) From the matrix, predict what the setup should do to incident right-circularly polarized light and to incident left-circularly polarized light.

(c) Check your prediction; you will first need to create right and left circularly polarized light using a separate polarizer and a quarter wave plate. You can do this by sending horizontally polarized light through a quarter wave plate set at $\theta = \pm 45°$.

11A.4   Elliptical Polarization

Create a source of unknown *elliptical* polarization by reflecting a linearly polarized laser beam polarized at $\theta = 45°$ from a metal mirror with a large incident angle ($\theta_i \approx 80°$), as shown in Figure 11.7. The electric field for elliptically polarized light spirals either in a right-hand or left-hand fashion while tracing out the shape of an ellipse. You can think of linearly polarized light and circularly polarized light as a very narrow or a very round ellipse, respectively.

(a) Demonstrate that the light is elliptically polarized by rotating a polarizer in the beam, noting that the intensity varies, but the beam cannot be extinguished.

(b) Demonstrate that a quarter-wave plate can be used to change the elliptically polarized light into linearly polarized light. Hint: position the quarter wave plate before the linear polarizer used in part (a), as shown in Figure 11.7. Adjust the orientation of both the quarter wave plate and the linear polarizer to find a combination of angles where the beam is completely extinguished.

(c) Determine the Jones vector of the elliptical polarization state just following the mirror. Hint: from the fact that the light is extinguished by the linearly polarizer, you can determine the Jones vector of the light coming through the wave plate. This linear state must equal the original (unknown) Jones vector $\begin{bmatrix} A \\ B \end{bmatrix}$ operated on by the wave plate (11.10). As you solve the matrix equation, it is helpful to know that the inverse of (11.10) is its own complex conjugate. Be sure to measure angles relative to the horizontal axis consistent with the angle convention used in Figure 11.6.

(d) Call the combination of angles found above (for the quarter wave plate and linear polarizer) combination A. You should be able to find a second, quite different combination that extinguishes the beam; call this combination B. Explain using polarization diagrams why there are two combinations that work. Repeat the process of finding the Jones vector of the reflected light, this time using the angles of combination B. You should get the same result as with

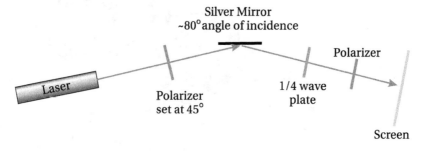

FIGURE 11.7   Schematic for 11A.4"Elliptical Polarization."

---

* $\dfrac{i}{2}\begin{bmatrix} 1 & i \\ -i & 1 \end{bmatrix}$ The factor i in front is an unimportant overall phase and can be dropped.

combination A, except for an unimportant overall complex phase factor (i.e., a factor of the form $e^{i\beta}$).

11A.5 Brewster's Angle and s- and p-Polarizations

The previous exercise illustrates that reflection from a surface can alter the polarization state of a beam. The natural basis for resolving the electric field components when reflecting from a surface is along what are called the *s* and *p*-dimensions. The *p-polarized* component of the electric field lies in the *plane of incidence*, which contains both the incident and reflected beams; *s-polarized* light has an electric field component that is perpendicular to that plane. The reflection can be quite different, depending whether the incident light is *p*-polarized or *s*-polarized. This includes significant differences in phase for reflection from metallic of multilayer mirrors. For reflection from a simple air-glass interface, the strength of the reflection can strongly differ for the two polarization components. A more complete discussion is available in chapter 3 of *Physics of Light and Optics*.

(a) As shown in Figure 11.8, use a power meter to measure the fraction of reflected power for *s*- and *p*-polarized light reflected separately from a flat glass surface at about ten different *angles of incidence*. Identify *Brewster's angle* where the reflection of *p*-polarized light vanishes. Hint: if the figure represents a top view of the experiment, then horizontal polarization corresponds to *p*-polarized light and vertical polarization corresponds to *s*-polarized light.

(b) Use a computer to plot your data with the theoretical curves given by

$$R_s = \left[\frac{\sin(\theta_t - \theta_i)}{\sin(\theta_t + \theta_i)} \right]^2 \qquad R_p = \left[\frac{\tan(\theta_t - \theta_i)}{\tan(\theta_t + \theta_i)} \right]^2$$

where θ_t is related to θ_i via Snell's law: $\sin\theta_i = n\sin\theta_t$. Note that Brewster's angle occurs when $\theta_i + \theta_t = \pi/2$ in which case $R_p = 0$ owing to an infinite denominator.

11.3 Gaussian Beams

In both scientific and industrial applications, it is often important to focus a laser to a very small spot, either to generate high intensity or to illuminate small spatial regions. In other applications, we may want the illumination to be fairly uniform over a wide area. In all of these situations, it is important to have an understanding of the structure of a basic *Gaussian laser beam*. It is sometimes necessary to "clean up" a beam to make it more like an ideal Gaussian beam.

Pre-lab question Q3 is a reminder that only certain very special configurations of electromagnetic fields are allowed by Maxwell's equations. One such configuration is a plane wave, which you likely encountered in a previous course. A more real-life solution to Maxwell's equations is a beam such as is

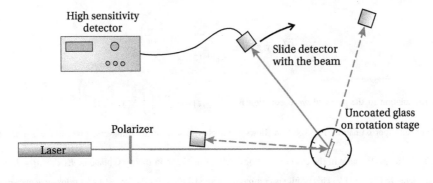

FIGURE 11.8 Experimental setup for11A.5 "Brewster's Angle and *s*- and *p*-Polarizations."

produced by many lasers. Maxwell's equations can be used to show that light, as it propagates in empty space, obeys the wave equation:

$$\nabla^2 \mathbf{E}(\mathbf{r},t) - \frac{1}{c^2}\frac{\partial^2}{\partial t^2}\mathbf{E}(\mathbf{r},t) = 0, \tag{11.16}$$

where $\mathbf{E}$ is the electric field vector and $\nabla^2 = \frac{\partial^2}{\partial x^2} + \frac{\partial^2}{\partial y^2} + \frac{\partial^2}{\partial z^2}$. A similar equation is obeyed by the accompanying magnetic field, and the initial conditions for $\mathbf{E}$ and $\mathbf{B}$ must be chosen consistently with Maxwell's equations. The different vector components of $\mathbf{E}$ must also be chosen mutually consistent with Maxwell's equations while they each obey (11.16). We consider a dominant vector component and write (11.16) as a scalar equation corresponding to that component.

Fortunately, laser light often travels as a beam with a nearly definite direction of travel, which we designate as the z direction. Furthermore, laser light is often virtually *monochromatic* with a single frequency ω. We are inspired by this to factor out pure plane-wave-like behavior and to write the solution we seek as

$$E(\mathbf{r},t) = \tilde{E}(\mathbf{r})\,e^{i(kz - \omega t)} \tag{11.17}$$

Where $k = \omega/c$ and $\tilde{E}(\mathbf{r}) = \tilde{E}(x,y,z)$ is a slowly varying function of position that is responsible for the laser-beam shape. The beam can change in strength as it spreads out or converges during propagation. Complex notation is used for convenience; only the real part of (11.17) is the physically relevant quantity.

Substitution of (11.17) into (11.16) gives

$$e^{i(kz-\omega t)}\frac{\partial^2 \tilde{E}(\mathbf{r})}{\partial x^2} + e^{i(kz-\omega t)}\frac{\partial^2 \tilde{E}(\mathbf{r})}{\partial y^2} + e^{-i\omega t}\frac{\partial^2 e^{ikz}\tilde{E}(\mathbf{r})}{\partial z^2} + \frac{e^{ikz}\tilde{E}(\mathbf{r})}{c^2}\frac{\partial^2 e^{-i\omega t}}{\partial t^2} = 0$$

which quickly reduces to

$$\frac{\partial^2 \tilde{E}(\mathbf{r})}{\partial x^2} + \frac{\partial^2 \tilde{E}(\mathbf{r})}{\partial y^2} + 2ik\frac{\partial \tilde{E}(\mathbf{r})}{\partial z} = -\frac{\partial^2 \tilde{E}(\mathbf{r})}{\partial z^2} \tag{11.18}$$

In the *paraxial* approximation, we ignore the right-hand side of (11.18) since it is small compared with other terms in the equation. This is justified by the fact that laser beams tend to propagate very directionally such that the field varies slowly in the z-direction, absent the factor e^{ikz} that was explicitly factored out.

An important solution to (11.18), representing a Gaussian laser beam, is*

$$\tilde{E}(x,y,z) = E_0\frac{z_0}{z_0 + iz}\,e^{-\frac{k}{2}\left(\frac{x^2 + y^2}{z_0 + iz}\right)} \tag{11.19}$$

---

* The general solution to the paraxial wave equation is $\tilde{E}(x,y,z) = -\frac{ik}{2\pi z}\int\int\limits_{-\infty}^{\infty}\tilde{E}(x',y',0)e^{i\frac{k}{2z}\left[(x-x')^2 + (y-y')^2\right]}dx'dy'$ which is

called the Fresnel approximation. This solution allows one to choose an arbitrary transverse shape of the beam, specified by $\tilde{E}(x',y',0)$ in the $z = 0$ plane. In the case of solution (11.19), this shape in the $z = 0$ plane is $\tilde{E}(x',y',0) = E_0 e^{-\frac{k}{2}\left(\frac{x'^2 + y'^2}{z_0}\right)}$, which is depicted in Figure 11.9. This profile is commonly labeled TEM_{00} for the lowest-order transverse electric and magnetic Gaussian field mode. Lasers can also be *multimode*, exhibiting more complicated structure.

One may confirm that (11.19) is a solution to (11.18) by direct substitution. This beam profile is shown in Figure 11.9; see footnote 3 for further details.

It is common to write this complex expression for the field in polar format, that is, as an amplitude times a pure phase factor. To this end, in the exponent of (11.19), we may write

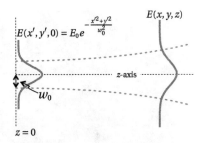

$$\frac{1}{z_0 + iz} = \frac{1}{z_0 \left(1 + z^2/z_0^2\right)} - \frac{i}{z + z_0^2/z}$$

The same factor out in front may be written as

$$\frac{1}{z_0 + iz} = \frac{1}{z_0 \sqrt{1 + z^2/z_0^2}} e^{-i \tan^{-1} \frac{z}{z_0}}$$

FIGURE 11.9 Diffraction of a Gaussian field profile.

With the aid of the above expressions and with a judicious introduction of new quantities, the full expression for the field, including the time and z dependence in (11.17), may be written as

$$E(\rho, z) = E_0 \frac{w_0}{w(z)} e^{-\frac{\rho^2}{w^2(z)}} e^{i\frac{k\rho^2}{2R(z)} - i \tan^{-1} \frac{z}{z_0} + i(kz - \omega t)} \tag{11.20}$$

where

$$\rho^2 \equiv x^2 + y^2 \tag{11.21}$$

$$w(z) \equiv w_0 \sqrt{1 + z^2/z_0^2} \tag{11.22}$$

$$R(z) \equiv z + z_0^2/z \tag{11.23}$$

$$z_0 \equiv \frac{k w_0^2}{2} \tag{11.24}$$

This formula describes the simplest and most common *laser-beam* profile.

As we analyze (11.20), consider the intensity profile $I \propto |E|^2$ as depicted in Figure 11.10:

$$I(\rho, z) \equiv I_0 \frac{w_0^2}{w^2(z)} e^{-\frac{2\rho^2}{w^2(z)}} = \frac{I_0}{1 + z^2/z_0^2} e^{-\frac{2\rho^2}{w^2(z)}} \tag{11.25}$$

By inspection, we see that $w(z)$ gives the radius of the beam anywhere along z; $w(z)$ indicates the radius at which the intensity reduces by the factor $e^{-2} = 0.135$ from its local axial value. At $z = 0$, the beam radius reduces to $w(0) = w_0$, as expected. w_0 is called the *beam waist*. The parameter z_0, known as the *Rayleigh range*, specifies the distance along the axis from $z = 0$ to the point where the intensity decreases by a factor of 2. Note that w_0 and z_0 are not independent of each other but are connected through the wavelength according to (11.24). There is a tradeoff: a small beam waist means a short *depth of focus*, or Rayleigh range.

We next comment on the phase terms that appear in the field expression (11.20). The phase term

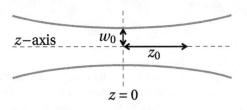

FIGURE 11.10 A Gaussian laser field profile in the vicinity of its beam waist.

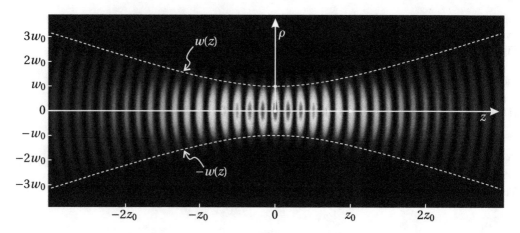

FIGURE 11.11 Real part of a Gaussian laser field at an instant in time. The radius of curvature of wavefronts is apparent.

$ikz + ik\rho^2/2R(z)$ describes curved wave fronts, where $R(z)$ is the radius of curvature of the wave front at z. The curvature of wavefronts is evident in Figure 11.11. At $z=0$, the *radius of curvature* is infinite (see (11.23)), meaning that the wave front is flat at the laser beam waist. In contrast, at very large values of z we have $R(z) \cong z$, and the wave front resembles a spherical wave emerging from the focus. Note that $kz + \dfrac{k\rho^2}{2z}$ is a parabolic approximation to the spherical shape $k\sqrt{z^2 + \rho^2}$. The phase $-i\tan^{-1} z/z_0$ is called the *Gouy shift*, which ranges from $-\pi/2$ (at $z=-\infty$) to $\pi/2$ (at $z=+\infty$). When light goes through a focus, it experiences this overall phase shift of π relative to a propagating plane wave. The Gouy shift is challenging to observe, since it affects only phase and not intensity.

Lab 11B   Laser Beams

11B.1   Focusing a Beam and f-Number

Far away from the focus (i.e., $z \gg z_0$), the radius of a laser beam expands along the edge of a cone:[*]

$$w(z) = w_0\sqrt{1 + z^2/z_0^2} \rightarrow w_0 z/z_0.$$

The *f-number* for a laser beam is defined to be the ratio of the cone height to base, as shown in Figure 11.12:

$$f^{\#} \equiv \lim_{z \to \infty} \frac{z}{2w(z)}.$$

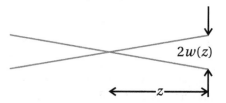

FIGURE 11.12   Dimensions for the definition of $f^{\#}$.

  (a) Show using the above equations and (11.24) that $f^{\#}$ dictates the beam waist w_0 according to

$$w_0 = \frac{2\lambda f^{\#}}{\pi}$$

  (b) Using lenses to expand a laser beam and then refocus it, create a focusing beam with $f^{\#} \sim 100$. Experimentally characterize the f-number and the resulting focal spot using attenuation filters

---

[*] For very wide cone geometries, the beam is often characterized by its *numerical aperture*, abbreviated NA, which is given by the sine of half the cone apex angle.

and a camera sensor to measure the beam diameter before and at the focus. Compare with the prediction in part (a). NOTE: you may be surprised at how distorted a beam can behave in comparison with the theoretical prediction. You may wish to intentionally introduce aberrations by tilting a lens or by inserting a low-quality piece of glass in the beam, perhaps with a fingerprint or dust on it. It is always good practice to directly measure your focus using a camera if its size is important in an experiment.

11B.2   The Airy Pattern and How to Clean Up a Beam

(a) Using lenses, create a collimated laser beam that is more than 1 cm wide. Place a $D \sim 5$ mm iris in the beam and then focus it using a $f = 1$ m lens. Using a camera and appropriate attenuation filters, record the *Airy pattern* that appears at the focus and confirm that it matches

$$I(\rho, f) = I_{\text{aperture}} \left(\frac{\pi D^2}{4\lambda f}\right)^2 \left[2\frac{J_1(kD\rho/2f)}{(kD\rho/2f)}\right]^2$$

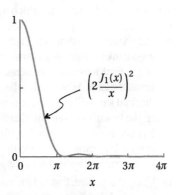

where I_{aperture} is the intensity of the beam at the aperture, ρ is the radial coordinate in the pattern, and J_1 is the first order Bessel function. The form of this function is shown in Figure 11.13.

FIGURE 11.13   Square of the *Jinc* function.

(b) Next *spatially filter* the beam by clipping off the rings of the Airy pattern. Place a pinhole in the focus, which transmits only the central portion of the Airy pattern (inside of the first dark ring, which occurs when the argument of J_1 is 1.22π). The arrangement is shown in Figure 11.14. Using a camera sensor, check that the emerging beam resembles an ideal Gaussian beam.

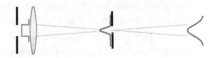

FIGURE 11.14   Spatial filtering of a Airy pattern.

(c) Show that the pinhole can also remove other distortions in the beam such as those caused by dust on a piece of glass inserted into the beam.

11B.3   The Mathematical Structure of Gaussian Beams

Consider the setup shown in Figure 11.16, where a diverging laser beam is collimated using an uncoated lens. A double reflection from the two surfaces of the lens, known as a ghost, comes out in the forward direction, focusing after a short distance (20 cm–30 cm). The collimated beam serves as a reference to reveal the phase of the focused beam through interference. Because the weak *ghost beam* concentrates near its focus, the two beams can have similar intensities for optimal *fringe visibility* (Figure 11.15).[*]

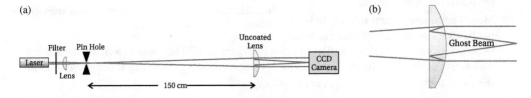

FIGURE 11.15   (a) Setup for 11B.3 "The Mathematical Structure of Gaussian Beams." (b) Enlarged view of the ghost beam.

---

[*] J. Peatross and M. V. Pack, "Viewing the Mathematical Structure of Gaussian Laser Beams in a Student Laboratory," *Am. J. Phys.* **69**, 1169 (2001).

The ghost beam $E_1(\rho,z)$ is described by (11.20), where the origin is at the focus. The collimated beam may be approximated by a plane wave $E_2 e^{ikz+i\phi}$, where ϕ is the relative phase between the two beams. The net intensity is then $I_t(\rho,z) \propto |E_1(\rho,z) + E_2 e^{ikz+i\phi}|^2$ or

$$I_t(\rho,z) = \left[I_2 + I_1(\rho,z) + 2\sqrt{I_2 I_1(\rho,z)} \cos\left(\frac{k\rho^2}{2R(z)} - \tan^{-1}\frac{z}{z_0} - \phi \right) \right]$$

where $I_1(\rho,z)$ is given by (11.25). We now have a formula that retains both the radius of curvature $R(z)$ and the Gouy shift $\tan^{-1} z/z_0$, which are not present in the intensity distribution of a single beam (11.25).

(a) Create the setup shown in Figure 11.15, and use a camera sensor to study the ghost beam of the final lens in the presence of the collimated beam. Hint: back reflections from the final lens will help you to center the lens on the beam and to ensure that it is straight. It is good to use relatively long focal lengths in the telescope setup to keep the beam relatively small at the final lens (i.e., < 1 cm).

(b) Using the camera, find the focus of the ghost beam and determine the f-number. Use it to predict values for w_0 and z_0. Hint: at the lens, the ghost beam has the same diameter as the collimated beam. See also the formula in Lab 11.B.1 (a) and Equation (11.24). You can measure the diameter of the collimated beam with a ruler, but a more accurate measurement can be made with the camera. If the beam is too wide to fit on the camera sensor, you can measure its width inside the telescope and adjust the result in proportion to the distance from the pinhole located at the telescope beam waist.

(c) Using a camera and appropriate attenuation filters, look at the ghost beam in the neighborhood of its focus as it interferes with the collimated beam. Check that w_0 and z_0 are plausibly similar to the prediction. NOTE: depending on parameters used, w_0 may be difficult to resolve, given the pixel size of the camera.

(d) Move the camera several times z_0 in front of the focus as well as behind the focus

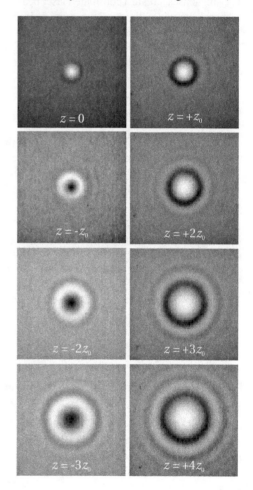

FIGURE 11.16 You should observe interference rings similar to these.

to observe interference rings similar to those in Figure 11.17. The rings arise from the curved wavefronts of the ghost beam. If you count N fringes out to a radius ρ, then the phase term in the above formula $k\rho^2/2R(z)$ has varied by $2\pi N$.

(e) Observe the effect of the Gouy shift $\tan^{-1} z/z_0$, which varies over a range of π. Notice that that the ring pattern inverts as you move the camera sensor from before the focus to after the focus (i.e., the bright rings exchange with the dark ones). Hint: you can make a subtle adjustment to the tilt of the lens. This causes the phase between the two beams to vary so that you can make the center of the interference pattern as dark as possible either far before or far after the focus for better observation.

FIGURE 11.17 Justin Peatross on a family trip to Mount Kilimanjaro in Tanzania.

Justin Peatross is a Professor of Physics at Brigham Young University where he does research in high-intensity laser physics. He earned his PhD at the University of Rochester in 1993. Prof. Peatross is interested in fundamental processes such as nonlinear Thomson scattering from relativistic free electrons and laser high-order harmonic generation from atoms. He holds patents for a biomedical scanner, a teleprompter design, and a three-dimensional display. He is coauthor of the textbook *Physics of Light and Optics*. Professor Peatross enjoys the outdoors including hiking, canyoneering, and spelunking. He has run over 20 marathons. He and his wife Cindy are the parents of seven children.

FIGURE 11.18   Michael Ware in his backyard.

Michael Ware teaches in the Department of Physics and Astronomy at Brigham Young University. His experimental research background is in quantum optics, but he also enjoys dabbling in the theory of classical optics and writing textbooks. He collaborates with Justin Peatross in experiments that use lasers to drive electrons in a high-intensity focus and then use single-photon detection techniques to study the radiation from these electrons. Michael has had a life-long fascination with pipe organs, both to play and to appreciate as a mechanical marvel and a beautiful instrument. He is the father of six children and loves to spend time with his family (especially in the outdoors) (Figure 11.18).

12

Vacuum

Walter F. Smith

CONTENTS

(to the tune "Listen to the Mockingbird")
Listen to the vacuum pump, listen to the vacuum pump.
The vacuum pump is pumping night and day!
Listen to the vacuum pump, listen to the vacuum pump.
[getting gradually quieter]
It's pumping all the atmo s p h e r e a w a y!

– Author unknown

12.1 Introduction

Physics research in a wide variety of subfields requires vacuum. Vacuum is needed to create ultracold atoms for Bose-Einstein condensation or quantum computing, to create plasmas for nuclear fusion, to conduct particle physics experiments such as those at the Large Hadron Collider (LHC), and to keep surfaces clean at the atomic level so their physical and chemical properties can be studied.

The degree of vacuum is measured by the pressure of gas inside the vacuum chamber. The SI unit for pressure is the Pascal, where $1 \text{ Pa} = 1 \text{ N/m}^2$. This is named for Blaise Pascal, who is mostly famous for his work in math, but also had key insights into the ideas of pressure and vacuum. However, for historical reasons, most scientists use the Torr, named for Evangelista Torricelli, the inventor of the barometer and a student of Galileo. You will also often find pressure measured in pounds per square inch, abbreviated psi. These units are summarized in Table 12.1.

Regulators for compressed gas cylinders have a pressure displayed in "pounds per square inch gauge" (psig).* Psig is defined as the pressure relative to one atmosphere, so a reading of 14.7 psig means that the pressure is really 2 atm.

Three properties of vacuum are important for physics research. One is the "mean free path," the average distance a gas molecule travels before colliding with another. The mean free path is inversely proportional to pressure. At 1 atm, it is about 70 nm. At 10^{-5} Torr, it's about 5 m.

As an example of why this is important, perhaps you want to coat a silicon wafer (the "substrate") with a thin layer of gold, and your process requires that the gold atoms come straight down onto the substrate. (This is a common requirement for lithographic patterning.) One method is "thermal evaporation," as shown in Figure 12.1. The substrate is suspended face down above a "boat," a thin metal strip made from a metal such as tantalum that melts at an extremely high temperature. The boat has a small cup-shaped area in the middle, and you put a few small pieces of gold into this. You pump the air out of the vacuum chamber, then heat the boat by running a large electric current through it. This melts the gold; due to

---

* The labeling usually just reads "psi," even though it should read "psig."

TABLE 12.1

Common units for pressure

Unit	Named for	Abbreviation	1 atmosphere equals
atmosphere	The Greek words "atmos" (vapor) + "sphairia" (sphere)	atm	1 atm
Torr	Evangelista Torricelli	Torr	760 Torr
Pascal	Blaise Pascal	Pa	101325 Pa
Pounds per square inch	Marvin Inch[a]	psi	14.7 psi
bar	The Greek word "baro" (weight)	bar	1.013

[a] Just kidding. Actually, "inch" is derived from the Latin "uncia", meaning one twelfth.

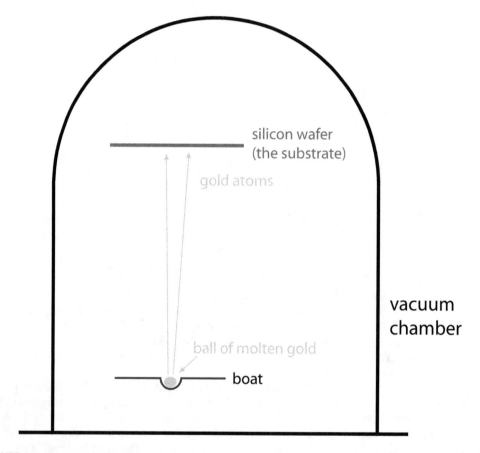

FIGURE 12.1   A thermal evaporator.

surface tension, it forms a sphere in the boat. Like any liquid, the molten gold evaporates. Of course, it is quite hot, so the evaporating atoms have a great deal of thermal energy compared with the gravitational potential energy difference between the boat and the substrate. Therefore, if they don't collide with gas atoms on the way, the gold atoms travel in straight lines up to the substrate, giving the desired directional deposition. Since the distance between the boat and the substrate is about 0.3 m, a vacuum of 10^{-5} Torr is enough to ensure that the gold atoms, as they travel from boat to substrate, will rarely collide with gas atoms.

The second important property of vacuum is thermal conductivity. You are familiar with thermos flasks, double walled containers with vacuum between the walls for insulation. In fact, as you reduce the pressure of a gas starting at 1 atm, the thermal conductivity doesn't change much until you get down to about 10 Torr. There are two conflicting factors at work. As the pressure is lowered, the mean free path increases, so gas atoms can more efficiently carry heat from one area to another. However, there are fewer molecules to carry the heat. Once the pressure falls below 10 Torr, the lack of gas molecules starts to have a stronger effect and the thermal conductivity drops. This change in thermal conductivity is important for the example of the thermal evaporator: because the pressure is low, little heat is transferred from the boat to the substrate, so heat damage to the substrate is minimized.

The final important property of vacuum is the frequency of air molecule impacts on a surface. Scientists creating multilayer semiconductors, perhaps for high-performance solar cells, require that each layer remains pristine until the next is deposited. Usually, the materials involved react quickly with oxygen, so it's essential to work under very good vacuum. A rule of thumb is that, if each molecule that strikes a surface sticks to it, then at a pressure of 10^{-6} Torr, the surface will be coated with a full layer of molecules in only 1 second. That would completely ruin such a process. Other scientists need to study uncontaminated surfaces for periods of hours, so they require vacuum of 10^{-10} Torr or better.

The background sections for this chapter are available at ExpPhys.com. They describe the various types of vacuum pumps, the gauges that are used to measure pressure, and the idea of "pumping speed" which is critical to vacuum system design. The lab for this chapter is also there. In it, you will create a simple vacuum system, and in the process learn the most common assembly methods, such as ConFlat flanges, KF flanges, and Swagelok connections. You will also practice how to find leaks using inexpensive equipment.

13

Particle Detection

Joseph Kozminski

CONTENTS

Science makes people reach selflessly for truth and objectivity; it teaches people to accept reality, with wonder and admiration, not to mention the deep joy and awe that the natural order of things brings to the true scientist.

– Lise Meitner

13.1 Introduction to Radioactivity

13.1.1 Introduction[1]

Radioactivity is the spontaneous emission of ionizing radiation or particles from the decay or de-excitation of unstable nuclei. Typically, these nuclear processes occur on energy scales of MeV (millions

---

[1] G. Knoll, *Radiation Detection and Measurement*, 4th ed., John Wiley and Sons (2010).

of electron volts). While the term radioactivity often has negative connotations, humans are constantly exposed to low-level radioactivity from natural sources such as rocks, cosmic rays, and even some kinds of food like bananas and table salt. The three most common types of radioactive emissions are:

- Alpha (α) particles: high energy helium (^{4}He) nuclei spontaneously emitted by nuclear decay processes ($^A_Z X \rightarrow ^{A-4}_{Z-2} Y + a$, where X, Y, A, and Z are the initial nuclide, final nuclide, mass number, and proton number, respectively) typically in elements with a high atomic number. Alpha particles are not very penetrating and can be easily stopped by a few centimeters of air, a sheet of paper, or the epidermis (the protective outermost layers of the skin). However, they can be very dangerous when inhaled or ingested because of the large amount of energy they deposit over a short distance in sensitive tissue.
- Beta (β) particles: high energy electrons or positrons emitted in nuclear processes involving either $n \rightarrow p + \bar{V}_e + e^-$ (β^- decay) or $p \rightarrow n + \nu_e + e^+$ (β^+ decay) in the nucleus. These decays produce nuclear isobars, i.e., nuclides with the same mass number. A β^- decay increases proton number (Z) by one and decreases neutron number (N) by one while a β^+ decay decreases Z by one and increases N by one. These particles are more penetrating than alpha particles and can penetrate the skin. Plastic or metal plates are needed to stop beta particles.
- Gamma (γ) rays: high energy photons from de-excitation of energetic nuclei in a process like $^A_Z X^* \rightarrow ^A_Z X + \gamma$. These are particles of electromagnetic radiation that are more harmful than X-rays (which are produced from electron transitions to inner shells, not nucleon de-excitation) and are even more penetrating than beta particles. Lead, concrete, or other metals like iron or steel are needed to stop gamma rays.

There are other, rarer forms of radioactive emission as well. For example, some isotopes undergo spontaneous neutron emission, and neutrons can be emitted in spontaneous or induced fission processes such as in nuclear reactors. Neutrons are extremely penetrating and require a substantial amount of concrete or water, often doped with a neutron absorber like ^{10}B, to stop them. However, since spontaneous neutron emission is a very rare process, it will not be considered in this chapter.

13.1.2 Activity

The strength of a radioactive source is defined by the number of decays per second it undergoes, known as its *activity*. The unit Curie (Ci) is used to measure activity where 1 Ci is 3.7×10^{10} decays/s, roughly the activity of 1 g of radium. Sources of around a μCi or less are deemed safe to handle and do not require licensing from the Nuclear Regulatory Commission (NRC) to purchase, store, and use. While the Curie is still commonly used, the Becquerel (Bq), where 1 Bq = 1 decay/s = 2.703×10^{-11} Ci, is the SI unit of activity.[1]

Another way to measure radioactivity is by measuring the *dose*, or amount of energy deposited per unit mass of the radiation's target. Radiation dose is commonly given by the unit *rad*, where 1 rad = 0.01 Joules of energy deposited per kilogram while the SI standard for radiation dose is the Gray (1 Gy = 1 J/kg = 100 rad).

A third measure is the internal radiation dose rate (also called the exposure rate), $\dot{D}$, due to intake of a radioactive material, given by

$$\dot{D} = \frac{kAE}{m} \tag{13.1}$$

where A is the activity (in Becquerel) of the source, E is the energy of the decay particle in MeV, m is the mass of the organ at risk in grams, and $k = 1.6 \times 10^{-8}$ rad·g/MeV is a conversion factor.[2] The dose,

[2] J. J. Bevelacqua, *Contemporary Health Physics: Problems and Solutions*, 2nd ed., Wiley-VCH (2009) p577.

D, (in rads) is then the dose rate times the exposure time. This equation is typically used for calculating alpha and beta dose rates since these particles deposit their energy over a fairly short distance. Gamma dose rates are more complicated to calculate due to their penetrating nature and ability to deposit energy outside the organ in question.

When gamma radiation travels through air to an exposed object, dose rate calculations have to take into account a $\dot{D} \propto 1/r^2$ factor, with r being the distance from the source. The dose rate for gamma rays traveling through air can then be written as

$$\dot{D} = \Gamma \frac{A}{r^2} \tag{13.2}$$

where Γ is the gamma ray dose rate constant, or gamma factor. The gamma factor is the dose rate in rad/hr one meter from a one Curie point source of a given radionuclide. Solid angle of the exposed area must also be considered, as does the geometry of the source if it is not a spherical point source. More information on calculating dose rates can be found in [2]. Because beta and alpha particles interact more readily with air than gammas, they attenuate more quickly than indicated by an inverse square relationship.

Radiation dose, however, does not give an accurate view of the biological impact of exposure to radiation. Instead, biological damage is given by the unit *rem*, which is the number of rads times the radiation weighting factor, w_R, for the type of radiation ($w_R = 1$ for γ rays and β particles, $w_R = 20$ for α particles, and a continuous function of neutron energy for neutrons).[3] This reflects the approximation that α particles are 20 times more damaging than γ rays or β particles of the same energy. However, the type of tissue exposed must also be considered as some tissue types are more sensitive to radiation than others. Sometimes the unit Sievert (Sv) is also used where 1 Sv = 100 rem. Table 13.1 provides a summary of radiation-related units. The typical average radiation dose for a person in the United States per year from natural sources is 310 mrem and 620 mrem from all sources, human-made and natural.[4] Table 13.2 provides a list of some common radiation doses received in the United States.

TABLE 13.1

Table of common units used when working with radiation

Units of activity	
Becquerel (Bq)	1 Bq = 1 decay/s
Curie (Ci)	1 Ci = 3.7 × 10^{10} decays/s
Units of dose	
Gray (Gy)	1 Gy = 1 J/kg
Radiation absorbed dose (rad)	1 rad = 10^{-2} Gy

[3] A. D. Wrixon, "New ICRP Recommendations," *Journal of Radiological Protection*, **28**, 161–168 (2008).
[4] US Nuclear Regulatory Commission, "Doses in Our Daily Lives," (2017), www.nrc.gov/about-nrc/radiation/around-us/doses-daily-lives.html.

Concept Tests

1. Considering a point (or spherically symmetric) source that can emit radiation uniformly, why must the inverse square relationship be applied to gamma radiation? Answer below.[5]

2. What relationship would alpha and beta particles from a uniform alpha or beta emitter follow in vacuum? Why? Answer below.[6]

3. How would the dose rate compare for a student whose finger is 10 cm from the source compared with holding the source at its center, 1 mm away from the source? Answer below.[7]

TABLE 13.2

Common radiation doses received in the United States[4]

Source	Dose (mrem)
Live within 50 miles of a nuclear power plant for a year	0.0009
Live within 50 miles of a coal-fired power plant for a year	0.03
Dental X-ray	1.5
Transatlantic flight	2.5
Chest X-ray	10
Annual cosmic radiation (sea level)	26
Average annual terrestrial radiation dose from the ground	43
Average annual dose from radon in the air	200
Average US annual natural background dose	310
Average total US annual dose	620
Full body CT	1000
Annual nuclear worker dose limit	5000

13.1.3 Safety

While the sources used in many undergraduate lab experiments are exempt μCi sources, it is always a good idea to use proper radiation safety practices. The general rule for working with radioactive materials is ALARA, "as low as reasonably achievable[8]." That is, every effort should be made to minimize exposure to ionizing radiation. The way to do this is through considering time, distance, and shielding. Time near or in contact with the source should be minimized since radiation dose increases linearly with exposure time. Distance from the source should be maximized since intensity, and therefore dose rate, follows an inverse square relationship (or stronger for β and α particles) with the distance from a source. For sources stronger than the μCi level, appropriate shielding should be placed between the experimenter and the source. The type of shielding needed varies by type of radiation, but, in general, intensity decreases exponentially with the thickness of the absorber material as will be discussed in more detail later.

Exempt disk sources should be handled by the edges. Standard laboratory practices like no eating or drinking and proper hand washing should also be enforced when dealing with radioactive sources. While hand washing will not remove radiation damage, it can remove radioactive contaminants (dust, etc.) that might be on your hands. While exempt disk sources are not hot enough to irradiate dust and surrounding surfaces to detectable levels, it is good to get in the habit of hand washing after working with radioactive materials. There are often local rules or instructions that must be followed when handling and using radioactive sources as well.

The only significant danger from exempt sources in the labs described would come from ingesting (or crushing up and inhaling) the sources. NEVER open the source enclosures or chew, eat, swallow, or ingest the sources!!!

---

[5] If you think about a uniform point source of light, the light energy is spread uniformly on the surface of a sphere with a radius of the position of the observer so the intensity at a distance r is the power over the surface area which scales as $1/r^2$, the surface area of a sphere being $4\pi r^2$. A uniform gamma source would likewise emit the gamma rays uniformly in all directions, so the count rate is analogous to the intensity, and it scales as $1/r^2$ since the gammas spread out over a surface of radius r.

[6] Alpha and beta particles count rates would also experience a $1/r^2$ dependence in vacuum for the same reason as the previous question. These particles, especially alphas, interact with and lose energy to the air more readily than gamma particles so the count rate falls off faster than $1/r^2$ in air.

[7] 1 mm is 100 times closer to the source than 10 cm. Since the dose rate falls off as $1/r^2$, the dose rate at 1 mm would be 10^4 times higher.

[8] Centers for Disease Control and Prevention, "ALARA – As Low as Reasonably Achievable," (2015), www.cdc.gov/nceh/radiation/alara.html.

When dealing with stronger sources, these and other protocols, like using monitoring badges, must be in place, and there are specific training programs that users must complete.

13.2 Detecting Radiation

In order to detect ionizing radiation, the high energy particles, including photons, must interact with the detector system and deposit energy in a medium suitable for detection. Common media used for the direct detection of ionizing particles are gases and semiconductor devices, in which the energy deposition frees charges in the active volume of the detector that can be collected by their interaction with an electric field in the detector. Solid or liquid scintillators coupled to a photomultiplier tube (PMT) or various kinds of semiconductor photodiodes or photomultipliers can also be used to detect ionizing radiation. In these materials, ionizing radiation excites molecules in the scintillator, which then de-excite and give off small light signals with wavelengths, ideally, near the peak acceptance wavelength of the detector. Though there are a variety of kinds of radiation detectors available, this section will focus on Geiger-Mueller (GM) tubes (colloquially known as "Geiger counters") and scintillators coupled to PMTs or to the increasingly popular and less costly silicon photomultipliers (SiPMs).

13.2.1 GM Tubes[1,9]

There are general purpose radiation detectors like GM tubes, that can detect several types of radiation, and there are specialized detectors that can only detect one type of radiation, and often only in a particular energy regime. Some detectors like the GM counter can only count particles while other detectors can yield other important information like particle energy.

A GM counter contains a simple gas-filled tube with an anode wire running down the center and the metallic tube wall acting as the cathode. (See Figure 13.1.) The primary fill gas is typically a noble gas, which does not form negative ions. A charged particle passing through the tube breaks the gas atoms into positive ions and "secondary" electrons. Because of the high potential difference between the anode and the cathode, the secondary electrons produced in the ionization process accelerate toward the wire and ionize the gas as well, producing even more secondary electrons and creating an avalanche. Because the tube contains a 5–10% concentration of a quench gas (either an organic compound or a halogen molecule), consisting of larger, more complex molecules that absorb energy, there is a limiting point for the Geiger discharge that ultimately ends the avalanche and makes every signal pulse roughly the same height, regardless of the energy of the particles initiating the avalanche. The disadvantage of the GM counter is that information about the original particle is lost, and it can only count particles passing through. However, the signal is large, often on the order of volts, so the readout electronics can be relatively simple, and the signal does not require external amplification.

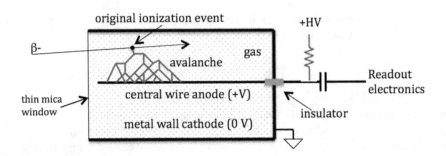

FIGURE 13.1 Geiger Muller tube showing an ionization event and the resulting avalanche of secondary electrons.

---

[9] W. Leo, *Techniques for Nuclear and Particle Physics Experiments: A How-To Approach*, 2nd ed., Springer Verlag (1994).

The GM tube has a high efficiency, nearly 100%, for counting charged particles that make it into the tube. Alpha particles do pose some challenges since they are easily absorbed by the window if it is too thick. For alpha counting, a GM counter with a very thin window, often made of mica, is needed. Gamma rays, on the other hand, are not counted directly by a GM counter since they are not charged particles. Instead, gamma rays are detected through interactions with the metal wall. Gamma rays can penetrate the tube from any direction, and the efficiency for gamma counting depends on the probability that the gamma ray produces a secondary electron in an interaction with the tube wall coupled with the probability that the secondary electron makes it out of the wall to the gas in the tube. Since only secondary electrons produced in the inner millimeter or two of the wall can make it into the fill gas, the gamma ray counting efficiency is only on the order of a percent. Geiger tubes are typically not used for neutron counting unless gases with high neutron-capture cross sections are used in the tube.

The GM tube can only be used when count rates are relatively low, because of the rather long recovery time after a discharge before a second discharge can occur. During this dead time, which is typically on the order of 50–100 μs, any particles entering the GM tube cannot be observed. Even after this dead time there is often a several hundred μs recovery time when the next pulse cannot be resolved; thus, a so-called dead time measurement actually accounts for the entire recovery time of the tube. For practical purposes, GM tubes are best used when count rates are less than a thousand per second. Correcting for dead time losses, even for count rates of a few hundred events per second, is important[1].

There are several ways to measure the dead time of a GM tube.[9,10] The two-source technique is one of the most common, though it can only resolve dead time to 5–10% accuracy[9]. If the true count rate is r, the dead time is τ_d, and the detector measures N counts in time T, then the amount of time the GM tube is dead during T is $N\tau_d$, and the number of particles not counted is $rN\tau_d$. The actual number of particles passing through the GM tube during T is then

$$rT = N + rN\tau_d \tag{13.3}$$

Solving for the true count rate in Equation 13.3 yields

$$r = \frac{R}{1 - R\tau_d} \tag{13.4}$$

where $R=N/T$ is the measured count rate. For the two-source technique, the rates of two sources are measured first separately and then together. If the two sources individually have true count rates r_1 and r_2, the true count rate of the combined measurement is $r_{12}=r_1+r_2$. Equation 13.4 can then be used to relate the true count rates

$$\frac{R_{12}}{1 - R_{12}\tau_d} = \frac{R_1}{1 - R_1\tau_d} + \frac{R_2}{1 - R_2\tau_d} \tag{13.5}$$

where R_1 and R_2 are the measured count rates of the individual sources and R_{12} is the measured count rate of the combination of the two. Equation 13.5 can be solved for the dead time:

$$\tau_d = \frac{R_1 + R_2 - R_{12}}{2R_1R_2} \tag{13.6}$$

when the approximation $\tau_d^2 \sim 0$ is made. The full solution can be found in [9].

There are other kinds of gas detectors such as ionization chambers and proportional counters, which use lower applied voltages and have smaller dead times but also have smaller signals, requiring more sophisticated amplification and readout. More information on ionization chambers and proportional counters can be found in [1, 9].

---

[10] C. Meeks and P.B. Siegel, "Dead Time Correction via the Time Series," *American Journal of Physics*, **78**, 589 (2008).

Concept Test

4. What is the maximum count rate a Geiger counter with a 100 μs dead time should encounter in order to lose less than 1% of the particles passing through due to dead time? Answer below.[11]

13.2.2 Scintillator-Based Detectors[1,9]

Scintillation-based detectors sense light produced in a material (the "scintillator") when ionizing radiation passes through. Unlike GM tubes, these detectors can measure the energy of the particle.

There are two main types of scintillator materials: organic liquids and plastics, and inorganic crystals like alkali halides. The type of scintillator must be chosen carefully for the application, taking into account factors such as type of particle being detected, required signal decay time, and particle energy range. While inorganic alkali halide crystals like sodium iodide tend to have a slower response time than organic scintillators, they have high light yield and linear response over a wide range of energies. These crystals are typically used for gamma ray spectroscopy due to the high Z values of the constituent atoms compared with organic scintillators. That is, these crystals have a shorter radiation length over which the gamma rays interact with it and lose energy to the material. Organic scintillators, on the other hand, tend to have a shorter decay time but lower light yield and are best for charged particle detection and spectroscopy.

Both types of scintillators work by the process of fluorescence, in which the molecules in the material that are excited by the incident particle promptly de-excite, yielding visible light. Specific scintillator properties can be found in [1, 9, 12]. The scintillator has to be chosen carefully, not only for the particle detection application, but also to ensure the peak wavelength of the emitted light is near the peak acceptance wavelength of the light detector.

Since the light signals from the scintillator are small, sensitive detectors must be used to detect the light, convert the light signal to an electronic signal, and amplify it. There are a number of detectors that can be used with scintillators, but two of the most common are the PMT and the SiPM.

A PMT (Figure 13.2) has a thin outer window to which the scintillator, or a light guide from the scintillator, is coupled by optical grease and an inner photocathode with a photosensitive material

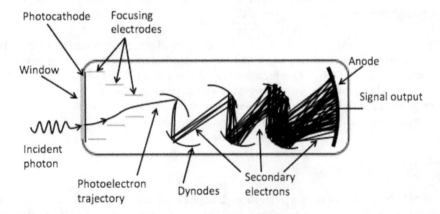

FIGURE 13.2 Photomultiplier tube with an incident photon ejecting a photoelectron from the photocathode. The signal is amplified as secondary electrons are ejected from a series of dynodes as the potential is stepped from high negative voltage to zero at the anode.

---

[11] The ratio of particles lost to total particles is $\frac{rN\tau_d}{rT}$. Setting this ratio less than 1% and solving for the measured count rate ($R < N/T$) yields $R < 100$ counts/s.

[12] M. F. L'Annunziata, *Handbook of Radioactivity Analysis*, Academic Press (2012).

that efficiently absorbs the incident photons and ejects low energy photoelectrons. Since the light signals are small, often only a few hundred photons, the electrical signal from the photocathode needs to be amplified significantly. To do this, the photoelectrons are accelerated across a potential gap toward a metal plate at a higher potential than the photocathode called a dynode, which produces secondary electrons when struck by the accelerated photoelectrons. These secondary electrons are likewise accelerated across a positive potential difference toward a second dynode, which ejects more secondary electrons in the collision. This progresses through a dynode chain with the final avalanche of 10^7–10^{10} electrons being collected at the anode as a current pulse, which can be further amplified and shaped by the data acquisition electronics. This amplification in the PMT is typically very linear such that the output current is directly proportional to the number of photoelectrons produced, and the electron multiplication process happens very quickly, on the order of tens of nanoseconds, preserving much of the timing information from the original light signal as well.[1]

Due to significant improvements in the technology, solid state PIN photodiodes have become more prevalent for scintillation counting due to their high quantum efficiency, ability to detect smaller scintillation signals, and their compactness. Photodiodes are semiconductor devices with a *p-n* junction that detect light when a photon (with energy higher than the band gap energy) creates an electron-hole pair in the depletion region. The electric field across the depletion region drives electrons to the *n* side and holes to the *p* side, generating a current in the photodiode. Typically, the number of electron-hole pairs produced is equal to the number of photons detected. While standard photodiodes do not have enough gain to detect small light pulses, they can be used for applications like calorimetry where light pulses typically contain at least several hundred photons.[1,13]

To resolve smaller light signals, devices with internal amplification like avalanche photo diodes (APDs) or SiPMs must be used. APDs have an applied electric field in the depletion region that cause impact ionizations at the p-n transition in which electrons collide with other electrons creating an avalanche of electron-hole pairs and, therefore, a stronger signal. Unless operated in a highly regulated environment where larger gains are possible, APDs have a gain of 50 to 200 over PIN photodiodes and tend to have better energy resolution than PMTs.[8,13]

Light signals down to the single photon level can be detected using APDs in Geiger mode (GM-APD), where amplification is self-sustaining due to an applied electric field higher than the breakdown point. This amplification enhancement comes from secondary avalanches initiated by holes and secondary photons in addition to electrons, and the avalanche must be stopped by a quench resistor. To minimize dead time and recovery time after the quench, GM-APDs can only have areas of a few thousand square microns.[8,13]

GM-APDs are often used in parallel in an array, called a silicon photomultiplier (SiPM). Here, an incident photon triggers an avalanche in a GM-APD pixel, which proceeds to the breakdown voltage, and is quenched. Then, the GM-APD is recharged to the bias potential. The signals from the individual pixels are integrated to produce a summed signal proportional to the photon intensity. SiPMs have several advantages over PMTs, including high gain with low noise from small bias voltages, excellent energy and timing resolution, and the ability to be used in magnetic fields[12].

Concept Test

5. In a PMT, the photocathode is typically at high negative potential and the anode at zero with the dynodes providing intermediate voltage steps. What is the direction of the electric field in the PMT and explain why the photoelectron and secondary electrons accelerate toward the anode. Answer below.[14]

[13] D. Renker and E. Lorenz, "Advances in Solid State Photon Detectors," *Journal of Instrumentation* **4** 04 P04004 (2009).
[14] The electric field points from the anode to the photocathode in the direction of decreasing potential. Electrons (or any negative charge) accelerate opposite the direction of the field lines.

13.3 Interactions with Matter

Different types of radiation interact through different physical processes with matter; however, the general trend for most beta particles and gamma rays (and neutrons) is similar – the intensity of radiation that gets through a sheet of material decreases exponentially with its thickness. Since alpha particles are blocked by very thin absorbers (e.g., a sheet of paper), the relationship between absorber thickness and count rate for alpha particles is not easily studied.

While low energy beta particles lose energy more rapidly than a decreasing exponential model would indicate due to collisional processes, most beta particle attenuation curves have a nearly exponential shape, which is a mostly empirical relationship that assumes electrons lose energy by radiation only[1,9]. A simple model can be developed by letting the probability of interaction dP over a small thickness of absorber dx be

$$dP = \frac{dx}{L_{\text{rad}}} \tag{13.7}$$

where L_{rad} is the radiation length, the mean length for a given material to decrease the number of particles passing through by $1/e$. Multiplying both sides by the count rate R gives the differential change in count rate, which is negative since electrons are absorbed and the count rate decreases:

$$-dR = RdP = R\frac{dx}{L_{\text{rad}}} \tag{13.8}$$

Solving this differential equation for an absorber of thickness t between the source and the detector yields

$$R = R_0 e^{-t/L_{\text{rad}}} \tag{13.9}$$

where R is the count rate after the absorber, and R_0 is the count rate without absorber. For beta particles, $1/L_{\text{rad}}$ is sometimes called the absorption coefficient.

Gamma ray absorption also has a decreasing exponential relationship with respect to thickness due to three physical processes: photoelectric absorption, Compton scattering, and pair production. A linear attenuation coefficient, μ, which is dependent on the energy of the gamma ray, can be defined as the probability per unit path length for a gamma ray to be removed via one of these processes. The count rate is

$$R = R_0 e^{-\mu t} \tag{13.10}$$

where again R_0 is the count rate without absorber, R is the count rate with an absorber of thickness t.

Often when dealing with gamma rays, the linear attenuation coefficient is not the most convenient parameter since it depends on the density ρ of the absorber. The "mass attenuation coefficient," μ/ρ, which is a function of the atomic number of the absorber and gamma ray energy, is often used instead.[15] In terms of this new parameter, the count rate is

$$R = R_0 e^{-\left(\frac{\mu}{\rho}\right)\rho t} \tag{13.11}$$

where the quantity ρt is called the "mass thickness," or area density, of the absorber and is measured in g/cm^2. When doing radiation studies, the thickness of an absorber is often reported in terms of mass thickness instead of physical thickness.

---

[15] National Institute of Standards and Technology, "X-Ray Mass Attenuation Coefficients: NIST Standard Reference Database 126," (2004), www.nist.gov/pml/x-ray-mass-attenuation-coefficients.

The mass attenuation coefficient depends on the energy of the gamma ray. Therefore, the graph of count rate vs. absorber thickness for radiation from isotopes that emit gamma rays strongly at multiple energies (such as ^{60}Co) is not linear on a semilog plot; the count rate is given by

$$R = R_{0,1}e^{-\left(\frac{\mu_1}{\rho}\right)\rho t} + R_{0,2}e^{-\left(\frac{\mu_2}{\rho}\right)\rho t} + \cdots + R_{0,N}e^{-\left(\frac{\mu_N}{\rho}\right)\rho t} \tag{13.12}$$

where the indices $1\ldots N$ represent gamma rays emitted at N different energies.

Concept Test

6. Show that integrating Equation 13.8 yields Equation 13.9. Answer below.[16]

13.4 Counting Statistics

Animals have a fairly well-defined lifespan. For example, once a human survives the first year of life, it is quite likely they will live to at least 55 years, and certain that they will die by age 125.[17] It is very different for particles that may undergo radioactive decay. Radioactive processes are statistical in nature; that is, each unstable nucleus of a particular isotope has a probability λ of decaying per unit time. If there are $\mathcal{N}$ unstable nuclei in a given sample, then the rate of change in $\mathcal{N}$ is given by

$$\frac{d\mathcal{N}}{dt} = -\lambda\mathcal{N} \tag{13.13}$$

It can be shown that this decay probability is related to the half-life of the isotope by $\tau_{1/2} = \ln(2)/\lambda$. This is left as an exercise in Problem 7b.

In nuclear decay processes, the probability that a particular atom in a sample decays is typically very small, but the number of atoms in the sample is very large such that the mean decay rate is $\bar{R} = \mathcal{N}\lambda$ or the mean number of decays in a time window Δt is $\bar{N} = \mathcal{N}\lambda\Delta t$. As long as the mean decay rate remains relatively constant (i.e., $\mathcal{N} \gg \bar{N}$), Poisson statistics can be used to describe the distribution of decays per given time window.[18] (This is a limiting case of binomial statistics which are discussed in more detail in [18].)

The Poisson distribution is a discrete distribution that gives the probability of v decays in a given time window for a mean number of decays, $\bar{N}$:

$$P(v;\bar{N}) = \frac{\bar{N}^v e^{-\bar{N}}}{v!} \tag{13.14}$$

---

[16] First, separate the variables $\frac{dR}{R} = -\frac{dx}{L_{rad}}$. Then integrate the rate from R_0 to R and x from zero to the thickness t:

$$\int_{R_0}^{R} \frac{dR}{R} = -\int_{0}^{t} \frac{dx}{L_{rad}}$$

$$\ln\left(\frac{R}{R_0}\right) = -\frac{t}{L_{rad}}$$

$$R = R_0 e^{-t/L_{rad}}$$

[17] Social Security Administration, "Actuarial Life Table," (2015), www.ssa.gov/oact/STATS/table4c6.html.
[18] J. R. Taylor, *An Introduction to Error Analysis: The Study of Uncertainties in Physical Measurements*, 2nd ed., University Science Books (1997).

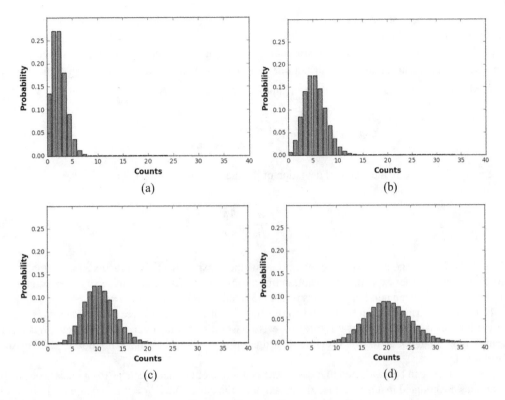

FIGURE 13.3 Poisson distributions for (a) $\bar{N}=2$, (b) $\bar{N}=5$, (c) $\bar{N}=10$, and (d) $\bar{N}=20$.

Figure 13.3 shows the Poisson distributions for several values of $\bar{N}$. The Poisson distribution is an asymmetric distribution about $\bar{N}$, and the sum of the individual count probabilities is 1 so the distribution is normalized. The variance, σ^2, in a statistical distribution is the difference between the mean of the squares of the random variable and the square of the mean of the random variable. In this case,

$$\sigma^2 = \overline{v^2} - \bar{v}^2 \tag{13.15}$$

As you can show in Problem 5, the mean number of counts is

$$\bar{v} = \sum_v vP(v,\bar{N}) = \bar{N} \tag{13.16}$$

and the mean of the squares is

$$\overline{v^2} = \sum_v v^2 P(v,\bar{N}) = \bar{N}^2 + \bar{N} \tag{13.17}$$

Therefore, the variance is $\sigma^2 = \bar{N}$, and the standard deviation of the Poisson distribution is

$$\sigma = \sqrt{\bar{N}} \tag{13.18}$$

It should be noted that this standard deviation only applies to a directly measured number of counts. It cannot be applied to any calculated function of number of counts such as count rates or sums or differences of counts.[1] Standard uncertainty propagation methods covered in Chapter 4 must be used to find the uncertainties on such calculated values.

As $\bar{N}$ becomes large, the Poisson distribution becomes more symmetric (as shown in Figure 13.3) approaching a Gaussian, or normal, distribution with a mean of $\bar{N}$ and a standard deviation of the Poisson standard deviation, $\sqrt{\bar{N}}$. That is, the Poisson approaches the Gaussian distribution

$$G(\nu, \bar{\nu}, \sigma) = \frac{1}{\sigma\sqrt{2\pi}} e^{-(\nu-\bar{\nu})^2/2\sigma^2} = \frac{1}{\sqrt{2\pi\bar{N}}} e^{-(\nu-\bar{N})^2/2\bar{N}} \tag{13.19}$$

If n experiments are run, the standard deviation of the mean of the Poisson distribution is[10]

$$\sigma_{\bar{\nu}} = \frac{\sigma}{\sqrt{n}} = \sqrt{\frac{\bar{N}}{n}} \tag{13.20}$$

where $\bar{N}$ is the average number of counts in one of the n experiments. In practice, for example in a particle physics collider experiment, it is often impractical to do many consecutive small trials, and the uncertainty on the mean number of counts is just the standard deviation since $n = 1$.

The Poisson distribution can be used when the mean number of events (decays, in this case) is small, the events are independent, and the mean rate is constant over the observation window. When the mean is more than about 10, the Gaussian distribution is a reasonable approximation to the Poisson distribution.

An experiment can be conducted to measure the count rate of emitted particles from nuclear decay. If N particles are counted in a time T in a given trial, then the count rate is $R = N/T$. When n trials, each of equal time length T, are conducted, the average count rate is given by the average number of counts $\bar{N}$ divided by the time interval:

$$\bar{R} = \frac{\bar{N}}{T} = \frac{\sum_{i=1}^{N} N_i}{nT} \tag{13.21}$$

In a counting experiment with n successive trials, the uncertainty of the mean number of counts is given by the Poisson uncertainty $\delta\bar{N} = \sqrt{N/n}$. As $\bar{N}$ gets very large, the Poisson standard deviation of the mean approaches the Gaussian standard deviation of the mean. Thus, the average count rate is:

$$\bar{R} = \frac{\bar{N} \pm \sqrt{\bar{N}/n}}{T} \tag{13.22}$$

When doing a radiation counting experiment, the background radiation must be measured and subtracted from the measurements. In this case, the signal rate is the total measured rate minus the background rate:[18]

$$\bar{R}_s = \bar{R} - \bar{R}_B \tag{13.23}$$

If dead time corrections need to be taken into account, this should be done before the background subtraction in Equation 13.23. In the experiments in this chapter, the background counts should be so low that the background rate need not be corrected.

Concept Test

1. To determine the background radiation rate, a scientist uses a Geiger counter to collect data in a single 45-minute run, during which 3860 counts are recorded. What is the background rate, including uncertainty? Answer below.[19]

13.5 Homework Problems

1. Classify each radioactive decay as alpha, beta plus, beta minus, or gamma decay and give the missing decay products. Some may have several step decay processes to classify.

   a. $^{90}_{38}Sr \rightarrow ^{90}_{39}Y^* + \underline{\quad} + \underline{\quad}$

   b. $^{210}_{84}Po \rightarrow ^{206}_{82}Pb + \underline{\quad}$

   $^{137}_{55}Cs \rightarrow ^{137}_{56}Ba^* + \underline{\quad} + \underline{\quad}$

   c. $\qquad\downarrow$

   $\qquad ^{137}_{56}Ba + \underline{\quad} (662 \text{ keV})$

   d. $^{204}_{81}Tl \rightarrow ^{204}_{80}Hg^* + \underline{\quad} + \underline{\quad}$

   $^{60}_{27}Co \rightarrow \underline{\quad}^* + e^- + \bar{\nu}_e$

   $\qquad\downarrow$

   e. $\qquad \underline{\quad}^* + \gamma (1173 \text{ keV})$

   $\qquad\downarrow$

   $\qquad \underline{\quad}^* + \gamma (1332 \text{ keV})$

2. The exposure rate from a 5 Ci ^{60}Co point source is 256 R/hr at 0.5 m from the source. The Roentgen (R) is an old unit of exposure still commonly found in radiation data tables.

   a. What is the exposure rate 2 m from the source?

   b. A typical exempt ^{60}Co source is only 1.0 μCi. What is the exposure rate 10 cm from an exempt ^{60}Co source in R/hr?

   c. The rad is more commonly used now as a unit for exposure, but the conversion from R to rad depends on the material exposed. If skin (soft tissue) is exposed, the conversion is 1 R=0.96 rad. Calculate the biological damage (in mrem) for a student working 10 cm from a 1 μCi ^{60}Co source for a 2 hr lab period, and compare this with the average total radiation dose a typical person receives per hour in the United States.

3. Ignoring the safety warning, a student ingests a 0.1 μCi ^{90}Sr source and somehow doesn't choke on the source holder. The electron in the ^{90}Sr decay product has an energy of 546 keV, and the decay product of the ^{90}Sr beta decays quickly (relative to ^{90}Sr) with an electron energy of 2280 keV.

---

[19] The background rate is:

$$\bar{R} = \frac{N \pm \sqrt{N}}{T} = \frac{3860 \pm \sqrt{3860}}{2700 \text{ s}} = 1.43 \pm 0.02 \text{ ct/s}$$

a. Assuming a total decay energy of the two electrons, calculate the internal dose rate $\dot{D}$ for an 850 g stomach in mrad/hr.

b. Given that a typical person in the United States receives about 620 mrem of radiation per year, how does the dose rate from the ingested ^{90}Sr compare with the amount of radiation a human receives on average per hour in the United States?

4. A patient undergoes radiation therapy, which uses 512 keV photons from a ^{137}Cs source. (a) In order to protect the patient from excess radiation, they drape the patient with a lead apron around the exposure area. If they want to block 99.9% of the gamma rays, how thick would the lead apron need to be? (b) In a medical setting, the typical apron thickness is at most 1.0 mm of lead-equivalent (i.e., an apron that will block as much as 1.0 mm of lead). What fraction of gamma rays are absorbed? (c) If the apron were made of iron instead of lead, what thickness would be needed for the same level of protection. Data for this problem can be found in the NIST x-ray attenuation database: www.nist.gov/pml/x-ray-mass-attenuation-coefficients. Table 3 on this website gives the mass attenuation coefficients for various elements. Select the element of interest and find the mass attenuation coefficient nearest the energy of the photon.

5. Evaluate the summations in Equations 13.16 and 13.17 to show that $\bar{v} = \bar{N}$ and $\overline{v^2} = \bar{N}^2 + \bar{N}$.

6. In a radioactive counting experiment, counts are recorded in 2.0 s intervals for 100 consecutive experiments. The data are binned in Table 13.3 showing the number of occurrences for each number of counts that was observed.

a. Plot a histogram of the data.

b. Calculate the total number of counts recorded during the 100 experiments.

c. Calculate the mean number of counts, $\bar{v}$, per 2.0 s interval.

d. Calculate the uncertainty of the mean number of counts per 2.0 s interval.

e. Calculate the mean count rate (i.e., counts per second) and its uncertainty.

f. These data are well described by a Poisson distribution (which could be confirmed by doing a χ^2 calculation, being careful to combine the last few bins into a ≥ 5 bin to avoid low statistics effects in the asymmetric tail). Calculate the expected number of occurrences of zero counts recorded if you were to run 500 consecutive experiments with 1.0 s intervals.

g. Calculate the expected number of occurrences of five or more counts if you were to run 500 consecutive experiments with 1.0 s intervals.

TABLE 13.3

Number of occurrences for each number of counts

v counts	Occurrences
0	14
1	31
2	28
3	17
4	6
5	2
6	1
7	0
8	1

7. A medical physics student is studying a weak radioactive source used for PET scans by recording the number of decays measured by a liquid scintillation counter in 30-minute intervals. Unfortunately, he forgot to record what sample he was working with in his lab notebook. Using the data provided in Table 13.4, work through the following steps to determine what the source is.

a. From Equation 13.13 show that decay data can be modeled by a decreasing exponential distribution:

$$f(t) = Ae^{-\lambda t}$$

where λ is a positive constant and A is a normalization constant. Integrate (from $t=0$ to ∞) to find the normalization constant A.

b. Show that the half-life is given by $\tau_{1/2} = \ln(2)/\lambda$.

c. The student measured the background to be 4 ± 2 counts per 10 minutes. Add a column to the table with background-subtracted counts and a column for the uncertainty on the background-subtracted counts. (*Note: you cannot just take the square root of the background-subtracted counts.*)

d. Make a plot of (background-subtracted) counts vs. midpoint time, including error bars.

e. Linearize $N(t) = N_0 e^{-\lambda t}$ and make a semilog plot of ln(counts) vs. Midpoint time. Be sure to derive the uncertainty on ln(counts) to correctly transform the error bars and apply the new error bars to your linearized plot.

f. Perform a fit to find the slope of the line and the y-intercept and uncertainties on these quantities. What is the physical meaning of the slope and y-intercept (in terms of N_0 and λ)?

g. From the fit parameters, determine $N_0 \pm \delta N_0$ and $\tau_{1/2} \pm \delta \tau_{1/2}$.

h. Which isotope(s) given in Table 13.5 does the measured half-life agree with (within uncertainty).

TABLE 13.4

Number of counts from unknown isotope per 30 min window, given at the midpoint of the counting window

Midpoint time (min)	Counts
15	414
45	352
75	293
105	240
135	198
165	171
195	133
225	119
255	100
285	81
315	69
345	62
375	54
405	42
435	46
465	34
495	33
525	27
555	21
585	22

TABLE 13.5

Possible isotopes with half lives

Isotope	$\tau_{1/2}$
^{11}C	20.3 min
^{18}F	1.828 hr
^{45}Ca	162.6 day
^{64}Cu	12.7 hr
^{66}Ga	9.49 hr
^{68}Ga	68 min
^{89}Zr	78.5 hr
^{94}Tc	52 min
^{129}Ba	2.16 hr
^{132}I	2.295 hr

Lab 13A Experiment on Counting Statistics

Learning goals: deepen understanding of Poisson statistics and their relationship to Gaussian statistics. Learn use of GM tubes. Deepen understanding of different types of radiation and how radiation rate depends on distance and absorber thickness.

13A.1 Objectives

1. Determine the background radiation in the lab using Poisson statistics.
2. Measure GM tube dead time.

3. Make measurements to examine one of the following dependencies:
   - Count rate vs. distance.
   - Count rate vs. attenuator thickness.
4. Plot the distributions of counts at each distance or each attenuator thickness, fitting them to both Gaussian and Poisson distributions using the measured average and standard deviation, for each distance/attenuation.
5. Make a plot of rate vs. distance or rate vs. attenuator thickness, including error bars, and discuss the behavior of the plot compared to theoretical expectations.

13A.2 Safety

Never break open, crush, ingest, or inhale the source. While these are exempt sources, it is good to practice ALARA: handle the source as little as possible; set up the experiment such that the source is as far as reasonably possible from people (e.g., far side of the lab table); and, if some shielding is available, use it. Also, follow any local rules and instructions provided for safe handling and use.

13A.3 Experiments

This set of experiments provides a useful set of counting experiments, which are a good introduction to more advanced nuclear and particle physics experiments. Experiment 13A.3.1 provides an application of the Poisson distribution in determining background, which will need to be subtracted in later experiments. Experiment 13A.3.2 is designed to give practice in working with Poisson and Gaussian distributions. Experiment 13A.3.3 is a measurement of the dead time of the radiation counter. It requires somewhat specialized sources for the best results and may be omitted, in which case count rates should be kept below the low 100s of counts per second to minimize dead time effects. Experiments 13A.3.4 and 13A.3.5 are opportunities to measure the behavior of radiation vs. distance and vs. absorber thickness, respectively.

For a first "look" at particles, it would be instructive to construct a cloud chamber. A cloud chamber allows for an observation of particle tracks in the clouds as high energy particles pass through. That is, the ionizing particles remove electrons from the molecules in the cloud. These ions become nucleation sites for the surrounding molecules, which are attracted to ions and condense to form droplets. As ionizing radiation moves through the cloud, it leaves a trail of these droplets. Cosmic rays (high energy muons, which are more massive cousins of electrons) that are continuously streaming to Earth can be observed at a fairly low rate. Putting a beta or alpha source in the cloud chamber shows much more activity. Instructions for building a cloud chamber can be found at [20].

13A.3.1 Background Measurement

The radiation counter or GM tube should be set up according to its supplied instructions for data collection. If it has multiple settings, use the ×1 setting. While measuring the background, the radioactive source should be at least 1.5 m away. While the background rate should be determined as accurately as possible, it is recommended to measure it to, at worst, 5% accuracy (i.e., $\frac{\sigma_{\bar{N}}}{\bar{N}} < 5\%$, where $\bar{N}$ is the average number of counts per measurement when n measurements are taken, and $\sigma_{\bar{N}}$ is the standard error of the mean). (Hint: see Equation 13.20.) Perform background radiation trials until you have the above precision on the background rate. If you are running an experiment in which your counts come from short, consecutive intervals, then each interval is a trial.

You should subtract the background in the following experiments using Equation 13.23.

20 S. Charley, "How to Build Your Own Particle Detector," *Symmetry Magazine* (2015), www.symmetrymagazine.org/article/january-2015/how-to-build-your-own-particle-detector.

13A.3.2 Poisson and Gaussian Distributions

This examination of Poisson and Gaussian Distributions could be done on its own or using data collected from Experiments 13A.3.4 or 13A.3.5 later. In this experiment, you should perform experiments, varying the distance between the radioactive source and sensor, and time interval for each trial, in order to achieve trials with average counts within the following ranges: <4 counts, 5–10 counts, and >50 counts. The sources should be oriented with the labels facing away from the detector. The collection time per trial does not matter for this experiment, nor does the distance from the detector. The goal is to run many consecutive trials, so that when you create the histogram below, you'll have a reasonable number of counts in each bin. You'll need about 250 trials for each version, e.g., 250 trials for the version with <4 counts per trial, another 250 trials for the version with 5–10 counts per trial, etc..

Using the data, compute the mean number of counts per trial. Then, find the standard deviation by using Poisson statistics and by Gaussian statistics separately. The standard deviation using Gaussian statistics is given by

$$\sigma = \sqrt{\frac{\sum_i (\nu_i - \bar{\nu})^2}{(n-1)}} \tag{13.24}$$

where ν_i is the number of counts in the ith trial, $\bar{\nu}$ is the mean number of counts, and n is the number of experiments, or trials run. Most analysis programs, such as Excel, Python, and MATLAB®, have a function to calculate this standard deviation.

Plot a histogram of occurrences vs. counts for each data distribution. Using the calculated mean, plot a histogram of the theoretical Poisson distribution and a Gaussian fit using the calculated mean and the Poisson standard deviation for each set of data as well, preferably all on one graph.

Also, calculate the χ^2 of the Poisson fit to the observational data:

$$\chi^2 = \sum_\nu \frac{\left(N_\nu^{obs} - N_\nu^{th}\right)^2}{N_\nu^{th}} \tag{13.25}$$

where N_ν^{obs} and N_ν^{th} are the is the measured and expected numbers of occurrences of ν counts, respectively. In general, if χ^2 is less than or approximately equal to the number of terms in the summation, there is good agreement between the observed and expected distributions. There are look up tables to find the probabilistic significance of the agreement.[18] A word of warning: one must be careful when dealing with the tails of the Poisson distribution since these low statistics bins can make χ^2 unreasonably large. Ideally, there should be at least four to five occurrences in each bin when doing the χ^2 calculation so bins on the tails are often made wider than those near the peak.

Questions:

1. How do the Gaussian and Poisson uncertainties compare for your distributions in each window?
2. Based on your fits, where would you recommend using the Poisson distribution, and when can you start using the Gaussian distribution? Explain.

13A.3.3 Measurement of GM Tube Dead Time

A Resolving Time Source Set has two active half-disk sources and a half-disk blank. The half disks are beta sources because of the high efficiency of detection of beta particles by a GM detector. The goal of this measurement is to have relatively low count rates for the individual half-disk measurements and a higher count rate (a couple hundred counts/second) that would require some correction for the combined count rate. It is critical to always position the sources in the same places, which is why there is a blank to set in the place of the source not in use. When the sources are placed together, there is also some chance for interaction between the sources, inducing decays, which is part of the reason this measurement is only accurate to 5–10%. Once the positioning of the sources relative to each other and the GM tube is determined, the individual count rates, R_1 and R_2, and the combined count rate R_{12} should be measured and the uncertainties on each computed. The dead time (i.e., the full recovery time of the tube) can be computed referring to Section 13.2.1.

13A.3.4 Measuring Count Rate vs. Distance

To measure count rate vs. distance, a gamma source should be placed at varying distances from the GM tube/radiation counter. The closest measurement should be far enough from the GM tube such that the count rate is in the low hundreds of counts per second to avoid the need for substantial dead time corrections. However, if the dead time experiment was conducted, the corrected count rate can be determined. Measurements should be made over a range of positions from this initial position until the count rates drop to near the background rate.

Once data are collected, the mean count rate and uncertainty at each distance should be calculated, the background subtracted, and count rate vs. distance (R vs. d) plotted. Determine how the count rate depends on the distance to the detector. Making a plot of residuals vs. position is recommended.

1. Does the curve fit agree with what you would expect for a gamma source? If not, what are some possible explanations for the discrepancy?
2. Is the measured distance the distance from the detector? Explain. If not, how might you improve your fit?
3. If this were run with a beta source instead of a gamma source, would you expect the same relationship between count rate and distance? Explain. Testing this would be a good extension to the lab.

13A.3.5 Measuring Count Rate vs. Absorber Thickness

Using several known sources of varying types (alpha, beta, and gamma if possible), qualitatively compare the effectiveness of different absorbers; e.g., a sheet of tissue paper, a piece of paper and multiple sheets of paper, plastic, aluminum foil, a thicker sheet of aluminum, a sheet of lead, or any other materials on hand.

To measure the count rate vs. absorber thickness, the source must be placed at a fixed distance from the GM tube with enough space between for the absorbers to fit. The count rate without absorbers should also be as high as possible but limited to the low hundreds of counts per second to minimize dead time effects. Count rates should be measured for several (at least five if possible) thicknesses of absorber. Do this measurement for a gamma source and a beta source.

Once the data are collected, plot the count rate vs. absorber thickness and determine the relationship between count rate and absorber thickness. This is a good opportunity to make a semilog plot, i.e., ln(*counts*) vs. thickness, and see if there is a linear relationship on this plot.

Questions:

1. Based on your observations of the effectiveness of different kinds of absorbers, which type of radiation might be best for treating surface tumors (e.g., skin cancer), and which might be better for treating deeper tumors?
2. Why does the last sentence of the activity recommend making a semilog plot and checking for a linear relationship? What does a linear relationship indicate? Explain/show.
3. Is either fit exponential? If any are not, why might this be the case?
4. For an exponential relationship, how would you extract the mass attenuation coefficient and its uncertainty from the semilog plot? If a gamma source is being used, does the measured mass attenuation coefficient agree with the known value for your absorber in the NIST x-ray attenuation coefficient database: www.nist.gov/pml/x-ray-mass-attenuation-coefficients.

Joseph Kozminski is Professor and Chair of Physics at Lewis University in Romeoville, IL. He is an experimental high energy physicist by training and is a member of the Mu2e collaboration at Fermilab. He also has an interest in Physics Education Research, especially around pedagogy, laboratory practices, and skill development in undergraduate labs. He chaired the subcommittee that wrote the *AAPT Recommendations for the Undergraduate Laboratory Curriculum* (2014) and is on the Board of Directors of the Advanced Laboratory Physics Association (ALPhA). He is an avid Chicago Cubs fan and enjoys hiking, golfing, and reading from a range of genres. He has also made a foray into politics, being elected to the Board of Education of his local public school district. He believes scientists need to be more active and engaged in government and policymaking (Figure 13.4).

FIGURE 13.4 Joseph Kozminski at Fermilab.

Part III

Fields of Physics

14

Development and Supervision of Independent Projects

Melissa Eblen-Zayas

CONTENTS

I think experimental physics is especially fun, because not only do you get to solve puzzles about the universe or on Earth, there are really cool toys in the lab... I had to learn to cleave optical fiber, machine a lot of parts, do a lot of plumbing. Are you feeling the fun? I had to measure the pulse durations and the frequency spectrum. Not all of the measurements showed what we expected. We had to figure out the problems and then a way around them. That was the fun part.

— Donna Strickland (2018 Nobel Prize in Physics)

14.1 Introduction

The advanced lab is designed to develop your skills in experimental physics, including a deeper understanding of instrumentation, experimental methods, and data analysis techniques. However, successful experimental physicists must do much more: to carefully plan a measurement; to know what tools they have available and how to use them; to be patient when things inevitably do not go as planned; to keep a detailed record of their work; to thoughtfully address unexpected problems that arise; to weigh the amount of time something will take and what will be learned from investing that time; to be detail-oriented; to think twice before doing, but not become paralyzed into inaction; and to learn from mistakes and failed experiments. All of these skills come from experience and spending time in the lab. Some of these skills are only developed when you have the opportunity to work through the experimental process from conception of an idea to completion of a project. However, traditional advanced lab activities do not always give you the opportunity to develop an experiment from initial idea to design and implementation.

Engaging in an independent project in an advanced lab course allows you to build a subset of skills that are often missing in other parts of the undergraduate curriculum. Even students who engage in undergraduate research experiences often do not have much opportunity to participate in research project goal setting, and they are not asked to develop contingency plans if things do not go as expected.* These activities, however, are key aspects of experimental research, and a student-driven independent project in the advanced lab allows you to experience these activities in the curriculum.

Because an independent project is one of the few opportunities to develop your skills in experimental design, it is important that you have realistic expectations for what success looks like in this realm. Your goal is not to try to get "good data" as quickly and easily as possible. One of the keys to successful experimental projects is to devote sufficient time to planning. Often, one must invest at least as much time in planning as in carrying out the project. If you plan to spend five weeks in the lab on a project, it would be reasonable to spend at least five weeks doing preparatory work – brainstorming, exploring the literature, developing and revising project plans, and identifying what items needs to be purchased, assembled, and built. The purpose of a project like this is to go through the process of doing real experimental work, with a particular focus on experimental design and moving a project from idea to implementation. Some experiments fail, some succeed – it is the nature of science. If a dedicated effort does not produce results as anticipated, that does not mean your project is not a success. You will learn more about experimental physics (and yourself) if you challenge yourself to take a risk with your project. You will also likely have a better experience if you judge your success based on whether you learn something from the process, gain experience using scientific reasoning to figure out the next steps in the project, and walk away from the independent project more comfortable with experimental design.

This chapter provides one approach to tackling advanced lab independent projects when you are given the opportunity to design and carry out your own experimental project. Unlike other lab activities where the physics content that the labs explore is central, this chapter is focused on helping you make the most of the often messy process of designing and carrying out an experiment. Section 14.2 walks you through some of the key elements you need to consider as you plan your project. Section 14.3 contains pointers designed to help with project execution. However, because of the broad parameter space that independent projects can occupy, this chapter only scratches the surface of what you need to consider when undertaking an independent project.

14.2 Project Proposal

Before experimental work is undertaken, physicists often have to sell their ideas through writing a research proposal aimed either a hiring committee or a funding agency. Although the audience for proposals is usually external, proposal writing benefits the authors by forcing them to clarify their thinking and to outline their plans. Unless your instructor provides directive guidance for your project, developing a proposal is an excellent place to begin your work on a project. Each proposal should include four key elements: research goals; literature review; work plan; and equipment and infrastructure. These elements are not independent of each other. For example, developing research goals and carrying out the literature review are often intertwined, requiring an iterative approach. Although this section presents each element separately, keep in mind that the process of developing a research proposal is fluid.

14.2.1 Research Goals

An important first step in engaging in an advanced lab independent project is choosing a topic to explore. This involves taking an inventory of your interests and skills, identifying what you hope to gain from the project, understanding the apparatus and instrumentation that is available to you, and exploring the literature to understand how your work can contribute to the conversation among physicists. When choosing a topic, think about the courses you have taken. Is there a concept you might enjoy exploring in a lab that you studied in one of your previous classes? Is there a topic that you have not studied before that would be accessible to you through the independent project? For example, one topic that is rarely covered in the undergraduate curriculum, but lends itself well to exploration in the advanced lab, is granular

---

* N. G. Holmes and C. E. Wieman, "Examining and contrasting the cognitive activities engaged in undergraduate research experiences and lab courses," *Phys. Rev. Phys. Ed. Res.*, **12**, 020103 (2016). https://doi.org/10.1103/PhysRevPhysEducR es.12.020103

materials.* In addition, think about the nature of the work that interests you. Are you more interested in the challenges of experimental design, hands-on fabrication, and tinkering, or are you more interested in the intersection of experimental and computational work? Your answer might help you decide on the type of project you want to try. For example, some classical mechanics experiments may appear more straightforward in terms of the experimental design, but they can allow in-depth data collection and analysis – perhaps with computational models – that allows the investigation of small deviations from theory. On the other hand, in condensed matter physics, experiments are often messy because it is difficult to know the true character of the samples being studied. In such cases, experimentalists must be clever with their experimental design to avoid unforeseen complications that could make well-controlled data collection challenging, and clear results may not emerge within the constraints of the time available.

Your experiment need not be entirely original. You can repeat an experiment that has already been published. The confirmation of previous results plays a crucial role in the advancement of science. You can also think about taking a previous experiment and tweaking it a little bit: changing a different variable, doing more precise measurements, investigating the role of something that the authors ignored. The *American Journal of Physics*, in particular, is a treasure trove, but you can also find inspiration in leading research journals like *Physical Review Letters* and *Science*.

EXAMPLE OF EXPANDING ON PREVIOUS WORK

Let's consider a group interested in doing a project in the general subject area of fluids and waves. Through a quick literature search, the group finds the 2002 *American Journal of Physics* article by Luna, Real, and Durán describing an undergraduate experiment to measure the speed of sound in a liquid by diffraction of HeNe laser light by the liquid. This previously published experiment has measured the speed of sound in distilled water. The group might want to start by trying to reproduce the results of this study, but then could consider a whole host of follow-up experiments to test how the speed of sound in a liquid depends on different properties of a liquid: using liquids with different densities and viscosities; changing the impurity concentration in the liquid by dissolving salt in the water; changing the temperature of the liquid and measuring the speed of sound; and so on.

Luna, D. A., Real, M. A., & Durán, D. V., "Undergraduate experiment to measure the speed of sound in liquid by diffraction of light," *American Journal of Physics*, **70**, 874 (2002).

Once you have identified a topic that is of interest to you, the next step is to articulate your research goals as clearly as possible. A well-written research question allows the reviewer to understand specific variables, controls, and what quantities will be measured. Also, how you articulate your research goals impacts the scope and development of your work plan. Because it is often difficult to assess how the project will fit within the academic calendar, you should consider developing several connected research goals. Depending on the time available and initial results, you can then make choices about subsequent research results.

EXAMPLE OF RESEARCH GOALS

Consider a proposal to study the diffraction of light produced by HeNe laser as it goes through an aperture. Here are two different ways of wording the same research goal:

"We will explore how the shape of an aperture impacts the diffraction pattern."

"We will explore how the width of a parallel slit impacts the intensity distribution in the diffraction pattern."

The second articulation of the research goal is much stronger than the first because it clearly specifies what will be explored.

---

* James Kakalios, "Resource Letter GP-1: Granular physics or non-linear dynamics in a sandbox," *American Journal of Physics*, **73**, 8 (2005). https://doi.org/10.1119/1.1810154

14.2.2 Literature Review

The purpose of research is not simply asking and answering questions in a void, but rather to contribute to the scholarly conversation in the field. As mentioned earlier, engaging in the scholarly conversation might involve revisiting the work of others in order to determine if the results are the same when carried out in a somewhat different context by different researchers. Or it might involve asking a more complex question than was previously explored or seeing if an apparatus can be simplified so that researchers with fewer resources can explore the questions at hand. In order to shape one's questions and figure out how one's work fits with the scholarly conversation, a physicist must know how to find and evaluate the work of others. That means understanding how to search the literature, both to get a sense of the big picture and to uncover specific information related to the research questions you have formulated. If you don't have a sense of the big picture, starting with textbooks, general overview articles (e.g., in *Physics Today*), or journal review articles on the topic can be helpful.

Once you have an understanding of the key concepts and questions in your proposed research area, then dive into the primary literature. Online databases (such as Inspec or Web of Science) are helpful tools for searching and navigating the literature. Contacting a librarian at your institution to find out what databases are available and to gain some insights into the search process using those databases is an excellent first step. The librarian can help you understand how to identify and use key search terms and determine appropriate boundaries for your search. If you are beginning with an already published article as the basis for your independent project, using the controlled indexing terms from that article, and additional criteria of your choosing, can help you identify other relevant articles. Following the citation trail can allow you to identify papers that cite your selected paper as well as the papers cited by your paper.

Once your initial search has helped you find articles connected to your research questions, the next step is to identify which are most relevant for your work. You likely don't have the time to read through all of the papers you identify with a fine-tooth comb. Make an initial pass through the articles by reading the abstracts and looking at the figures. Are the figures the result of experimental data that has been collected or computational models of the relevant system? What variables does the particular research explore? What assumptions are being made? From the articles yielded by your initial search, select a subset of the most relevant articles to read more thoroughly.

Making an annotated bibliography of your selected sources can help focus your understanding and shape your proposal. Each entry in the annotated bibliography should summarize the methods of investigation and the key contributions that the article makes to the scholarly conversation. However, your annotation should go beyond a summary of content and assess the relevance of the resource to your research project:

- Are you interested in employing or modifying a particular technique that was used by the authors?

- Do you think the model for the physical system described is particularly insightful or problematic?

- How convincing do you think the results or conclusions are? What is left unanswered in your mind or what could be improved?

When reading your annotations, readers should not only understand the key aspects of the article that is referenced, but they should also gain insight into how you see the article connected to and informing your thinking about your project.

14.2.3 Work Plan

One pitfall of eager experimentalists is to dive right into doing things in the laboratory without sufficient planning. For the independent project, developing a feasible work plan is a crucial element of the proposal because it provides a framework for structuring your approach once the actual laboratory work time begins. When developing a work plan, aim for modular design that allows for assembly and testing of various aspects of the physical apparatus and/or the relevant computer code in parallel and separate from each other.

It is tempting to begin your work plan by focusing on the primary measurements you want to make and how you will process and analyze the results that you get. However, in most experiments, there are many additional steps that need to be considered while planning your primary measurements. For example, it is often useful to begin your project work in lab by engaging in some feasibility analysis. This involves choosing preliminary tests that will help you understand limits of instrumentation, timeline, etc. In addition, these tests might be needed to explore underlying assumptions, which if not true might undermine all your work. Your feasibility analysis might include setting up a makeshift apparatus or taking some measurements to determine if a design concept is valid. After the feasibility testing, consider what baseline measurements or calibration measurements need to be made. Are there tests you need to carry out to understand offsets, backgrounds, non-linearities, scaling factors, etc., that might impact your measurements and data analysis? Perhaps you want to run a control experiment where you use the apparatus with a dummy sample to determine the impact of noise, environmental conditions, etc., or perhaps you want to send a voltage signal of known size through your instrument chain to see if it behaves as expected. Your work plan should include the feasibility analysis you will carry out and how those results will inform your primary measurements.

Your work plan should also lay out your proposed approach for collecting and analyzing data. Clearly identify the quantities that you are able to measure. Then consider the physical models that you want to test. What must you do with the quantities you measure in order to compare them with the models that you are exploring? How will you handle outliers? How will decide whether your data is appropriate for your goals or whether you will need to iterate the design of your apparatus or your experimental methods? Thinking ahead about your approach to analysis helps avoid subconscious bias, including confirmation bias. In addition, you should decide on a data management plan: where will you store your data, what file naming conventions will you use, how will you keep track of raw versus processed data, and who has access to and responsibility for the data?

Because it is impossible to foresee all the experimental twists and turns that will arise during your project, you can help yourself by developing a branched work plan that acknowledges uncertainties and provides alternate paths for proceeding depending upon the outcomes of the first portion of your project.

EXAMPLE OF DEVELOPING A BRANCHED WORK PLAN

Let's consider a project group that is developing a proposal based on the paper in the *Review of Scientific Instruments* by Rutgers, Wu, and Daniel that describes the use of vertically falling soap films to study two-dimensional (2D) fluid flow. This group has identified their initial research objective as classifying how 2D fluid flow patterns around a finite planar aerofoil change with varying angles of attack.

The initial elements in this work plan would include a design for assembling a simplified version of the soap film apparatus described in the article (perhaps without the pumps) as well as setting forth the criteria that the group would use as part of their feasibility analysis (e.g., benchmarks with regards to reproducibility of film thickness and flow velocity based on key parameters of the apparatus). In addition, the initial elements of the work plan would develop a strategy for image capture and processing to measure the fluid flow. After these initial steps, the group could begin exploring the primary research objective.

Because building the soap film apparatus may take several iterations with intervening design modifications, a branched work plan outlines how the project might move forward depending on both the time available and the initial results. For example, such a plan might provide some criteria that the experimenters would use to decide whether to revise the initial measurements to increase the quality of initial results, either through additional data or higher precision data. In addition, the branched work plan would outline the next steps of the exploration if the preliminary results meet the standards that the group members set for themselves. Additional explorations might involve exploring how changing aerofoil thickness or characteristic length impacts the generation of lift.

M. A. Rutgers, X. L. Wu, W. B. Daniel, "Conducting fluid dynamics experiments with vertically falling soap films," *Review of Scientific Instruments*, **72**, 3025 (2001).

14.2.4 Equipment and Infrastructure

One key consideration when developing your project proposal is what instrumentation is available to you and what infrastructure you might need (e.g., fume hoods, vibration isolation). Begin by identifying all of the instruments you might use or apparatus that you might need to build. In particular, talk with your instructor or the lab manager in your department to identify if there is equipment available of which you might not be aware. Because of concerns about time constraints and reliability, projects are often best served by using or modifying existing apparatus or instrumentation, particularly if the apparatus or instrumentation is complicated. If you have a particular penchant for engineering and your instructor approves, you may choose to build your own apparatus,* but be warned that it will likely take longer than you anticipate. You may find that what you had intended to be one step along the way toward your goals – building something – becomes most of your project, as you prototype, test, iterate, and finalize the design of your apparatus.

One common error that students make when identifying equipment and infrastructure needs is to overlook key instrumentation details such as resolution, noise floor, response time, etc. Some questions to ask yourself as you consider the instrumentation and infrastructure available, and whether it will meet the requirements of your project, include:

- What is the expected size of the signal you want to measure? How much will that signal size change? Compare that with the resolution and noise of the instrumentation.

- What is the time scale over which the system will respond? If it is very short, will your apparatus or instrument be able to measure changes over such a short time period? If it is long (days or weeks), how will that limit the data you will be able to collect during the course of your project?

- Might the signal from your experiment push the apparatus or equipment outside of its normal operational boundaries (e.g., saturating a photometer or overloading an electrometer)? In such cases, is there a chance for damaging your equipment? How will you know if you are operating outside of operational boundaries and how can you try to avoid that?

- Are there environmental considerations that could create unanticipated signals or disruptions (e.g., building vibration, ambient light, changes in temperatures/humidity levels)?

CAUTIONARY EXAMPLES OF BUILDING APPARATUS

Assessing the possible challenges you might encounter in your experimental project can be difficult, but one of the places where the actual project progress most often deviates from the work outlined in the project proposal is building apparatus from scratch. Here are two examples of pitfalls:

1. Even if there are clear schematics to follow, apparatus is time-consuming to construct.

   One topic that lends itself to exploration in advanced lab independent projects is exploring the conditions under which a pendulum system exhibits chaotic behavior. Commercially built apparatus are available from Pasco and Klinger, but they come with a price tag that can be beyond the budget of an advanced lab project. Several *American Journal of Physics* articles describe chaotic pendulums that have been used for the intro or the advanced lab. Although proposing to build one of these pendulums may seem a straightforward first step of a project because the papers clearly describe the apparatus, sorting out all the details of the pendulum design, data collection, and processing often takes more time than anticipated, and many groups find the process of building a chaotic pendulum from scratch consumes most of the project work time in a course.

R. DeSerio, "Chaotic pendulum: the complete attractor," *American Journal of Physics*, **71**, 250 (2003). P. Laws, "A unit on oscillations, determinism and chaos for introductory physics students," *American Journal of Physics*, **72**, 446 (2004).

---

* If you do choose to build your own apparatus, an excellent resource on the topic is: J. H. Moore, C. Davis, M. A. Coplan, and S. C. Greer, *Building Scientific Apparatus*. Cambridge University Press (2009).

2. Making modifications to an apparatus may have unforeseen impacts on other aspects of the experimental design.

A group decides to measure the heat capacities of solids using a calorimetric approach described in the 2011 *American Journal of Physics* article by Mahmood, Anwar, and Zia. The basic concept is to use a force sensor to measure the mass of vaporizing nitrogen before and after a solid block is dropped into a Styrofoam cup to determine the mass of vaporization and use that to calculate the molar heat capacity of the solid block. The group makes some alterations to the experimental set-up described in the article, but in doing so, the group doesn't account for the fact that the modified set-up produces a smaller mass change. The force sensor used in the article no longer has sufficient resolution, and therefore, the modified set-up also requires selection of a different force sensor.

W. Mahmood, M. S. Anwar, and W. Zia, "Experimental determination of heat capacities and their correlation with theoretical predictions," *American Journal of Physics*, **79**, 1099 (2011).

In addition to developing a list of the equipment you will need, drawing diagrams of your experimental set-up can also help with planning. Different types of diagrams play different roles. Schematic diagrams show all the details of the circuitry, mechanics, etc. of a system, but when you are just beginning to design an experiment, you may not know all of those details. The block diagram provides a high-level overview of the major components of an experimental set-up in a way that shows meaningful relationships between concepts and components, and developing such a diagram as part of your proposal can help clarify your thinking about your experimental set-up and what equipment you will need. Figure 14.1 shows how a block diagram for a Mossbauer drive control circuit is helpful in understanding flows, interactions, and causal relationships, without all the nitty gritty details of the circuit within the box. In addition to a functional block diagram, you might also want to consider other representations of your system. A topographic representation shows components of a system, their locations, and interconnections, but

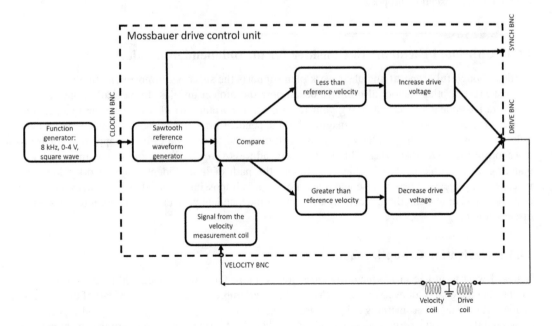

FIGURE 14.1 Example of a functional flow diagram of a system for controlling the drive of a traditional Mossbauer spectroscopy system. This diagram provides a few details about structural elements of the system (e.g., relevant BNC connections), but it mostly includes information that highlights the functional behavior of this system. It does not show the details of the circuit within the drive control box.

TABLE 14.1

Elements of a project proposal

Project proposal elements	What this looks like in an excellent proposal
Research goals	Specific, well-defined research goals that are presented in a clear and compelling manner. Relevant physics concepts are clearly and concisely discussed.
Literature review	Authors have thoroughly researched the available literature to understand broader context, previous work, and how it connects to and informs the proposed work.
Work plan	Provides detailed outline of first steps. Equipment, techniques, and methods to be employed are clearly identified (and supported with diagrams if appropriate). Possible alternatives for next steps, depending on how the project unfolds, are clearly presented.
Equipment and infrastructure	List of equipment to be built, bought, or borrowed is clear, and includes considerations of relevant specifications and possible limitations. Relevant faculty and staff are consulted.

doesn't show the functional relationships. Thinking about various representations of your systems will help with understanding equipment needs and with troubleshooting your set-up.

14.2.5 Summary

After having carefully worked through the four elements of a project proposal outlined above, the final step is to combine all the elements in a clear narrative that allows someone outside your group to understand your research goals, your path to achieving those goals, and what you will need in order to get there. Convince the reader of your proposal that you have carefully thought through your ideas and approaches, and that in some small way, your project will contribute to the broader scholarly conversation. Because it is easy to become wrapped up in your project, asking a colleague to read over your proposal before finalizing it will help identify if there are holes where you have assumed too much background knowledge of the topic or too much familiarity with your work plan. Table 14.1 summarizes what to look for in an excellent proposal.

14.3 Additional Elements to Consider for an Independent Project

While a thoughtful proposal and clear work plan supports the successful implementation of an independent project, other factors also contribute to how the project unfolds. In particular, physics is a collaborative endeavor, and managing your team is an important part of managing the independent project process. Another facet of managing the independent project process is to engage in weekly reflections on your progress relative to your work plan and to have an effective approach for troubleshooting problems that arise. This section will address three topics – group dynamics, weekly planning, and troubleshooting – that can smooth the path of independent project work. However, independent projects are so diverse and dynamic that it is impossible to develop a one-size-fits-all set of recommendations, and there are many elements of independent project implementation that are not discussed here.

14.3.1 Navigating Group Dynamics

You will want to utilize group members' strengths to move the project along within given time constraints, but the goal is for everyone to learn new skills and improve existing skills as part of the process. While certainly there is something to be said for letting a member of your group who is good at programming take the lead on writing code, that individual should also welcome the opportunity to share their expertise, become a teacher, and help another group member become better at coding. Be careful not to sideline group members who are a little slower in carrying out particular aspects of the project, but who are looking to strengthen their skill set in those areas.

To address the diversity of skills that group members bring, you are encouraged to take an inventory of both the strengths and areas for development of each group member. One way to do this is to develop a skills inventory, or competency matrix, that lists key skills that are important for the project you are undertaking along one axis of a matrix. These competencies should include relevant technical skills, data analysis skills, communication skills, team management skills, and project management skills. Along the other axis of the matrix list each group member. Each group member can then identify three to four skills that are strengths with a + and three to four skills that they want to develop with a ^ sign in the relevant cells of the matrix. After the matrix has been filled out, group members should engage in a discussion about how to capitalize on strengths and provide opportunities for the development of new skills. In addition, group members should consider how they will manage aspects of the project that rely on competencies that are not a focus (either a strength or an area of development) for any group member. Throughout the course of the project, the group should revisit this matrix to make sure that individuals are developing the skills that they want.

Beyond regularly revisiting the skills inventory, make time periodically to reflect on and talk about how each group member is contributing, how decisions are being made, and whether there are ways to make your group function more equitably. Members should monitor the group to make sure that they aren't allowing subconscious biases to creep into the group functioning. As a group, consider if you are automatically letting the group caretaking roles (e.g., schedule organizing, note taking) fall to women. Or when there is uncertainty about what to do next, does your group tend to defer to the person who has received the highest grades in other physics classes? Be aware of the ways your group may unintentionally operate that lead to unequal workload or leadership opportunities.

In addition to sharing the workload, talking about your thinking and approaches with your group members can be a powerful way to minimize unnecessary stumbles and find ways to overcome difficulties. To that end, making sure that all the members of the group articulate their thinking as they work on the project is important. One of the most powerful questions that group members can ask each other* is: "What are you doing, why are you doing it, and how does this help?" Asking a colleague to explain, clarify, or justify their reasoning to other members of their group helps ensure robust consideration of multiple ideas, enhances everyone's understanding, and tries to avoid foreseeable pitfalls. This is not designed to be adversarial, but rather with the goal of having all group members articulate their thinking so that the group as a whole can arrive at the best possible approach. Ask questions of your group members with a true spirit of curiosity and a desire to learn, rather than to grill or demonstrate your skepticism of your partners; the former approach can promote engagement and collaboration, while the latter will likely cause some group members to shut down.

14.3.2 Weekly Planning

A solid proposal should help your group navigate your time in the lab as the project unfolds. Nevertheless, deciding how to proceed on an independent project at any given point in time can be daunting. At the beginning of each week, I suggest that your group take stock of how things are going and set concrete goals. Begin each week by taking stock of last week's challenges and successes. Here are some questions your group might want to reflect on:

- What were the roadblocks that you faced in your work last week? How did various members of the group respond and contribute in an effort to address and overcome (or develop workarounds for) those roadblocks?
- What were some of your successes in the past week? How did various members of the group contribute to those successes?

---

* Kevin L. Van De Bogart, Dimitri R. Dounas-Frazer, H. J. Lewandowski, and MacKenzie R. Stetzer, "Investigating the role of socially mediated metacognition during collaborative troubleshooting of electric circuits," *Phys. Rev. Phys. Educ. Res.*, **13**, 020116 (2017). https://doi.org/10.1103/PhysRevPhysEducRes.13.020116

- List the main contributions to the group functioning and to the project work that each group member has made in the past week. Does everyone feel like others are contributing? Does everyone feel like their contributions are being acknowledged by others? Check with the skills inventory. Are group members finding that they are able to strengthen their skill set in areas where they expressed interest?

After taking stock of the previous week, the next step is to plan for the week ahead so that everyone is on the same page. If you have developed a modular project design, consider in what parallel aspects of the work you will engage. Also, just as it helps to be specific in proposal research goals, it is helpful to be specific about your weekly goals. A poor weekly goal would be: "This week we hope to collect data." What data will you collect? How will you evaluate if the data indicates there is a problem with your experimental set-up, with your theoretical model, or if it addresses the questions you hoped it would? What will you be doing in parallel with data collection? Who is working on what? How will you help each other and check in on each other's work?

14.3.3 Troubleshooting

When working on an independent project in the advanced lab, it is likely that you may be the only local expert on the topic of your particular project. Whereas in curricular labs the instructor likely knew what to expect and what might go wrong, and therefore could help with troubleshooting, in the domain of independent projects, your instructor may not know your experimental design and instrumentation well enough to help you solve the problems that will inevitably arise. To that end, it is important that you have a clear approach to follow when troubleshooting your experimental set-up, as discussed in Section 2.7.

14.3.4 Summary

While planning and preparation are critical to the success of independent projects, reflection and flexibility during the project execution are also important. Regularly taking stock of how your group members are experiencing the collaborative efforts is necessary to ensure progress toward project goals. Similarly, adjusting the project work plan, after reflecting on particular successes and challenges, is an expected part of doing real experimental work. Remember that you should not measure success by how smoothly the project unfolds, but rather by how much you learn about experimental physics (and yourself) in the process.

Melissa Eblen-Zayas is a Professor of Physics at Carleton College. Her research interests focus on understanding the electronic and magnetic properties of correlated electron materials, where the interactions of electrons with each other are strong. These strong electron interactions give rise to materials with unusual properties such as materials where a colossal change in the resistance is produced in response to applied magnetic fields or materials in which electric fields can be used to change magnetic domains. She has also served as the Director of the Perlman Center for Learning and Teaching at Carleton College. When not in the lab or the classroom, Melissa enjoys spending time outdoors – bicycling, hiking, and camping (Figure 14.2).

FIGURE 14.2 Melissa Eblen-Zayas hiking in Blue Mounds State Park, Minnesota.

15

Condensed Matter Physics

Walter F. Smith

CONTENTS

I am drawn to condensed matter physics because it is simultaneously useful and fundamental, simultaneously mundane and fantastical, simultaneously large and small, and most of all, some of the phenomena that are observed in condensed matter systems are just flipping cool.

– Inna Vishik* (condensed matter physicist at University of California, Davis)

15.1 Introduction

"Condensed matter" means liquids, solids, and things in between (like jello). It includes solid state physics, which is responsible for the transistor, integrated circuits, and virtually every device powered by electricity. It has strong connections to many other fields, including materials science, biophysics, and chemistry. It is the basis of nanoscience and nanotechnology. The majority of physicists are in the areas of condensed matter, and by far the most publications are in condensed matter.

Many condensed matter physics experiments probe the electronic, magnetic, and optical properties of samples, understanding them using ideas from quantum mechanics, statistical physics, and electromagnetism.

The field is far too vast to provide representative experiments in each subfield. Instead, I have chosen a typical example from the area of electronic properties: measurement of the Johnson noise of a resistor. This is a very challenging experiment, which requires you to integrate ideas from Sections 6.1–6.7 and 7.1–7.3, as well as to think carefully about experimental design. If you're able to succeed at the 5% target level of accuracy, you will know that you've accomplished something noteworthy.

---

* Taken from Prof. Vishik's delightful set of pieces on Quora: www.quora.com/profile/Inna-Vishik.

15.2 Equivalent Noise Bandwidth for a Measurement Chain

(Please review the first discussion of equivalent noise bandwidth and noise spectral density in Section 6.7.)

A "measurement chain" consists of the sequence of instruments used to make a measurement. Often, this includes one or more amplifiers and one or more filters. You often need to calculate the equivalent noise bandwidth of the entire chain, either to find how much amplifier noise to expect at the output or (if you are measuring the noise output from a sample) to find the noise spectral density from the measured rms noise amplitude at the output.

The definition of the "equivalent noise bandwidth" B, which appears in the Johnson noise formula, is as given in Equation (6.18):

$$ B = \int_0^\infty \left(\frac{V_{rms,out}}{V_{rms,in}} \right)^2 df, $$

where $V_{rms,\,in}$ is a function of frequency, and is the rms amplitude at the input of the measurement chain, while $V_{rms,\,out}$ is the rms amplitude at the output of the measurement chain, "referred to the input," i.e., divided by the nominal gain.

Measuring *B*

Method 1: Entire Chain

One approach to determining B experimentally is simply to measure V_{out}/V_{in} for your *entire measurement chain*, including the amplifier at the gain you will use for the actual experiment and all filters you will use, then applying the above equation (e.g., using the trapezoid rule for numerically integrating your data), with appropriate correction for the gain. If the gain is large, you will need to use a voltage divider to create a small enough signal to avoid overloading your amplifier. You can accurately measure the input voltage to the voltage divider, and accurately measure the resistors of the divider to determine the division ratio. Make sure to choose the resistors used for the divider so that its input impedance is high enough and output impedance low enough that you can ignore loading effects.

There are two variants of this method. In the first, you apply sinusoidal voltages (e.g., from a function generator) at a series of frequencies and measure V_{out}/V_{in} for each. Space the frequencies appropriately; if V_{out}/V_{in} is varying slowly with frequency, you can use a larger frequency spacing. Then, apply a numerical integration technique such as the trapezoid rule to your data, and use Equation (6.18) (remembering to refer output voltages to the input) to find B.

In the second variant, you use a white noise generator (if available) to apply a white noise signal to the input of the measurement chain, measure the rms amplitude at the output, and use Equation (6.17):

$$ V_{rms} = v_n \sqrt{B}, $$

where V_{rms} is the measured amplitude at the output (referred to the input), and v_n is the noise spectral density produced by the white noise generator. For 1% accuracy, make sure the noise generator creates noise in a bandwidth that is at least 100 times the f_{3dB} of the lowest frequency low-pass filter in your chain. Also be sure that the apparatus you use to measure the rms amplitude at the output (such as an analog interface unit for a computer) has a bandwidth and sampling rate adequate to accurately measure a signal at 100 times the f_{3dB} of the lowest frequency low-pass filter in your chain. This is the easiest and also the most accurate method to measure B if you have the required equipment.[*]

---

[*] Dedicated white noise generator units are expensive and are not found in most research labs. The white noise generator that is built into some general purpose signal generators is often of low quality, creating noise that is not truly white. It may be possible for you to use the digital-to-analog (D/A) interface unit that is built into the computer interface unit you are using to create a white noise generator; this depends on the particular hardware and software you have. Sometimes there are frustrating timing issues that make it impossible to make measurements with the A/D at the same time that you are creating the white noise with the D/A.

Method 2: Values of f_{3dB} for Each Filter or Effective Filter

A different approach is to measure the f_{3dB} for each filter in the measurement chain and the bandwidth for each amplifier in the chain. You can model the effect of the amplifier bandwidth as a low-pass filter. Then, using the ideas below, you can combine these measured frequencies to calculate B.

For a low-pass filter with $f_{3dB} = f_{lo}$, it is easy to show that $\dfrac{V_{out}}{V_{in}} = \dfrac{1}{\sqrt{1 + \left(f / f_{lo}\right)^2}}$,

while for a high-pass filter with $f_{3dB} = f_{Hi}$, $\dfrac{V_{out}}{V_{in}} = \dfrac{1}{\sqrt{1 + \left(f_{Hi} / f\right)^2}}$.

In all of the following, I ignore loading effects (i.e., I assume there are buffers or the equivalent between the filters). For a combination of two low-pass filters, V_{out}/V_{in} is then just the product of two expressions such as those above. These can be integrated with Mathematica to obtain the following results:

For a single low-pass filter: $B = \dfrac{\pi}{2} f_{3dB}$

For two low-pass filters in series (no loading) with f_{3dB}'s of f_1 and f_2: $B = \dfrac{\pi}{2} \dfrac{f_1 f_2}{f_1 + f_2}$

For three low-pass filters in series (no loading): $B = \dfrac{\pi}{2} \dfrac{f_1 f_2 f_3 \left(f_1 + f_2 + f_3\right)}{\left(f_1 + f_2\right)\left(f_1 + f_3\right)\left(f_2 + f_3\right)}$

For four low-pass filters in series (no loading):

$$B = \frac{\pi}{2} \frac{f_1 f_2 f_3 f_4 \left[\alpha + \beta\right]}{\left(f_1 + f_2\right)\left(f_1 + f_3\right)\left(f_1 + f_4\right)\left(f_2 + f_3\right)\left(f_2 + f_4\right)\left(f_3 + f_4\right)}$$

where $\alpha \equiv f_1^2 \left(f_2 + f_3 + f_4\right) + f_2^2 \left(f_1 + f_3 + f_4\right) + f_3^2 \left(f_1 + f_2 + f_4\right) + f_4^2 \left(f_1 + f_2 + f_3\right)$ and

$\beta \equiv 2\left(f_1 f_2 f_3 + f_1 f_2 f_4 + f_1 f_3 f_4 + f_2 f_3 f_4\right)$.

To take into account the effects of a high-pass filter, so long as its f_{3dB} is much lower than that of any low-pass filter in the circuit, just subtract off $\dfrac{\pi}{2} f_{Hi}$ from the above results.

Lab 15A Quantitative Measurement of Johnson Noise

Learning goals: In this lab, you will synthesize concepts you've learned from preceding labs, including low-interference measurement techniques, amplifier noise, and Fourier analysis. You will have a genuine experience of designing and debugging an experiment. It is a significant challenge to make the measurements at the accuracy required; this will only be possible if you understand your equipment deeply. You will perform a full uncertainty analysis, requiring you to understand how to combine uncertainties from several sources, including sources for which an absolute tolerance is quoted.

Pre-lab reading

- H. Nyquist, "Thermal Agitation of Electric Charge in Conductors." *Physical Review* **32**, 110–113 (1928), available free on the web; you are only required to read through equation 1. (You're welcome to read the rest, but it is not as important for this lab.) Note that Nyquist uses the symbol E to represent the voltage noise density.
- It is especially important for this lab to read through this entire lab write-up before beginning.
- This lab will take at least two lab sessions.
- Sections 6.1–6.7, 7.1–7.3, and 15.2.

Post-lab reading: (preferably read before lab, but definitely before writing your report.)
- 3.2–3.3

Pre-lab questions: There are three of these distributed through the text below.

Safety: Depending on your method, you may be using a hot oil bath. The oil is between 100°C and 150°C, and will burn you if you spill it. Treat it as you would hot oil in cooking, e.g., from frying bacon. You should wear protective gloves when carrying the oil from one place to another and ensure that anything you put into the oil is completely dry. (Any water on the surface of an object put into the oil will vaporize and cause spattering.)

Introduction: In 1926, J. "Bert" Johnson, then 39 years old, was working in the research labs of AT&T. He was investigating a "steady rustling noise" that was heard on some telephone receivers and was about to make a discovery that is still central to all low-noise electrical measurements. With remarkable ingenuity and careful experimentation, he found that the noise was a *universal characteristic of all objects!* He showed (see the top graph in Figure 13.2.1) that the square of the rms amplitude of the voltage noise scales linearly with the resistance of the object. It doesn't matter what material is used, what the shape of the object is, or whether electrical current in the object is carried by electrons in a solid or ions in a liquid solution. He also found that the power, V_{rms}^2/R, is proportional to temperature (see the lower graph in Figure 15.1), and that the slope of this graph is related in a simple way to Boltzmann's constant k_B.

He consulted with his colleague, Harry Nyquist, a theorist. Nyquist explained Johnson's results by considering the normal modes of electromagnetic waves on a transmission line connecting two resistors (see pre-lab reading). Johnson's paper and Nyquist's paper were published back-to-back in the same issue of the *Physical Review*. Theorists refer to the phenomenon as "Nyquist noise," but experimentalists call it "Johnson noise."

It took at least 30 years for the full implications of this discovery to be realized. It led to the "fluctuation-dissipation theorem," a very deep statement that the more a system can dissipate energy into heat (e.g., the way a resistor converts electrical energy to heat), the greater are the fluctuations in the system (e.g., Johnson noise). Another example of this theorem is in the motion of a small particle in a fluid: the greater the viscous resistance to movement is, the more random motion of the particle there will be due to Brownian motion.

Pre-Lab Question 15A.1

Use Nyquist's Equation 1 to find a relation between the rms voltage measured in an experiment, the resistance, the temperature, and the equivalent noise bandwidth B of the experimental apparatus. Note that Nyquist's formula gives "the square of the voltage within the interval dv," where v is frequency. This is the input voltage to our measuring apparatus, which has a transfer function $F(v) \equiv \dfrac{V_{out}(v)}{V_{in}(v)}$. So, the resulting square of voltage in the interval dv at the output of our measuring apparatus is $F^2 4Rk_BT\,\mathrm{d}v$.

For any low-noise electrical measurement, the experimenter must consider the effects of Johnson noise. In some experiments, the Johnson noise from a device is used to measure its temperature; this is called "noise thermometry." The Boltzmann constant is known to a precision of about 1 part per million. Although that may seem like a high precision, it is poor compared with our knowledge of other fundamental constants. For example, Plank's constant is known to about 0.01 parts per million, and the speed of light is known to about 0.02 parts per billion. Researchers at the National Institute of Science and Technology are using Johnson noise, combined with a voltage standard based on superconductivity, to determine k_B to a higher precision*.

In this exercise, you will emulate their efforts, albeit in simpler form. You will measure the Johnson noise of a resistor quantitatively and use it to deduce the value of Boltzmann's constant. You will have

---

* See Pollaro *et al.*, *IEEE Trans. Instrum. Meas.*, **62**, 1312–1317 (2013).

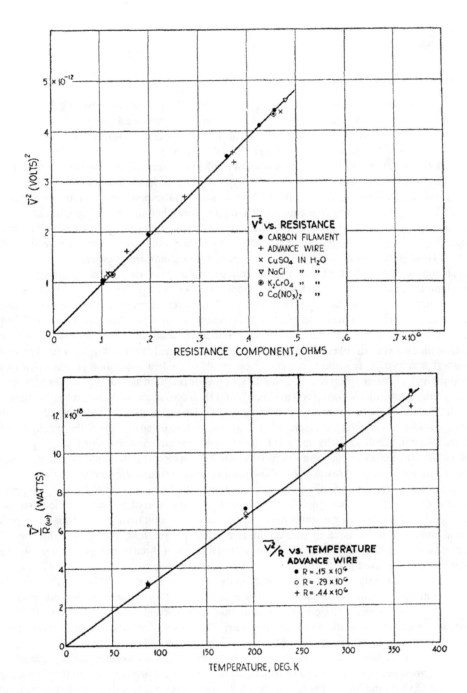

FIGURE 15.1   Two figures from Johnson's paper (*Phys. Rev.*, **32**, 97–109 [1928]).

available to you a shielded resistor (you choose the value), any other small connector boxes containing resistors or capacitors that you care to make, a thermometer to measure room temperature, a bath of hot oil (with a thermometer), your amplifier, an oscilloscope, a signal generator, a well-calibrated multimeter, and your computer which includes an analog-to-digital converter. It's up to you to figure out how to use these items to make an accurate measurement of k_B. If you are careful, you should be able to measure k_B to an accuracy of better than 5%. (Because the Johnson noise amplitude is proportional to $\sqrt{k_B}$, this means you need to measure the noise amplitude to an accuracy of better than 2.5%.) Some of the following discussion refers to changing temperature, but your method may involve measurements made at only one temperature.

Experimental Considerations

Bear in mind the goal of measuring k_B to an accuracy of 5%, which requires measuring the noise amplitude to an accuracy of 2.5%. This requires you to be quite careful about all aspects of your experiment. Bear in mind that, because uncorrelated sources of uncertainty add in quadrature, it is usually only important to consider the one or two largest sources of uncertainty.

Before doing any experiment, every scientist must think and plan carefully to make sure they are optimizing the signal-to-noise ratio, are performing any needed controls, are ready to record all important variables and observations, etc. It is usually best to run some trial experiments first, to see roughly how things are working and to refine your equipment and procedures, before you take serious data.

You should connect the output of the amplifier both to your computer interface unit and to an oscilloscope. The scope gives you immediate feedback on whether things are working properly, and also provides a good ground, while your computer is superior for quantitative analysis.

You should use the computer to acquire the Johnson noise. Your data acquisition program should run in a loop for a user-specified number of iterations, capturing a large, user-selectable number of data points each iteration and calculating the rms noise amplitude each iteration. The distribution of rms values thus obtained will be an important part of your uncertainty analysis. You should arrange things so that the full loop with all iterations completes in two minutes or less. (Obviously your lab period is much longer than this, but you will likely need to do the experiment several times to develop a good procedure.) To ensure that everything is working correctly, you should use a data acquisition program that includes a histogram and a Fourier spectrum. You should set up your program so that the Fourier spectrum has good frequency resolution, allowing you to check for 60 Hz interference, and also probes frequencies up to at least 25 kHz. (Digital instruments often broadcast interference at about 20 kHz.) In order to improve the signal-to-noise, arrange for the displayed Fourier spectrum to be an average of the spectra for all the iterations. You will use this Fourier analysis to find various rms amplitudes of signals.

You will need to consider which value of resistor, when connected to your amplifier, will give the most accurate result for k_B. A good resistor will allow you to ignore certain noise sources that are difficult to quantify. Even after this optimization, you may not assume that amplifier noise is negligible, nor may you assume that interference is negligible. (You should, of course, use what you've learned to minimize interference.) You may, however, assume that these factors are constant in time, as long as the impedance to ground connected to the input of your amplifier is constant. You may also assume that the voltage and current noise densities of your amplifier are within a factor of three of the spec values. You should consider how to minimize the effects of $1/f$ noise with your amplifier.

Also, note that the resistance of your resistor depends weakly on temperature. This means that, in principle, the contribution from the current noise of your amplifier is also temperature dependent. However, if you've chosen your resistor properly, the contribution of the current noise should be small. Since the temperature dependence is weak, and since the contribution is small anyway, you may ignore the temperature dependence of the contribution of the amplifier current noise.

Because you are trying to make rather precise measurements, don't assume that any aspect of any piece of equipment necessarily has a value exactly equal to its spec. For example, you should not assume that the gain of your amplifier or the f_{3dB} of the filters you engage are exactly equal to the spec values. You should plan to measure the actual values if they will affect your final result for k_B by more than 0.3%. You will need a high gain to measure the Johnson noise accurately; experiment a bit to determine what gain you'll want in the final measurement. To measure this gain accurately, you need to start with a signal that's large enough to measure accurately (at least 0.25 Vrms), then use a voltage divider to divide this down to a small enough signal for the input of the amplifier. Your instructor may provide the divider, or you can build it yourself. (Again, don't assume that the resistors used for this voltage divider have the spec values of resistance!)

If you are measuring the amplitude of a sinusoid that includes significant noise, you can do it most accurately by using the peak height of your Fourier transform. Make sure you are using the flattop window, that your Fourier spectrum shows "magnitude" (rather than a power spectrum), and that your data acquisition window covers at least 50 periods of the sinusoid. It would be smart to check that your spectrum peak heights work the way you think they do.

You may find it helpful for measurements of gain and values of f_{3dB} to write a modified version of your data acquisition program that acquires two differential channels of data, e.g., one that monitors the input to the voltage divider and one that monitors the output of the amplifier. It's easiest to do this by acquiring the set of data for one channel first, and then the other, rather than attempting to acquire them simultaneously.* If you are using LabVIEW, you should use the "sequence" structure (looks like a film strip) to ensure that the data taking on the first channel is done first, followed by the data taking on the second channel.

You *can* probably assume that the gain of the amplifier that is built into your interface unit has the spec value of gain. You should consult the manual and verify that the uncertainty in this gain is negligible compared with the ~1% uncertainty you'll be getting for the amplitude of the Johnson noise.

Your measurement setup will include an "unintentional filter" created by the output impedance of your resistor (which is equal to its resistance) and the parallel combination of the capacitance of the cables you have used on the input of your amplifier and the input capacitance of your amplifier. The f_{3dB} of this filter is weakly temperature dependent because the resistance of your resistor is weakly temperature dependent.

You will need to find an accurate value for the equivalent noise bandwidth B that appears in the equation for Johnson noise. The bandwidth is influenced by several factors: (1) the unintentional filter; (2) the effects of any filters you choose to intentionally engage in your amplifier; (3) the bandwidth of your amplifier; and (4) the bandwidth of the amplifier that is built into the analog-to-digital converter (ADC) you are using to take data with your computer. It is likely that the f_{3dB} of the unintentional filter is much lower than the bandwidth of either amplifier, so you may use the spec values for those bandwidths in your calculations, rather than measuring them. You may assume that the f_{3dB} for any filter you choose to engage intentionally is within 50% of the spec value. Depending on your experimental method and how strongly that filter affects B, that might or might not mean that you need to measure the f_{3dB} experimentally.

For your measurement of Johnson noise, you'll connect one end of your resistor to your amplifier input and ground the other end. However, you will certainly need to experimentally measure the f_{3dB} of the unintentional filter described above; this will require that you temporarily disconnect the "bottom end" of your resistor from ground. To make it easy to connect or disconnect this end of the resistor to ground, use a "shorting cap," which connects the inner and outer conductors of a coax connector to each other, as shown in Figure 15.2. When the shorting cap is attached, the bottom end of the resistor is connected to the box, which you will want to have connected to ground.

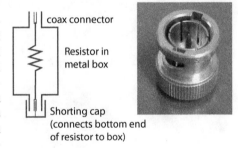

FIGURE 15.2 A coax "shorting cap" (right) can be used to connect one end of the resistor to the box. If the cap is removed, that end can instead be connected to an instrument.

Pre-Lab Question 15A.2: Why Should the Box Be Grounded?

Despite all the precautions you will take to avoid interference, some will still get into your experiment, and because of the high amplification needed, this will be significant. *To minimize the effect, you should intentionally engage a 1 kHz low-pass filter.* Likely, this will still leave a significant amount of

---

* This approach also allows you to keep the sampling rate at the maximum value. Most interface boxes include only two A/Ds. A one-channel differential measurement uses both of these. So, to make "simultaneous" measurements on multiple differential channels, the interface box rapidly switches the inputs to the A/Ds from one channel to another. This is called "multiplexing." One result is that, if you are measuring two channels simultaneously, you must lower the sampling rate for each channel by a factor of two.

interference, and you'll need to take this into account in your experimental design. For example, you may wish to account for interference by measuring the height of interference peaks in your Fourier transform and "subtracting in quadrature."

You should think explicitly about how you're modeling your instruments. For example, if your model for your amplifier includes the assumption that the gain is independent of frequency, you may wish to check this experimentally. This is just one example of the ways in which you model your instruments, whether consciously or unconsciously. Try to think about all of these and whether you need to test if that aspect of your model is justified.

If your method involves measurements at two different temperatures, make sure to measure the resistance of your resistor at both temperatures.

Pre-Lab Question 15A.3

After carefully reading the discussion above, devise a possible method for measuring k_B. Write this method up in brief form, as a series of bullet points. (If you are working with a partner or as part of a group, you should talk with the other group members, jointly devise a method, and jointly write it up.)

At the beginning of the lab period, get started on assembling any connection boxes you may need (see ExpPhys.com for general instructions on box assembly and soldering), on making any needed modifications to your data-taking program, and on running a rough version of the experiment to see how things are working. Before taking your serious data, you should discuss your method with your instructor.

Uncertainty Analysis

For your report, you should perform a thorough uncertainty analysis, and report your value for k_B with a 95% confidence interval. Be explicit about which sources of error and uncertainty you're ignoring because they are negligible compared with other sources.

16

Biophysics

Mason Klein

CONTENTS

When you have seen one ant, one bird, one tree, you have not seen them all.

– E.O. Wilson

16.1 Introduction

Biophysics is a young field, inherently interdisciplinary, and covers a wide range of research topics. A biophysicist may be interested in the physics behind the folding of proteins, long molecules that arrange themselves into complex structures and that participate in nearly every process within cells. Or the physics of colloids and other objects dispersing through liquids and gels – with the right modifications they can self-assemble into crystals or nano-scale machinery. Or the physics of the cytoskeleton, tubules, and filaments that function as scaffolding within cells to maintain their shape and structure. Or the physics of ion channels in neurons, cells that transmit electrical signals to other neurons and form complex neural networks that move and store information like an electrical circuit does. Thermodynamics, classical mechanics, fluid dynamics, and electromagnetism are all important in understanding biological systems at such scales.

To a student or anyone new to thinking about biological phenomena at this level, the relevant physics can feel different from sometimes more intuitive findings in mechanics, for example. Imagine being a molecule or suitably small object living at the micro or nanoscale: your neighbors slam into you at hundreds of meters per second, from all directions, and this happens a trillion times per second; moving through essentially any fluid when you are this small is like swimming through honey where viscous drag is the dominant force; your life is defined by random thermal motion and your trajectory feels chaotic, with only probabilities to provide any sort of predictability (not unlike quantum mechanics!). A key question then: how does so much structure and order arise in an environment so strongly driven by randomness and dissipative processes?

In addition to studying the physics of biological systems, a biophysicist can use knowledge, ideas, or tools from physics to help observe and understand such systems. They might develop new imaging techniques that allow us to observe living cells beyond the diffraction limit. They might work in advancing radiology to treat diseases in humans or build medical devices to measure physical properties of blood or tissue. They might sequence DNA by measuring electrical currents as the long molecules pass through a nano-scale pore in a membrane. They might consider larger-scale systems like population dynamics of organisms; or an even larger scale with the Earth as an engine taking in ordered energy from the high temperature sun and emitting energy into random thermal motion of the low temperature surrounding space; or a much smaller scale where quantum mechanics is needed to understand how photon absorption energy is transferred efficiently in photosynthesis.

Biophysics can mean studying the physics of biological systems, or it can mean applying ideas and techniques from physics to studying biological systems. In the next sections, you will primarily be doing the second thing, but in a tractable macro-scale model system (the fruit fly larva) that allows you to work with physical ideas that are extremely important at the nano-scale of biological systems.

Lab 16A   Navigation in the *Drosophila* larva

Broadly speaking, biophysics is the science of applying physics concepts and tools to investigate problems in biology. Biology is itself a very large field, with topics spanning the molecular scale (as small as 10^{-9} meters) to entire ecosystems (as large as our own planet, 10^7 meters). In this section you will be working somewhere in between, at the level of a small organism (10^{-3} meters), *Drosophila melanogaster*, also known the common fruit fly. While *Drosophila* is historically the most widely studied organism in genetics, and thus well-suited for understanding the functions and properties of proteins (molecular level) or neurons (cellular, neural circuit level), here we will focus on the mechanical side of things – that is, the animal's *physical* response to stimuli. The key question you will seek to answer: How does an animal use simple "rules" in its movement to accomplish a complicated task?

The primary goal of this chapter is to develop a quantitative understanding of the motion of navigating insects. You will assemble an apparatus that measures the behavior of groups of larvae as they crawl up a stimulus gradient toward one side of a two-dimensional (2D) square arena, and then use video analysis to determine the basic parameters of their motion. In the next chapter you can seek to understand how the motion of crawling insects – and other systems – fits into the context of a classical physics model, namely the random walk. You can use the parameters of motion as the basis for computer simulations of behavior in Lab 16B, and through these simulations explore how adjustments of the probabilities of heavily random decisions can lead to directed, purposeful navigation.

Chemical Sensing and Response

To ensure survival, most animals that are able to move must be proficient in finding a healthy and sustaining food source while avoiding more toxic environments. For example, an animal uses its olfactory system to sample the surrounding air for chemical cues, detects changes in chemical concentration, and moves either up or down a concentration gradient (i.e., toward or away from the source), depending on whether the odor is attractive or repulsive. The process of *chemotaxis* refers to the movement of an organism in response to a chemical stimulus, and in this module you will attract populations of larvae to an apparent food source and seek to understand *how* a larva can successfully navigate by modulating the properties of its crawling. The *Drosophila* larva is a good fit for the video recording and analysis that you will perform. Specifically, the creature is macroscopic, slow-moving, and has a short life cycle, all traits that minimize the barrier of entry into the world of animal behavior analysis.

You most likely have not used living animals in physics experiments before, but aside from some extra gentleness required during handling, you will investigate the fruit fly larva as you would any other physical system. Think of it like a mystery component in an electrical circuit, or the more general concept of a "black box" that we can only view in terms of inputs and outputs. The larva is a system that has *inputs*: various stimuli such as light, sound, mechanical touch, temperature, or the chemical concentration gradient you will deliver. And it has *outputs*: movement of its muscles causing it to crawl forward, turn, stop, eat, dig, roll, and so on. Like an impenetrable black box, in the limited time of this lab module you will not be able to view the inner components or logic, although neuroscientists interested in brain circuitry, for example, do study the animal at this level as well.

Hardware Assembly

You will use a CCD camera to record the motion of larvae crawling on a flat, square surface, approximately 20×20 cm. Use an empty square plastic dish to mark the space on a flat table. Then, use standard laboratory posts, clamps, and claws to hold the camera pointing down at the square area in a bird's eye view. Attach the lens in front of the camera and note that there are two adjustments possible: the aperture and the focus. Arrange the LED bars in a square, with the lights pointed toward the center, approximately 3–4 cm above the crawling arena. The lights should be either red or infrared (IR), in order to be outside the visible spectrum for fly larvae. Depending on the lighting conditions in your workspace, you may find it useful to turn off overhead lights when you run experiments. It can also be useful to stick black tape over reflective surfaces like the metal rods and clamps above the camera.

Connect the camera to a computer with acquisition software and make sure you can see a live image on the monitor. Physically, there are three things you can adjust: (1) the distance between the camera and the arena, which determines your field of view; (2) the aperture size, which determines how much light reaches the camera; and (3) the focus (i.e., the distance between the lens and the CCD), which you will set to give you the sharpest image. In software, you can adjust the exposure time, which is the duration for the CCD to collect light before saving each image. The inverse of this time is referred to as the frame rate. An exposure time of approximately 200 ms (frame rate of 5 Hz) should be suitable for most purposes.

As a medium for larvae to crawl on, you will use agar gel. Without moisture these animals will dry out and cease moving. To make large batches of gel, you can use an autoclave machine if your institution has one. Otherwise, gels can be made one at a time using a microwave oven (preferably one with a door that you can see through), with the following protocol:

- Pour 200 mL of water into a 500 mL Pyrex flask
- Measure out 5 grams of bacto agar, add to flask
- Measure out 1.5 grams of charcoal; add to flask (optional, but helpful for image contrast)
- Cap the top of the flask with plastic wrap and poke a couple small holes in it
- Set microwave timer to 3–4 minutes (on HIGH), microwave for ~90 seconds
- Continue to microwave, but look through the door and watch for the mixture bubbling up to the top. If the liquid/bubbles reach near the top, interrupt the microwaving, open the door, shake the flask, and return it, then continue microwaving (be sure to have oven mitts or something similar, as the flask will be hot). Note: not all microwaves are the same, so if you find your agar has not dissolved, or that you are evaporating large amounts of water, you might need to adjust the times above.
- When the heating reaches the point where the bubbles are large and pop on their own without starting to boil over, the agar is dissolved, and the flask is ready. Pour the liquid into a square plastic bioassay dishes, push any bubbles toward the edge (or pop them) with a pair of tweezers or something similar, then place the top lid on the plate. Let cool for approximately 30 minutes. Store in a refrigerator if available. When removing from a refrigerator, it is generally best to wait 10–15 minutes for the gel to reach room temperature; *Drosophila* are not mammals and move quite slowly (if at all) when placed on a cold surface.

For control experiments that do not involve an odor stimulus, the same gel can be used several times. For odor experiments, one gel per experiment is recommended. [Side note: you will want to be careful maintaining the top plates of the square dishes – clean them gently with alcohol and avoid anything abrasive, as small scratches can render them less transparent and affect the image quality during odor gradient experiments.]

[Optional] You can also make a set of gels in small, round Petri dishes. These may be convenient for staging your experiments and doing initial observations. Follow the recipe above but exclude the charcoal and pour small amounts of the liquid (enough to cover the bottom) into Petri dishes. These will solidify quickly and can be stored in the refrigerator or at room temperature.

Computer Software

In addition to the software for image acquisition (generally whatever comes with your camera), you will want the following programs or their functional equivalents. These examples are all free and open source, and run on Windows, Mac OS, or Linux. For the download links for each of these, please see the entry for this section on ExpPhys.com.

- Fiji. This is a distribution of ImageJ ("FIJI" is a recursive acronym for "Fiji Is Just ImageJ"), an open source image processing program. This can open almost any image format, view your larva behavior recordings, and also save the movies into other formats.
- Ctrax. This is an open-source machine vision program that can track the positions and orientations of walking flies or larvae, while maintaining their identities. The tracking steps are displayed visually with a tracking wizard, where the user can see steps in the tracking settings. While tracking, it clearly displays the trajectories. You can then export the trajectories into a generic spreadsheet type format (.csv), and from there you can analyze your crawling data in more detail.
- any2ufmf. An open source program for converting movies into "ufmf" (micro fly movie format), which Ctrax can read.

If automated tracking causes you too many problems, either due to software issues or the image quality of your movies, most of the analysis described below can be completed with just Fiji, or any program that can play your recordings back as a movie. The analysis just becomes a bit more "manual" than you might prefer.

Odor Stimulus Delivery

Drosophila larvae are highly motivated by food, so a robustly attractive odor can serve to draw the animals specifically toward one side of your arena. In both this experimental module and the following computational module, you will investigate how the larvae modulate their behavior based on whether they are moving toward or away from the odor source.

A reliable attractant is the chemical ethyl acetate (EA). The smell is characteristically sweet and might remind you of glue or nail polish remover; in fact it is an ingredient of both of these things. You should dilute your sample in water to be between 0.1% and 1% concentration. EA is not very soluble in water, but it will dissolve at this low concentration range. Pure ethyl acetate is flammable and should be stored as you would standard laboratory alcohols (read the material safety data sheet if you are unsure how to handle it).

To create a chemical concentration gradient of EA, place small (~1 cm) pieces of filter paper near the left or right edge of your arena and add a small (approximately 30 μL) drop of the diluted EA to each. Put the lid back on the square dish. The experiment should be started within a few minutes of adding the EA drops.

An Alternative Stimulus

Based on the set of existing hardware in your laboratory, or your own personal preference, you could use stimuli other than chemical concentration to investigate animal navigation. Temperature, for example, is another convenient type of stimulus: (1) it is of nearly universal importance for mobile organisms to find an ideal temperature, so navigation is quite robust; and (2) it is relatively straightforward to control, either with a feedback loop between your sensor and heating/cooling elements or with reasonably large reservoirs. If, say, a low and a high temperature could be maintained on each side of a rectangular plate, you end up with a smooth one-dimensional gradient (dT/dx) across one direction.

If you choose to use temperature or another stimulus, what follows should work essentially the same way – you would make similar observations about crawling motion, and try to piece together the fundamental modulations of behaviors that the animals use to achieve their goals. Note also that a thermal gradient experiment would not require a lid over your behavior arena, and agar gels can be reused several times.

Fly Maintenance and Larva Selection

Fruit flies are generally straightforward to maintain, and the larvae can be safely handled with a small paint brush. Laboratory flies usually live in small vials with food at the bottom and a breathable (but inescapable) cap or cotton ball at the top. The population is maintained by occasionally (every 3–4 weeks) moving the flies onto a new food source. To transfer the adult flies: (1) remove the cap from the empty new food vial and have the vial ready; (2) tap the old vial gently but repeatedly onto the table, which knocks the flies to the bottom, preventing them from flying away; (3) remove the cap from the old vial, continue tapping the old vial, and move the new vial upside-down over the top of the old vial; (4) flip this pair of vials over so the new one is on the bottom, and tap the pair of vials so that the flies fall to the bottom of the new vial; (5) move the old vial away, continue tapping the new vial, and replace the cap on the new vial; and (6) discard the old vial (if maintaining the population in the "off-season") or keep it if using for larva experiments.

Allowed to inhabit the vial, flies will mate and the females will lay numerous eggs on the food. If you then move the adult flies into a new vial, the old vial will contain larvae within a day or so. Briefly, the larval life cycle looks like this:

- 24 hours as an embryo/egg
- 24 hours as a first instar (L1)
- 24 hours as a second instar (L2)
- 48 hours as a third instar (L3)

After this the larva turns into a pupa, and some days later ecloses and an adult fly emerges. Larval "instars" refer to periods between molting (they shed their skin – and teeth! – like snakes or many other animals). For these experiments you will probably want third instar larvae ("L3s") during the first half of that stage. If you move the adult flies to a new vial every day, and save the old vials, one of the vials will contain the L3s you need. For example, if adult flies inhabit a vial from Monday to Tuesday, then early L3s will be ready on Friday. Note: depending on laboratory conditions, you might find that these time frames don't work out exactly; the important things are to choose larvae that you can see on your camera pretty well, and that the larvae are approximately the same size as each other.

To extract larvae from a vial, the simplest method is to squirt water into the vial and dump the contents into an empty Petri dish. You will see larvae swimming in the liquid and you can remove them with a small paint brush. An alternative method is to remove the food in chunks with a small spatula, and then spread it out and sift through it with the brush. Either way, at this stage the larvae should be rinsed off in a few large droplets of clean water in succession, and then can be moved onto a transfer plate or your large arena gel for recording crawling. The rinsing step is particularly important for odor-based experiments because you want to remove the scent of the food they have recently been crawling in.

Control Experiments

Before you start formal image recording and image processing steps, first try to get a sense of the physical motion of the animals. Find a few larvae that are actively crawling and place them onto one of your gels and watch them one at a time, either by eye directly or through a magnifying glass or laboratory microscope.

- **Q1: Describe qualitatively what you see, but include details. Include a physical description of the animal and discuss the apparent movement of its muscles, especially as it crawls forward and turns. Do you notice other behaviors?**
- **Q2:** Focus specifically on what happens when the animal changes its crawling direction. **Discuss what happens to its body shape when it turns and what happens to the velocity of its forward motion.**
- **Q3: Estimate the crawling speed of your larvae in m/s (or perhaps more appropriately mm/s or μm/s). Does each larva have the same speed? Does each larva appear to maintain a constant speed?**
- **Q4:** You may notice a longitudinal wave-like motion of muscle contraction during forward crawling. The motion is called *peristalsis* and works on the same principle as the human intestines, for example. **Estimate the period of the peristaltic waves.**

Next, you will use your full apparatus to perform video recordings of the crawling motion. Initially this will be done without a stimulus. Retrieve and rinse 10–20 larvae, place them on a transfer plate (if you are using one), and then place them near the middle of your square gel, which should be positioned directly underneath the camera. Adjust the aperture and focus to give a sharp image that doesn't saturate the camera. The lid for the square plate is optional for these control experiments, but will be required for the odor gradient experiments you will perform later. Choose the recording time (sometimes inputted as the number of frames, depending on software), typically around 10 minutes, and start your recording. When the recording finishes, remove the larvae with a brush and do not reuse the same animals. You will eventually want to record several of these experiments to later compile better statistics of behavior.

View your images as a movie by opening the movie file(s) in Fiji. (If you are using the default recommended Mightex camera, your movie was saved as a folder of .jpg files – to open in Fiji, go to Open→Import→Image Sequence, then find the folder you saved in.) Drag the progress bar around to watch the paths of the larvae and consider both the net motion of the population and the trajectories of individual animals.

- **Q5:** To follow an individual path, focus on one animal at a time as you watch your movie. **Sketch the trajectories for a few of your animals, and discuss what they look like, and how the individual trajectories vary.**
- **Q6: Estimate the average velocity of the population during your experiment and explain your reasoning. What about the average speed?**

By looking more carefully at the features of some of these trajectories, we can characterize the motion and behavior more precisely and quantitatively. Typically, the paths of individuals are made up of periods of approximately straight crawling punctuated by abrupt turns (although there are exceptions), which is why behavior in simple animals is sometimes treated as a random walk, which you might explore in the next module. You can start by examining the following properties of the tracks, but feel free to consider other features that you find noteworthy. For now, only examine one experiment, and you can incorporate additional control experiments in the next section.

- **Q7:** INTERVALS BETWEEN TURNS. For each larva in your single experiment, identify its abrupt turning events, and estimate the distance traveled between each turn that you see (note: if you hover your cursor over a location in the movie, the x,y coordinates will be displayed in the top menu in Fiji). In a single experiment, you will likely have more than 100 turns total. Compile the list of the distance intervals, and **compute the mean, standard deviation, and standard error. Note that the raw units you extracted are in pixels (for distance) and you will want to convert those to more familiar units.**

- **Q8:** TURN ANGLE. For each abrupt turn you observe, estimate the size of the turn (in degrees or radians). As with the turn intervals, compile a list and **compute the mean and uncertainty.**

- **Q9:** VELOCITY. For each trajectory, first approximate its average speed by estimating the total distance traveled (not displacement) and divide by the total elapsed time. **Do all the larvae travel at the same speed?** You can also examine the average *velocity* of each larva: determine the positions (x,y coordinates) of each larva at the start of your movie, and their positions at the end, and use that to compute the average velocity in the x-direction and the y-direction (note that by convention, Cartesian coordinates for images have the origin in the upper left, with $+y$ pointing downwards). Compile these values for all animals and **find the mean and uncertainty. Is this consistent with your initial estimate above?**

An "error analysis" interjection here. When measured values deviate from the mean, you might be used to thinking of this as "error," where the value of some quantity was not measured to be its true or correct value, and you might think this reflects some failure in the experimenter or the equipment. In the case of these animal behavior measurements, the inconsistency is actually an extremely important property of the behavior itself. Variance (standard deviation squared) is often the metric used to describe this. The animals really do behave differently from one another, particularly over shorter time intervals, and the size of your "error bars" describes the range of behavior the animals exhibit under the conditions you have set up. This is characteristic of processes (in many areas of both physics and biology) that are driven at least partly by randomness – these are known as stochastic processes, and you might delve into this more in the next module.

Having familiarized yourself with how features of 2D crawling behavior can be extracted by eye, you will next employ machine vision software to perform some inspection and analysis of the larva trajectories. In particular, the image processing routines should give you the position of each larva as a function of time. From introductory physics courses, you are likely quite familiar with what you can do once you are given the functions $x(t)$ and $y(t)$ for moving objects, and should be able to extract a variety of interesting properties of the trajectories.

From Fiji you can save your crawling images to a movie format (.avi is a common one). From here, run the "any2ufmf" program on your movie file. As its name suggests, it saves your movie as a .ufmf (micro fly movie format). Using the program Ctrax, open the .ufmf file. Run the Tracking Wizard, which will guide you through selecting some tracking settings, including background subtraction, and the thresholds for counting larvae. After proceeding through the wizard steps, tracking analysis will commence, and will show you a graphical overlay of the objects (larvae) that the program finds. Notice that when larvae occasionally collide, the tracking program will be unable to identify the two individuals, but will resume tracking after they move apart. You could personally probably track the two animals and maintain their identity, an example of the fact that automated image processing can do things quickly, but is not necessarily better at all tasks.

After the tracking finishes, the program will have saved an annotation file (.ann extension), and you can also export the results as .m (if you happen to be using MATLAB®) or .csv, which you can open in MS Excel or other spreadsheet programs. If you are continuing with .csv (the most common option), the data will appear as a large number of columns. Each track is described by six columns of data, in the order [ID, x, y, w, l, theta], and the sequence continues across the width of the sheet. ID is the identifying label for the larvae (an integer, arbitrarily assigned), x and y give the position (in pixels), l and w are the length and width of the animal (technically the major and minor axes of the ellipse that surrounds the animal), and "theta" is the current orientation direction (with respect to the positive x-axis, in radians, with range $[-\pi, +\pi]$). To examine a trajectory of any individual, plot a y-column vs. an x-column. In this visual format, it should be more convenient and faster to identify intervals between turns and turn sizes.

- **Q10:** Taking at first a small number of trajectories, make a new column and compute the speed at each time point (row), which will be the displacement divided by the time interval. (Note: you might want to make this new column in a separate sheet or paste to another program altogether.) **Plot speed as a function of time. It might look noisier than you expected. Why do** you think that is? Try zooming in on a few short time intervals (~30 seconds): can you see a periodic structure? If so, what is its physical origin? What is the mean speed and how does this compare with your estimates from earlier?

- [Optional] If you are finding the plots too noisy to understand, you can run a smoothing routine on your x and y values. In MATLAB®, run the function smooth(vec), where vec is a vector of your x or y values. In Excel, run Data→Data Analysis, and choose the "moving average" option. Other programs should have equivalents to this.

- **Q11:** With the same trajectories you used to look at speed vs. time, you can also look at the crawling direction (θ) vs. time (t). You can either use the "theta" from the .csv file directly, or compute it yourself from the x and y data using basic trigonometry (note: use ATAN2 in Excel, not ATAN, so that it can give you directions in all four quadrants). **Why does θ sometimes appear to jump between near $+\pi$ and near $-\pi$? What does a relatively abrupt jump in theta signify physically, and what does a steady increase or decrease signify physically? Look at the speed vs. time and theta vs. time graphs next to each other – is there a connection between the two variables? What typically happens during larger direction changes?**

- **Q12:** Using either the by-eye method described earlier, or the x(t), y(t) data you have now, find the distance intervals between turns for all your control data. **Compute the mean and uncertainty, and also make a histogram of the values.**

- **Q13: Do the same thing for turn size, including a histogram of its values.** These histograms can be used as probability distribution functions for the Monte Carlo simulations in the next module.

Chemotaxis Experiments

The goal here is to record crawling behavior movies as in the previous section, but this time with an ethyl acetate (EA) chemical concentration gradient in your arena, and be sure to include the top plate on the plastic square dish. Your animals should now be exhibiting *navigation* behavior, moving toward targets rather than merely searching a space. View your images as movies in Fiji and confirm that the animals generally crawl toward the filter paper with your ethyl acetate drops.

- **Q14:** Follow the trajectories of a few individuals, one at a time, by eye. **Are they always moving toward the odor source? Considering each abrupt turn as a "decision," roughly how often are they making the "correct" decision?**

- **Q15: From an ecological standpoint, would it be better for individual animals to be perfect navigators, or to exhibit some degree of random behavior?** Feel free to speculate, but explain your reasoning. It may help to come back to this question later.

- **Q16: Estimate the average velocity of the population during your experiments and explain your reasoning.**

Next you will want to quantitatively analyze your new navigation behavior data. This will be very similar to what you did for the control, odor-free crawling data, with the following important difference. When you are extracting the distance between turns, the turn size, or the speed, divide your results into two categories: (1) larva is crawling toward the EA squares or (2) larva is crawling away from the EA squares. For sections between abrupt turns that spend time crawling toward both sides, use the direction immediately prior to the turn to assign a category.

- **Q17:** Find the average x-direction velocity for each larva (x-displacement divided by total crawling time), and **compute the overall mean and uncertainty for all your chemotaxis experiments put together. How does the number compare with what you found for the control data?** Note that if most of the animals reach the filter paper targets early in the experiment, it might be better to compute the x-velocity over the shorter time windows leading up to this.

- **Q18:** For category 1 (toward EA) and category 2 (away from EA) compute the mean and uncertainty of the speed, and if you have enough data, plot the histogram for each category on the same graph. How do the two compare? How do they compare with the control results? Note: you might want to compute the speed of individual segments of a track and average, since the tracks will often be split between the two categories.

- **Q19:** For category 1 and category 2, compute the mean distance between successive abrupt turns. How does this value compare between the two categories? How do they both compare with the control results?

- **Q20:** For category 1 and category 2, compute the mean turn size (magnitude). How does this value compare between the two categories? How do they both compare with the control results?

- **Q21:** Based on your findings from the last three questions, discuss *how* larvae are able successfully navigate toward higher concentrations of ethyl acetate.

In the next lab you can use your results to explore more deeply the mechanics of how robust navigation arises in a simple system with limited perceptive and motor capabilities. To do this you will generate simulated tracks of your own, and test how modulating parameters like turning rate, turn size, and crawling speed enhance or diminish an animal's performance of navigation tasks.

If you have additional time, there are a number of directions you could take this experimental module, especially regarding the hardware design and stimulus delivery. Some suggestions are listed here, but feel free to generate your own ideas and try them out as well!

- So far you have been using what is called a "wild type" strain of *Drosophila*, which is a laboratory-raised strain that is meant to act similarly to a wild fly without any significant mutations. If you have the strain *Orco*[1] available, try running chemotaxis experiments with it. "Orco" is short for "olfactory receptor cofactor," and is missing this gene (i.e., unable to manufacture the protein in its chemical-sensing neurons). Thus, the larva should be unable to smell anything. Do the larva behave as you would expect?

- The ability to respond to olfactory cues is extremely important for larvae (and most other animals), but there are many other important stimuli too. Come up with a way to deliver a different kind of stimulus, then observe and describe the resulting behavior. You could use an ice water bath and a resistive heater to establish a spatial temperature gradient, for example. Or perhaps aim a lamp (not red) at your gel from an angle and see if larvae crawl toward or away.

- As noted earlier, most animals must have proficiency in searching for food. As a search continues without success, animals may alter their movement strategies. You could try starving your larvae for several hours (leave them on an agar-only transfer plate), and then measure their behavior, either with no stimulus or in a chemical gradient. In terms of the quantitative metrics you used earlier, does anything change? If so, do you think these changes would be reasonable for an intelligent searcher?

- If you happen to have the hardware for delivering a temperature gradient and a chemical concentration gradient, why not try both together? You could place them at odds with each other, or in the same direction, or even perpendicular. How would you determine the relative importance of each type of stimulus? This type of experiment could study what is called multisensory integration, where animals must process many inputs and decide on a physical output action. This could become overwhelmingly complicated if taken too far, but would place us on the road toward a better understanding of how these animals would need to behave in the "real world."

Lab 16B   Biophysics: Modeling and Stimulating Behavior

Modeling the world around us with consistent and simply stated ideas is a great strength of physics. Complex systems like biological organisms might not appear to have straightforward equations that predict their internal or external states (e.g., probably no $PV = NkT$ for an insect), but there are a number of ideas from basic physics that show up remarkably often. For example, simple electrical circuits with batteries, resistors, and capacitors can describe brain connectivity and communication. When trying to understand the essential physics of a situation, it is often useful to determine the fundamental components of a process. In this lab, we will attempt to do this for the crawling insect system used in the previous lab. Specifically, what are the fundamental changes to physical trajectories that animals can make in order to successfully navigate toward a desirable environment?

Ideally you have completed the previous experimental lab already, and have data of your own to use as the basis for our simulations here. It's perfectly fine if you are using this chapter as a standalone laboratory about modeling and simulations. If this is the case, data can be provided to you, although you might want to look over the previous chapter as well. Once we build a foundation for simulating motion, many sections and questions will present you with quite open-ended tasks. You are strongly encouraged to explore your own ideas.

Random Walks

A random walk is a path through mathematical space that is made up of a sequence of steps whose magnitudes and/or directions are assigned randomly. These walks emerge and can serve to explain behaviors in a wide range of systems throughout the natural and social sciences: the path of a molecule in an ideal gas, the Brownian motion of pollen grains in water, stock prices in the market, and so on.

In the simplest conception of the random walk, we can consider a 1D path along the x-axis, which starts at the origin. Then at each step, we (randomly, like a coin flip) add either +1 or −1. Thinking of the steps as time advancing (a sequence of times $t_0, t_1, t_2, \ldots$), we can write this as

$$x(t_i) = x(t_{i-1}) + (1 \text{ or} - 1),$$

and $x(t)$ becomes the 1D trajectory for a single object performing a random walk. An example of such a random walk ($n = 1000$ time steps) can be seen in Figure 16.1 (left panel). As its reputation would suggest, it appears to be a fairly aimless path, although it would be difficult to draw conclusions based on a single trajectory. Plotting $N = 10$ such trajectories at once (middle panel) gives us a somewhat better sense of the range of possible outcomes. Every step is completely random (or as random as the programming language can accomplish), and as a result the paths can end up in a wide range of locations.

This range can be quantified a bit better if we increase the number of paths we are generating, to $N = 10,000$, and build a histogram of the resulting $x_{n=1000}$ positions (right panel). We see what amounts to an essentially symmetric distribution of outcomes, centered at $x = 0$, which is in fact the most likely result. For sufficiently large N, the mean of the distribution is not very interesting, since movement to the left or right is equally likely, but its *width* is important, as a way to describe the range of possible locations. You can explore this facet a bit more in the questions below.

- **Q1:** In the figure, each trajectory has 1000 steps. **Q1: Why does the histogram fall off to essentially zero before reaching even ± 100?** (Note: in this specific simulation, 12 of the 10,000 tracks fall outside the range shown.)
- **Q2:** Try writing code to generate your own 1D random walk trajectories! The $x(t_i)$ equation above serves as the algorithm. The specific syntax depends on which programming language you are using, but the main tools you need are for-loops, and a random number generating capability. **Try generating ten trajectories and plotting them like the middle panel in the figure. Do they match that figure? Would you expect them to?**
- **Q3:** Generate a large number of trajectories and save the final positions of each into a single array. **What is the mean and standard deviation of this "data"?** Then plot a histogram like the one in the right panel in the figure. There is probably a histogram function at your disposal already, but if not, you can choose your bins, loop through all your final x_n position values, and increment the number of entries in each bin as you go. **Try doing a fit to a Gaussian function – how does it look?** [Note: to prevent computer memory problems, you might find

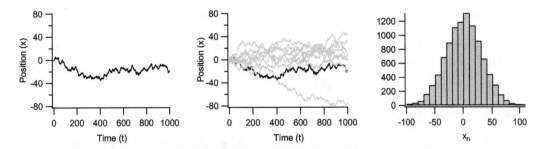

FIGURE 16.1 Left: position vs. time for a single random walk. Center: multiple random walks. Right: histogram for final positions (after 1000 steps) for 10,000 random walks.

it helpful to delete most of your trajectories and save only the final x_n values, especially if you choose to use a very large n and/or N].

- **Q4:** As a somewhat more open-ended task, examine the connection between the final position (x_n) distribution and the number of steps (n). Keeping N the same, make histograms for, say, $n = 10, 100, 1000,$ and $10,000$, and plot them on the same graph or as a sequence of graphs (keep the horizontal scaling the same if you are using separate graphs). **What do you observe happening over time? Does this phenomenon remind you of anything else in physics?**

- **Q5:** We will continue developing these simulations throughout this section. **At this point though, if you completed the experimental module, what are some similarities and differences between these 1D random walk trajectories and the actual recorded trajectories from the crawling larvae?**

Diffusion

As you most likely discovered for yourself in the previous section, the greater the number of steps n in your simulated 1D random walk, the farther away from the starting point your "walkers" will end up. Your simulated trajectories were generated one at a time, but you could imagine them all running together: N walkers start at the same point, and as a population spread out. Most stay near the starting location, some move far away. The more time they are given to walk, the greater the average absolute distance from the start. Importantly, the mean displacement is near zero, but that number obscures the spreading phenomenon, so it makes more sense to consider the *square* of the displacement.

- **Q6: As a quick exercise, compute the mean of the square of the final x-positions, x_n^2, for your data in the previous section.**

This is closely linked to another physical phenomenon called *diffusion*. Diffusion is an idea used in numerous scientific disciplines: ions moving across cell membranes, CO_2 spreading in air, and so on. In a general sense, diffusion refers to the idea of objects spreading out from a location with a high concentration of the object to regions of lower concentration. Classically, the idea was used to describe the motion of atoms and molecules, but as we will start to see in this module, the tools can be applied to macroscopic objects like the organisms you might have been wrangling in the previous lab.

To be a bit more precise, a *diffusion coefficient* quantifies the rate that the square of the spread distance increases with time. Its dimensions are length²/time (m²/s in MKS units) – we have so far just dealt with an abstract number line, but length and time scales for the larva tracks will be important later). As an instantaneous quantity, we can write this as $D = \dfrac{1}{2}\dfrac{d}{dt}\langle x^2 \rangle$, where x is the displacement from zero and $\langle\ \rangle$ denotes averaging over the population. For large numbers of steps, this can be approximated as $D = \dfrac{1}{2}\dfrac{\langle x^2 \rangle}{t}$, with t the duration of the trajectories. The 1/2 is a factor related to the dimensionality; it is 1/2 for 1D, 1/4 for 2D, and 1/6 for 3D motion.

The diffusion coefficient can also be related to the width of the distribution. You can find the derivation of this for random walks in many places, and the result is that the *variance* of the final positions of the objects is $\sigma^2 = \dfrac{t}{\Delta t} L^2 = 2Dt$ (L is the distance moved at each step; the factor of 2 is a 4 for 2D motion and a 6 for 3D motion). Or, $D = \dfrac{L^2}{2\Delta t}$.

- **Q7: Compute D for your longest 1D simulated random walk, using both methods described above. Since these initial simulations were both on an abstract number line, L and Δt are both 1. Do the two agree exactly? If not, what is the reason, and what could you do differently in their simulation to make them match more closely?**

- **Q8:** Notice that the variance of the position distribution is directly proportional to the elapsed time ($\sigma^2 \sim t$). Variance has the same units as the square of the position units, meaning that the average absolute value of the displacement scales like $\sqrt{t}$. Perhaps the $\sqrt{t}$ scaling is not intuitively obvious, but **how could you have guessed that it was something slower than $\sim t$ scaling? What kind of motion would have the absolute displacement directly proportional to t?**

Two-Dimensional Random Walks

To bring us closer to simulating, and hopefully gaining some insight about, the crawling behavior of *Drosophila* larvae you perhaps dealt with in the previous lab, let us extend the capability of our random walk simulations into two dimensions. The simplest way to implement this would be to have each single step have four outcomes (step in $+x, -x, +y$, or $-y$) instead of two. But in preparation for being able to handle larva turning better, we treat the direction changes a bit more generally.

For the one-dimensional case, trajectories were generated by $x(t_i) = x(t_{i-1}) + (1 \text{ or } -1)$. Now we will have three things to keep track of: the x-position, the y-position, and the current crawling direction. We define the crawling direction θ as the angle the path heading makes with the positive x-axis, and use a range of $[-\pi, +\pi]$. A new heading, for now, will be chosen randomly within that range at every time step (later we will choose the new heading based on deviations from the previous heading, which is slightly more difficult to implement, but more realistic for these particular crawling animals). Let us now also introduce more explicitly a movement speed v_0 (constant for now) and a time step duration Δt. Thus, our new algorithm for generating trajectories would look like

$$\theta_i = \text{random value in range} \left[-\pi, +\pi \right]$$

$$x(t_i) = x(t_{i-1}) + v_0 \cos(\theta_i) \Delta t$$

$$y(t_i) = y(t_{i-1}) + v_0 \sin(\theta_i) \Delta t$$

where θ_i is generated uniquely for each time step. These equations probably remind you of the kinematics material from the first few weeks of your first physics course. Figure 16.2 shows an example of one simulated 2D trajectory (left panel), and then ten such trajectories (middle panel), for $n = 1000$ steps.

The right panel is a histogram of the displacement from the origin for $N = 1000$ tracks each with $n = 10,000$ steps. Note that although this histogram is counting a value with the same dimensions (length) as the 1D histogram, it's not quite the same: this is not centered around 0 (the mean is ~90) and is not symmetric. In the following exercises, you will build your own 2D simulated trajectories, and in the next section you will simulate larva behavior more directly.

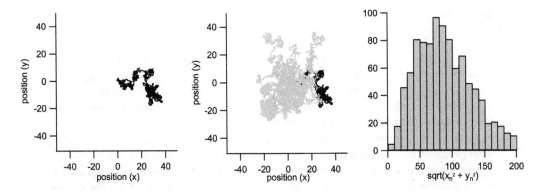

FIGURE 16.2 Left: a two-dimensional random walk. Center: ten random walks. Right: a histogram of 1000 random walks, each 10,000 steps long.

- **Q9:** Larvae, like all organisms, are not really two-dimensional. **Is it appropriate to treat everything as 2D in our behavior analysis and simulated trajectories?**

- **Q10:** Generate your own set of 2D random walk trajectories, employing the algorithm above in the programming language of your choice. **Try producing your version of the figure panels above and comment on the similarities and differences.**

- **Q11:** You hopefully stored the final coordinates (x_n, y_n) for all your tracks. **Plot two separate histograms, one for x_n and one for y_n. Do these look like your original 1D histogram for histogram for x_n? If you use the same n and N values, do you expect your x_n 2D motion to have the same width as it did for 1D motion? Why or why not?**

- **Q12:** Look at one of your 2D simulated trajectories next to one of your larva crawling trajectories from the previous lab (if you have one). **What properties of the two paths appear to be different? If you "zoom in" on your simulated track, do they seem more alike? What does that tell you?**

Simulated Control Experiments

Your procedure for generating 2D simulated random walk tracks is the foundation for simulating the larva behavior from the experimental lab (or you can obtain example data and work with that). There are two important changes to implement before moving forward: (1) the *time* between steps is not constant, and we will model the intervals between turns as a Poisson process (described below); and (2) the change in *direction* between steps is not completely random, and we will choose the *change* in heading from a Gaussian distribution of possible angles.

As to the first part, in your observations of crawling animal trajectories, the intervals (temporal or spatial) between turns were not constant. We choose to treat it here as a *Poisson process*. Simply stated, this means that a repeating event has a constant probability of occurring in a given time interval; of particular note is that this probability is *independent* of whether the event has occurred recently. Suppose in a rainstorm, an average of 100 raindrops hit your roof every second, and the raindrops arrive independently of one another. During a given one-second interval, you will not *always* observe 100 (λ) rain drops hitting, and the likelihood of a given number of drops (k) showing up follows a Poisson distribution, $\lambda^k e^{-\lambda}/k!$, which has the unique property that its mean is equal to its variance, and looks like a Gaussian for large enough λ. The width of this distribution is where shot noise arises, in electronics and photon counting experiments, for example.

More pertinent to our task at hand, if an event (like a crawling larva turning, or a flipped coin landing on heads) has a fixed probability of occurring during a time interval (i.e., is a Poisson process), then the distribution of the number of time intervals between repeated events is exponential (like $e^{-t/\tau}$). It turns out that if the probability of the event occurring during the interval Δt is p, then the mean number of intervals between events is $1/p$, and the decay rate of the exponential $1/\tau$ (if τ has units of steps) is also equal to p.

- **Q13:** Take a look at your results for the control data in the experimental lab. In particular, look at your histogram for the distance between turns in a track. Dividing by the average crawling speed would give you a histogram of the same shape, and this would roughly be the distribution of time intervals between turns. **Based on the information presented above, could one describe larva turning as a Poisson process, at least approximately?**

- **Q14:** Even if the agreement isn't perfect, we can estimate the probability p for your larva turn behavior data. Let's take our time interval to be $\Delta t = 1$ second. **Compute p by (1) calculating the mean amount of time between turns and (2) by fitting your time interval or path length distribution to an exponential and extracting the time constant. Do the two methods agree?** Keep this value for p handy, as we will code our own Poisson simulation coming up shortly.

The second change to our 2D random walk simulation we want to make is in the change in direction between time steps. As you likely observed, larvae do not choose a completely random direction every time they turn, but their new direction is related to their previous heading. You recorded the change in angle $\Delta\theta$ for many turns and crafted a histogram to show the distribution. Here we will use this distribution as a sort of weighted average for randomly choosing a $\Delta\theta$ for each turn in a simulation.

We model the $|\Delta\theta|$ distribution as a Gaussian function (a.k.a. normal distribution), of the form $\left(\dfrac{1}{\sqrt{2\pi\sigma^2}}\right)e^{-\frac{(x-x_{\mathrm{avg}})^2}{2\sigma^2}}$. The variance is σ^2 and the approximate width is 2σ. The factor in front is to force the area under the curve to be 1, important when using it as a probability distribution function (PDF). This leaves the distribution with only two parameters (σ and x_{avg}). We use this distribution partly because it is so universal, and partly because it's mathematically more convenient. At this point you can fit your empirical distribution to a Gaussian and extract x_{avg} and σ. If your data is too noisy, feel free to estimate both parameters. Note that most likely your Gaussian is not peaked at exactly $\pi/2$ (although it could be close), so the mean of the data that makes up your histogram does not necessarily align with the peak of the curve.

An approximated example of a Gaussian distribution for the turn sizes $|\Delta\theta|$ is shown in Figure 16.3. The left panel is a distribution of turn size with peak value $\pi/3$ and $\sigma = \pi/6$. Our goal is to choose a number for $|\Delta\theta|$ from 0 to π, where the probability of choosing a particular value is weighted by the shape of the curve. Hence the function is termed a *probability distribution function*. You may have been introduced to the idea when learning about the Maxwell-Boltzmann distribution in statistical mechanics. For this specific example, a value near $\pi/3$ would be the most likely turn size, and a value near π the least likely.

Procedurally, how can your simulation pull a value from the distribution each time? The middle panel is what is called a cumulative distribution function (CDF). This is the integral of the PDF from 0 to x.

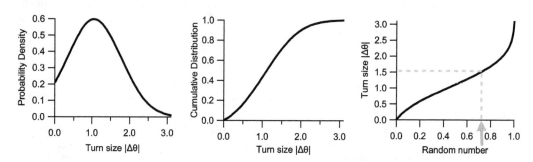

FIGURE 16.3 Left: distribution of turn sizes. Center: cumulative distribution function. Right: same function as center plot, but with axes flipped, illustrating the conversion of a random number into a turn size.

Now we have a one-to-one unique mapping between a number in the [0,1] range and the turn size. The CDF increases the fastest when the PDF is largest. So you will notice that values near the peak of the PDF cover more of the [0,1] range on the vertical axis than values near π. It is sometimes more convenient to flip the axes (right panel). In your program, the function will have discrete values and this is easily done. Now a random number chosen from [0,1] on the horizontal axis corresponds to a specific angle on the vertical axis. In programming terms, you have your turn size values (a vector or array with values evenly spaced between 0 and π) and your CDF values (between 0 and 1, but not evenly spaced). To choose a turn size, take your random number and find the linearly interpolated $|\Delta\theta|$ value.

Taking our two changes to the original 2D random walk, your algorithm can look something like the steps shown below. Let p be the probability for turning during a given time step Δt.

Choose a random number in the range [0,1]

If this value is less than p, then a turn is made:

- Generate another random number from [0,1] and use it to select a $|\Delta\theta|$.
- Randomly make the number positive or negative.
- $\theta_i = \theta_{i-1} \pm |\Delta\theta|$

If the value is greater than p, then no turn is made:

- $\theta_i = \theta_{i-1}$

Then do the same thing as before to update positions:

- $x(t_i) = x(t_{i-1}) + v_0\cos(\theta_i)\Delta t$
- $y(t_i) = y(t_{i-1}) + v_0\sin(\theta_i)\Delta t$

Notice that this has incorporated both of our changes, the first in deciding each step whether to change the crawling direction, and the second in how the change in direction is chosen. Note that the $\pm$ part of the turn is (for now) another random choice, with + corresponding to a left turn, and − to a right turn.

[Note: Your CDF can be generated manually, so to speak, by taking the sum of your PDF at each angle bin in your histogram. If you are using a Gaussian fit, it can also be done analytically. For a Gaussian

$$\text{PDF}(x) = \left(\frac{1}{\sqrt{2\pi\sigma^2}}\right) e^{-\frac{(x-x_{\text{avg}})^2}{2\sigma^2}}$$, its definite integral from 0 to x is:

$$\text{CDF}(x) = \frac{1}{2}\left[\text{erf}\left(\frac{x-x_{\text{avg}}}{\sqrt{2}\sigma}\right) - \text{erf}\left(\frac{0-x_{\text{avg}}}{\sqrt{2}\sigma}\right)\right]$$

You might want to divide by $\text{CDF}(\pi)$ to ensure that your CDF is normalized properly.]

Your simulation program should now be able to generate some pretty nice looking larva trajectories – let's find out!

- Using your empirical values for p and for the x_{avg} and σ in the turn size PDF, generate a set of simulated trajectories. You might want to choose fewer times steps this time, so that they roughly correspond to the actual duration of your experiments. If you did not complete the preceding experimental lab, feel free to use the data provided, or try out your own p and x_{avg} values.

- **Q15: Plot some of your simulated trajectories next to some of your empirical ones. What similarities and differences do you observe at this point?**

- **Q16:** Think about some limiting cases of parameters in your simulations. **What would you expect if the mean turn size x_{avg} were very small? What if it were near $\pi/2$? Similarly, what happens when p is made very large or very small?** Try out some or all of these (i.e., generate new trajectories as you tweak these parameters). **How should each of these changes affect the diffusion coefficient? Pick one parameter (or more if have time) and** make your own plot of **D** vs. ___ (each point on the graph will come from a batch of simulated trajectories).

- **Q17:** Using a very large number of trajectories averaged together, and with realistic parameters, what is the diffusion coefficient? Would you expect this to agree with your empirical trajectories? If not, what is still "missing" from your simulations?

These kinds of explorations (the third bullet point above especially) are extremely powerful as a general technique in science. You are able to test the effect of individual components of behavior on the resulting movement of the population. It would be essentially impossible to breed or create real larvae with a specific desired p or σ, for example, but you can "observe" their navigation easily with your simulations. The general technique we are employing here is called the *Monte Carlo method*, widely used in physics and many other disciplines. The idea is to use random sampling many times to arrive at some quantitative conclusion or result. Monte Carlo methods are especially appropriate here, where there is a very high uncertainty in the behavior of individual animals.

Simulated Chemotaxis: Elements of Navigation Strategy

Your work with simulated trajectories so far has been, to within the uncertainty due to random sampling, *symmetric* with respect to direction. That is, the spreading out of trajectories from the origin would look the same along the x-axis, y-axis, or any tilted axis in between. This aspect probably does match your control data reasonably well, but you also have a set of odor gradient experiments where the outcomes do not match this idea at all. Individual trajectories may appear to still have a lot of "randomness," like the random walks you generated, but there is clearly a net movement of the population toward a more preferred odor concentration.

The goal of this section is to explore in greater detail *how* larvae modify the conditions of a standard random walk to move toward improved conditions, running a *biased random walk*. Further Monte Carlo simulations will investigate the relative importance of the fundamental components of navigation strategy by isolating individual behavior modulations and observing the outcome.

- **Q18:** Look back at your results from your chemotaxis experiments (or the data provided to you) and estimate how the properties of turning (the distance or time between them, and their size) depend on the crawling direction. The simplest way is to compare "category 1" and "category 2" situations, although if you have enough data you could break up the direction into finer categories. **Determine values for p (probability of turning in a 1-second Δt window) for each category and do the same for x_{avg} and σ for the turn size distributions.**

- **Q19:** Armed with a separate set of parameters [p, x_{avg}, σ] for $+x$ direction crawling and $-x$ direction crawling, modify your simulation code to allow for either. Changes will be minimal. Importantly, before you start the commands in your previous algorithm for each time step, check the current crawling direction (most easily by looking at the sign of $\cos(\theta_{i-1})$), and then assign the parameters accordingly. **Run your simulations to generate chemotaxis trajectories and compare them with the empirical trajectories. Are these two minor modifications enough to produce navigation?**

- **Q20:** As we proceed to determine the key features that lead to navigation in these animals, it is useful to have a figure of merit in mind – that is, a quantitative way to ask how efficiently the animals navigate. The most straightforward is probably to take the average x-position (*not* the square in this case) at the end of the trajectories. If the two sets of trajectories have the same duration, they can be directly compared this way. **Compute this for your simulated tracks and for your empirical ones. Is this a fair comparison? If not, do you have other ideas?**

- [Optional] At this point you have two distinct, independent (at least as coded in your simulations) behavioral modulations that cause directed navigation. Try to add a couple more. One could be that speed depends on crawling direction. Another could be a bias to turn toward a preferred side of the arena when crawling approximately perpendicular. For the latter, you could adjust the way the ± on the turn angle is decided – rather than 50/50, you could make it 60/40 for −1 (right turn) when the crawl direction lies within some range near $\pi/2$, and then 60/40 for +1 (left turn) when the crawl direction is near $-\pi/2$.

- **Q21:** Whether or not you added the above optional behavioral modulations, let's consider the degree to which each type of modulation affects the trajectory outcome. Try enabling or disabling each of the biases one at a time, simulate trajectories, and compute your figure of merit (see above). **For realistic (i.e., your empirically determined) conditions, which type of bias is the most significant?**

- **Q22:** Play around with the behavior biases even more. Set them to more extreme limits (e.g., turning the correct direction 100% of the time) and observe what happens to your paths. **If you think about a more realistic, ecological setting for this type of motion, would "perfect" navigation be a desirable thing? Why or why not?**

- **Q23:** As a more open-ended exercise, try adding new features to your simulations. Think about properties of the empirical tracks that your simulated tracks don't have. [Examples could include extra time spent while turning, the fact that the trajectory shape between turn events are not straight lines, variance in crawling speed both between animals and during a single animal's track, etc.]. **Describe in as much detail as you can the effect of your new addition(s).**

- [Optional] Try giving your simulated animals a slightly different task, rather than a general goal of moving to the +x or −x direction. For example, you could add a finite vertical line as a barrier that animals "bounce" off and investigate how optimal behavioral biases might now be different.

- [Optional] Through all of this, we have treated each animal as a completely independent actor. What if there is some kind of interaction between nearby animals? To include this, you would probably want to build your trajectories in parallel rather than in series. Perhaps once two animals become close, you could add a component of attraction or repulsion. If you make the effect quite strong, what would you expect the new trajectories to look like?

- [Optional] So far, motion has been in two dimensions. What if it were done in 3D instead? What kinds of new behaviors would you now have access to?

- [Optional] Our organism of choice has been the *Drosophila* larva. It has a rather simple behavioral repertoire, which is great when we are first getting used to modeling. But maybe think about another animal? Choose one you like, look up some facts about, and try writing a simulation of its motion. How is it similar to or different from the crawling fly larva? [Note: this could complement the previous question if you wanted an animal that is more 3D, due to jumping or flying, for example].

An important property of many biological systems, and more physical ones than you might think, is randomness. In these past two modules you have ostensibly been studying directed motion -- how animals manage to navigate toward better conditions. You probably noticed how imperfect this process actually is. The crawlers do the "wrong" thing very frequently. This is quite interesting, and in fact makes the animals more robust to finding what they need, even if the process becomes less efficient.

Lab 16C   Biomechanics: Modeling Physical Actions

Not all forms of using physical ideas to model biological phenomena have to involve the extremely high variance and randomness of crawling animals making decisions. Familiar ideas like Newton's laws can be used to help us understand how physical actions come about. This has applications in mechanical design and robotics, and can work in both directions: a good physical model could create something not seen in nature (the flight of a 100,000 kg airplane, for example), and sometimes an organism accomplishes a task better than human mechanical designs, and can help improve those designs.

This lab is meant to be an open-ended opportunity for you to model some physical actions and learn about the conditions under which types of motion become possible. A simple, two-masses-and-one-spring model of jumping behavior in 2D will be presented, just to provide a rough foundation. After that, it's very much up to you! Feel free to use the programming language of your choice, even a spreadsheet program if you're feeling adventurous.

We start by imagining two masses (m_1 and m_2) connected by a massless spring of rest length ℓ_0. Initially the spring is compressed (or stretched) to length d_0 (the distance between the masses), and the masses are oriented such that the spring makes an angle θ_0 with the vertical direction. Let Δt be the time step of our simulation, and N be the number of steps we want it to run.

We will compute the accelerations ($\vec{a}_1, \vec{a}_2$), velocities ($\vec{v}_1, \vec{v}_2$), and positions ($\vec{r}_1, \vec{r}_2$) of both masses as functions of time: at each step we use the forces (due to gravity and the spring) to determine acceleration, and go from there. Our initial conditions are:

$$\vec{r}_{1,0} = (0,0)$$

$$\vec{r}_{2,0} = (d_0 \sin\theta_0, d_0 \cos\theta_0)$$

$$\vec{v}_{1,0} = \vec{v}_{2,0} = (0,0)$$

$$\vec{a}_{1,0} = \frac{1}{m_1}\left[k(d_0 - \ell_0)\hat{r}_0 - m_1 g\hat{y}\right]$$

$$\vec{a}_{2,0} = \frac{1}{m_2}\left[-k(d_0 - \ell_0)\hat{r}_0 - m_2 g\hat{y}\right]$$

Where the unit vector $\hat{r}$ is $(\vec{r}_2 - \vec{r}_1)/d$, pointing along the length of the spring. From here we simply let the system evolve and see where the two masses go! The process iterates (from $i=1$ to N) like this:

$$\vec{v}_{1,i} = \vec{v}_{1,i-1} + \vec{a}_{1,i-1}\Delta t$$

$$\vec{v}_{2,i} = \vec{v}_{2,i-1} + \vec{a}_{2,i-1}\Delta t$$

$$\vec{r}_{1,i} = \vec{r}_{1,i-1} + \vec{v}_{1,i}\Delta t + \frac{1}{2}\vec{a}_{1,i-1}(\Delta t)^2$$

$$\vec{r}_{2,i} = \vec{r}_{2,i-1} + \vec{v}_{2,i}\Delta t + \frac{1}{2}\vec{a}_{2,i-1}(\Delta t)^2$$

$$\vec{a}_{1,i} = \frac{1}{m_1}\left[k(d_i - \ell_0)\hat{r}_i - m_1 g\hat{y}\right]$$

$$\vec{a}_{2,i} = \frac{1}{m_2}\left[-k(d_i - \ell_0)\hat{r}_i - m_2 g\hat{y}\right]$$

Where d_i gets updated to be $|\vec{r}_2 - \vec{r}_1|$ and $\hat{r}_i$ is as defined above.

Note that in order to prevent the masses from falling through the floor, we would need to impose another condition on the positions $\vec{r}_1$ and $\vec{r}_2$. For example, if for either of the masses $v_{y,i} < 0$ and $y_{i-1} \leq 0$, then we could force $\vec{v} = 0$ and $\vec{r}_i = \vec{r}_{i-1}$. This particular workaround would make a mass that would otherwise pass through the floor instead of resting on it. If you wanted to allow it to continue sliding (without friction) in the horizontal direction, then only force the vertical velocity to zero, and $y_i = y_{i-1}$.

To see your result, be sure to save the positions as you go, and then plot them at the end. Figure 16.4 shows a schematic of the situation on the left, conditions where there is insufficient energy stored to

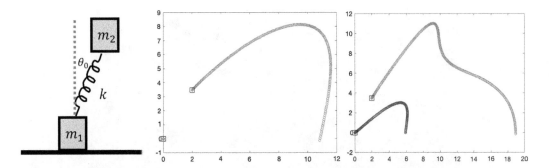

FIGURE 16.4 Left: the system to be modeled. Center: trajectory of m_2, starting on the left, with insufficient energy in the system for m_1 to leave the ground. Right: trajectory of m_2 (top) and m_1 (bottom) with enough energy for the masses to make a "jump."

raise the bottom mass off the ground at all in the middle ($m_1 = m_2 = 1$, $\ell_0 = 10$, $g = 10$, $k = 5$, $\theta_0 = \pi/6$, $d_0 = 4$, $\Delta t = 0.01$, $N = 3000$), and then conditions where the masses "jump" on the right (same conditions except $k = 10$). Note that m_1 is initially located at the origin, and m_2 above and to the right.

- **Q1:** Implement the above code yourself in the software of your choice. There is a lot of parameter space to cover, so pick a few parameters and vary each systematically while holding the others fixed. **Qualitatively, what is the effect you see as you change your parameters? Do the limiting cases make sense?**

- **Q2:** Roughly speaking, what would you say are the conditions that must be met for this object to jump off the ground?

- **Q3:** Two masses and a spring doesn't seem like much of an animal, as far as modeling goes, but the system can be surprisingly informative. As an exercise, you could treat your system vertically ($\theta_0 = 0$), and consider the maximum height reached by one of the masses. **From an energy perspective, what determines the height of a jump?** There is a result found in nature (not entirely true, but by order of magnitude pretty close) that all animals jump the same height, regardless of their mass. **Based on what you see in your simulation, does that seem possible? Why or why not?**

From here feel free to take this simple jumper system further or explore any kind of motion you want. You could try a more sophisticated version of this thing above, using more masses or springs, or choose to include air drag or floor friction and see if and how that affects your results. You could build a ridiculous contraption with many masses and springs and see what it does. Part of what makes simulations both very powerful and very fun is that, when you have an idea of the underlying physics/rules, you can perform almost any "experiment" you want!

FIGURE 16.5   Mason Klein (right) and his daughter (left) with an ice cream cone (not much left).

Mason Klein is an Assistant Professor of Physics and Biology at the University of Miami. Following a winding research trajectory through chaos and pattern formation in plasmas, mixing and de-mixing of vibrated sand particles, and slow light in warm atomic vapor, his research currently focuses on the behavior and brain function of fruit fly maggots. A broader goal of behavioral neuroscience is to understand how sensory inputs ultimately lead to behavioral outputs. Prof. Klein's lab studies how crawling larvae navigate through environments featuring an array of stimuli like temperature, mechanical agitation, gravity, light, and odors, plus the animals' own internal preferences. Longer term goals include characterizing the neural circuits that lead to navigation and understanding the nature of randomness in behavior. He also enjoys spy novels, baseball statistics, and whatever his favorite organism (his one-year-old daughter) comes up with (Figure 16.5).

17

Non-Linear, Granular, and Fluid Physics

Nathan C. Keim

CONTENTS

Nature can produce complex structures even in simple situations, and can obey simple laws even in complex situations.

– Nigel Goldenfeld and Leo P. Kadanoff

17.1 Introduction

We live in a world of breathtaking complexity and variety that often seems to come from nowhere. We add cream to coffee and witness a turbulent display of whorls and plumes. A potato chip bag is smooth, pliant, and as thin as human hair, yet crumpling it makes a pointy, noisy mess that seems to willfully refuse to be any smaller. Living matter can assemble itself and function with no direction from without, and surprisingly little from within. The solid earth quakes without warning, and one can walk on sand and yet pour and stir it like a liquid.

It seems unlikely that we will ever understand these examples in quite the same way that we understand a helium atom or even a bottle of helium gas, even though they all obey the same known laws of physics. But fields like non-linear, fluid, granular, and soft matter physics use physical reasoning to help make sense of the world's complexity and surprises. These fields are in dialogue with biology, chemistry, engineering, and geophysics, but they are also true to the tradition of physics, especially in the way that a small number of themes, ideas, and lessons seem to underlie a staggering array of phenomena. Such themes include:

- *Mesoscopic structure.* This refers to building blocks with sizes in the vast middle ground between atomic length scales (microscopic) and the size of the system (macroscopic): chunks of rock, rod-shaped molecules in liquid crystals, or the bundles of macromolecules that make up the cytoskeleton. The interactions among these components are of similar strength as among atoms, but instead they apply to much larger volumes, often making a material "soft" – easier to deform and change. Large deformations can in turn bring out the non-linear properties of matter such as buckling, creasing, fracture, and transitions from rigid to flowing. And by being larger and slower than the structures and dynamics of atoms and small molecules, mesoscopic physics can be easier to observe and manipulate. For example, by using colloidal particles in liquid instead of atoms, one can see the dynamics of crystal defects and phase changes in incredible detail. These building blocks and their interactions can also be customized far more than atoms, so that, for example, they assemble spontaneously into complex structures – a route to new types of matter and machines.

- *Spontaneous pattern and structure formation.* Softness may allow structures to arise from a featureless background: for example, the creases and cones in a crumpled sheet, the many large and small eddies of cream in a coffee cup, or tiny micelles made of surfactant molecules. These structures are often beautiful, but they can also give a material its distinctive properties and texture.

- *Physics far from equilibrium.* Statistical and thermal physics show us that even though the microscopic state of a system might seem random and constantly changing, simply by exploring many such microscopic states, the system tends to relax toward a well-defined equilibrium. But sometimes this exploration is too slow, or too costly in terms of energy, and the system is stuck in one state. This leads to materials with memory of their history, as when two buckets contain the same volume of the same kind of sand, yet because they were prepared differently, one weighs 7% more. "Active matter" in which energy is injected at the mesoscopic scale – swimming bacteria or contracting motor proteins – represents another frontier far from equilibrium.

This lab will touch on all three of these themes. A beautiful, dramatic structure will arise spontaneously from two featureless fluids. When we add mesoscopic components – a small amount of polymer molecules – they will dramatically change the behavior as they are driven farther and farther from equilibrium.

Further Reading

- D.T.N. Chen, Q. Wen, P.A. Janmey, J.C. Crocker, and A.G. Yodh. Rheology of soft materials. *Annual Review of Condensed Matter Physics* **1**, 301–322 (2010).
- N. Goldenfeld and L.P. Kadanoff. Simple lessons from complexity. *Science* **284**, 87–89 (1999).
- L.S. Hirst. *Fundamentals of Soft Matter Science*. CRC Press (2012).
- N.C. Keim, J.D. Paulsen, Z. Zeravcic, S. Sastry, and S.R. Nagel. Memory formation in matter. *Reviews of Modern Physics* **91**, 035002 (2019).
- S.R. Nagel. Experimental soft-matter science. *Reviews of Modern Physics* **89**, 025002 (2017).
- S.H. Strogatz. *Nonlinear Dynamics and Chaos: With Applications to Physics, Biology, Chemistry, and Engineering*. Westview (2015).
- T.A. Witten. *Structured Fluids: Polymers, Colloids, Surfactants*. Oxford University Press (2010).

Lab 17A: Drop Pinch-Off

FIGURE 17.1 Sequence of images from the pinch-off of water-glycerol solution in air. A drop is falling from the tilted nozzle at the top, which has diameter 4.0 mm. The bright spots in the middle of the drop are artifacts due to the light positioned behind the drop. The dark smudge on the left is from the laser used to trigger each photograph.

Introduction

As a drop of water falls from a faucet, one body of liquid separates into two. This everyday event is actually a beautiful and violent non-linear phenomenon, shown in Figure 17.1. In this lab, you will use high-speed imaging techniques to capture this event and use some basic fluid dynamics to model it. You will also use these techniques to reveal the presence of polymer molecules in a second fluid.

Objectives Shared with Other Areas of Experimental Physics

Video and stroboscopic techniques; image analysis; high-speed imaging; power-law and exponential dynamics.

Goals of This Experiment

- Construct simple models of low-viscosity drop pinch-off in Newtonian and polymeric fluids.
- Study an example of a viscoelastic fluid.
- Reason (with data) about power-law dynamics and dynamical singularities.
- Use a stroboscopic technique to "slow down" a very fast process.

Time Requirements

Typically two or three three-hour lab sessions. The polymer solution should be stirred unattended for two hours, then an additional 24.

Safety Precautions

- The chemicals used in this experiment are relatively safe, but you and your instructor should discuss their specific hazards and safety requirements. Use disposable gloves and goggles, preferably splash goggles, when mixing fluids and setting up the tubing. Label any containers you use and do not pour waste down the drain.

- Do not look directly into intense light sources. If you are using a laser, do not turn it on until you have read the safety labels and checked with your instructor about what precautions apply.

- This lab may use a strobe light, which may cause discomfort or nausea, and can have more serious effects in people with certain medical conditions.

Readings

- *OpenStax, Chemistry*. OpenStax CNX. February 1, 2020. https://openstax.org/details/chemistry (Relevant sections of this are specified later in this chapter.)
- *OpenStax, University Physics, University Physics Volume 1*. OpenStax CNX. February 1, 2020. https://openstax.org/details/books/university-physics-volume-1 (Relevant sections of this are specified later in this chapter.)
- X. Shi, M.P. Brenner, and S. Nagel. A cascade of structure in a drop falling from a faucet. *Science* **265**, 219–222 (1994).
- Material Safety Data Sheets (MSDS) for glycerol and polyethylene oxide.

Suggested Additional References

- J.C. Burton, J.E. Rutledge, and P. Taborek. Fluid pinch-off dynamics at nanometer length scales. *Physical Review Letters* **92**, 244505 (2004).
- J.C. Burton, J.E. Rutledge, and P. Taborek. Fluid pinch-off in superfluid and normal ^{4}He. *Physical Review E* **75**, 036311 (2007).
- J. Dinic, L.N. Jimenez, and V. Sharma. Pinch-off dynamics and dripping-onto-substrate (DoS) rheometry of complex fluids. *Lab on a Chip*, **17**, 460–473 (2017).
- J. Eggers. Nonlinear dynamics and breakup of free-surface flows. *Reviews of Modern Physics* **69**, 865–929 (1997).
- A. Morozov and S.E. Spagnolie, in *Complex Fluids in Biological Systems*, pp. 3–52. Springer (2015).
- A. Sack and T. Pöschel. Dripping faucet in extreme spatial and temporal resolution. *American Journal of Physics* **85**, 649 (2017).
- J.B. Segur and H.E. Oberstar. Viscosity of glycerol and its aqueous solutions. *Industrial & Engineering Chemistry* **43**, 2117–2120 (1951).
- V. Tirtaatmadja, G. McKinley, and J. Cooper-White. Drop formation and breakup of low viscosity elastic fluids: Effects of molecular weight and concentration. *Physics of Fluids* **18**, 043101 (2006).
- The user manual for your camera.

Introduction to Fluid Dynamics

In the first half of this lab you'll be working with Newtonian fluids, a type of fluid with which you are already very familiar (you're breathing one right now!). This is the simplest type of (real) fluid to describe mathematically. Before going further, you should be familiar with the concept of viscosity and its units, Pa s, and the concept of surface tension and its units, N/m (or J/m^2). If you can't find these topics in one of your physics textbooks, try an introductory chemistry book. One good option is the free online text *OpenStax Chemistry*. There you can read Section 10.2, "Properties of Liquids." You can stop when you reach the discussion of adhesive forces. It may be helpful to know that "IMFs" stands for intermolecular forces.

You should also be familiar with three other basic fluid dynamics concepts: hydrostatic pressure, conservation of volume, and Bernoulli pressure. One good source for these topics is *OpenStax University Physics*, Volume 1, Sections 14.1, 14.5, and 14.6.

In nearly all situations, we treat a fluid as *continuous*, meaning we do not think about individual atoms or molecules, and *incompressible*, which in this context means that the density ρ is uniform throughout. The velocity and pressure are generally *not* uniform, however; they are fields that depend on position and time: $\vec{v}(\vec{x},t)$ and $p(\vec{x},t)$.

Exercise

1. Consider a cylindrical tube of fluid parallel to the x axis with initial radius $R=R_0$. It extends from $x=-L/2$ to $x=L/2$. Something is causing R to contract, expelling fluid out the ends of the tube. Sketch graphs (qualitative only) of
    a. The x component of fluid velocity as a function of x, $v_x(x)$. Where is it positive, negative, or zero? Is $v_x(x)$ a continuous function?
    b. The density $\rho(x)$.
    c. The pressure $p(x)$. Where along the x axis is it highest?
    d. Surface tension and Laplace pressure

Surface Tension

What happens where one fluid meets another? If the fluids don't mix (like water and air), there is an interface or surface that separates them. Unlike a solid boundary, this surface does not have a fixed shape; it can bend, expand, contract, and undulate as it pleases. Why then does a small drop of water sitting on a table have such a smooth, rounded shape? As you saw in the reading, surface tension is the energy associated with creating or destroying a unit of interfacial area. Imagine a sphere and a cube that each contain 1 cm^3 of liquid. Which has the lower interfacial energy?

Because making a surface smaller will lower its interfacial energy, surface tension exerts forces and can do work. It's often useful to model this effect as *Laplace pressure* – a pressure exerted on a fluid by its interface. This is $p_{\text{Laplace}}=\gamma\kappa$, where γ is the surface tension, and κ is the *mean curvature* of the surface. For example, a liquid-filled sphere of radius R has $\kappa=1/R$; its contents are always under Laplace pressure γ/R, in addition to pressure from the outside atmosphere and hydrostatic pressure. As the sphere gets smaller, the contents are under greater pressure. A cylinder of radius R has mean curvature $\kappa=\dfrac{1}{2}\dfrac{1}{R}$, because its surface only curves in one direction.

Exercises

2. Two spherical soap bubbles are connected by a very thin horizontal straw that allows air to flow between them. The bubbles are of equal radii $R_1=R_2=2$ cm. Calculate the air pressure in each bubble relative to atmospheric pressure, Δp_1 and Δp_2. Take the surface tension of a soapy air-water interface to be 0.036 N/m.

3. Suppose the right bubble starts out slightly smaller, $R_2<R_1$. Which bubble is under greater pressure? Because of this pressure imbalance, will the air inside the straw be pushed *toward* the smaller bubble or *away* from it? As the air flows, which bubble will gain air and get bigger, and which one will get smaller?

4. Continuing the previous part: as R_2 gets smaller, and R_1 gets bigger, what will happen to these bubbles' pressures? The flow of air through the straw is driven by the difference between these pressures; will it slow down or speed up? As this process continues toward the limit $R_2 \to 0$, what will happen to the pressures and the air flow?

What you have just sketched out, in which some physical quantities are expected to become infinite, is an example of a *singularity*. In the real world, these quantities grow large but don't actually diverge, because there is always some other feature of the system that cuts them off. In this case, the nonzero diameter of the straw keeps R_2 from reaching zero.

A Simple Model of Drop Pinch-Off

You will need to do the preceding exercises (and be reasonably confident in your answers) *before* reading this section. Next, go to the paper by Shi et al. Read the first page and the paragraph that is continued on the second page.

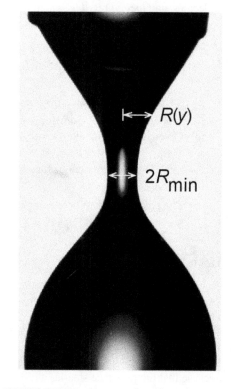

In the first part of this lab, you will study the pinch-off of an aqueous solution much like the one in Figure 1B of Shi et al. You will use photography or video to measure the radius of the fluid neck (h in Shi et al., R in this lab) at its thinnest point (Figure 17.2), as a function of time: $R_{min}(t)$. You might expect the equations that describe the shape of the neck to be difficult to solve, and you would be right (see Equations 2–4 of Shi et al.). But using this reading and what you already know about physics, you can correctly predict the behavior of $R_{min}(t)$ when it is between 10 and 100 μm – and possibly outside this range.

Exercises

5. First, decide what will be the ingredients in your prediction – what quantities do and don't matter to the dynamics when $10 \lesssim R_{min} \lesssim 100$ μm. Use the following list of possibilities:

   a. Neck minimum radius R_{min}
   b. Nozzle diameter D
   c. The size of a water molecule
   d. Acceleration of gravity g
   e. Surface tension γ (in other papers, σ)
   f. Liquid density ρ
   g. Liquid viscosity η

FIGURE 17.2 A photograph from Figure 17.1, annotated to define R and R_{min}.

You will be using a solution of 50% glycerol and 50% water (by mass), which at room temperature has viscosity $\eta = 0.00600$ Pa s, or 6.0 times that of pure water. This liquid has the same surface tension and density as pure water, to within about 10%.

As suggested by the Shi et al. paper, you will decide what's important by comparing the expected R_{min} – roughly 10 or 100 μm – with other *length scales* in the problem. This is straightforward for the first three items, which are all length scales: if a length scale is much larger or much smaller than R_{min} (by a factor of roughly 10 or more), whatever is happening at that scale should have very little to do with the instantaneous dynamics of R_{min}. For instance, you should conclude that the molecular nature of water is irrelevant.

For the last four quantities, you can *construct* length scales to compare with R_{min}. For instance, the capillary length L_γ describes the competition between surface tension and gravity. You may have noticed that a tiny drop of water on a table is round (surface tension dominates), while a large puddle of water is flat (gravity dominates). "Tiny" and "large" have precise meanings because of L_γ (which you can derive yourself by setting Laplace pressure equal to hydrostatic pressure). This also applies to the pinching drop: if $R_{min} \ll L_\gamma$, you know that surface tension is much more important than gravity to the dynamics. Shi et al. also define L_η: if surface tension acts on a large body of liquid, we must account for the liquid's inertia (via ρ); otherwise viscosity matters more (via η).

Now, decide which items on the list above should and shouldn't be important to the pinch-off dynamics of a water drop in air. Show quantitative reasoning whenever possible. You should find that **only three quantities matter** (hint: one of them is R_{min}).

6. There's a fourth important quantity: Δt, the time remaining until the neck reaches zero radius at time t_{pinch}. We define $\Delta t = t_{pinch} - t$, so that Δt counts *down* to zero; as R_{min} gets smaller, Δt gets smaller.

   Use dimensional analysis to find an expression that contains R_{min}, ρ, and γ, and that has units of time. You can use exponents, square roots, etc. as needed. Set your expression equal to Δt, so that you have one equation with all four key quantities.

7. Use your equation to make a prediction for $R_{min}(\Delta t)$. This is your prediction. Really! You have ignored a lot of complexity (see again Equations 2–4 of Shi et al., which are themselves approximations), so you shouldn't be surprised if your prediction is a little too small or too large. In anticipation of this, you should multiply your expression for $R(\Delta t)$ by a dimensionless prefactor α, which for now we'll assume to be 1.

Polymers and the Maxwell Model

Many fluids have more mechanical properties than just viscosity, density, and surface tension. They contain large molecules, or other ingredients like particles or cells, that are much larger than a solvent molecule but much smaller than any macroscopic length scale. As the fluids flow, these *mesoscopic* components can stretch, jam, break and heal, or even swim, imbuing these *complex fluids* with a staggering variety of behaviors. Examples include mayonnaise, toothpaste, fluids used in oil drilling, and your own saliva and mucus. Newtonian fluids lack these mesoscopic components, though this is not a precise definition.

In the second part of this lab, you will uncover how a tiny amount of polymer additive, put under stress at the pinch-off singularity, can radically change the behavior of a fluid. You will model the fluid as a simple kind of *viscoelastic* fluid that combines the viscosity of the solvent with the elastic stretchiness of the polymer molecules.

Each dissolved polymer molecule is a long, flexible chain of atoms, subject to constant bombardment by molecules of solvent. A consideration of entropy suggests that at equilibrium, the molecule will be gathered into a loose, random, roughly spherical mess, where the largest number of microstates is possible. On the other hand, if the molecule is stretched into more of a straight line, this has much lower entropy. If it is left alone, a stretched molecule will gradually relax back to a more disordered state, at a rate characterized by a *relaxation time* λ, which has units of time. This change in entropy is associated with a change in free energy, meaning that the relaxation does mechanical work. A similar idea explains the elasticity of a rubber band. The extension of a polymer molecule in one dimension (how much it is stretched out) may be quantified with the dimensionless number A, where $A = 1$ corresponds to the unextended, equilibrium state.

The simplest model of fluid with polymers is the Maxwell model, named for James Clerk Maxwell of electrodynamics fame. In this model, when a fluid is stretched very slowly, the molecules always have plenty of time to relax, so they don't do much. But stretching a fluid quickly will stretch the polymer molecules much faster than they can relax, and the result is a kind of elastic tension (like a rubber band). We decide which limit we are in by comparing the *extensional strain rate*, $\dot{\epsilon}$, with the inverse relaxation time $1/\lambda$ (each has units of s^{-1}). This comparison is expressed as the dimensionless number $\dot{\epsilon}\lambda$, also known as the Weissenberg number.

Moving closer to the experiment at hand, let's approximate the fluid neck as a cylinder of radius R. You showed earlier that the fluid in a cylinder gets stretched out as R decreases. In this case,

$$\dot{\epsilon} = -\frac{2}{R}\frac{dR}{dt}$$

Exercises

1. Let's revisit your prediction for $R_{\min}(\Delta t)$, but with dissolved polymer of relaxation time λ. Assume that the neck is roughly cylindrical near its thinnest point, and derive an expression for $\dot{\epsilon}(\Delta t)\lambda$ during pinch-off. Your answer should be in terms of Δt and constants such as λ, not other variables like $R_{\min}$. From this, speculate as to whether the polymer will have a significant effect during pinch-off, as $\Delta t \to 0$. Hint: $\dfrac{dR}{dt} = -\dfrac{dR}{d\Delta t}$.

2. (Optional but recommended.) The Maxwell model for polymer extension A is the linear differential equation $\dot{A} = \left(2\dot{\epsilon} - \lambda^{-1}\right)A + \lambda^{-1}$. Show that this model has the following behaviors:

   a. When $\dot{\epsilon}\lambda \ll 1$, $A(t)$ relaxes to equilibrium (molecules unstretched), with a timescale λ.
   b. When $\dot{\epsilon}\lambda \gg 1$, $A(t)$ will grow very large (molecules stretched a lot).
   c. When $\dot{\epsilon}\lambda$ is neither large nor small (so, not a good basis for approximation), and $A(t)$ starts out being $\gg 1$, $A(t) \simeq A_0\exp\left[t\left(2\dot{\epsilon} - \lambda^{-1}\right)\right]$.

Part I: Low-Viscosity Newtonian Pinch-Off

To do this experiment, you will need some kind of "leaky faucet" to slowly produce the drops, with a way to adjust the rate, and something to catch the drops. The light source and camera will be on opposite sides of the drop. (This explains why there are there bright spots inside the drop shape in Figure 17.2.)

The physics of drop pinch-off sets two basic requirements:

- Spatial resolution: To get good data you will need to resolve the width of the neck when it is ~20 μm, or even smaller. Your instructor will have provided you with this equipment. In general, the closer the camera is to the drop (while still in focus), the higher the magnification will be.
- Time resolution:
  - The camera cannot record images instantaneously. It has to wait for a certain *exposure time* in order to gather enough photons at the sensor. If the subject of your photo moves too much during this time, the result will be *motion blur*. Your predictions should tell you that small=fast, which means you will need to take each picture very quickly; 10 μs is a reasonable goal.
  - In this experiment, a lot can happen in 1 ms or even 10 μs. The more precisely you can slice the pinch-off process into short time intervals, the better your results.

There are approximately two ways to achieve the required time resolution. Your instructor has probably already chosen one of these for you:

- Use a high-speed camera that can record video at 10,000 frames per second or more. These cameras typically require a very bright, continuous light source.
- Set a long exposure time on the camera (0.1 s or even 10 s) and use a *strobe light*, which gives a short, intense flash of light. To achieve good time resolution, something will have to trigger the strobe at precisely the right moment.

The Trigger and Strobe

High-speed video cameras that are suitable for this experiment have only been available since the 1990s, and for most of that time they have cost US$100,000 or more. Even today, these cameras typically have much lower resolution than the one in your phone. Yet people have been making high-resolution, high-speed observations of fluids since the 1880s! How is this possible? Discoveries about electricity led to the use of electric sparks (dielectric breakdown of air) for illumination – the forerunner of the strobe light. But even so, sparks and strobes reveal single instants. How did people assemble many such snapshots to discover an entire sequence of events in the blink of an eye – what today we call a "movie"?

Experiments like the one you are doing have a very important feature: because the dynamics depend only on the boundary conditions (i.e., the nozzle) and the fluid properties, drop breakup is exactly repeatable. If the falling drop triggers the strobe in a way that is also exactly repeatable, each picture should look like the next. Triggering when the drop crosses a laser beam is one way to do this. If we then introduce a delay d between when the laser beam is blocked and when the strobe fires, adjusting d will let us image different moments in the pinch-off process. Do this at regular intervals in d, and you will have a movie that is the envy of all but the best high-speed cameras.

The recipe to do this is as follows:

1. Use a strobe light that is fast enough (if it has an intensity adjustment, dimmer means faster) and has an external trigger input.
2. Send a laser beam across the path of the drop, where it just starts to fall rapidly. Have a photodiode convert the laser light to a voltage. Send the output to a circuit that compares the photodiode signal with some adjustable threshold voltage.
3. Use a delay generator, or any circuit that fires the strobe after an adjustable delay d.

Preparing the Newtonian Fluid

This experiment should work with any low-viscosity Newtonian liquid, including water (see the discussion of viscous and capillary lengths for a definition of "low"). However, using a water-glycerol mixture will let you make a more direct comparison with the polymeric fluid in Part II.

Prepare a solution of 50% glycerol and 50% deionized water (by mass). Wear safety goggles, preferably splash-proof, and gloves. Roughly 200 mL should be more than enough for one session of the experiment, and you will probably need much less if you are using a high-speed camera. The mixing goes fastest if you start with water and add glycerol gradually. Note that glycerol does not evaporate, so you will need to clean up spills using a damp paper towel.

Since you are studying a capillary phenomenon (i.e., involving surface tension), you need to beware of *surfactants* – molecules like soap, detergent, and fatty acids that change the surface tension and dynamics of a liquid. Residues of these chemicals are everywhere, and dissolved minerals in the water also play a role. For best results use deionized water, wipe and rinse glassware and stirrer bars with an organic solvent (such as acetone or isopropanol), and wear clean disposable gloves. Surfactants are also carried on dust particles in the air, so cover your samples when possible.

If you are also doing Part II, now is a good opportunity to start mixing that sample as well.

Initial Setup

You should first get to the point where you can take focused pictures of a drop hanging from the nozzle (known as a "pendant drop"). The camera, drop, and light source should be arranged in a horizontal line. If you will be using a strobe light for the experiment, start out by using a continuous source like the room lights instead.

Once it's time to start the dripping, make sure that each pinch-off event does not interfere with the next. A good rule of thumb is to keep the drops about 2 s apart. Be on the lookout for bubbles getting stuck in the tubing or fittings; these can reduce the flow rate, and (even worse) cause it to slow down dramatically as the fluid level gets low. You can dislodge bubbles by opening all the valves momentarily and/or squeezing the tubing.

Using a Strobe

The camera shutter needs to already be open (i.e., it must be taking a picture) when the strobe fires. You may have been given a triggering circuit that takes care of this for you, but you can also do it by hand, using a remote control so you don't shake the camera. Either way, set the exposure time to "Bulb" if your camera allows it, which means you that you or your circuit have direct control over whether the shutter is open or closed.

Once you have a steady drip and can take pictures, set up the laser beam, perpendicular to the camera's line of sight. By moving the nozzle, laser, and/or photodiode up and down, you can make the strobe fire earlier or later in the pinch-off process. Configure your triggering circuit to have zero delay, so that you will be seeing the earliest moment that you can record. Make this the moment when the neck radius is roughly 1/2 of the drop radius. **Remember that this technique relies on the time interval between drops, and the way each drop hangs from the nozzle, staying nearly constant during setup and data-taking.** To troubleshoot, start with the advice about bubbles in the previous section.

Once you are happy with your setup, you are ready to take a "movie." Experiment with different values of delay in your triggering circuit: establish a lower bound for the delay, where the drop is connected but a neck has formed, and an upper bound, where the drop is separated but still in the picture. Keep pushing these bounds toward one another, until they are just several ms apart. Within this range, you will be able to observe $R_{min}(t)$ change noticeably as you take pictures that are spaced by 25–100 µs.

While they won't be needed for your analysis, be sure to take some pictures *after* pinch-off.

Using a High-Speed Camera

Once you are ready to take data, you should record at least several drops because of the variation in when each movie starts (more on this in the analysis section).

Your camera likely allows you to get a higher framerate by reducing the size of the image (using only a portion of the image sensor). Remember, though, that you must be able to see the thinnest part of the neck in order to measure R_{min}. Some sensors perform better when the image is short and wide, so you may even want to rotate your camera onto its side.

Overview of Image Analysis

The goal of image analysis for this experiment is to measure the minimum radius of the neck, $R_{min}(t)$, up until the very last moment that the neck is unbroken.

There are two basic approaches. Measuring the size of the neck by hand is straightforward but can be tedious. In research today, analysis of movies is almost always automated with a computer. After skimming the sections below, come to a decision with your instructor as to which method you will use.

Obtaining Good Images

Image analysis is easier and works better when you start with good images. Five minutes of adjustments in the lab can save hours of frustration on the computer. This is why you should look over this section before you finish setting up your experiment. It is also a good reason to revisit the experiment once you have given the analysis a try.

Whatever imaging method you use, getting enough light to the camera sensor may be a challenge. Start with the aperture on the lens open all the way (the lowest f-stop number). You can also turn up the gain or "ISO" control on your camera to make sure that the sensor is as efficient as possible, but don't overdo it – you will make the image brighter, but you will also amplify the noise. The background should be light gray, **not bright white**. Otherwise the fine details of pinch-off will be washed out. Since you will be recording and measuring many images, make sure that your camera doesn't change any settings between pictures by disabling all automatic features including autofocus (most cameras have a "manual" mode that helps).

Your illumination may also be uneven. Experiment with positioning and aligning the light, and with adding translucent materials like wax paper that diffuse the light. If the bright spots inside the drop are too large, you need to make the light source "smaller" by moving it farther away, or by covering up parts of it with tape or heavy paper.

Finally, you need to know the scale of your images – the conversion from distances in pixels on your image to μm on the drop. One easy way to do this is to make sure that the end of the nozzle is always in the picture. Then you can compare the nozzle's size on the image with its diameter in real life.

Analyzing Images by Hand

The software of choice for this is ImageJ (free download, https://imagej.nih.gov/ij/). You can open many images as a "stack," which is sort of like a movie (in the File menu, look for Import → Image sequence, or check out the copious documentation and tutorials). If your images are high-resolution, be sure to use the "Virtual Stack" option so that the computer does not try to load all of them into memory at once. You can then zoom in (use the "-" and "=" keys) and use the line tool to quickly measure distances on each image. (Holding the shift key while drawing a line will keep it horizontal.)

Analyzing every image you have is slow and will give you more data than you need. Instead, look at just ten or so frames, spread out over the range of times you are interested in. Then start doing the analysis in the next section using just these data points. This will tell you which additional frames to measure in order to get a better representation of your experiment.

Automated Image Analysis

For purposes of this lab, "automated" does not mean "time-saving." But it will ultimately give you more precise results and produce a tool that you can apply to other movies. That is why it is preferred by researchers today. Since your image is already structured as rows of pixels, each corresponding to a different height y, you can measure $R(y,t)$ by considering the image one row at a time. Take the minimum of $R(y,t)$ for that frame to obtain $R_{min}(t)$.

Because your program is not as smart as you are, the quality of your images is tremendously important. Artifacts like laser light can ruin a picture. If you used a color camera, it recorded separate red, green, and blue images. (In many computing environments, the image will actually be a three-dimensional array, with the third axis selecting the color channel.) Each of the three channels will have a different amount of laser light.

Data Analysis

Once you have your R_{min} data, converted to physical units (μm and μs are appropriate), you are ready to analyze these data in the context of your model.

Your prediction was that R_{min} should be a function of the time remaining (i.e., counting down) until the moment when $R_{min}=0$ and the neck disconnects. Specifically, it should be proportional to $\Delta t^{2/3}$, where $\Delta t = t_{pinch} - t$, and t_{pinch} is the precise moment of disconnection – the time of the singularity. To proceed, we should put ourselves in this frame of mind. Identify the last frame in which the neck is connected. Now imagine playing the movie backward, starting from this frame. As you do this, the thinnest part of the neck grows thicker as a power law of time. However, before long, the singularity's effect on the dynamics appears to fade. The shape of the neck starts to become more symmetric between top and bottom. The growth of the neck radius slows down and deviates from its original power law. If you were to patiently wait to the end of the movie, you would see the neck grow so large that it has nearly the same diameter as the drop itself, and it would be almost motionless. It would bear no resemblance to the beginning of your movie.

Now, plot R_{min} vs. $\Delta t = t_{pinch} - t$ on log-log axes, where t_{pinch} is the time of the first frame that shows the neck disconnected. Now is a good time to add error bars to your plotted points, based on the known sources of random error in the R_{min} and/or Δt of each data point. (What is uncertain about t_{pinch}?) The exact error bars you estimate matter much less than a clear explanation of how you arrived at them. Comparing the error bars with the actual scatter in your data will help you think about aspects of the experiment that could be improved and aspects that are already good.

On the same axes, plot your prediction for R_{min} from the pre-lab exercises. Estimate the value of the dimensionless prefactor α. (On log-log axes, parallel lines have the same power law exponent but different prefactors.) Where does the model agree with your data? Where doesn't it? Why? Try a few other values of the exponent to get a sense of what is an acceptable range of agreement.

At this point, you can either continue to Part II of this lab, or first delve a little deeper into the present analysis.

What you've done so far should have shown you that in order to better analyze the power-law dynamics associated with the singularity, you will have to ignore most of your data! You need to choose t_{start} and consider times only in the range $t_{start} \le t < t_{pinch}$. To do this, you should consider your observations in three ways:

- A plot on linear axes of R_{min} vs. t.
- A plot on log-log axes of R_{min} vs. $\Delta t = t_{pinch} - t$. (Here, t_{start} will be the rightmost limit of your data.)
- The *shape* of the neck at various times before pinch-off.

Once you have decided on t_{start}, extract the data for the power-law regime; you will work with only these data from here on. The most robust way to fit your model to these data is to use the fitting function $R_{min} = B(t_{pinch} - t)^{2/3}$, where B is the only free parameter, and t_{pinch} is the time of the first frame after disconnection, as before.

A better way is to make t_{pinch} another free parameter. (You will want to supply the computer with a guess.) This is especially valuable if you are analyzing a movie from a high-speed video camera. You can also let the computer vary the exponent.

As you are finishing this part of the lab, there is one final twist that you may have been lucky enough to notice: very close to the singularity, the top of the drop develops a dimple, and the thinnest part of the neck is actually inside it! You can't see R_{min} anymore, but the 2004 paper by Burton et al. uses an electrical measurement of liquid metal to show that the power law persists down to the nanoscale.

Part II: Polymeric Fluids

Making a Polymeric Fluid

You will be making about 200 mL of a solution that is 49.9% glycerol, 49.9% water, and 0.2% polyethylene oxide (PEO) by mass. Wear safety goggles, preferably splash-proof, and gloves. Start by measuring out each substance, with the water and glycerol in separate beakers. Put the beaker of water on a stirrer plate and set it to stir rapidly. If the stirrer plate can also heat, set the thermostat to 40°C to speed up dissolution. Gradually add the PEO granules, a bit at a time. Cover the solution and let it stir at low speed for at least 2 hours or until the PEO appears fully dissolved. Next, turn off the heat, slowly add the glycerol, then stir for about 24 hours more.

Once your sample is ready, take a few minutes to compare it with the Newtonian fluid and record your observations. For example, dip your (gloved) thumb and forefinger into a liquid and then rub them together. Can you tell the difference?

The Experiment

This experiment is almost identical to the Newtonian one, though the results will not be! As before, be sure to explore a wide range of delay values to be sure you are not missing anything. When it comes time to take data for analysis, you will be covering a longer time and will require less precision than in the Newtonian experiment, so incrementing the delay by 100 μs should work well.

Data Analysis

Extract $R_{\min}(t)$ in a similar way as in the Newtonian analysis, discarding late times when the neck is broken or not easily measured. Put your data into units of μm and μs, and plot $R_{\min}$ vs. t on linear axes.

Visually, you may have noticed that at some times, the polymeric and Newtonian pinch-offs are hard to distinguish (just like the fluids themselves). Is this ever true in your data? Load the data from your previous analysis and put them on the same plot. You will need to move the Newtonian curve forward or backward in time to bring it into better alignment with the polymeric curve.

The portion of your movie in which the neck resembles a long, slender thread is known as the *filament thinning regime*. The thread is described by an approximately uniform $R(t) \approx R_{\min}(t)$. This obviously has to do with the polymer, and we would like to model it. First, make a new plot with $R_{\min}$ on a logarithmic scale. The filament thinning regime should now be much straighter, hinting that an exponential function is appropriate for that portion of your data. You may also notice that at late times, pinch-off speeds up again and deviates from an exponential. You can ignore those points for now.

Go ahead and fit the exponential segment of your data using the exponential function $R = R_0 \exp(-t/t_{\text{filament}})$. What does t_{filament} mean? Recall that in the Newtonian case, surface tension (Laplace pressure) was driving the collapse of the neck, and the inertia of the liquid was the only thing slowing it down. But here, something else must be stepping on the brake: a stress (force per unit area) due to the polymer. For reasons that aren't important here, this is known as the *polymeric stress difference* $\Delta\tau_p$. We can surmise that for $\Delta\tau_p$ to actually do something, it must be about the same strength as Laplace pressure:

$$\Delta\tau_p \approx \sigma\kappa = \frac{\sigma}{R(t)}$$

Conveniently, $\Delta\tau_p \propto A$, the extension of polymer molecules that you modeled in the pre-lab exercises. In one of those exercises you considered the case where $\dot{\varepsilon}\lambda \approx 1$, and A is already $\gg 1$ (i.e., the pinch-off process has already stretched the molecules considerably). Using that result, you can show that

$$R(t) \sim \exp\left[-t\left(2\dot{\varepsilon} - \lambda^{-1}\right)\right] \approx \exp\left(-t/t_{\text{filament}}\right)$$

and

$$t_{\text{filament}} \approx (2\dot{\varepsilon} - \lambda^{-1})^{-1} \approx \left(-\frac{4}{R}\frac{dR}{dt} - \lambda^{-1}\right)^{-1} \approx 3\lambda$$

Use these results to measure λ of the fluid.

Finally, one notable feature of the Maxwell model is that it allows $A \to \infty$. Why and when could this be a problem? Do you see any evidence for this in your data?

Optional Analysis

Use the equations to show that $\dot{\varepsilon}\lambda$ is a constant. Compute $\dot{\varepsilon}(t)$ from your data and compare it with your prediction. What does $\dot{\varepsilon}\lambda$ do *before* the filament thinning regime?

Suggested Extensions to This Lab

1. Ultimately, Newtonian drop pinch-off is just forces acting on a mass of fluid. Like most dynamical processes it should depend on initial conditions – as set, for example, by the size and tilt

of the nozzle. Yet initial conditions appear nowhere in your model, suggesting that pinch-off is *universal*. Devise experiments to test this idea – what aspects of pinch-off can you show are *not* universal?

2. Find ways to vary the properties of the fluids you have studied. The work you did before the lab, and the readings, will let you make predictions about some effects that you can test experimentally. (Note: avoid using pure glycerol – it is highly hygroscopic, meaning that it rapidly absorbs water vapor from the air.)

3. Now that you have studied the birth of a drop, investigate its life and death. What happens as it falls, and as it impacts a liquid pool or a dry surface?

4. Visualizing fluid flow is both a scientific and an artistic pursuit. Modify the experiment to make images that are more beautiful and more informative. Some ideas to try: milk, food coloring, smoke, multiple light sources, colored lights.

5. As you increase the liquid flow rate, so that drops follow each other closely and each pinch-off influences the next, the dripping is known to become chaotic. Identify and study the transition to chaos, preferably by measuring the time interval between drops. What do your high-speed photographs or movies reveal?

Nathan C. Keim is an Associate Research Professor of Physics at the Pennsylvania State University, before which he was Assistant Professor of Physics at Cal Poly in San Luis Obispo. He and his group study soft matter physics, including experiments to study how disordered solids like mayonnaise and sand deform and flow. One particular focus is how these and other materials retain memories of how they were deformed in the past, that can later be read out. He and his wife (also a physicist) enjoy hiking and kayaking (Figure 17.3).

FIGURE 17.3 Prof. Nathan Keim.

18

Atomic and Molecular Physics

Robbie Berg and Glenn Stark

CONTENTS

Anything worth doing is worth doing twice, the first time quick and dirty and the second time the best way you can.

– Arthur Schawlow

18.1 Field Overview

Optical spectroscopy has proved to be an invaluable tool for probing atomic and molecular systems. The historical role of atomic spectroscopy in revealing quantum effects and validating quantum theory is well-documented, from the first interpretation of the Balmer series in hydrogen to the measurement of the Lamb shift in hydrogen, spectroscopy was the proving ground for the emergence of a comprehensive understanding of quantum effects. Small effects can reveal deep truths, so techniques that advance our ability to resolve closely spaced features are highly prized. Since its invention more than 60 years ago, the laser has been central to high-resolution spectroscopy because the light it produces can be made highly monochromatic. Ten physicists have shared Nobel Prizes since 1981* for ground-breaking work involving laser-based precision spectroscopy and its applications. Experimental methodologies and novel instrumentation are continually being developed, and today atomic and molecular physics play central roles in research efforts that address a wide range of fundamental and applied questions.

Current research efforts in atomic and molecular physics include (but are not limited to) the following:

(a) Ultracold matter: following the development of laser-based atom cooling and trapping techniques and the subsequent discovery of Bose-Einstein condensates, ultracold matter studies seek to answer a broad array of questions regarding the collective behavior of atoms in weakly interacting gases.

(b) Precision measurements: ultra-precise measurements of fundamental constants of nature, searches for the electric dipole moment of the electron, studies of anti-matter, and the development of ever more precise atomic clocks share common experimental approaches in atomic physics.

(c) Quantum computing: the internal states of atoms are being explored as candidates for "qubits" – the fundamental logical units of quantum computers.

Lab 18A Circular Dichroism in Atomic Rubidium

18A.1 Learning Goals

- To further develop a working understanding of the polarization properties of light, with an emphasis on the relationships between linearly and circularly polarized light.
- To review the physics of the normal Zeeman effect in an atomic system, including the polarization requirements for transitions between magnetic substates and the resulting phenomenon of circular dichroism.
- To gain further experience with optical alignment techniques, and to apply those techniques in the context of manipulating an infrared laser beam.
- To gain experience recording absorption spectra using a high-resolution tunable diode laser system.
- To learn and implement "best practices" safety techniques for the manipulation of an infrared laser beam. Specific instructions are outlined in the "Laser safety" section of this write-up.

18A.2 Introduction

In 1845, Michael Faraday discovered that when linearly polarized light passes through certain types of glass immersed in a magnetic field, the plane of polarization of the light is rotated. Now known as "Faraday rotation," this phenomenon is due to slight differences in the indices of refraction for right-handed and

---

* The Nobel Lecture by 1981 Prize winner Arthur Shawlow prominently features saturated absorption spectroscopy, the topic of Lab 18B.

left-handed circularly polarized light ("RCP" and "LCP" light) that are magnetically induced in the glass. Faraday was soon able to replicate this effect with several transparent solids, liquids, and gases.

Intimately related to Faraday rotation is the phenomenon of *circular dichroism*, which describes the differences in *absorption* of RCP and LCP light by a solid, liquid, or gas. Circular dichroism finds wide application in the study of the structures of biological molecules that exhibit chirality (e.g., the double helix of nucleic acids). In this laboratory, you will measure an *atomic* effect induced by a magnetic field: the circular dichroism of rubidium (Rb) vapor. The Rb circular dichroism signal is dramatically enhanced by observing the differences in RCP and LCP light absorption in the vicinity of a strongly allowed Rb transition. The physics behind this "resonant" effect is best understood by the Zeeman effect – the splitting of atomic energy levels in a magnetic field. The underlying physical mechanisms of resonant circular dichroism are key elements in a variety of modern experimental techniques in atomic physics, including saturated absorption spectroscopy, frequency stabilization of laser light, and atom cooling and trapping.

In the following experiment, you will observe a resonant circular dichroism signal in atomic Rb, relying on a clever manipulation of the polarization of a linearly polarized laser beam. By measuring the relationship between the strength of the applied magnetic field and the frequency shifts of the RCP and LCP light, you will test the prediction of the Zeeman effect.

18A.3 Pre-lab Reading

To prepare for this laboratory, you will need to familiarize yourself with three pieces of physics:

1. The polarization properties of light. Please read Section 11.2 "Polarization." The relationships between linearly and circularly polarized light are particularly relevant.
2. Atomic energy levels and notation. You will be monitoring transitions between the $5S_{1/2}$ and $5P_{3/2}$ energy levels in the Rb atom. A basic presentation of atomic energy level structure can be found in many intermediate-level modern physics texts. Sections 7.3 and 8.2 of *Modern Physics for Scientists and Engineers*, 4th edition, by Thornton and Rex, describes the angular momentum of atomic states and atomic energy level notation. Though not directly relevant to the measurements in this lab, you should satisfy yourself that you understand the meanings of all of the symbols in "$5\,^2S_{1/2}$" and "$5\,^2P_{3/2}$."
3. The Zeeman effect. For the purposes of this lab, an understanding of the "normal Zeeman effect" will suffice. Section 7.4 of *Modern Physics for Scientists and Engineers*, 4th edition, by Thornton and Rex, covers the basics of the Zeeman effect.

To prepare yourself on the experimental side, you should also read Sections 10.1–10.4, 10.6, and 10.7.

Suggested Additional Reading

Benumof, R. Optical pumping theory and experiments. *American Journal of Physics* 33(2), 151–160 (1965). Contains a good overview of the Zeeman effect in Rb.

Corwin, K. L., Lu, Z.-T., Hand, C. F., Epstein, R. J., and Wieman, C. E. Frequency-stabilized diode laser with the Zeeman shift in an atomic vapor. *Applied Optics* 37(15), 3295–3298 (1998).

Patterson, L. H., Kihlstrom, K. E., and Everest, M. A. Balanced polarimeter: A cost-effective approach for measuring the polarization of light. *American Journal of Physics* 83(1), 91–94 (2015).

Van Baak, D. Resonant Faraday rotation as a probe of atomic dispersion. *American Journal of Physics* 64(6), 724–735 (1996). A must read.

18A.4 Laser Safety

The laser you use in this experiment emits light at a wavelength of 780 nm, which is in the near infrared region of the electromagnetic spectrum. The laser beam is well-collimated, with a power that likely exceeds 10 mW (consult your instructor for the specifics of your infrared laser). *This beam is bright enough to damage your eyesight instantly. To protect your eyes, you must wear appropriate laser safety goggles when operating the infrared laser.* Goggles are specially designed to greatly attenuate the 780 nm laser beam while remaining mostly transparent to visible wavelengths. Here are some additional safety procedures to follow when working with the infrared laser.

- The work area in which the laser is mounted and over which the laser beam propagates must be enclosed by appropriate barriers that confine any stray light to the working area.
- Remove all jewelry and metal objects from your hands and wrists.
- Never bring your eyes down to the height of the laser beam.
- Always block the beam before adding or removing any optical component from the beam.
- Be very careful with wrenches and screwdrivers to be sure they don't cross the active laser beam.
- An infrared (IR) viewer, camera, or IR-sensitive card should be used to monitor the location of the laser beam(s) and all reflections. Whenever optical components are moved or adjusted, the laser beam and all reflections should be checked both before and after the adjustments.

18A.5 Pre-lab Questions

The following questions use the polarization notation introduced in Chapter 11 as well as the mathematics of Jones matrices. As in the examples in Chapter 11, we will consider a light wave propagating in the $+z$-direction.

1. As discussed in the development of Equation (11.14), right-circularly polarized light (RCP) is expressed as:

$$\vec{E}(z,t) = \frac{E_0}{\sqrt{2}} \left[\cos(kz - \omega t)\,\hat{x} + \sin(kz - \omega t)\,\hat{y} \right]$$

   or

$$\frac{E_0}{\sqrt{2}} \begin{pmatrix} 1 \\ -i \end{pmatrix},$$

   where $\vec{E}$ is derived from the real part of this Jones vector.

   Construct similar mathematical expressions for left-circularly polarized light (LCP).

2. In the upcoming experiment, your laser beam will be vertically polarized. Show that vertically polarized light (linearly polarized along the y-axis) can be written as a linear combination of RCP and LCP.

3. The vertically polarized laser light will pass through a quarter-wave plate oriented at 45° to the vertical before impinging on a polarizing beam splitter. The combined effect of these two optical elements is to physically separate the right-circularly polarized and left-circularly polarized components of the initial beam (this is why you were asked in Question 2 to show that a vertically polarized beam can be considered as a superposition of RCP and LCP light). Now you are asked to find the effect of the quarter-wave plate on the RCP and LCP components:

   Using the geometry of Figure 11.2, with the fast axis of the quarter-wave plate oriented at 45°, show that RCP light becomes vertically polarized and that LCP light becomes horizontally polarized. It is easiest to use the Jones matrices introduced in Chapter 11. This manipulation of the polarization of the laser light is illustrated in Figure 18.10. (Note that in Chapter 11, it was shown that a quarter wave plate with its fast axis at 45° converts linearly polarized light to circularly polarized light. That's still happening here, but we're thinking about it in a different way. The horizontally and vertically polarized beams at the output of the quarter wave plate in this problem are 90° out of phase, i.e., the combination can also be thought of as circularly polarized light.)

18A.6 Background

Absorption Spectroscopy

The circular dichroism measurements that you make here, and also those in the saturated absorption lab that follows, are examples of *absorption spectroscopy*. The basic idea is shown in Figure 18.1. Light from a tunable laser passes through a glass cell filled with a vapor of Rb atoms. The frequency of the laser beam ν_{laser} is scanned and the transmitted intensity is monitored by a photodetector. Assume that each atom has two distinct quantum states – a ground state and an excited state, separated by an energy $h\nu_0$. The energy of the incident photons is $h\nu_{\text{laser}}$, so when $\nu_{\text{laser}} = \nu_0$ some of the incident light will be absorbed and there will be a dip in the transmitted intensity. Thus, the absorption spectrum reveals information about the spacing of the energy levels in the atoms.

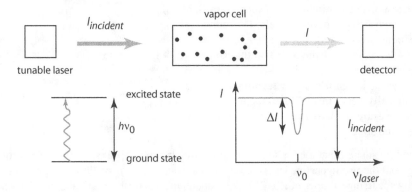

FIGURE 18.1 Absorption spectroscopy in a two-level system. A tunable laser beam passes through a vapor cell and the transmitted intensity is monitored. When $\nu_{\text{laser}} = \nu_0$ the resonance condition is satisfied and the light can be absorbed by the atoms.

Hyperfine Structure of Rubidium

Atomic Rb has a strongly allowed transition from its ground state (the $5S_{1/2}$ state) to the excited $5P_{3/2}$ state at 780 nm in the near infrared (Figure 18.2).*

A careful examination of this transition reveals that both atomic states are, in fact, more complicated than is shown in Figure 18.2. Both the electrons and the nucleus of Rb have magnetic dipole moments, and the magnetic interaction between these dipole moments, which is very weak relative to the Coulomb forces in the atom, results in "hyperfine structure" – each electronic state of the atom actually consists of multiple closely spaced energy levels. Figure 18.3 shows the "hyperfine structure" of the $5S_{1/2}$ and $5P_{3/2}$ states. There are two hyperfine sublevels associated with the $5S_{1/2}$ state and four associated with the $5P_{3/2}$ state. Note that the energy level spacings in Figure 18.3 are not to scale. As the term "hyperfine" suggests, the splittings

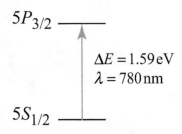

FIGURE 18.2 Ground and excited states of the Rb atom. A transition from the ground $5S_{1/2}$ state to the $5P_{3/2}$ excited state is induced by the absorption of a photon with an energy of 1.59 eV. This corresponds to a wavelength of 780 nm in the near infrared.

within each electronic state are orders of magnitude smaller than the 1.59 eV splitting between the two electronic states.

---

* Standard spectroscopic notation for atomic energy levels includes a superscript indicating the "multiplicity" of the level. All relevant Rb energy levels are "doublets"; the complete notation for the levels in Figure 18.2 is $5\,^2S_{1/2}$ and $5\,^2P_{3/2}$. For notational simplicity, the superscripts are not included in this lab write-up.

The details of hyperfine structure will be explored in more depth in Lab 18B. However, there are a few specifics that are of relevance to this circular dichroism lab:

a) The range of frequencies through which the laser can be scanned encompasses transitions originating from both hyperfine levels of the ground state, and those two sets of transitions are easily distinguished from one another.

b) The hyperfine splittings in the excited $5P_{3/2}$ state are about 30 times smaller than those of the $5S_{1/2}$ ground state. As a result, your Rb absorption spectra will not distinguish between transitions to different $5P_{3/2}$ hyperfine levels; the individual transitions are masked by "Doppler broadening" due to the motions of the atoms in the Rb vapor. Doppler broadening is discussed in Lab 18B, where you will learn a clever technique to spectroscopically peer "inside" the Doppler broadening to measure the hyperfine energy splittings in the excited state.

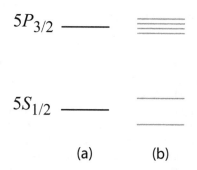

(a)   (b)

FIGURE 18.3 Hyperfine splitting. An energy level diagram showing the ground and excited states of the Rb atom (a) and the hyperfine structures of those two electronic states (b). The energy splittings between the hyperfine levels in the $5S_{1/2}$ state are about five orders of magnitude smaller than the 1.59 eV splitting between the electronic states, and the splittings in the $5P_{3/2}$ state are an order of magnitude smaller than that.

Points (a) and (b) suggest that a laser scan will show two absorption peaks, with the absorption features originating from the two hyperfine levels of the ground state. There is one last complication! There are two naturally occurring isotopes of Rb, [85]Rb (72% abundance) and [87]Rb (28% abundance). The nuclei of these isotopes have different magnetic dipole moments, and, hence, the isotopes have slightly different hyperfine structures. This results in a laser scan showing four transitions rather than two. Figure 18.4 displays a typical absorption spectrum of a mixed-isotope sample of Rb.

The Zeeman Effect and Circular Dichroism in Rubidium

When a Rb atom is placed in a magnetic field, each hyperfine level of the ground and excited states is split into multiple levels – this is referred to as the Zeeman effect. The presence of the hyperfine levels in Rb leads

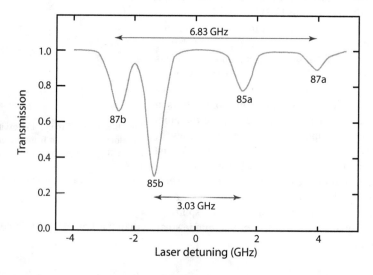

FIGURE 18.4 A $5S_{1/2} - 5P_{3/2}$ absorption spectrum of a mixed sample of [85]Rb and [87]Rb at 780 mn. The four features correspond to transitions originating from the two hyperfine levels of the ground states of the two isotopes. In reality, each feature consists of multiple closely spaced transitions to the hyperfine levels of the $5P_{3/2}$ state, but these features cannot be resolved.

to quite complicated Zeeman splittings. However, for the purposes of this lab, we can illustrate the basic principle using the simplified model shown in Figure 18.5, which only takes into account the orbital angular momentum of the electrons ($l=0$ for the 5S state and $l=1$ for the 5P state). The intrinsic angular momenta (the "spins") of the electrons and the nucleus, which lead to the fine structure and hyperfine structures of the atomic energy levels, are ignored.

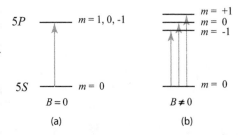

(a)   (b)

As Figure 18.5 indicates, in the presence of a magnetic field, three transitions become observable. One transition (the $m=0 \rightarrow m=0$) is unshifted in energy relative to the 5S $\rightarrow$ 5P transition when $B=0$. The ($m=0 \rightarrow m=+1$) transition is shifted to higher energy, while the ($m=0 \rightarrow m=-1$) transition is shifted to lower energy. A critically important aspect of the Zeeman effect for the observation of circular dichroism follows from these properties of atoms and photons:

FIGURE 18.5 A simplified energy level diagram for the 5S and 5P states of Rb. In the absence of a magnetic field, the magnetic sublevels of the 5P state are degenerate (a); when a magnetic field is applied, the sublevels shift according to the values of the quantum number m and the degeneracy is lifted. Three closely spaced transitions are observed. This is the so-called "normal Zeeman effect."

a) When an atom absorbs a photon and makes a transition, the total angular momentum of the system (atom plus photon) must remain constant, as must the component of the total angular momentum in any direction.

b) When light propagates in a particular direction, each photon carries one unit of angular momentum either parallel to or anti-parallel to the direction of propagation. Photons associated with right-handed circularly polarized light carry -1 units of angular momentum along the propagation direction (that is, their angular momentum vector points opposite to the direction of propagation) and photons associated with LCP light carry $+1$ units of angular momentum along the propagation direction.

c) The quantum number m determines the component of the atomic angular momentum along the direction defined by the propagating light. So, for example, an absorption transition from (5S, $m=0$) to (5P, $m=+1$) results in the atom gaining one unit of angular momentum in the propagation direction.

If LCP light is used to excite the Rb atoms when they are exposed to a magnetic field, only one transition conserves angular momentum: the (5S, $m=0$) to (5P, $m=+1$) transition, which is slightly higher in energy than the 5S to 5P transition in the absence of a magnetic field. RCP light can only produce transitions to the 5P, $m=-1$ state.

Consider a single absorption feature in the Rb spectrum displayed in Figure 18.4. The top trace of Figure 18.6 shows that single feature, with a central frequency of ν_0, in the absence of a magnetic field. When a field is applied, RCP light is absorbed at a lower frequency and LCP light is absorbed at a higher frequency. The frequency shifts are proportional to the strength of the applied magnetic field. The difference between the RCP and LCP absorption spectra constitutes resonant circular dichroism.

18A.7 Experimental Setup

A schematic of the experiment is shown in Figure 18.7. For safety, the entire setup should be surrounded by an enclosure that will effectively confine any stray laser beams to the optical table.

The arrangement of optical elements in Figure 18.7 will be explained in the *Procedure* section. But first, we suggest that you complete the "warm up" optical-alignment exercise below.

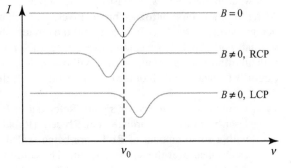

FIGURE 18.6 Circular dichroism and the Zeeman effect. Top trace: a single absorption feature in the Rb spectrum in the absence of a magnetic field (with a central frequency of ν_0). Middle and bottom traces: in the presence of a magnetic field, RCP and LCP light are absorbed at lower and higher frequencies, respectively.

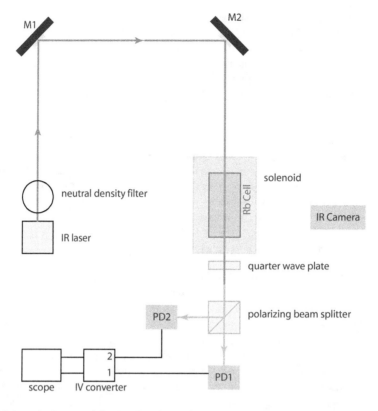

FIGURE 18.7 Schematic diagram of circular dichroism optical setup.

18A.8 Procedure

The careful alignment of optical elements is critical to the success of laser-based experiments. This skill is developed through experience (and time – be patient!). Before proceeding with the alignment of your infrared diode laser beam, please complete or review Lab 10A.2 "Precision Optical Alignments" in which a pair of mirrors mounted on adjustable mirror mounts are aligned to reflect a helium-neon laser beam through a pair of iris diaphragms. You will soon use this alignment technique with your 780 nm infrared laser.

In what follows, you will construct, one step at a time, the optical apparatus shown in Figure 18.7. Along the way, you'll introduce progressively more complex optics that take advantage of the polarization properties of the laser light to reveal the circular dichroism signal in a Rb spectrum. The focus of your lab work should be on gaining experience with optical alignments – it is a non-trivial matter to work with an infrared laser beam, so there will be challenges in even the simplest steps. Be patient and work carefully to understand all of the experimental details that follow.

1. *Observing a fluorescence signal*: Referring to Figure 18.7, you will first direct the infrared laser beam to pass through your Rb cell. (Rubidium cells are first evacuated and then filled with a small amount of Rb before being sealed, such that an equilibrium vapor pressure is established over the solid rubidium. The equilibrium vapor pressure at room temperature is 2.6×10^{-7} Torr, high enough to produce easily observable spectroscopic features.) When the frequency of the laser is properly adjusted, the $5S_{1/2} - 5P_{3/2}$ transitions in ^{85}Rb and ^{87}Rb are excited. After absorption of a laser photon, each excited atom will emit a photon (on a time scale of about 30 nsec) and fall back into its ground state. The photons from the collection of excited Rb atoms are emitted isotropically, that is, they are emitted in equal numbers in all directions. You can directly observe this "fluorescence" with the aid of an infrared-sensitive

camera. For this exercise, in addition to the infrared laser, the apparatus consists only of neutral density filters, the two mounted mirrors that were used in the warm-up exercise, the Rb cell, and the IR camera. The cell should not yet be housed inside a solenoid – that comes later.

Experimental Details

(a) Your instructor will provide instructions for powering up your IR laser. *Before turning the laser on, discuss laser safety issues with your instructor and take all necessary precautions,* including (i) wearing appropriate laser goggles; (ii) removing all jewelry from hands and wrists; and (iii) having beam stops available to intercept the beam. *Place a #2 neutral density filter directly in front of the laser output aperture – this will reduce the laser power by a factor of 100. This will protect the photodiode from being exposed to potentially damaging photocurrents. (Power levels in excess of a few mW are capable of inducing damaging photocurrents in a typical photodiode.)*

(b) Using IR-sensitive cards or your IR camera to monitor the beam alignment, make the appropriate mirror adjustments so as to direct the beam through the two windows of the Rb cell. Block the beam with a beam stop after it passes through the cell.

(c) Observe the cell on a monitor with your IR camera. It is highly unlikely that the initial laser frequency will correspond to a Rb transition, so you will not initially see evidence of fluorescence. To observe fluorescence, you will need to tune the laser frequency as described next.

The fundamentals of an "external cavity diode laser," such as the one used in this lab, are presented in *Laser Spectroscopy 1, Basic Principles* (Demtroder, 2014). The frequency of the laser output can be shifted by adjusting the position and angle of a diffraction grating that forms one end of the resonant cavity of the laser – this is accomplished by applying a voltage to a piezoelectric transducer to which the grating is attached. You can create a linear frequency scan by feeding a "voltage ramp" created by a function generator into the appropriate input of your laser controller. For this purpose, we recommend an initial ramp frequency of 1 Hz, an amplitude sufficiently large to make the laser frequency scan several GHz, and a "symmetry" setting of about 0.8 (meaning that the ramp spends 80% of the overall period rising and 20% of the overall period falling). Use an oscilloscope to monitor the voltage ramp; it should resemble the ramp illustrated in Figure 18.8.

(d) Once the voltage ramp is established, the laser frequency will scan, once per second, through a small spectral range. Your instructor will provide you with the operating parameters of the laser (i.e., laser current) that result in an initial laser frequency that is close to the Rb transition frequencies. Your job is to adjust the voltage ramp parameters (ramp amplitude and ramp offset) until you observe fluorescence with the IR camera. There is no magic formula for finding the correct voltage ramp settings. You'll likely need to hunt around a bit for the proper combination of ramp amplitude and offset. But once you find this combination, you'll be in for a visual treat! If you observe fluorescence, declare victory and move on.

2. *Recording an absorption spectrum*: To fine-tune your voltage ramp so as to encompass all Rb transitions of interest, you'll need to begin monitoring the absorption of the laser light by the Rb atoms.

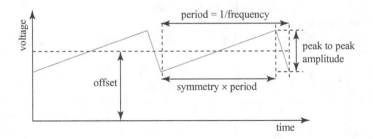

FIGURE 18.8 Voltage ramp signal for scanning the laser frequency.

Experimental Details

Place a photodiode directly behind the Rb cell so that it intercepts the IR laser beam (PD1 in Figure 18.7). Leave plenty of room between the cell and the photodiode so that you can later insert additional optical elements.

(a) At this point you may also want to add the solenoid to your apparatus. The Rb cell should be housed at the midpoint of the solenoid and must be completely enclosed by the solenoid. (Note that you'll no longer be able to observe fluorescence with the IR camera.)

(b) Connect the photodiode to one channel of your current-to-voltage converter and monitor the resulting signal on your oscilloscope. Trigger the oscilloscope on the rising slope of the voltage ramp (you should observe the voltage ramp and the PD signal on two channels of your scope). To optimize your PD signal, you may need to adjust the amplification factor of your current-to-voltage converter and, of course, the oscilloscope settings.

(c) Assuming that you were successful in observing fluorescence from the Rb cell, you should see, over the course of each laser frequency scan, one or more absorption dips as the laser frequency comes into resonance with a Rb transition. As described earlier, there are a total of four closely spaced transitions that can be observed; two from each Rb isotope. Now, with the aid of the absorption signal on the oscilloscope, you should make additional adjustments to the voltage ramp parameters (amplitude and offset) so that all four absorption features are observable. You should aim to reproduce, as best as possible, the spectrum shown in Figure 18.4. The relative amplitudes and spacings of the four features in Figure 18.3 can be used as a guide in identifying the features that you are observing. When you are satisfied with your spectrum, electronically save the oscilloscope trace.

3. *Adding the polarizing beam splitter:* Now you will begin to take advantage of the polarization properties of the IR laser beam. In what follows, we assume that the linear polarization of your IR laser beam is in the vertical direction. Your instructor can verify this assumption. The optical component to be added is a polarizing beam splitter. Most polarizing beam splitters are cubical, formed by cementing two right-angle prisms along their hypotenuses. Light impinges normal to one face of the beam splitter; a dielectric coating on one of the hypotenuses produces a strong polarization dependence in the reflected and transmitted beams. With the orientation of the coated surface as shown in Figure 18.9, the beam splitter transmits light with a horizontal polarization (referred to as "*p*" polarization, in the plane defined by the propagation direction of the light and the normal of the reflecting surface) and reflects light with a vertical polarization (referred to as "*s*" polarization, perpendicular to the plane).

Experimental Details

Place the polarizing beam splitter in the optical path in front of photodiode #1 (see Figure 18.7). Adjust its position and orientation so that the IR laser beam strikes normal to the front surface.

(a) Assuming vertically polarized laser light, <u>all</u> of the IR beam should reflect at 90°; there should be no transmitted light. Check that this is so with your IR-sensitive card or your IR camera.

Place a second photodiode (PD2 in Figure 18.7) to intercept the reflected laser light and observe the absorption signal from PD2 on your scope. You'll now have three signals displayed on your scope – the voltage ramp (also used to trigger the scope), PD1 (this should be zero, or very close to zero), and PD2 (the Rb

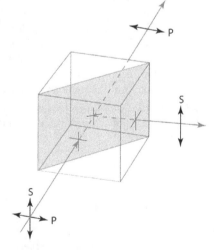

FIGURE 18.9 Polarizing cubical beam splitter. With the coated common hypotenuse oriented vertically, horizontally polarized light is transmitted and vertically polarized light is reflected.

absorption spectrum). Note that the only effect of the beam splitter so far is to reflect the vertically polarized laser light. That's not terribly interesting! The fun starts when you add the next optical component…

4. *Adding the quarter-wave plate*: The last optical element to be added is a quarter-wave plate (QWP), situated in front of the polarizing beam splitter (PBS) and with its fast axis oriented at 45° to the vertically polarized laser light. Understanding what this accomplishes is crucial to understanding how circular dichroism will be observed with your apparatus. The key points are: (i) the vertically polarized laser light that passes through the Rb cell can be considered as a 50–50% superposition of RCP light and LCP light [see Pre-lab Question 2]; (ii) when RCP light passes through a QWP, the beam becomes linearly polarized at an angle of 45° to the fast axis of the QWP [see Pre-lab question 3]; and (iii) LCP light also becomes linearly polarized, with its polarization direction perpendicular to that of RCP light. Figure 18.10 illustrates the changes in polarization in the RCP and LCP components of the laser light. The QWP "converts" the RCP and LCP components into linearly polarized light in the vertical and horizontal directions. The polarizing beam splitter then separates the two linearly polarized components, sending

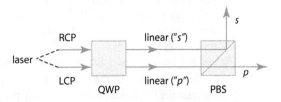

FIGURE 18.10 A schematic representation of the effects of a QWP and a PBS on a linearly polarized laser beam (vertical polarization). The combination of optical elements separates the RCP and LCP components of the laser light so that they can be monitored independently. Note that the "s" and "p" beams entering the PBS are not physically separated; they are separated in the schematic for clarity.

one to PD1 and the other to PD2. In this way, you have separated the RCP and LCP components of the laser light – and this is what is needed to observe circular dichroism.

Experimental Details

(a) You'll need to orient the fast axis of the QWP at 45° to the vertical. The fast axis is usually marked on the optical element, so a simple way to accomplish the orientation is to place the QWP in a rotation mount and use the rotation mount settings as a guide.

(b) A sensible check of the orientation is as follows: (i) with the QWP fast axis oriented vertically (parallel to the laser light polarization), there should be no effect on the polarization properties of the light (convince yourself of this!) and all of the laser light should continue to reflect from the PBS onto PD2. As you rotate the fast axis away from the vertical, the intensity of the light falling on PD2 should decrease and the intensity of light falling on PD1 should begin to increase. When the fast axis is oriented at 45° to the vertical, half of the laser light will be falling on PD2 and half on PD1. Beyond a 45° orientation, the light will begin again to be preferentially reflected to PD2. So, a *minimum* in the signal from PD2 should coincide with a 45° orientation of the fast axis.

(c) With the QWP alignment and orientation complete, you should see two identical versions of the Rb absorption spectrum on your scope, originating from the RCP and LCP components of the laser light. Finally, you're ready to observe circular dichroism!

5. *Observing circular dichroism*: A current passing through the solenoid that surrounds the Rb absorption cell will produce a uniform magnetic field that is parallel to the propagation direction of the laser light. This field induces small shifts in the energy levels of the Rb atom (the Zeeman effect), and the absorption spectra of the RCP and LCP components of the laser light will be shifted in frequency relative to one another. You will measure the frequency shifts as a function of the strength of the magnetic field and compare these shifts with a simplified model of the Zeeman effect.

Experimental Details

(a) To quantitatively interpret the frequency shifts that you'll be measuring, you'll need to know the strength of the magnetic field inside the solenoid as a function of the current flowing in the solenoid (it is the current that you will directly control and monitor). For starters, you can estimate the strength of the B field using the standard expression for a solenoid field (which assumes a very long solenoid length relative to its diameter). More accurate values of the B field might be provided by your instructor, or you might be asked to use a Hall probe to directly measure the field as a function of current. Field strengths on the order of 10 to 100 Gauss (1×10^{-3} to 1×10^{-2} T) should produce easily measurable frequency shifts.

You'll also need a reliable frequency scale to quantify the shifts. For this purpose, you'll use the known spacings of the four Rb absorption features that you are observing. Referring to Figure 18.4, the inner two absorption features originate from ^{87}Rb and are separated in frequency by ~3040 MHz. The outer two features, originating from ^{85}Rb, are separated in frequency by ~6830 MHz. The rightmost two features (the stronger from ^{85}Rb and the weaker from ^{87}Rb) are separated in frequency by ~1300 MHz.

Apply a B field in the range 1×10^{-3} to 1×10^{-2} T with an appropriate solenoid current. Because the frequency shifts are directly proportional to the strength of the B field, you might initially opt for a field at the larger end of the range. As you increase the field, you should see a clear difference emerging in the horizontal positions of the Rb features associated with absorption of RCP and LCP light. It might be helpful at this point to zoom in on a subset of the four absorption features (using the laser-controller voltage ramp and/or the oscilloscope time base) – this will give you a more sensitive frequency scale. The two strong rightmost features in Figure 18.4 are good candidates for this strategy. When you are satisfied with the quality of your RCP and LCP spectra, electronically save both spectra for later analysis.

(b) Repeat your frequency-shift measurement for two or three more B field values and save the data.

6. *Observing a weak circular dichroism signal (optional)*: Observing and measuring the RCP and LCP frequency shifts becomes progressively more difficult as the strength of the applied magnetic field is lowered. With the current experimental arrangement, it's quite difficult to see these frequency shifts when the field strength is below about 1×10^{-3} T. Time permitting, you might want to try a simple alternative arrangement for monitoring the two photodiode outputs that significantly enhances the sensitivity of your apparatus to small circular dichroism signals. This approach is an example of the "null measurement technique," in which small differences in relatively large signals are measured by *subtracting* one signal from the other. Null measurements are a common and powerful tool in many branches of experimental physics.

In Figure 18.11, the two photodiodes are internally wired to have opposite output polarities. PD1 is wired so that when illuminated by light, current flows *out of* the center pin of the output connector while PD2 is wired so that when illuminated by light, current flows *into* the center pin. Thus, if you connect the two photodiode outputs to a T-connector, as shown, the current that flows into the current-to-voltage converter will be proportional to the *difference* of the light intensities falling on the two photodiodes. (If your photodiodes are not wired as described above, see your instructor for possible workarounds.)

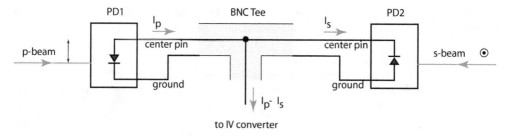

FIGURE 18.11 Photocurrent subtraction. Use a BNC Tee-connection to take the difference between the photocurrents from the p-polarized and s-polarized beams.

In the absence of a magnetic field, the signals from the two photodiodes should be almost identical (there will be small differences in the output voltages due to optical alignment issues and to variations in the photodiodes' responses to light). This means that the difference between the signals will be very close to zero. When a weak magnetic field is applied, you will be looking for *changes* in this signal. You can take advantage of the small amplitude of the "background" signal by amplifying the signal – this leads to an improved sensitivity to small frequency shifts.

Experimental Details

(a) With no external magnetic field applied, connect the outputs of PD1 and PD2 to a T-connector and use one channel of your current-to-voltage converter to produce an output voltage that reflects the difference in the two photodiode signals. Monitor the signal on your scope. Adjust the gains of the current-to-voltage converter and the scope appropriately.

(b) To optimize the sensitivity of this "null" arrangement of the two photodiode outputs, the signal that you are observing should be as close to zero volts as possible. Small differences in the two outputs are inevitable, but you should think about how to minimize those differences. For example, a microscope slide inserted before one of the photodiodes will reduce the light intensity falling on the photodiode by about 8%. For finer control, a variable neutral density filter, if available, might be helpful in balancing the two outputs.

(c) Now observe the difference signal when the solenoid magnetic field is set to the lower end of the range of fields that you explored earlier. Does the shape of the new output signal make sense? Construct on paper a qualitative plot of the difference between the RCP and LCP spectra and compare this with the signal on the scope – hopefully they are consistent!

(d) Finally, begin lowering the current in the solenoid (hence reducing the magnetic field) while monitoring the scope signal. Determine the smallest magnetic field that results in a measurable difference spectrum. You should find that this minimum field strength is significantly smaller than the field strengths used earlier. Voilà!

18A.9 Analysis

While the main intention of this lab is to familiarize you with optical alignments and with the manipulation of the polarization properties of light, there are follow-up analyses that lend insight into the circular dichroism signals that you've recorded.

1. According to the discussion of the Zeeman effect and circularly polarized light, which set of absorption features – those associated with RCP light or LCP light – should be shifted to higher frequencies and which to lower frequencies? Go back to your recorded spectra and verify that the signs of the frequency shifts are consistent with the Zeeman effect analysis. Refer back to Figure 18.10 to remind yourself how the polarization properties of the light are manipulated by the quarter-wave plate and the polarizing beam splitter.

2. The frequency shifts in RCP and LCP light that you measured are explained by the Zeeman effect. Figure 18.5 presents a highly simplified energy level diagram of the $5S$ and $5P$ states in Rb that ignores many of the details of the Rb energy levels structure (e.g., hyperfine structure) but is nevertheless quite useful in interpreting the frequency shifts. In this question, you are asked to check for the general consistency of your experimental frequency shifts with this model.

When the intrinsic angular momenta of the electrons and the nucleus in an atom are ignored (as in the simple model in Figure 18.5), the shift in energy of an atomic level in the presence of a magnetic field is given by

$$\Delta E = \mu_B B m,$$

where μ_B is the Bohr magneton (9.27×10^{-24} J/T) and m is the quantum number describing the projection of orbital angular momentum along the z-axis. A more exact expression includes the

"Landé-g factor," which is a unitless number on the order of unity that is determined by the electron orbital angular momentum, electron spin, and nuclear spin quantum numbers for each atomic energy level – we will ignore the Landé-g factor in this approximate analysis.

(a) Plot your measured frequency shifts as a function of the strength of the applied magnetic field. Remember that one feature is shifted up in frequency and the other is shifted down in frequency, so you'll want to divide the measured differences in frequency by two before plotting. Determine the slope of a line that best represents your data.

(b) Relate the measured slope to the prediction of your simple Zeeman-effect model. Don't expect exact agreement! Because of the approximations being made and the uncertainties in your measured frequency shifts and magnetic field strengths, you should only expect general agreement with the prediction (perhaps within a factor of two or so).

3. A more careful theoretical analysis would call for a more careful analysis of experimental uncertainties. Make reasonable estimates of the uncertainties in your measured frequency shifts and in the strengths of the magnetic field inside the solenoid. Are these uncertainties small enough to justify expecting decent agreement with the simplified theory?

Lab 18B Doppler-free Saturated Absorption Spectroscopy in Rubidium

18B.1 Learning Goals

- To further build upon experience with optical alignment techniques, and to apply those techniques in the challenging context of co-aligning a pair of counter-propagating infrared laser beams.
- To correlate the observed location of features in Doppler-free spectra with the hyperfine structure of rubidium.
- To use a balanced photodetector to perform background suppression and noise cancellation in order to observe subtle features in a saturated absorption spectrum.

18B.2 Introduction

In this experiment you will use laser light to make high-resolution absorption measurements on a gas composed of the two naturally occurring isotopes of rubidium, ^{85}Rb (natural abundance 72%) and ^{87}Rb (natural abundance 28%). You will learn the technique of "Doppler-free saturated absorption spectroscopy," which can reveal energy differences smaller than the natural Doppler broadening of spectral lines, allowing, in the case of Rb, the hyperfine structure of energy levels to be observed. You will apply the Doppler-free technique to the $5S_{1/2}$ to $5P_{3/2}$ transition at 780 nm in the infrared. Saturated absorption spectroscopy, first developed in the 1970s, and for which the 1981 Nobel Prize was awarded, is now a common high-resolution spectroscopic tool in a variety of laboratory settings, including apparatuses used to produce "optical molasses" and Bose-Einstein condensates. These techniques have led to many important applications, for example, improving the accuracy of atomic clocks, which has led to dramatic improvements in the accuracy of global positioning systems.

18B.3 Pre-lab Reading

1. *The Spectrum of Atomic Hydrogen*, in Scientific American (March 1979) by Hänsch, Shawlow, and Series.* This is a good summary of the energy level structure in hydrogen including fine structure, the Lamb shift, and hyperfine structure. It also treats Doppler broadening and saturated absorption spectroscopy.

---

* Hänsch, T. W., Schawlow, A. L., and Series, G. W. The spectrum of atomic hydrogen. *Scientific American*, **240**, 3, 94–111 (1979). An excellent semi-quantitative discussion of high-resolution laser spectroscopy, including saturated absorption spectroscopy, authored by leading experts.

2. A chapter on atomic physics from a modern physics text, for example, Chapter 8 in Thornton and Rex's *Modern Physics for Scientists and Engineers (Fourth Edition).**

3. (Optional) For more detailed treatments see Chapter 18 of *Atomic and Laser Spectroscopy*,† which treats the hyperfine interaction and includes Rb as a prominent example. Chapter 3 of Demtroder's *Laser Spectroscopy 1* monograph‡ treats saturated absorption and his *Laser Spectroscopy 2* monograph§ provides a detailed exposition of Doppler-free techniques.

Suggested Additional Reading

Budker, D., Orlando, D. J., and Yashchuk, V. Nonlinear laser spectroscopy and magneto-optics. *American Journal of Physics* **67**(7), 584–592 (1999).

18B.4 Laser Safety

The safety precautions needed when using the 780 nm infrared laser at the heart of this experiment are listed in the *circular dichroism* lab earlier in this chapter (Section 18A.4). It is imperative that these precautions be followed carefully.

18B.5 Pre-lab Questions

1. Show that the full width at half maximum (FWHM) of the 780 nm spectral line of ^{85}Rb at 297 K due to Doppler broadening is 513 MHz. Refer to the discussion on Doppler broadening in the *Background* section (below) of this laboratory.

2. In the absence of Doppler broadening, the linewidth of an atomic transition is determined by the Heisenberg uncertainty relation,

$$\Delta E \approx \frac{\hbar}{2} \cdot \frac{1}{\tau},$$ where τ is the excited

state lifetime. For the $5P_{3/2}$ excited state, $\tau = 28 \times 10^{-9}$ s = 28 ns. How does the resulting "natural linewidth" compare with the Doppler-broadened linewidth?

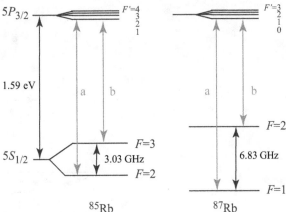

18B.6 Background

The nuclei of ^{85}Rb and ^{87}Rb have different magnetic dipole moments and hence different angular momenta. This results in a hyperfine structure that is different for the two isotopes, as can be seen in Figure 18.12. We can probe the hyperfine splitting of both the ground and excited states by measuring an absorption spectrum of the

FIGURE 18.12 Energy level diagram of the hyperfine structures of the $5S_{1/2}$ and $5P_{3/2}$ states in ^{85}Rb and ^{87}Rb. (Not to scale.) The quantum number F refers to the total angular momentum of the atom, including electron orbital angular momentum, electron spin, and nuclear spin. Because the hyperfine splittings are so small compared with an electron volt, it is conventional to state them in units of frequency, with the understanding that you have to multiply these frequencies by $h = 4.136 \times 10^{-15}$ eV·s in order to obtain energies in electron volts. The "a" and "b" labels reference the absorption feature labeling in Figures 18.4 and 18.23.

---

* Thornton, S. T., and Rex, A. *Modern Physics for Scientists and Engineers*, 4th ed. Cengage Learning, 2012.
† Corney, A., and Deslattes, R. D. *Atomic and Laser Spectroscopy*. Clarendon Press, 1977.
‡ Demtroder, W. *Laser Spectroscopy 1: Basic Principles*, 5th ed. Springer, 2014.
§ Demtroder, W. *Laser Spectroscopy 2: Experimental Techniques*, 5th ed. Springer, 2016. See Chapter 2.3 for a saturated absorption overview.

$5S_{1/2}$ to $5P_{3/2}$ transition. The energy spacings shown in Figure 18.12 are not to scale – in both isotopes the hyperfine splitting of the $5S_{1/2}$ ground state is five orders of magnitude smaller than the 1.59 eV energy difference between the $5S_{1/2}$ and the $5P_{3/2}$ levels. And the hyperfine splitting of the $5P_{3/2}$ state is an order of magnitude smaller still.*

Doppler Broadening

Resolving such closely spaced energy levels with a spectroscopic measurement is challenging because of "Doppler broadening" of spectral lines. There is a distribution in the velocities of the atoms that make up a gas in thermal equilibrium. And due to the Doppler effect there will be a corresponding distribution in the frequencies of light that can be absorbed or emitted in an electronic transition. The result is a *Doppler broadening* of the spectral lines, as described in Figure 18.13.

The expected "lineshape" of a Doppler-broadened transition is a Gaussian function of frequency. The FWHM of this lineshape is given by:

$$\Delta\nu_{1/2} = 2\frac{\nu_0}{c} \cdot \sqrt{\frac{2k_B T \ln 2}{M}}$$

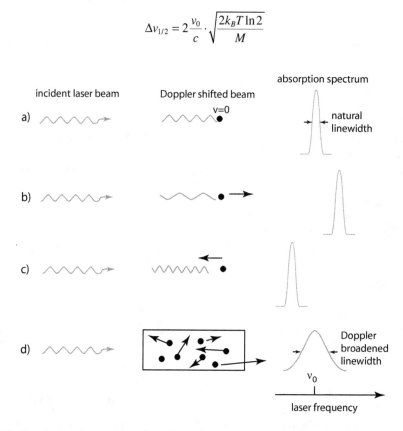

FIGURE 18.13 Doppler broadening. A laser beam is incident on a gas of atoms from the left. Atoms in the gas have a distribution of velocities, so at any given moment some are moving toward the incident laser beam and others are moving away. The Doppler shift depends on the component of an atom's velocity parallel to the laser beam. **(a)** The atom has no velocity component along the beam, so there is no Doppler shift and the absorption reaches its maximum value when $\nu_{\text{laser}} = \nu_0$. The width of the peak, the "natural linewidth," is ~1/lifetime of the excited state. **(b)** The atom is moving away from the beam so the apparent frequency of the light is lower than in (a) and the absorption reaches its maximum when $\nu_{\text{laser}} > \nu_0$. **(c)** The atom is moving toward the beam so the apparent frequency of the light is higher than in (a) and the absorption reaches its maximum when $\nu_{\text{laser}} < \nu_0$. **(d)** For an ensemble of atoms, the absorption spectrum is obtained by summing the contributions from all the atoms. The result is an absorption spectrum whose width is greater than the natural linewidth.

---

* The smaller splitting of the *p*-state can be explained qualitatively on the grounds that an *s*-state has a significantly greater probability of penetrating close to the nucleus than a *p*-state and hence the nuclear magnetic field "seen" by the electron is much greater for an *s*-state.

where ν_0 is the resonance frequency of the transition, T is the absolute temperature of the gas, M is the mass of the atom, c is the speed of light, and k_B is Boltzmann's constant. For the Rb $5S_{1/2}$ to $5P_{3/2}$ transition at room temperature $\Delta\nu_{1/2}=513$ MHz. Figure 18.4 in the circular dichroism lab illustrates the Doppler broadening in the four features of the $5S_{1/2}$ to $5P_{3/2}$ transition of a mixed sample of ^{85}Rb and ^{87}Rb. The FWHM of the features is less than the ground state hyperfine splittings but larger than the hyperfine splittings of the $5P_{3/2}$ states, so the hyperfine splittings of the upper states are entirely obscured.

But do not despair! Fortunately, laser light, in addition to being highly monochromatic, has another property that can exploited to overcome this problem – it is *intense*. This allows for nonlinear (two-photon) effects which lie at the heart of the Doppler-free technique you will employ in this lab.

Saturated Absorption Spectroscopy

Saturated absorption, which relies on the high intensity of a laser beam, forms the basis for an extremely clever technique for getting around the resolution limits introduced by Doppler broadening. The basic idea is illustrated in Figure 18.14. Two counter-propagating laser beams are sent through a glass cell containing Rb vapor. The beams are derived from a single tunable laser so they both have the same frequency. The absorption of a low intensity beam, the "probe beam," is monitored by a photodiode as the laser frequency is scanned. The intensity of the second beam, the "pump beam," is much higher than the probe intensity.

Consider the distribution of Rb atoms in the cell and how their energy states are affected by the presence of the pump and probe beams. In the absence of any external light, almost all (very close to 100%) of the Rb atoms are in the ground $5S_{1/2}$ state. Somewhat surprisingly, the population of Rb atoms in the ground state is barely affected by the presence of the probe beam alone. This is because, while the probe beam does cause transitions from the ground state to the excited state (hence the absorption spectrum!), each excitation is quickly followed by the emission of a 780 nm photon as the excited Rb atom falls back into the ground state (this process is referred to as "spontaneous emission"). As long as the rate of absorption of probe-beam photons is significantly less than the rate of spontaneous emission (or, alternatively, the average time between absorptions is much greater than the average time for spontaneous emission), the ground state population will remain essentially unchanged.

On the other hand, the presence of the much stronger "pump" beam can strongly affect the population of ground state atoms: the pump beam can significantly deplete the number of atoms in the ground state if the rate of absorption of pump-beam photons becomes comparable with the rate of spontaneous emission. The absorption rate is directly proportional to the intensity of the beam, so there will be, roughly speaking, a threshold pump-beam intensity above which the population of ground-state atoms is noticeably depleted.

Before explaining how the technique of saturated absorption spectroscopy takes advantage of this effect, we need to note some complicating factors in the relationship between energy level populations and laser beam intensity. A careful analysis of your experiment requires some exposure to these issues:

(a) While we've discussed the absorption of light from a laser beam and the spontaneous emission of light from an atom in an excited state, there is in fact a third fundamental process at play: "stimulated emission", in which an atom in an excited state is induced to make a downward transition by the presence of the laser light and emits a photon. In stimulated emission, the

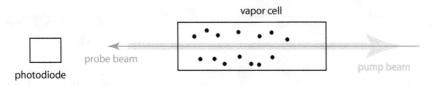

FIGURE 18.14 Saturated absorption spectroscopy. Two laser beams with identical frequencies, the "pump" and "probe" beams, counter-propagate through a Rb vapor cell. For atoms with no velocity component in the direction of the beams, the high-intensity pump beam depletes the population of ground-state atoms that are available to absorb probe-beam photons.

emitted photon has the same direction, frequency, and polarization as the photons in the laser field. (As you likely know, stimulated emission is at the heart of the physics of lasing.)

(b) Stimulated emission plays a critical role in "saturating" an absorption transition. An analysis of the populations in the ground and excited states as a function of the intensity of the incoming laser beam that includes all three radiative processes, shows that, at very high laser intensities, a limit is reached in which the populations of the ground and excited states become equal.* At this point, the combination of absorption and stimulated emission processes results in no net absorption of light!

(c) Long before this extreme situation is reached, there are noticeable effects associated with the depletion of the population in the ground state by the pump beam. The "saturation intensity," I_{sat}, of the pump beam identifies the intensity above which saturation effects are significant. This intensity is determined by the radiative lifetime of the upper state of the atomic transition. The lifetime of the $5P_{3/2}$ levels of Rb is ~28 nsec, leading to a saturation intensity of ~1.6 mW/cm^2. It is important to know the numerical value of I_{sat} since it will serve as a guide in setting the optimal intensities of the laser beams for acquiring the highest resolution spectra.

(d) The standard analysis that leads to the numerical value of I_{sat} treats the atom as a two-level system. However, due to the hyperfine splittings of the $5S_{1/2}$ and $5P_{3/2}$ levels, this is an over-simplification and the treatment above is not the full story. For the case of a multi-level system like Rb, there is an additional mechanism called *optical pumping* which can lead to an intensity-dependent depletion of the ground state. This is often the dominant mechanism for ground state depletion and typically "turns on" at intensities lower than I_{sat}.

Returning to the experimental setup: a high-intensity pump beam (with $I \approx I_{sat}$) will noticeably deplete the population of Rb atoms in the ground state, which means that the absorption of light by the probe beam will be diminished. Here is a key point: the probe beam will experience a reduced absorption only when the pump and probe beams are both absorbed by the *same* atoms. And this will only happen for atoms that have no Doppler shift. Only atoms that happen to have $v_z = 0$ (where z is the laser propagation direction) will experience pump and probe beams at the same frequency. And only this subset of atoms will have their absorption saturated by the pump beam when the laser frequency equals the transition frequency. The result is a sharp Doppler-free "*Lamb dip*" in absorption (it is a peak in transmission) when $\nu_{laser} = \nu_0$ (Figure 18.15). The width of the dip is determined by the natural width of the transition or the line width of the laser, whichever is larger. The obscuring effect of Doppler broadening has been defeated!

Each of the four Doppler-broadened absorption features in Figure 18.4 masks multiple transitions from one of the $5S_{1/2}$ hyperfine levels to the closely spaced $5P_{3/2}$ levels (see Figure 18.12). A scan of the laser frequency over a Doppler-broadened feature will reveal each of those transitions, and a Doppler-free

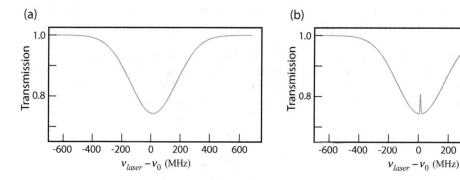

FIGURE 18.15 Doppler-free Lamb dip. (a) Doppler-broadened transmission spectrum as the laser frequency ν_{laser} is scanned in the vicinity of the resonance frequency ν_0. (b) In the presence of an intense counter-propagating pump beam there is a small but sharp peak in transmission (corresponding to a dip in absorption) at $\nu_{laser} = \nu_0$.

---

* See Demtroder, W. *Laser Spectroscopy 1: Basic Principles*, 5th ed. Springer, 2014, chapter 3.

saturated absorption spectrum will appear as in Figure 18.16.

Background-free Measurements

The Lamb dips shown in Figure 18.16 appear as small features riding on a Doppler-broadened background. For more effective observations, it is helpful to introduce a "reference beam" that travels through the Rb cell, parallel to the probe beam, but without overlapping the pump beam (Figure 18.17). The probe and reference beams can then be directed onto two identical photodiodes. By taking the difference of the probe and reference signals, a "background free" saturated absorption signal can be obtained.*

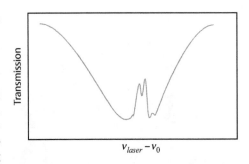

FIGURE 18.16 Resolving the $5P_{3/2}$ hyperfine splitting. A transmission spectrum of the ^{85}Rb $5S_{1/2}$ ($F=3$) to $5P_{3/2}$ (F') transition in the presence of a saturating pump beam. Analysis of the Doppler-free features can reveal the hyperfine structure of the excited $5P_{3/2}$ state.

18B.7 Experimental Setup

The optical alignment in this experiment is challenging: co-aligning pump and probe beams that are the width of a pencil lead over the length of the Rb cell is not easy. Plus, the beams are invisible! However, it's the right level of challenge for an advanced atomic physics lab: hard enough to be very satisfying when you succeed, but not so hard as to be painful. It is possible to complete the measurements described here over a span of 10–20 hours, starting "from scratch" on an empty optical breadboard. This "build your own" aspect of the lab greatly increases what you'll take away from the experience. And if you're coming to this having just completed the circular dichroism lab, the optical alignment will go substantially faster since there's a fair amount of overlap between the two optical arrangements.

Laser characteristics: To perform the saturated absorption measurements below requires a laser with the following characteristics:

- Narrow bandwidth (~10 MHz)
- Tunable range ~ 10 GHz around 780.24 nm
- At least 1 mW power

Proper *intensities* of the laser beams (pump, probe, and reference beams) are quite critical for recording optimal saturated absorption spectra. The pump beam intensity at the Rb cell should be roughly equal to $I_{sat} = 1.6$ mW/cm^2, while the probe and reference beam intensities should be less than 1% of this.† The online materials explain how to measure power and intensity. It is very likely that the power from your laser (commercial or home-built) will need to be reduced by neutral density filters.

A schematic of the entire experiment is shown in Figure 18.18.

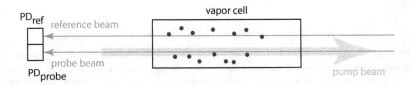

FIGURE 18.17 Background subtraction scheme. The difference in transmission recorded for the probe and reference beams isolates the Doppler-free components of the spectra.

---

* Note how this is similar in spirit to the balanced polarimeter you may have created in the circular dichroism lab (part 6 of Lab 18A,8). In both cases, you are able to enhance a small effect by taking the difference between two similar signals.
† B. E. Sherlock and I. G. Hughes, "How weak is a weak probe in laser spectroscopy?" *Amer. J. Phys.* **77**, 111–115 (2009).

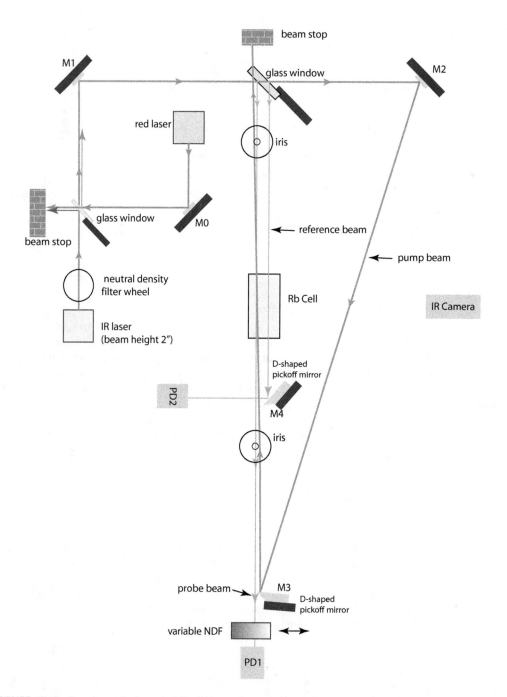

FIGURE 18.18 Experimental schematic. M0 – M4 are mirrors on kinematic mounts. PD1 and PD2 are photodiodes. Not shown: the photodiodes are connected to a dual-channel current-to-voltage converter. The converter outputs are connected to a digital oscilloscope. Data from the scope can be uploaded to a desktop computer for analysis. The computer is also used to display the image from the IR-sensitive camera. A function generator supplies a ramp voltage for tuning the IR laser.

18B.8 Procedure

0. *Instructors should make sure that the IR laser is mounted on the optical table so that the beam travels at constant elevation.* If not, place shims under the laser to level it. Optical alignments will be significantly simplified if all the optical components are mounted so that their centers are at the beam height. It is important to note at the outset the amount of "real estate" on the optical table that will be needed to implement the setup shown in Figure 18.18. In particular, note that (as explained in step 4 below) in order to maximize the overlap between the pump and probe beams it is desirable to have a large distance (ideally at least 1 m) between the glass window at the top of the setup – used to create the probe and reference beams – and mirror M3.

1. *Using a low-power visible laser for initial alignment*: The fact that 780 nm is in the infrared makes optical alignments difficult. A low-power visible laser (e.g., a helium-neon laser), carefully co-aligned with the IR laser, greatly simplifies the early stages of the optical alignment. Figure 18.19 illustrates a scheme for making a visible laser beam coincide with the IR beam.

   You can achieve this alignment using an algorithm similar to the one used in Lab 10A.2:

   Step A: Adjust the M0 kinematic mount so that the visible laser hits the glass window at the same location as the IR beam, as shown in Figure 18.19. Note that there are reflections from both surfaces of the glass window. Position a beam stop as shown so as to block the unwanted reflections. Use an IR viewer card to verify that the visible and infrared beams are overlapping a short distance from the slide. (Point *A* in Figure 18.19.)

   Step B: Adjust the orientation of the glass window so that the beams overlap when a viewer card is placed in front of a distant beam stop. (Point *B* in Figure 18.19.)

   Repeat steps A and B until the two beams maintain their overlap as you move a viewer card along the entire path.

   It is worthwhile taking the time to do this initial alignment carefully, so that, as far as you can determine with the IR viewer card, the beams maintain their overlap "perfectly" as the viewer card moves along the path. This way you can do the bulk of the remaining alignment using visible beams, relying on the fact that the IR beams are being similarly steered by the

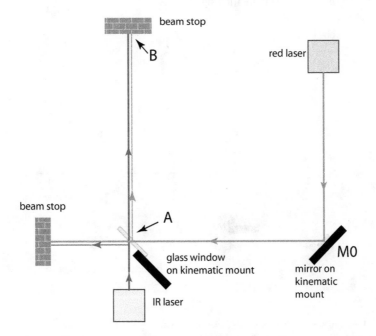

FIGURE 18.19   Co-aligning a low-power visible laser with the IR beam.

mirrors and beamsplitters. Once you are satisfied that you have co-aligned beams, you can block the IR beam and work only with the visible beam.

2. *Use iris diaphragms to mark the beam path.* Install a pair of iris diaphragms in the positions shown in Figure 18.18 in order to establish the desired path of the probe beam through the Rb cell. (Tip: you can see when the beam is centered in the diaphragm opening by holding an IR viewer card immediately behind the opening in the diaphragm.)

3. *Use a glass window to create the probe and reference beams.* Install M1 and a glass window as shown. As the laser beam crosses the air/glass interface on the front surface of the window, about 3% of the incident power will be reflected, creating the probe beam. As the beam exits the window, another 3% of the power will be reflected, creating the reference beam. The transmitted light will constitute the pump beam; for now, block that beam with a beam stop. Adjust the orientations of M1 and the glass window so that the probe beam passes through the centers of both iris diaphragms. With the diaphragms open, the reference beam should also pass through. Temporarily block the visible laser and use an IR camera (or viewing card) to verify that the IR probe and reference beams are coincident with their visible counterparts. If not, return to step 1 and tweak the co-alignment of the infrared and visible beams.

4. *Install M2 and M3 as shown so that the pump beam overlaps with the probe beam as much as possible in the region midway between the iris diaphragms, where the Rb cell will be installed. This is a tricky yet critical step!* Maximizing the overlap between the probe and pump beams as they travel through the Rb cell means that we desire as *small an angle as possible* between the pump and probe beams. The angle cannot be zero, since the probe beam needs to clear M3 as shown in the close-up view of Figure 18.20. For this reason, it is desirable to have a large distance (ideally at least 1 m) between the window at the top of the setup – used to create the probe and reference beams – and M3. As shown in the inset, using a D-shaped "pickoff mirror" for M3 allows for minimal spacing between the probe and pump beams, which leads to a small value of θ. You should aim for $\theta < 1.0°$.

   The path that the pump beam travels in reaching the Rb cell is considerably longer than the path followed by the probe beam. Due to beam divergence, the pump beam will be bigger in diameter than the probe beam, facilitating overlap.

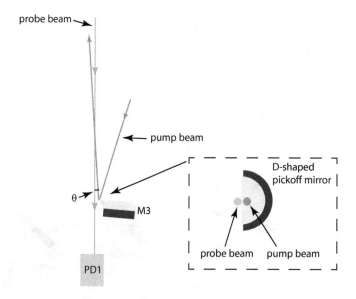

FIGURE 18.20 Detail of step 4. The goal is to make θ as small as possible. As shown in the inset, for this purpose a D-shaped "pickoff mirror" for M3 is particularly useful.

A piece of translucent vellum paper taped to a movable mount is useful here, since you can see both the pump and probe beams even though they are impinging from different sides. Try blocking and unblocking the pump beam while looking at the translucent paper; you should be able to tell when the two beams are optimally overlapped.

Block the red laser and check the overlap of the IR beams using the IR camera and vellum paper. Make final adjustments to mirrors M2 and M3 to maximize the spatial overlap of the IR probe beam with the IR pump beam. *Again, this is challenging. But it is also critical to obtaining a good Doppler-free absorption spectrum.*

5. *Install the Rb cell and direct the probe beam onto a photodiode.* Place the cell midway between the diaphragms as shown, so that both the probe and reference beams pass through the cell. Install PD1 and connect it to one channel of your current-to-voltage converter. Observe the photodiode output on an oscilloscope to verify that you can see the signal from the IR beam on the oscilloscope.

6. *Recording a Doppler-broadened absorption spectrum.* You will first tune the IR laser to encompass the four $5S_{1/2}$ - $5P_{3/2}$ transitions of interest and record a Doppler-broadened absorption spectrum using the probe beam. For this purpose, temporarily block the pump beam from passing through the Rb cell. The sequence of steps – setting up a "voltage ramp" for the laser, using fluorescence observations to properly tune the laser frequency, and recording an absorption spectrum – are described in the circular dichroism lab – please refer to those instructions to get started.

View both the PD1 and voltage ramp signals on the oscilloscope. Adjust the Amplitude and Offset settings on the function generator to so that you can see all four absorption features during the rising portion of the voltage ramp.

Use the neutral density filter wheel to vary the laser intensity. At high intensities (above I_{sat}) the absorption should begin to "saturate" as described earlier, which results in smaller fractional absorption signals. Can you see this effect?

Upload your oscilloscope data to your computer – both the photodiode channel and the voltage ramp channel.

7. *Set the pump and probe intensities.* For optimal saturated absorption spectra, you'll want $I_{pump} \approx I_{sat}$. You could measure the intensities directly, but you can do pretty well with the following simpler method. The general idea is to monitor the absorption features produced by the *pump beam* and observe when intensity-induced changes in the fractional absorption of the features begin to occur. Here's how to do it:

- Block the probe and reference beams. *Place an OD2 neutral density filter directly in front of PD2* and position the PD2 and OD2 filter to monitor the intensity of the pump beam just after it exits the Rb cell. (The neutral density filter will protect the photodiode from being exposed to damaging power. Powers in excess of a few mW are capable of damaging a typical photodiode.) Use the neutral density filter wheel to vary the pump intensity and observe (on the oscilloscope) how $\Delta I/I_{incident}$ changes (see Figure 18.21.) You'll have to vary the current-to-voltage converter gain and/or the vertical sensitivity on the scope to keep the spectrum viewable.

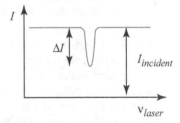

FIGURE 18.21 Monitor the fractional change in absorption depth, $\Delta I/I_{incident}$.

- You should observe that at sufficiently low intensities the ratio $\Delta I/I_{incident}$ is approximately independent of $I_{incident}$. But as you increase the pump intensity, the transition will begin to saturate so that $\Delta I/I_{incident} \to 0$. We find that setting the pump-beam intensity so that the ratio $\Delta I/I_{incident}$ is approximately half of its low-intensity value is a good way to set the value of I_{pump}.

- The probe-beam intensity: The reflectivity of the glass window used to create the probe and reference beams depends on both the polarization and angle of incidence of the laser beam. Assuming that the laser is polarized perpendicular to the plane of incidence and that the angle of incidence is 45°, the reflectivity should be ~ 0.05. Note that, due to beam divergence, the pump-beam diameter may be significantly larger than the probe-beam diameter at the position of the Rb cell. Therefore, measuring the relative powers in the pump and probe beams may not be an accurate measure of their relative intensities. You can lower the probe beam intensity by adding a neutral density filter in the probe path.

8. *Do you see Doppler-free features?* Adjust the voltage ramp signal and oscilloscope settings to "zoom in" on the ^{87}Rb $F = 2$ to $F' = 1, 2, 3$ features. Repeatedly unblock and block the pump beam. Do you see sharp Doppler-free features appear when the pump beam is unblocked? These features should disappear when you block the pump beam. The features may be subtle until you optimize the beam paths. If you don't see these features, the first thing to check is the degree of overlap between the pump and probe beams as they traverse the Rb cell. The greater the overlap, the more pronounced the Doppler-free features will be.

   Upload your oscilloscope data – both the probe-beam photodiode channel and the voltage ramp channel. Do this for the cases in which the pump beam is unblocked and when it is blocked. Time permitting, you can also explore the Doppler-free signals that appear within the other three $5S_{1/2}$–$5P_{3/2}$ transitions.

9. *Background subtraction.* The Doppler-free features that you observed in the previous step appear as small structures imposed on the Doppler-broadened background, as in Figure 18.16. For more detailed observations, it is necessary to make use of the reference signal to perform a "background subtraction".

   Install M4 and PD2 and direct the reference beam onto PD2, as shown in Figure 18.18. (Use the optical alignment laser as an aid if necessary. It will make things easier if you use another D-shaped pickoff mirror for M4.) The photodiodes have different polarities with respect to their cases, making it simple to measure the difference in their signals. The details of the subtraction technique are presented in the circular dichroism lab. The resulting signal produces a "background free" saturated absorption spectrum.* Inevitably, the strengths of the Doppler-broadened background signals from the probe and reference beams won't quite match, but you can significantly improve the background subtraction by installing a continuously variable neutral density filter on a translational stage in front of whichever photodiode is producing the bigger signal, as shown in Figure 18.22.

10. *Record a high-quality spectrum of one Doppler-broadened feature.* Adjust the amplitude and offset of the voltage ramp to "zoom in" on one of the four Doppler-broadened features. The lowest-energy peak, associated with transitions from the $F = 2$ ground state of ^{87}Rb, is a good candidate, as its Lamb dip features are more pronounced than those of the other Doppler-broadened peaks. Keep in mind that your goal in analyzing this spectrum will be to correlate the saturated absorption spectrum with the energy level structure shown in Figure 18.12.

    To optimize your spectrum, try varying the intensity of the pump beam (with the

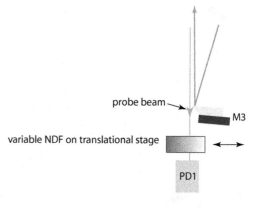

FIGURE 18.22 Fine tuning the background subtraction. Translating a continuously variable neutral density filter in the beam path provides a simple way to vary the beam intensity.

---

* See, for example, Figure 13 in *A narrow-band tunable diode laser system with grating feedback, and a saturated absorption spectrometer for Cs and Rb.* MacAdam, K. B., Steinbach, A., and Wieman, C. *Amer. J. Phys.* **60**, 1098–1111 (1992).

filter wheel) and the intensity of the probe and reference beams relative to the pump beam by inserting different neutral density filters into the probe path. Display the ramp voltage and/or Michelson fringe signal (see below) on other scope channels so that you will be able to obtain a frequency scale for your spectrum. Many digital oscilloscopes have the ability to average a number of measurements, which can lead to an improvement in the signal to noise ratio in your spectra. It may be helpful to take advantage of this feature, although since the laser frequency tends to drift, there is a limit to how many spectra you can usefully average. *Upload your best data for analysis.*

11. *Frequency calibration.* The oscilloscope data you saved in the previous step is a plot of transmission vs. time. But what you really want is a plot of transmission vs. laser frequency.

   We start by noting that the frequency of the laser light increases with increasing ramp voltage. Therefore, the left-most of the four Doppler-broadened features in your data corresponds to the lowest laser frequency and hence to photons with the lowest energy. This feature is associated with transitions that originate in the $F=2$ hyperfine state of the ^{87}Rb ground state (see Figure 18.12). Similarly, the rightmost feature is associated with the most energetic photons and corresponds to transitions that originate in the $F=1$ hyperfine state of the ^{87}Rb ground state. Examining the energy level diagram in Figure 18.12 we see that the difference in laser frequency between these two features is about 6.83 GHz.

   To a good approximation the changes in laser frequency $\Delta \nu_{laser}$ during a scan are linearly proportional to the changes in applied voltage V_{ramp}. With a bit of thought, you can use the measured ramp-voltage difference between the outermost two features along with this linear relationship to calibrate $\Delta \nu_{laser}$ as a function of the ramp voltage (this is why you've saved your ramp voltage signals) (see Figure 18.23).

12. *A better frequency calibration (optional).* With the analysis scheme suggested above, we are making use of a known energy-level spacing in our frequency calibration scheme, an approach that is routinely used in spectroscopy. This method has a drawback in that it assumes that the laser frequency

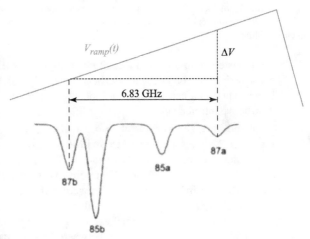

FIGURE 18.23 Frequency calibration. A Doppler-broadened transmission spectrum of the $5S_{1/2}$ to $5P_{3/2}$ transitions is recorded on one channel of an oscilloscope and the voltage ramp signal is recorded on another channel. Given that the 87a and 87b dips are separated by 6.83 GHz, we can calculate the relationship between V_{ramp} and $\Delta \nu_{laser}$.

varies linearly with the ramp voltage, which is only approximately true. An alternative method is to use a Michelson interferometer to provide a direct measure of the changes in laser frequency as the ramp voltage is applied.

   The Michelson interferometer is shown in Figure 18.24. A beamsplitter (BS1) is used to direct 50% of the IR laser beam into the interferometer.

   If the distances from the interferometer beam splitter, BS2, to the retro-reflecting mirrors differ by $x \equiv d_2 - d_1$ then there will be a *path difference* of $2x$ introduced between the recombined beams. This path difference leads to a phase difference ϕ in the recombined beams:

$$\phi = \frac{2x}{\lambda} \cdot 2\pi = \frac{4\pi x}{\lambda}$$

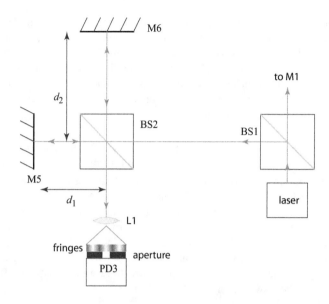

FIGURE 18.24   Michelson interferometer for frequency calibration. A fraction of the IR beam can be directed into a Michelson interferometer. As the laser frequency is scanned, the spacings of interference maxima can be used for frequency calibration.

If ϕ is equal to an integral multiple of 2π there will be constructive interference when the beams recombine, while if ϕ is equal to a half-integral multiple of 2π there will be destructive interference.

In the idealized version of the instrument, shown in Figure 18.24, the surfaces of M5 and M6 are exactly perpendicular and the reflecting interface in BS2 is oriented at exactly 45° to the mirrors. Therefore, the path difference will vary in a uniform fashion across the surfaces. Deviations from this ideal will result in an interference pattern consisting of *parallel fringes*, representing the constantly changing phase difference between the two beams (see Figure 18.25.) These are referred to as "fringes of equal thickness." The closer the two mirrors are to being parallel, the broader the individual fringes, and vice versa.

As the laser frequency is scanned, the fringe pattern will shift laterally. If the pattern is incident on a photodiode with a small aperture, as the pattern moves across the

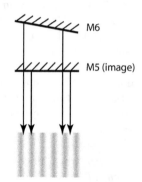

FIGURE 18.25   Parallel fringes of equal thickness. The spacing between the fringes depends on the angle between the surfaces of M5 and M6. In the diagram, the virtual image of M5 produced by the beamsplitter is shown.

aperture, the light intensity reaching the photodiode will oscillate. It is straightforward to show (see Section 18B.10 *Deliverables* below) that one full cycle of the oscillation corresponds to a change in the laser frequency that is determined by the path difference x:

$$\Delta\nu = \frac{c}{2x}.$$

So, for example, if $x = d_2 - d_1 = 0.15$ m, then one full cycle corresponds to a laser frequency shift of

$$\Delta\nu = \frac{3.0 \times 10^8 \text{ m/s}}{2 \cdot 0.15 \text{ m}} = 10^9 \text{ s}^{-1} = 1 \text{ GHz}.$$

Alignment Steps

Refer to Figure 18.24. As with the saturated absorption alignment, it is best to first align the interferometer optics with the aid of the visible laser. Once that alignment is complete, you can block the visible laser and unblock the co-aligned IR beam.

(a) Install BS1 and M5, angling M5 slightly so that the beam doesn't reflect back into the laser.

(b) Install BS2.

(c) Install M6 so that $d_2 - d_1 \approx 15$ cm. Adjust M6 so that the reflected beams from M5 and M6 recombine on a viewing screen. Install L1 to expand the interference pattern.

(d) Fine-tune the mirror orientations to produce a small number of parallel fringes. Install PD3 with an appropriate aperture and monitor the photodiode output on an oscilloscope.

(e) Scan the IR laser to observe fringe oscillations on the oscilloscope. Record Michelson fringes on one scope channel and the ramp voltage on another. What is $\dfrac{d\nu_{\text{laser}}}{dV_{\text{ramp}}}$? Is it constant? That is, does the laser frequency appear to vary linearly with ramp voltage, as we assumed above?

18B.9 Analysis

Your main analysis goal is to quantitatively connect the Doppler-free features in your measured absorption spectra with the hyperfine structure of Rb that is displayed in Figure 18.26. The analysis is complicated, in part, by the presence of so-called "cross-over transitions" in the spectra – these are explained below.

There are a number of *selection rules* that govern which transitions can be observed in an optical absorption measurement like the one that you perform in this lab. One of the rules is $\Delta F = 0, \pm 1$. This selection rule (which derives from conservation of angular momentum considerations) limits the number of allowed transitions from each $5S_{1/2}$ hyperfine level to three. The allowed transitions are indicated in Figure 18.26.

Crossover Resonances

In addition to the Lamb dips associated with the allowed transitions shown in Figure 18.26, additional Doppler-free dips can also be observed when the laser is tuned to a frequency exactly halfway between two resonant frequencies. The resulting feature is known as a *crossover resonance*. These features are often more prominent in the Doppler-free spectra than the ones associated with the allowed transitions.

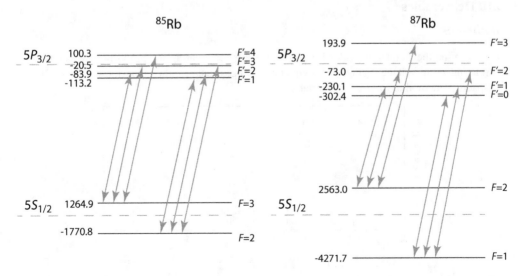

FIGURE 18.26 Allowed optical transitions between the $5S_{1/2}$ and $5P_{3/2}$ hyperfine levels. The energy shifts, relative to the theoretical energies of the levels in the absence of the hyperfine interaction, are given in MHz.

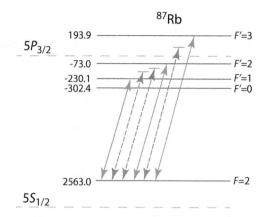

FIGURE 18.27 ⁸⁷Rb $F=2$ to F' transitions and crossover resonances. Allowed transitions are shown with solid lines, crossover resonances are shown with dashed lines.

When the laser is tuned to a frequency halfway between two resonant frequencies, there will exist a single non-zero (positive) value for ν_z, the atomic-velocity component parallel to the pump-beam propagation direction, that leads to a resonance in which the pump beam induces transitions to the higher-energy excited state and that also leads to a resonance in which the probe beam induces transitions to the lower-energy excited state. For example, if the laser is tuned halfway between the $F=2$ to $F'=2$ and the $F=2$ to $F'=3$ transitions in ⁸⁷Rb, there will be a particular value of $\nu_z>0$ such that atoms moving at this velocity will, for the pump beam, be in resonance with the $F=2$ to $F'=3$ transition, causing a depopulation of the $F=2$ ground state, *and these same atoms* will be moving in the same direction as the probe beam at the proper velocity to be in resonance with the $F=2$ to $F'=2$ transition. Crossover resonances for the ⁸⁷Rb $F=2$ to F' transitions are shown by dashed lines in Figure 18.27.

Exercise 1: Identify the features indicated by the arrows in the absorption spectrum shown in Figure 18.28 with the transitions shown in Figure 18.27. Since we can only measure shifts in the laser frequency, you will need to treat the zero of the frequency scale as an adjustable parameter. Note that the crossover features are often the most pronounced.

18B.10 Deliverables

Preliminaries

1. Determine the frequency scales for your spectra by calculating the proportionality constants relating ramp-voltage changes to frequency changes, using either the "known energy-level spacing" method or interferometer data.

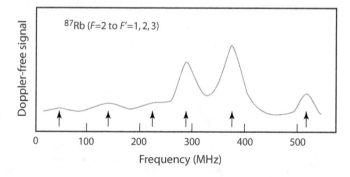

FIGURE 18.28 Doppler-free absorption spectrum for the ⁸⁷Rb $F=2$ to F' transition. Arrows indicate the observed absorption peaks. Some are due to allowed transitions, others to crossover resonances. Can you figure out which is which?

2. Measure the FWHM for each of the two Doppler-broadened lines of ^{87}Rb and compare with the theoretical result that you calculated in the pre-lab questions.

3. Make plots of your saturated absorption spectra both with and without a background subtraction.

4. (*Optional*) Show that one full cycle of the oscillation of the fringes in a Michelson interferometer corresponds to a change in the laser frequency that is determined by the path difference x:

$$\Delta \nu = \frac{c}{2x}.$$

The Main Course

Do a detailed analysis for one of the four sets of Doppler-free features in which you identify all hyperfine transition and all crossover resonances. If you are having difficulty with the identifications it's okay to look at published data, such as that given by MacAdam and co-authors.* The identifications can be quite challenging!

FIGURE 18.29 Prof. Robbie Berg, at "Light Play", a project he has worked on for a few years. This activity allows participants to explore light, shadow, and motion using a variety of simple materials along with programmatically controlled lights and motors. This project is a collaboration between the Tinkering Studio at the Exploratorium museum and the Lifelong Kindergarten Group at the MIT Media Lab.

Robbie Berg is Professor of Physics at Wellesley College. His research is centered on optical studies of new materials that hold great promise for creating new kinds of electronic devices. For example, he studies a particular type of atom-sized defect in diamond crystals called an "NV center", which is of interest because it is possible to optically monitor and manipulate the quantum state of a single center. On a different note, since 1996, he has collaborated on a series of projects with the Lifelong Kindergarten Group at the MIT Media Lab. He has helped develop new technologies that, in the spirit of the blocks and fingerpaint of kindergarten, expand the range of what people design and create—and what they learn in the process. For example, he helped design a new generation of "programmable bricks" called Crickets that enable kids to build all kinds of robotic inventions.

---

* MacAdam, K., Steinbach, A., and Wieman, C. "A narrow-band tunable diode laser system with grating feed-back, and a saturated absorption spectrometer for Cs and Rb". *Am. J. Phys.* **60**, 1098–1111 (1992).

Glenn Stark is a Professor of Physics at Wellesley College. His research centers on molecular spectroscopy in support of atmospheric and astrophysical research. His main focus is in studying the photoabsorption of vacuum ultraviolet light that leads to dissociation pathways in diatomic and triatomic molecules (Figure 18.30).

FIGURE 18.30   Prof. Glenn Stark.

19

Photonics and Fiber Optics

Jay Sharping and Walter F. Smith

CONTENTS

The beauty of it is, a scientist or engineer can utilize the fundamental principles of photonics and the tools offered by it in their own areas of work, or even enter a field distinctly different from their own, yet arrive at a fresh perspective of the underlying mechanics they set out to investigate. Photonics thus offers a way to bridge the gap between seemingly disjoint domains, facilitating fruitful interdisciplinary research in the process.

– Srikanth Sugavanam (photonics researcher at Aston University)

19.1 Overview

Photonics* refers to the field of science and engineering associated with controlling light energy (photons) in a manner analogous to that in which we control electrical energy or electrons. Photonics includes light generation, transmission, modulation, amplification, storage, and detection. It includes both classical and quantum physical concepts. Photonics expands on traditional optics to include a much broader range of frequencies extending from microwave frequencies well into the ultraviolet spectrum.

Photonics, like many optical technologies, is a field of study that enables other fields. For example, much of the research effort associated with photonics involves improving the performance and reducing the cost of optical communications technology, time and position metrology, optical sensing devices, biological devices, and high-power applications in manufacturing and defense. The intense demand for such *practical* improvements stimulates *fundamental* research in materials, light-matter interactions, and quantum optics.

A key feature of using light in all of these endeavors is that one has exquisite control over the state (classical and quantum) of the fields. Lasers provide high-power, directional beams of good spectral and polarization purity. If we are careful, we can manipulate the spatial mode (beam quality), temporal mode (pulse quality), and polarization state without introducing too much loss (i.e., destroying the photons). Because of the high frequency of optical waves, we normally have access only to the energy carried by the wave. But with the aid of an interferometer, we gain the ability to manipulate and characterize the amplitude and phase of the wave. Waveguides, including optical fibers, enable the spatial confinement of a beam in a well-defined volume and the transmission of light from one location to another.

---

* B. E. A. Saleh and M. C. Teich, *Fundamentals of Photonics* (Wiley, 2007).

19.2 Fiber Optics Fundamentals

Fiber Optic Terminology

We assume that you've learned the basics of fiber optics in a previous course. (For a refresher, please see the link to suggested reading on ExpPhys.com.) Light is transmitted through the "core." The surrounding "cladding" has a lower index of refraction, so that the light is kept in the core by total internal reflection. The cladding is surrounded by a coating for strength and protection, as shown in Figure 19.1. In the types of cables used in scientific applications, the core is surrounded by a "tight buffer," which provides even more protection. (In some telecommunications cables, multiple core/cladding/coating fibers are contained in a single buffer.) In patch cables that are used to connect different devices together, this tight buffer is surrounded by a protective outer jacket, typically 3 mm in diameter.

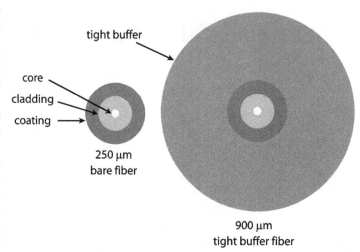

FIGURE 19.1 Bare and tight buffer fibers.

Types of Fiber Optic Cable

We define the direction along the length of a fiber optic to be z, and the perpendicular directions to be x and y. There are two primary types of fiber optics: "single-mode" and "multimode." The names refer to the normal modes for the electromagnetic field in the **cross-section** of the fiber (i.e., in the x-y plane). We refer to the distribution of fields in this plane as the "transverse state." The wave propagates primarily in the z-direction, which we also refer to as the longitudinal direction.

The situation is somewhat analogous to the two-dimensional infinite square well problem from quantum mechanics.* As you may recall, the energy eigenstates for that problem involves fitting an integer number of half wavelengths into the x-dimension of the well, and also an integer number of half wavelengths into the y-dimension. In the ground state, we fit one half wavelength in each direction, as illustrated in Figure 19.2a. In the next higher energy state, we fit two half wavelengths in one direction and one in the other. Figure 19.2b shows the version with

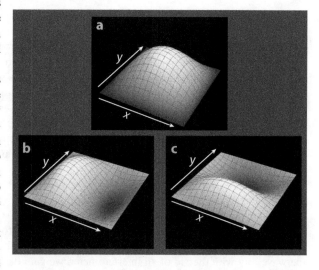

FIGURE 19.2 Wavefunctions for the two-dimensional square well.

* Hopefully, you also think it's cool that we can use your understanding of quantum mechanics as a tool to teach you by analogy about a classical system!

two half wavelengths in the *x*-direction and one in *y*. This state has the same energy as one with two half wavelengths in *y* and one in *x* (Figure 19.2c). Let's imagine fixing the energy of the particle to equal that of the ground state of a well of a particular size. (This is analogous to fixing the frequency, or equivalently the wavelength, of the light in a fiber optic.) Only one state of this energy can exist. A well of this size is analogous to a single-mode fiber: only one transverse state of the electromagnetic field can exist at this longitudinal wavelength.

Now consider a larger well, where for the energy under consideration we can fit two half wavelengths in one direction and one in the other. There are two states with the same energy. This is analogous to a multimode fiber; multiple transverse states can exist for the given longitudinal wavelength.

The fiber optic has a round cross-section, rather than square. This means that, for a large diameter cable, there can be many different transverse modes corresponding to a single wavelength, as shown in Figure 19.3. (An analogous thing happens in quantum mechanics.) The mode at the lower left is also the field distribution in a single-mode fiber; it is called the LP_{01} mode. ("LP" stands for "linear polarization.")

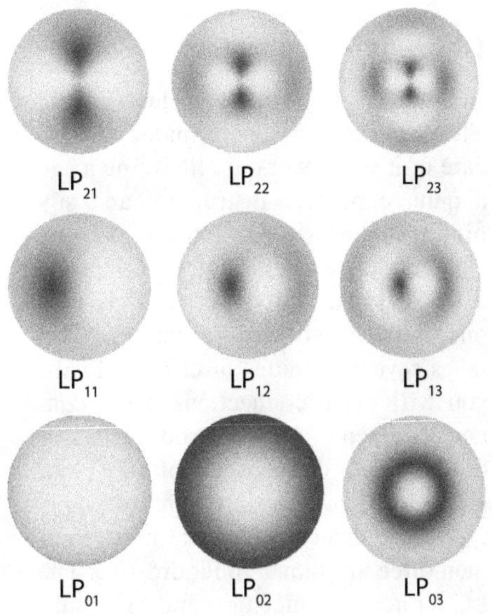

For any other mode, there is a small transverse component to the wavevector. In a somewhat simplified picture, this means that the wave bounces multiple times off the interface between the core and the cladding, as shown in Figure 19.4, whereas in the LP_{01} mode, the light essentially goes straight along the fiber. The multiple reflections lead to the three disadvantages of multimode fiber: (1) there is more leakage of energy out of the core, so transmission over very long distances is less efficient. For example, at 1550 nm (one of the most common wavelengths for telecommunications, and the one you will use for the labs in this book), the attenuation is less than 0.18 dB/km for single-mode fiber and 1 dB/km or more for multimode. (Recall that 1 dB corresponds to a change in energy of a factor of $10^{1/10}$, i.e., a change of about 20%.) (2) When the light exits the end of the fiber, it comes out with a wider angular spread. (3) The path lengths for the different modes are different (e.g., compare the three paths shown in Figure 19.4). So, a pulse that is short in time when it's injected gets spread out by the time it reaches the other end of the fiber. This means that the energy is spread out over a greater time, so the peak amplitude of the pulse is reduced, making it harder to detect. For these reasons, single-mode fiber is essentially the only kind used in telecommunications.

However, there are disadvantages to single-mode fiber. The core is quite small: 9 μm for the fiber used in telecommunications. This means that it is challenging to couple light from an external source (such as a laser) into the fiber; the light must be very tightly focused, and carefully aligned. In particular, this means that LED light sources (which are the most inexpensive type) cannot be efficiently coupled into single-mode fiber, but

FIGURE 19.3 Some of the transverse modes of a multimode fiber, shown at $t=0$. Light areas are positive amplitude and dark negative.

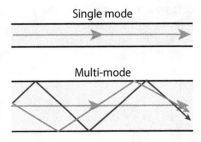

FIGURE 19.4 For a single-mode fiber, there is only a single path length. For a multimode fiber, the path length depends on the mode, causing pulses to spread out. (The degree of lateral motion is exaggerated in the image.)

can be into multimode. Single-mode fiber is designed for a particular wavelength, so in any application that uses multiple wavelengths, multimode fiber is a better choice. Multimode fiber is often less expensive and requires less precision in interconnects.

If the fiber is in the form of a cable (with a 3 mm outer jacket), then it's easy to tell the difference: single-mode cables are always yellow, while multimode cables are usually orange.

Fiber becomes birefrigent (i.e., index of refraction depends on polarization axis) when it is stressed, e.g., by bending. Therefore, the polarization is strongly affected in an essentially random way when light goes through a fiber. If this is problematic, you should use "polarization maintaining" (PM) fiber. The two main types are "PANDA" and "Bow Tie"; the names refer to the appearance in cross-section of members added to the cables to create intentional stress in the fibers, which overwhelms the accidental stress from bending. The two types are equally good. It's important to recognize that "polarization maintaining" only means that such fibers will maintain linear polarization that comes in parallel or perpendicular to the stress axis of the fibers; any other polarization will be essentially randomized. PM cables are usually blue.

Fiber Optic Connectors

In a pinch, you can make an electrical connection between two wires by simply twisting them together. There isn't even a remote analogy to this for fiber optics. Every connection has to be made with great care or it won't work at all! Doing a serious, low-loss splice between two fiber optic cables requires

a quite expensive instrument and significant expertise. For about US$7, you can purchase a mechanical splice for a single joint, which works fairly well once you've had some practice, but it takes several minutes of effort. Mostly, you will make connections using connectors, each of which introduces a loss of about 0.75 dB (a factor of $10^{0.75/10}$ or about 16% power loss). There are many types of connectors, but the most common three are shown in Figure 19.5. The LC ("Lucent connector") and SC ("subscriber connector") types are the most common in telecommunications and are superior in most respects. However, the FC ("ferrule connector") type is most common in science labs and so is the type we will focus on.

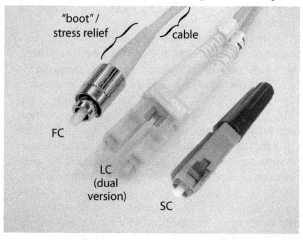

FIGURE 19.5 The three most common types of fiber optic connectors. See below for image credits.[†]

There are two main types of FC connectors: PC ("physical contact"), in which the fiber ends are flat, and APC ("angled physical contact"), in which the ends are angled at 8° to minimize back reflections. At this point, APC is superior. (When it was first introduced, it had higher insertion loss, but no longer.) However, PC is more common. It's easy to tell the difference: for APC, the "boot" color is always green, while for PC it is usually blue, but sometimes white, black, gray, or yellow. **Important:** don't mix APC and PC! If you do, there will be an air gap, because you have an angled fiber end on one side and a flat end on the other, resulting in huge losses. Also, nowadays when you buy PC connectors, you should buy the UPC ("ultra physical contact") version, which is superior to the original PC version.

Assuming you're buying FC connectors, this means you should opt for "FC/APC" when you can, otherwise "FC/UPC." If you were getting LC connectors instead, you would get "LC/APC" or "LC/UPC."

Concept test (answer below*): What is the designation for an FC connector with a green boot and an orange cable? Is the cable single-mode or multimode?

---

* Answer to concept test: FC/APC connector, multimode cable.

[†] LC connector image by Wikimedia Commons user Adamantios. SC connector image by Wikimedia Commons user Alei Phoenix. License for both images: https://creativecommons.org/licenses/by-sa/3.0/deed.en.

Unfortunately, there is one more aspect of connectors you must consider: key width. As shown in Figure 19.6, the inner metal part of the connector has a raised "key," which fits into a slot on the mating piece, allowing control of the rotation. This keeps the two fiber ends from rotating relative to one another (and perhaps damaging each other) when the connection is tightened. In "wide key" connectors, the key is 2.14 mm wide, while in "narrow key" connectors it is 2.02 mm wide. The types were perhaps introduced to prevent accidental connection between PC and APC connectors, but both kinds now come with both key widths. The width difference is too small to see by eye and is truly an annoyance. However, if you try to make a connection and the key won't fit the slot, now you know why. When purchasing new components, always choose narrow key.

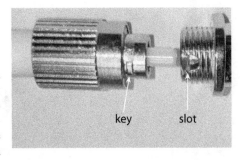

FIGURE 19.6 FC connector being inserted into a female connector.

Using Fiber Optic Connectors

Important: you need to use *much* more care when using fiber optic connectors than with electrical connectors. If you are not careful, you will get poor, likely unusable connections, and it is surprisingly easy to damage the end of the fiber. Please view the video on this at ExpPhys.com

Lab 19A Free Space Sagac Interferometer

This lab is available at ExpPhys.com.

Lab 19B Fiber Optic Sagac Interferometer

This lab is available at ExpPhys.com.

Jay Sharping received his Ph.D. from Northwestern University in 2003 and served as a postdoctoral scientist at Cornell University. Jay is currently an Associate Professor at the University of California – Merced where his research includes all flavors of parametric devices such as the development of novel pulsed light sources for applications in photonics and quantum optics (Figure 19.7).

FIGURE 19.7 Prof. Jay Sharping.

20

Experiments with Entangled Photons

Enrique J. Galvez

CONTENTS

Nature is always right. It is us who are wrong.

20.1 Introduction

Quantum mechanics is an amazing theory with tremendous predictive power. It is probably the most successful theory ever invented. It correctly verifies the results of measurements, but while it tells us what nature does, it does not tell us what nature is. Thus, we have to learn how to think about nature in the eyes of quantum mechanics. It gives us conflicting visions with what we are used to from our everyday experience. The set of experiments in this chapter leads us into the rabbit hole. It takes into confronting the most troubling concepts of quantum physics: superposition, wave-particle duality, and entanglement, to name a few. Thinking more deeply about them will bring about a better understanding of the subtle ways in which nature works.

This chapter describes three experiments to test and demonstrate the principles of quantum mechanics using correlated photons. These include the demonstration that the photon exists; production and measurement of entangled states of photons, including a measurement of a violation of a Bell inequality; and the interference of a single photon with itself and the quantum eraser. The experiments are presented in order of difficulty.

Lab 20A Spontaneous Parametric Down-Conversion and the Existence of Photons

This is the first of three experiments with single photons. They use an unusual source of light. It starts with a laser beam of short wavelength, which after interacting with a crystal produces pairs of photons that are correlated in many properties, such as energy, momentum, and polarization. The pairs are generated simultaneously from the conversion of a parent photon. The first and third experiments are called "heralded-photon" experiments because we use one photon to herald the presence of the other one. Part of the early exercises also explain why we go through all of this trouble instead of just using an attenuated source of light. *A first step that must be followed is to be familiar with aligning laser beams, so you should complete Lab 10A.2 "Precision Optical Alignments" before doing these labs.*

20A.1 Spontaneous Parametric Down-conversion

As a first step, we provide a general description of the source: spontaneous parametric down-conversion.

Basic Features

In this process, a pump photon of energy E_p is converted into two photons of energy E_1 and E_2 such that

$$E_p = E_1 + E_2, \tag{20.1}$$

or expressed in terms of the wavelength in vacuum via $E = hc/\lambda$,

$$\frac{1}{\lambda_p} = \frac{1}{\lambda_1} + \frac{1}{\lambda_2}. \tag{20.2}$$

In a birefringent medium, spontaneous parametric down-conversion occurs due to the interaction of the light and the medium. For down-conversion to occur, momentum must be conserved:

$$\vec{p}_p = \vec{p}_1 + \vec{p}_2. \tag{20.3}$$

Suppose that the wavelength of the light in vacuum is λ. In a medium with index of refraction n the wavelength the light travels at a speed c/n, where c is the speed of light in vacuum. As a consequence, the wavelength of the light in the medium is λ/n. The momentum of the photon in the medium will be $p = 2\pi n\hbar/\lambda$. If the two photons do not have the same energy (and therefore, wavelength), then by conservation of momentum in the transverse direction we get

$$0 = \frac{n_1}{\lambda_1}\sin\theta_1 - \frac{n_2}{\lambda_2}\sin\theta_2. \tag{20.4}$$

Photons come at different angles, as shown in Figure 20.1. However, in any medium the index or refraction depends on the wavelength. Conservation of momentum in the axial direction leads to

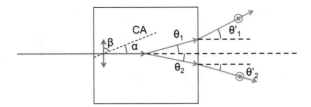

FIGURE 20.1 Angles that the photons form with the relevant directions of the problem. CA is the crystal axis, which forms an angle β with the electric field of the pump beam, and an angle α with the propagation direction. The angles formed by the down-converted photons with the incident direction inside the crystal are θ_1 and θ_2, and the angles that they form outside the crystal are θ_1' and θ_2'.

$$\frac{n_p}{\lambda_p} = \frac{n_1}{\lambda_1}\cos\theta_1 + \frac{n_2}{\lambda_2}\cos\theta_2. \tag{20.5}$$

If we consider the case where $\lambda_1 = \lambda_2$, then from Equation (20.4), $\theta_1 = \theta_2$ and so Equation (20.5) together with $\lambda_1 = \lambda_2 = 2\lambda_0$ reduces to

$$n_p = n_1 \cos\theta_1. \tag{20.6}$$

It is not possible to satisfy this relation in an isotropic medium. To satisfy it, we need a non-linear medium.

In these experiments we use beta barium borate (BBO) crystals, which have negative uniaxial birefringence. That is, they have two indices of refraction, extraordinary, for polarization parallel to the crystal axis, given by

$$n_e = \left(2.3753 + \frac{0.01224}{(\lambda/1\,\mu m)^2 - 0.01667} - 0.01516(\lambda/1\,\mu m)^2\right)^{1/2} \tag{20.7}$$

and ordinary, for polarization perpendicular to the crystal axis, given by

$$n_o = \left(2.7359 + \frac{0.01878}{(\lambda/1\,\mu m)^2 - 0.01822} - 0.01354(\lambda/1\,\mu m)^2\right)^{1/2}. \tag{20.8}$$

For example, the indices of refraction for pump-beam light of wavelength 405 nm are $n_{ep} = 1.568$ and $n_{op} = 1.6923$. Similarly, the indices for the down-converted photons at 810 nm are $n_{e1} = 1.5461$ and $n_{o1} = 1.6611$. A good exercise consists of graphing the two indices of refraction as a function of the wavelength from 400 nm to 900 nm. From the graphs one can see that for the index of a given kind (e or o), $n_p > n_1$, which implies that Equation (20.6) can never be satisfied if the polarizations of the pump and down-converted photons are the same. In addition, the extraordinary index can be tuned by having the polarization of the input light form an angle β with the crystal axis. It is usual to express the new index of refraction in terms of the complement of this angle: $\alpha = \pi/2 - \beta$, which is also the angle that the direction of propagation forms with the crystal axis. In this case, the index of refraction is given by

$$n_{e,\alpha} = \left(\frac{\cos^2\alpha}{n_o^2} + \frac{\sin^2\alpha}{n_e^2}\right)^{-1/2}. \tag{20.9}$$

When $\beta = 0$ (or $\alpha = \pi/2$; see Figure 20.1), the crystal axis is parallel to the electric field, so the index of refraction reduces to n_e. If $\beta = \pi/2$, the crystal axis is perpendicular to the field so the index must be n_o. For angles β in between 0 and $\pi/2$ the index of refraction is in between and specified by Equation (20.9).

Here is where the non-linear optics comes in. It is possible for the crystal to help with the emission of photons with a polarization such that Equation (20.6) is satisfied. In type-I parametric down-conversion, the down-converted photons have a polarization that is perpendicular to that of the pump photons. When this occurs, the indices of refraction satisfy Equation (20.6) at a specific angle α. For example, if the

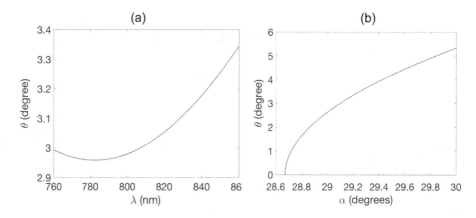

FIGURE 20.2 Graphs showing the angles at which down-converted photons are emitted: (a) as a function of wavelength for fixed $\alpha = 29.1°$, and (b) as a function of α for fixed $\lambda = 810$ nm.

angle between the crystal axis and the propagation direction is 29.1°, we have that $n_{ep,\alpha}(405 \text{ nm}) = 1.6602$. Down-converted photons with a polarization as shown in Figure 20.1 are perpendicular both to the pump photons and to the optic axis. Therefore, these photons experience the "ordinary" index of refraction. Using (20.6), the angle of propagation at 810 nm will be

$$\theta_1 = \cos^{-1}\left(\frac{n_{ep,\alpha}}{n_{o1}}\right), \tag{20.10}$$

which for our case is $\theta_1 = 1.81°$. Outside the crystal, the down-converted photons get refracted, so applying Snell's law, they come out at the angle

$$\theta_1' = \sin^{-1}\left(n_{o1}\sin\alpha\right), \tag{20.11}$$

Or $\theta_1' = 3.01$ for the conditions mentioned. For other wavelengths, down-conversion occurs at different angles. Figure 20.2a shows the emission angles, outside the crystal, of down-converted photons as a function of wavelength when $\alpha = 29.1°$. From the graph, one can obtain that if $\lambda_1 = 840$ nm, the emission angle outside the crystal is $\theta_1' = 3.17$. What is the wavelength of the partner photon if the pump photon is 405 nm?[*] Use Figure 20.2a to find the angle θ_2' at which the partner photon appears.[†]

The graph in Figure 20.2b shows the dependence of the angle of the down-converted photons as a function of the optic-axis angle α. As can be seen, changes by only a fraction of a degree have a corresponding effect on the angle of the down-converted photons.

In actuality, there may a slight difference between the directions of the pump photon outside and inside the crystal due to refraction. That is, the input crystal face may not be perpendicular to the input beam. The crystals already come cut to the desired angle when the input pump beam is perpendicular to the plane of the face of the crystal, so slight tilts do not alter significantly the results shown in Figure 20.2.

In summary, when the crystal is set to the appropriate orientation, it produces down-converted photons pairs. They emerge at angles that depend on their wavelength and the orientation of the crystal axis. The analysis given can be generalized to show that the down-converted photons are created in planes other than the horizontal plane. All the down-converted photons of a given wavelength form a cone about the pump-beam axis, but photons in each pair follow, by conservation of momentum, correlated paths in the cone.

Procedure

The first step in this experiment is to set up the apparatus to produce spontaneous parametric down-conversion. It is a weak process; in an average-size crystal only about 1 in $10^8 - 10^{10}$ photons get converted.

---

[*] Answer: $\lambda_2 = 782$ nm.
[†] Answer: $\theta_2 = 2.96°$.

TABLE 20.1

Parts for Lab 20A.1

Qty	Part	Description/comments	Photo example (Figure 20.3)
1	GaN diode laser	405 nm, > 20 mW, polarized.*	Laser pointer laser: (a)
1	Alignment laser	HeNe or equivalent of circular shape.	
1	Flip-mirror	(F) A mirror in a mount that can flip in and out of the way. It is mounted on a magnetic pedestal mount.	(b)
1	Plumb bob	Nylon thread and non-magnetic bob.	Example: (c).
1	Curved plate with 1 m radius of curvature	For help with the adjustment in the placement of collimators to detect partner photons.	3D printed: (e)
1	BBO crystal	Type-I, mounted, 29° phase matching angle, with optical axis on the horizontal plane.	Best mount: tilting and rotating: (d)
1	Mirror + translation stage	(D) For easy aligning of the pump beam.	
2	Fiber collection assembly	(DA and DB) Multimode fiber with mounted collimator, plus iris mounted on the collimator mount.	Held by magnetic mount: (e)
2	Band-pass filters	800–810 nm center wavelength, 40 nm bandwidth. Best mount is one that attaches to iris of collimator mount.	Attached to collimator (e)
2	Detectors	Avalanche photodiodes, fiber-coupled.	Modules: (f)
1	Beam dump	For safety, to stop the pump beam.	
1	Laser goggles	For safety, to block 405 nm during alignment.	
1	Electronics	To record singles counts and coincidences.	
1	Computer	Interfaced to the apparatus.	

* Note: this can be a fairly inexpensive "laser pointer," powered by a DC power supply, with polarization provided by an external polarizer. If you use this option, leave the laser turned on for the whole duration of the experiment; this will improve the stability of the power, since the laser will be at a relatively stable temperature. Also, don't be fooled by the "laser pointer look" of this instrument; it has high enough power to be dangerous, and should be treated accordingly.

That is okay because we are also detecting single photons and an ordinary laser source has too many. For example, a modest laser beam from a laser pointer with a power of 1 mW carries 3×10^{15} photons per second. Electronic systems have trouble distinguishing photons at rates higher than 10^8 per second. The parts that are needed are listed in Table 20.1. Photo references in the table correlate with the images in Figure 20.3. The parts' upper-cap labels refer to the schematic in Figure 20.5.

Pre-setup (Optionally Can Be Completed by the Instructor)

The down-converted photons are created in a cone about the pump-beam direction. The experiment uses the horizontal slice of the down-conversion cone that contains the pump beam. As a consequence, all the optical elements have to be contained in that plane.

1. The very first step in setting up the apparatus is to decide the height at which the photons will be traveling over the breadboard. Because of mounting constraints, we must make sure that three critical elements are at the same height above the breadboard: the pump laser beam, the alignment laser beam, and the down-conversion crystal. This height is typically 3 inches (or 7.5 cm).

2. The pump laser beam has to be set to be horizontally polarized.

3. The crystal has to be mounted on a mount so that its crystal axis is in the horizontal plane.

4. The best results are obtained when fiber collimators are mounted on a mirror mount, which is attached to a magnetic mount with a rectangular footprint (see Figure 20.3e). An iris or circular target placed concentric with the collimator helps with the alignment. Attaching the band-pass filter to the collimator assembly is convenient but not necessary.

FIGURE 20.3 Figures of examples of parts: (a) pump laser pointer; (b) flipper mirror; (c) plumb bob; (d) crystal mount; (e) collimator mount; (f) detectors.

5. Detector power supply needs be 5 V, 2 A. The high-current supply is needed for powering the thermoelectric cooler inside the detector unit.

6. The output of the detectors has to be connected to a counter/coincidence unit. Verify that the electric pulses from the detectors match the pulses expected by the electronic unit (they must use the same digital logic convention: either 5 V or 3.3 V).

7. A dedicated computer must be able to read the data from the electronic unit and display it in graphical form as well as by numeric format.

8. Other optical components needed for the experiment have to be mounted at the appropriate height. In most cases, pedestal mounts (see, for example, Figure 20.3b) are recommended because of their low footprint and stability.

Setting up the Alignment Beam

To set up the experiments, we first need to carefully position each component in a deliberate order, described below.

1. Set up the alignment laser as shown in Figure 20.4: mirror *A* reflects the beam parallel to the rows of holes on the breadboard keeping the same height. The beam must reflect off mirror *B* close to its edge, and the reflected beam must be parallel to the holes on the breadboard while keeping the same height over the breadboard. Follow the recommended dimensions.

2. The next step is to set up the path of the beams relevant for down-conversion: those of the pump beam and the down-converted photons, with the former forming angles of ±3° with the latter.

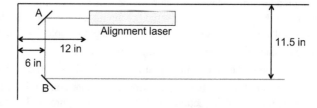

FIGURE 20.4 Arrangement for setting up the alignment beams in the apparatus. We start with an alignment laser and two mirrors, labeled A and B. Recommended approximate distances to the edge of the breadboard are shown.

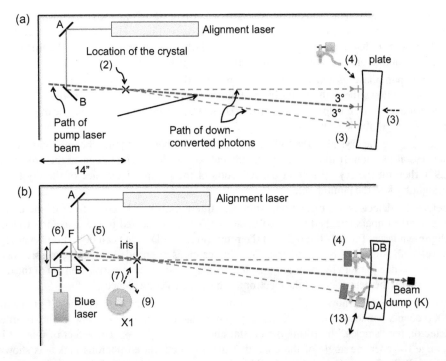

FIGURE 20.5 (a) Schematic of the paths of the pump photons (thicker central dashed line) and down-converted photons (thinner dashed lines above and below central line). (b) Schematic of the set up and alignment of the parts of the optical setup, which includes mirrors A, B, F (which flips), and D (on a translation stage). The numbers refer to steps in the procedure.

These paths are shown in Figure 20.5a. The path of the alignment laser, which uses mirrors *A* and *B*, as shown in Figure 20.4, mimics the path of one of the down-converted photons. Place a mark ("×") at the approximate location shown in Figure 20.5a exactly under the alignment beam. A small plumb bob (see, for example, Figure 20.3c) can help with this: make a mark under the bob when the bob's line, placed in the path of the pump beam, casts a shadow on a screen. The down-conversion crystal will be placed at this location later.

3. Make three cross marks (e.g., "+") at 1 m downstream from the location of the crystal, as shown in the figure. One mark is just 1 m downstream from the "×" mark. An easy angle for down-conversion experiments is $\theta_1' = \theta_2' = 3°$, so make "+" marks 1 m from the "×" mark in directions forming 3° and 6° from the direction of the alignment beam. It is useful to know that $\sin 3° = 0.052$. If available, place a plate with 1 m radius of curvature with the curved surface along the "+" marks. This way, the center of curvature of the plate is the location of the crystal.

4. Place the fiber collimator so that the light from the alignment beam reflected by mirror (B) enters the fiber. The base of the collimator must rest on the curved plate. There are techniques to do this alignment. The best one involves attaching an iris to the collimator mount marking its center and adjusting the unit so that the beam is centered on the iris. The collimator has to look into the beam. To accomplish this, place a mirror up against the collimator/iris, and tilt the mount until the alignment beam is reflected straight back (placing an iris in the path of the alignment laser beam can help with this). If the band-pass filter can attach to the iris/collimator, then mount it and use the reflection off the filter instead of the mirror. It makes it more convenient. This is an iterative process.

5. Mirror (F) should be mounted on a "flipper mount" (Figure 20.5b), which can flip in and out of the path of the alignment beam. Place mirror (F) such that it steers the light along the path of the second down-converted beam. This path must go over the location of the crystal and go above the corresponding "+" mark downstream. The use of irises placed at the two locations

helps significantly. The adjustment of the position and tilt of mirror (F) is challenging and requires patience. Once the beam follows the right path, we need to place a second collimator and adjust it so that the light enters the fiber, similarly to what was done with the first collimator. The collimator base must rest against the curved plate.

6. Now the pump laser is added to the table. Steer the light from the pump laser with mirror (D) so that it follows the path shown in the figures: halfway in between the paths of the down-converted photons. It helps to place mirror (D) on a translation stage, and the use of irises at the crystal and "+" mark locations.

7. Finally, place the crystal in the "×" location. The alignment and pump beams should cross at the crystal location. If the crystal is already cut so that the crystal axis is at the proper angle (29°), then tilt the crystal so that the reflection of the pump off the face of the crystal goes straight back to the pump laser.

8. Before the detectors are turned on, band-pass filters must be placed in front of the collimators. Optical fibers attached to the collimators must be connected to the detectors. Block the alignment laser. Turn off the lights in the room (green LED lights can be left on because that wavelength is blocked by the filters). These precautions are needed because excessive photon detections due to room lights or alignment laser may destroy the sensitive element of the detector. Turn on the detectors and the electronics that records coincidences.

9. Photon counts must now display on the computer. Best is if they are graphed. Counts above 2000 counts per second may indicate down-conversion. Maximize the raw counts from each detector, the "singles," by tilting the crystal about a vertical axis. The effect of this tilt is to change the aperture angle of the cone that the down-converted photons make, as shown in Figure 20.2b. The singles signals should be very sensitive to the tilt of the crystal at the sweet spot: they should peak dramatically at a certain tilt of the crystal and sharply go down for other tilts. If the singles from the two detectors do not peak at the same tilt angle, then the collimators are not at the right angles. To fix this, you adjust the tilt to peak at one detector, and displace by very small amounts the other collimator along the curved plate until the singles for the corresponding detector go through a peak.

At this point you have to check that you have real down-conversions (you may have down-conversion detections but not necessarily be detecting both partners; this also occurs because the overall efficiency in detecting a single photon produced by the source is 10% or less, so many partners get reflected or absorbed by optical components, or not detected by the detector of ~60% efficiency). The coincidence counts of the two detectors must be larger than the accidental coincidences (photons that arrive randomly at the same time but that are not partners). This can be calculated using probabilistic arguments. If the data is recorded over a dwell time t_d, the number of accidental coincidences over that time is given by

$$N_{\text{acc}} = \frac{N_1 N_2 \Delta T}{t_d}, \tag{20.12}$$

where N_1 and N_2 are the singles counts and ΔT is the maximum time-delay between pulses that is considered a coincidence. For example, a typical coincidence circuit has $\Delta T = 50$ ns. If the singles are 20,000 for each detector over a dwell time of 1 s, then we should expect an average of 20 accidental coincidences in that time. Optimal alignment should produce coincidences that are 10% of the singles counts. There are several tests that can be done to test this relation. If ΔT can be changed, test the relation as a function of that variable. If this quantity is fixed, block the laser and put a cloth cover over the collimators. Then turn some room lights on. Gradually uncover the collimators until the number of singles counts is of the order of the counts when the down-converted photons were present. In this case, all the coincidences are accidental. This can be used to calibrate the previous equation a bit more precisely.

10. Record singles and coincidences from detectors A and B for a given dwell time t_d. Calculate the accidental coincidences and compare them with the measured count. Propagate errors: you can consider the statistics to be Poissonian, and so the error in the number of recorded counts N, is $\sqrt{N}$.

20A.2 The Photon Exists

Once down-conversion pairs are produced, one type of experiment that can be done involves sending one photon through a particular optical setup and have the other one go straight to a detector, heralding the presence of the first one. The coincidence detection makes this source mimic the ideal quantum source: a single atom undergoing a single photon transition. Such a quantum source will emit a single photon.

Theory

A photon incident on a beam splitter provides a definite proof that photons exist. Contrary to common belief, the photoelectric effect and even the Compton effect do not constitute a proof that photons exist because they can be explained using classical waves and quantum detectors. Photons are whole items. We detect either the whole photon transmitted or reflected. The alternative is that light is made of classical waves that split at the beam splitter, which would trigger simultaneous recordings at both sides of the beam splitter. This would, of course, negate the existence of photons with energy $h\nu$ (with ν being the frequency of the photon).

We measure this quantum probability via an anticorrelation parameter, also known as the degree of second-order coherence, $g_2(0)$. If we have detectors B and C at the two outputs of the beam splitter, then

$$g_2(0) = \frac{\mathcal{P}_{BC}}{\mathcal{P}_B \mathcal{P}_C}, \tag{20.13}$$

where P_B and P_C are the probabilities of detection at detectors B and C, respectively, and P_{BC} is the probability of detecting photons in both detectors simultaneously. Thus, for the ideal single-photon source $g_2(0)=0$. The equivalent measure for classical waves is obtained from the Cauchy-Schwartz inequality, giving $g_2(0) \geq 1$. A more complete description can be found in the physics education literature.[*] With heralded photons, we account for the probabilities in the following way. If the heralding photon is recorded at detector A, then based on the detections at detectors A, B, and C,

$$\mathcal{P}_B = \frac{N_{AB}}{N_A} \tag{20.14}$$

$$\mathcal{P}_C = \frac{N_{AC}}{N_A} \tag{20.15}$$

$$\mathcal{P}_{BC} = \frac{N_{ABC}}{N_A}. \tag{20.16}$$

Note that the last equation involves triple coincidences. The degree of second-order coherence then becomes

$$g_2(0) = \frac{N_{ABC} N_A}{N_{AB} N_{AC}}. \tag{20.17}$$

We can estimate the uncertainty based on simple propagation of errors:

$$\left(\frac{\Delta g_2(0)}{g_2(0)} \right)^2 = \left(\frac{\Delta N_{ABC}}{N_{ABC}} \right)^2 + \left(\frac{\Delta N_A}{N_A} \right)^2 + \left(\frac{\Delta N_{AB}}{N_{AB}} \right)^2 + \left(\frac{\Delta N_{AC}}{N_{AC}} \right)^2 \tag{20.18}$$

---

[*] J.J. Thorn et al., *Am. J. Phys.* **72**, 1210–1219 (2004).

The Poissonian statistical errors in counts $\Delta N_i = \sqrt{N_i}$, thus we can obtain the error

$$\Delta g_2(0) = g_2(0)\left(\frac{1}{N_{ABC}} + \frac{1}{N_{AB}} + \frac{1}{N_{AC}} + \frac{1}{N_A}\right)^{1/2} \tag{20.19}$$

Hanbury Brown and Twiss first conducted this type of experiment in 1956, and in doing so, inspired the field of quantum optics.

When considering these correlated-photon experiments, one might ask: why go through all of this trouble of doing experiments with heralded photons if a simple laser beam attenuated strongly enough can easily provide an average of one photon moving through the apparatus at any given time? The answer is that there is something misleading in the term "average." Using an attenuated laser source, one gets $g_2(0) = 1$! A statistical source of light behaves the same way. Of course, we believe in photons, as no experiment has proven the contrary. However, an attenuated laser source mimics a classical wave by the detections, causing coincidences, and so failing the beam-splitting test. This is because photons coming out of a laser are in a coherent state, which is an infinite superposition of states where more than one photon is emitted simultaneously. As a consequence, when the laser light is incident on a beam splitter, some photons may be transmitted and others reflected, mimicking the classical wave. Therefore, we cannot do experiments with an attenuated laser and claim that it involves one photon at a time. In contrast, heralded photons do so, and the Hanbury-Brown-Twiss experiment proves it.

Procedure

To do this experiment, we only need to add a few components to the previous one. These involve mostly the hardware for detecting photons, following a third path involving the reflection from a beam splitter. The parts that are needed are listed in Table 20.2 and refer to Figure 20.6.

1. Assemble the apparatus shown in Figure 20.6 by adding a beam splitter to the previous setup. Align it so that its sends reflected photons at 90° from the incident direction.
2. Add and align a third fiber collimator (C) to collect the light reflected by the beam splitter. Use the same collimator alignment technique done before.
3. The data acquisition setup has to be adjusted to detect double coincidences between A and B, and A and C; and triple coincidences between A, B, and C.
4. Accumulate data for a period of about 1 minute.
5. Determine the degree of second-order coherence. Propagate errors as well. State your conclusions.

20A.3 Questions

1. If a 50 mW pump beam with a wavelength of 405 nm is used to produce down-converted photons, a fraction of $\sim 10^{-8}$ photons get converted. How much power is in the down-converted

TABLE 20.2

Parts for Lab 20A.2

Qty	Part	Description/comments
1	Beam splitter	(BS) Non-polarizing.
1	Fiber collection assembly	(DC) Multimode fiber with FC type connectors, mounted collimator plus concentric iris mounted on the collimator mount.
1	Band-pass filter	800–810 nm center, 40 nm bandwidth.
1	Detector	Avalanche photodiode, fiber-coupled.

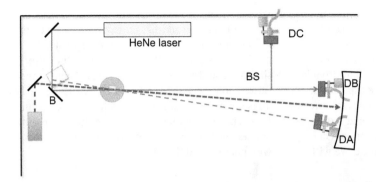

FIGURE 20.6 Apparatus to do the test that photons do not split at a beam splitter (BS). It requires three detectors A, B, and C, and the ability to detect double and triple coincidences.

light at 810 nm? (We note that a fraction of about 10^{-2} ends up headed toward the collimators, because they get produced into a cone.)

2. Down-converted photons are produced at random times but suppose that on average they are evenly separated in time. If 10^7 photons per second were to travel past the confines of the optical table, how far apart would consecutive photons be?

Lab 20B Entanglement and a Bell Test of Local Realism

The next experiment is a fundamental one in quantum physics. It is not your typical physics one where you measure something new or confirm a physical theory. Surely it has both those aspects of laboratory experimentation, but this experiment also settles a philosophical debate. Is nature realistic? That is, do objects have an inherent reality? Is a frog a frog? You would answer, "Of course it is!" Quantum mechanics says that too, but only because the frog has billions of atoms interacting with their surroundings. Suppose we settle for something simpler, say a single photon, and launch it by turning on a specially prepared source. The flying photon has definite properties: energy, momentum, and polarization, right? Well, that might not be strictly true. Quantum mechanics allows for superposition. A photon can be in a superposition of having two (or more) energies, say red and blue, or polarizations, horizontal and vertical. It can do so as long as the two possibilities are indistinguishable. That is, that once the photon is launched, it is impossible to distinguish those two properties. However, when we measure the property by something that distinguishes them, red from blue with a filter, or horizontal from vertical with a polarizer, the state of the photon snaps, randomly, into one of the possibilities. We measure this by doing repeated measurements of identical photons launched at different times. This is in contrast to what a realistic view would say: when the photons are launched they come out in a statistical mixture of being red or blue, or horizontal or vertical, but all along, once launched, the properties of the photon are the same and definite. Quantum mechanics says something very different: the reality of the color or the polarization of the photon is undefined until it is measured.

Let us take our argument one step further. Let us consider two photons, which can be horizontal or vertical, and launch them in a state where they are in a superposition of both being vertically polarized with both being horizontally polarized. In this case, the superposition carries further into two objects. Imagine that you and your twin sibling always dress the same (in principle you could have such a sibling – or maybe you do!). Some day you and your sibling go to class in a superposition of two outfits: both siblings wearing t-shirts and shorts, "and" dress shirt and pants. But these are special outfits, because neither you nor anybody else can tell either possibility. Maybe there is an undergarment that masks the texture and feel of the outfit, and there is a robe that covers it all, but neither can reveal the option, even to an X-ray machine or any other device (it is a thought situation after all). Then if the two possibilities are indistinguishable, you are in such a superposition: two outfits at the same time. Then both go to class, but in different buildings, and you pull the robe out so all classmates present, including yourself, can see what the outfit is. Quantum mechanics says that the uncovering of the outfit snaps into one of the possibilities, randomly. More strikingly, the outfit of your sibling, somewhere else on campus, snaps into the same one as you, under his or her robe, before being uncovered, instantaneously!

This argument was of much debate since 1935, until 1964 when John Bell devised a situation where these two distinct philosophical possibilities could be resolved by an experiment. Experimenters quickly set out to do the experiment. Today, after a long series of landmark experiments, the issue is all settled: quantum mechanics is correct. Nature is not realistic, and also "non-local" (the instantaneous snapping, or "spooky action at a distance"). However, it is stunning that one can do an experiment that tests this fundamental question. Our experiment here does precisely that: you will confirm that nature is not realistic and non-local. We will do this with photons and their polarization.

20B.1 Theory

In this experiment, we prepare the photon pairs in the entangled state

$$|\psi\rangle = \frac{1}{\sqrt{2}}\left(|H\rangle_1|H\rangle_2 + |V\rangle_1|V\rangle_2\right), \tag{20.20}$$

where we have used the Dirac notation to label the horizontal and vertical states of the photons 1 and 2. In the laboratory, we set up two crystals (already put together by the manufacturer) that are rotated 90° from each other. They are also very thin (0.5 mm or less). If we send a pump photon with its polarization at 45° to the horizontal, its horizontal component produces vertical pairs with one crystal, and its vertical component produces horizontal pairs in the second crystal. The orientation of the crystal is set up such that the pairs are created at exactly the same shallow angles. Thus, even if the photons are created in different crystals, it is inherently impossible to distinguish in which crystal they were produced, so we have a superposition. It is also known as an entangled state (or more technically, a non-separable state) because the state of the two photons cannot be factorized into a product of the state of each photon. This is the state of Equation (20.20).

One of the fundamental axioms of quantum mechanics is that a measurement projects the system into an eigenstate of the measuring device. The projection operator expresses a measurement leaving the system in a particular measuring state, which analytically is the outer product of the measuring state. A measurement on photon 1 using, for example, a horizontal polarizer is expressed by doing the projection of the initial state:

$$\widehat{P_H}|\psi\rangle = |H\rangle_1\langle H|_1\psi\rangle \tag{20.21}$$

$$= \frac{1}{\sqrt{2}}|H\rangle_1|H\rangle_2. \tag{20.22}$$

The measurement leaves the photons in the state where both are horizontally polarized. Analytically, it yields the final state multiplied by the probability amplitude of being in that state. The probability is the absolute value squared of the probability amplitude, which in the above case is 1/2. This is done without measuring the state of photon 2. It also brings the topic of non-locality, that a measurement on one particle instantaneously also projects the state of photon 2. We could have also obtained the joint probability by projecting the state of both photons. For example,

$$P_{HH/\psi} = \left|\langle H|_1\langle H|_2|\psi\rangle\right|^2 = 1/2 \tag{20.23}$$

$$P_{VV/\psi} = 1/2 \tag{20.24}$$

$$P_{HV/\psi} = 0 \tag{20.25}$$

$$P_{VH/\psi} = 0. \tag{20.26}$$

Let us consider the alternative to the entangled state: the mixed state. This state must be realistic and local. Consider the following reasoning: the photons are produced in only one crystal, both vertical if they are produced in one crystal, or both horizontal if they are produced in the other crystal. Since we do not know in which crystal they are produced, then we can conclude that they could both be either horizontally polarized or vertically polarized. Since only a small fraction of the photons from the incoming beam are down-converted, we can state that those produced in the first crystal do not affect those produced in the second crystal. Therefore, we could conclude that half the time the photons come out horizontally polarized, and the other half vertically polarized. Once the photons are produced, they are independent of each other: they get produced in a state of polarization, horizontal or vertical and continue to be so as they travel. Whatever happens to the partner, with the same polarization, does not matter. The photons in this mixed state are realistic because they have their polarization defined all along, and local because an arbitrary measurement on one photon does not correlate with a similar measurement on the other photon.

When photons in a mixed state reach two polarizers, one for each, we get the following probabilities:

$$P_{HH/m} = 1/2 \tag{20.27}$$

$$P_{VV/m} = 1/2 \tag{20.28}$$

$$P_{HV/m} = 0 \tag{20.29}$$

$$P_{VH/m} = 0. \tag{20.30}$$

This being again because they are both of either kind half the time. Note that the probabilities given for the mixed state are the same as those for the entangled state. However, a key distinction appears when we measure at other angles, such as diagonal or antidiagonal. These states are defined respectively by

$$|D\rangle = \frac{1}{\sqrt{2}}\left(|H\rangle + |V\rangle\right) \tag{20.31}$$

$$|A\rangle = \frac{1}{\sqrt{2}}\left(-|H\rangle + |V\rangle\right). \tag{20.32}$$

The probabilities of joint detection with the entangled state give:

$$P_{DD/\psi} \left|\langle D|_1 \langle D|_2 |\psi\rangle\right|^2 = 1/2 \tag{20.33}$$

$$P_{AA/\psi} = 1/2 \tag{20.34}$$

$$P_{DA/\psi} = 0 \tag{20.35}$$

$$P_{AD/\psi} = 0. \tag{20.36}$$

This is because if we express state ψ in the diagonal basis, we get

$$|\psi\rangle = \frac{1}{\sqrt{2}}\left(|D\rangle_1|D\rangle_2 + |A\rangle_1|A\rangle_2\right). \tag{20.37}$$

If the photons are in the mixed state, the results are different. The probability with the mixed state is:

$$P_{DD/m} = \frac{1}{2}\left|\langle D|_1 \langle D|_2 |H\rangle_1 |H\rangle_2\right|^2 + \frac{1}{2}\left|\langle D|_1 \langle D|_2 |V\rangle_1 |V\rangle_2\right|^2 = \frac{1}{4} \tag{20.38}$$

It can be shown that

$$P_{AA/m} = 1/4 \tag{20.39}$$

$$P_{DA/m} = 1/4 \tag{20.40}$$

$$P_{AD/m} = 1/4. \tag{20.41}$$

This difference is what is exploited in a Bell test of local realism. One problem with the state-vector formalism is that there is no state-vector expression for the mixed state.

Probabilities have to be calculated in an ad-hoc way, as shown in Equation (20.38) (that is, as a sum of probabilities). The density matrix formalism, not used here, is able to represent mixed states and so the probabilities can be calculated analytically in a more convenient way.[*]

One of the versions of Bell inequalities we present here goes by the name Clauser-Horne-Shimony-Holt (CHSH) inequality, after the names of the authors. We begin by defining a variable E that expresses the correlation between the polarizations of the two photons. If they are perfectly correlated, as is the case in state $|\psi\rangle$, then $E = 1$. If they are perfectly anticorrelated (see Question 3 in Section 20B.3 below), then $E = -1$. If they are uncorrelated then $E = 0$. The expectation value of the correlation can be defined for arbitrary angles θ_1 and θ_2 that the polarization axis of each state forms with the horizontal. At these angles E is

$$E(\theta_1, \theta_2) = (+1)P(\theta_1, \theta_2) + (+1)P\left(\theta_1 + \frac{\pi}{2}, \theta_2 + \frac{\pi}{2}\right)$$
$$+ (-1)P\left(\theta_1, \theta_2 + \frac{\pi}{2}\right) + (-1)P\left(\theta_1 + \frac{\pi}{2}, \theta_2\right), \tag{20.42}$$

(see Question 4 in section 20B.3below). For state $|\psi\rangle$ this simplifies to

$$E_\psi(\theta_1, \theta_2) = \cos\left[2(\theta_1 - \theta_2)\right] \tag{20.43}$$

The perfect correlation of the state $|\psi\rangle$ is manifested in Equation (20.43) because $E(\theta,\theta) = 1$ regardless of θ. In contrast, the mixed state gives a different correlation:

$$E_{mix} = \cos(2\theta_1)\cos(2\theta_2). \tag{20.44}$$

Therefore, one can find situations where the two expectation values give different results.

CHSH defined a variable S that depends on four angles. It is given by

$$S = E(\theta_1, \theta_2) - E(\theta_1, \theta_2') + E(\theta_1', \theta_2) + E(\theta_1', \theta_2'), \tag{20.45}$$

where θ_1 and θ_1' are two polarizer angles for photon 1, and similarly, θ_2 and θ_2' are two angles for photon 2. A local realistic theory satisfies[†]

$$|S| \leq 2. \tag{20.46}$$

Entangled states measured at certain angles will violate this inequality.

---

[*] E.J. Galvez, *Am. J. Phys.* **78**, 510–519 (2010).
[†] D. Dehlinger and M. Mitchell, *Am, J. Phys.* **70**, 903–910 (2002).

20B.2 Procedure

This experiment entails minor modifications of the previous one: the beam splitter and detector DC are removed, and the down-conversion crystal is switched for a different one. The parts that are needed are listed in Table 20.3 and refer to Figure 20.7.

As preparation for this experiment, all polarization elements have to be calibrated. Best results are obtained when their axes coincide with the zero axes of the rotational mounts. For example, the polarizer mount should read 0 when the transmission axis of the polarizer is horizontal. If this adjustment is not possible, then the horizontal or vertical settings should be recorded. Similarly, the angles of the waveplates' fast axis should be recorded. The down-conversion crystals and quartz crystal should also be square with the horizontal vertical axes.

The polarizers can be calibrated using the following method. Set the alignment laser beam to reflect off a piece of glass in a horizontal plane. When angle of incidence on the glass is set to the Brewster angle (approximately 56.6° for BK7 glass), the reflected light is polarized perpendicular to the plane of incidence (or vertically polarized). Placing a polarizer in the path of the reflected beam can be used to calibrate it. Once the polarizers are calibrated, placing optical elements in between crossed polarizers in the horizontal vertical settings helps in aligning the waveplates and crystal: when their axes are aligned with vertical or horizontal directions, no light should emerge from the second polarizer.

The instructions below assume that the pump beam is polarized horizontal, so make sure that this is the case. The experimental procedure involves the following steps:

1. Set up the apparatus as shown in Figure 20.7. Replace the crystal for the new one exactly in the same place and aligned as perpendicular to the pump beam as possible. You align this by making sure that the weak reflection of the pump beam off the face of the crystal heads straight back to the laser.

TABLE 20.3

Parts for Lab 20B

Qty	Part	Description/Comments
1	BBO crystal pair	(X) Two type-I BBO crystals 0.5 m thick, with 29° phase matching rotated 90° relative to each other and mounted on a rotation plus tilting (mirror-like) mount. Axes of the crystals must be in horizontal/vertical planes.
1	Half-wave plate	(V) 405 nm, zero order.
1	Compensating crystal	(U) Quartz, A-cut, 6–8 mm thick on a rotation mount, which can also be tilted about a vertical axis.
2	Half-wave plate	(H) 810 nm, zero order on a rotation mount.
2	Prism polarizer	(P) Glan-Thompson.

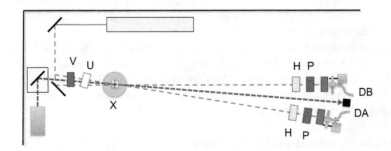

FIGURE 20.7 Apparatus for measuring entanglement and Bell Inequalities. It includes a half-wave plate (V) and quartz crystal (U) for the pump beam, and half-wave plates (H) and polarizers (P) for the down-converted photons. Down-conversion crystals are (X).

2. Polarizers (P) must be set to horizontal transmission. Half-wave plates (H) must be set to 45°, and half-wave plate (V) must be set to 0. This is set up to produce and detect down-conversion photons that are vertically polarized. The quartz crystal (U) must have its axis either vertical or horizontal.

3. Start the apparatus and tilt the crystal (X) about a vertical axis to maximize coincidences. A sign that pairs are being detected is that they have the same polarization. Thus, if you set one of the half-wave plates (H) to detect horizontally polarized photons (What angle would that be?), you should not detect any coincidences.

4. Set the half-wave plate (V) so that the pump beam is vertically polarized when it is incident on the crystal. Set the half-wave plates (H) to make sure horizontally polarized photons are detected. Tilt the crystal about a horizontal axis and maximize the down-conversion counts.

5. In Section 20B.1, we mentioned that when we produce the pairs, they are in an entangled state because we cannot distinguish from which crystal they were produced. However, in practice, the production in the two crystals is not symmetric: in one case, the pump beam creates pairs in the first crystal and the down-converted photons travel with a speed $v_{dc} = c/n_{dc}$ through the second crystal (c is the speed of light and n_{dc} is the index of refraction), taking a time $t_{dc} = d/v_{dc}$, where d is the crystal thickness. In the other case, the pump photon travels through the first crystal at a speed $v_p = c/n_p$ (n_p is the index of refraction), taking a time $t_p = d/v_p$, and the pairs are generated in the second crystal. For a given polarization state, in the crystal where the pairs are generated, the travel time is the same as that for the other polarization state in the other crystal, so they compensate; but in the crystal where the pairs are not generated, the differences in travel time are not the same for the two cases because $n_p > n_{dc}$. If we imagine the photons are wave-packets or pulses, the above difference amounts to a time delay $t_p - t_{dc} \sim 200$ fs, which introduces a temporal distinguishability (pairs produced in one crystal emerge sooner than those produced in the other crystal). The net effect is that it reduces the purity of the entangled state.

Moreover, this production of pairs in different crystals also translates into a phase shift δ, leaving the photons in the state:

$$|\psi'\rangle = \frac{1}{\sqrt{2}} \left(|H\rangle_1 |H\rangle_2 + |V\rangle_1 |V\rangle_2 \, e^{i\delta} \right). \tag{20.47}$$

Earlier we mentioned the correlations that are expected of this state when $\delta = 0$: when the polarizers are set to detect one photon in state $|D\rangle$ and the other photon in state $|A\rangle$, we should get no coincidences (Equations [33–36]). However, if $\delta \neq 0$ we can get coincidences. Worse yet, if $\delta = \pi$ we get that $P_{DA/\psi'} = 1/2$ (a maximum; see Question 2 in Section 20B.3 below).

The quartz crystal, oriented correctly, can be used to compensate for the temporal distinguishability and for setting δ to be 0. The quartz crystal's birefringence pre-delays in time one polarization component of the pump beam relative to the other. As a consequence, the pairs emerge at the same time, regardless of where they are created. It also introduces an additional phase shift. We can adjust δ via this phase shift by tilting of the quartz crystal (U) about a vertical axis.

Set up the pump beam to form 45° with the horizontal. Sometimes the crystal is more efficient at generating one set of pairs over the other. As a consequence, the coincidences for detecting photons in the states $|H\rangle_1 |H\rangle_2$ is not the same as when detecting them in states $|V\rangle_1 |V\rangle_2$. Adjust the angle of the waveplate (V) so that the coincidences for both states are about the same. This will involve an iterative procedure with the two polarization detection settings.

6. Set the half-wave plates (H) so that one photon is detected in state $|D\rangle$ and the other photon in state $|A\rangle$. If $\delta = 0$ there should be no coincidences. If there are, then very gently tilt (U) about a vertical axis until the coincidences are a minimum. If this minimum is not less than 25% of the coincidences that you get when you are set to detect $|H\rangle_1 |H\rangle_2$, then the quartz crystal may

be oriented to increase the delay between the two components of the pump beam instead of decreasing it to zero. If this is the case, then rotate the quartz crystal on its rotation mount by 90° and repeat the adjustment of δ to get a minimum of coincidences (zero ideally). If this is true, then you have an entangled state!

7. The measurement of the parameter S that provides the maximal violation of the Bell inequality of Equation (20.46), can be done for state of Equation (20.20) with the angles: $\theta_1 = 0°$, $\theta_1' = 45°$, $\theta_2 = 22.5°$, $a_2' = 67.5°$. Because of a lack of 100% detector efficiency, the correlation parameters have to be normalized as follows:

$$E\left(\theta_1,\theta_2\right) = \frac{N\left(\theta_1,\theta_2\right) + N\left(\theta_1 + \frac{\pi}{2}, \theta_2 + \frac{\pi}{2}\right) - N\left(\theta_1, \theta_2 + \frac{\pi}{2}\right) - N\left(\theta_1 + \frac{\pi}{2}, \theta_2\right)}{N\left(\theta_1,\theta_2\right) + N\left(\theta_1 + \frac{\pi}{2}, \theta_2 + \frac{\pi}{2}\right) + N\left(\theta_1, a_2 + \frac{\pi}{2}\right) + N\left(\theta_1 + \frac{\pi}{2}, \theta_2\right)}. \tag{20.48}$$

where $N(\theta_1,\theta_2)$ is the number of coincidences in a set amount of time when the polarizer angles are θ_1 and θ_2. Take the necessary measurements to calculate S in Equation (20.45) via Equation (20.48). If the accidental coincidences are significant, then they should be subtracted from the data. In that case, the singles counts should also be recorded.

8. An important quantity is the uncertainty in S, given by

$$\Delta S = \sqrt{\sum_{i=1}^{16}\left(\Delta N_i \frac{\partial S}{\partial N_i}\right)^2}, \tag{20.49}$$

where the uncertainties in the counts are due to statistical errors: $\Delta N_i = \sqrt{N_i}$. Note that the measurements involve four sets of four measurements. Measurements for each set define the value of one of the terms in Equation (20.45). Thus, if we label the 16 measurements, then

$S = E(N_1,N_2,N_3,N_4) - E(N_5,N_6,N_7,N_8) + E(N_9,N_{10},N_{11},N_{12}) + E(N_{13},N_{14},N_{15},N_{16})$, so the terms in the error formula are not so overwhelming:

$$\Delta N_i \frac{\partial S}{\partial N_i} = \sqrt{N_i} \frac{2\left(N_3 + N_4\right)}{\left(N_1 + N_2 + N_3 + N_4\right)^2}$$

for $i = 1, 2$, and

$$\Delta N_j \frac{\partial S}{\partial N_j} = \sqrt{N_j} \frac{-2\left(N_1 + N_2\right)}{\left(N_1 + N_2 + N_3 + N_4\right)^2}$$

for $j = 3, 4$. Other terms 5–16 are calculated the same way. Use this procedure to calculate the error in your determination of S. State your conclusions.

9. Now rotate the quartz waveplate about its horizontal axis by 90°. This will yield a slightly mixed state because the pairs will emerge from the down-conversion crystals at slightly different times, as mentioned earlier, and so they will be slightly distinguishable. That setting will produce a state that will not violate Bell. Verify this by repeating the previous 16 measurements and calculations.

20B.3 Questions

1. Derive Equation (20.37).

2. If the state $|\phi\rangle$ is given by

$$|\phi\rangle = \frac{1}{\sqrt{2}}\left(|H\rangle_1|H\rangle_2 - |V\rangle_1|V\rangle_2\right), \tag{20.50}$$

find the expression for the state in the diagonal basis.

3. If photons are in state

$$|\varphi\rangle = \frac{1}{\sqrt{2}}\left(|H\rangle_1|V\rangle_2 - |V\rangle_1|H\rangle_2\right), \tag{20.51}$$

(a) You can define a rotated basis $|H'\rangle$ and $|V'\rangle$, shown in Figure 20.8, where

$$|H'\rangle = \cos\theta|H\rangle + \sin\theta|V\rangle \tag{20.52}$$

$$|V'\rangle = -\sin\theta|H\rangle + \cos\theta|V\rangle. \tag{20.53}$$

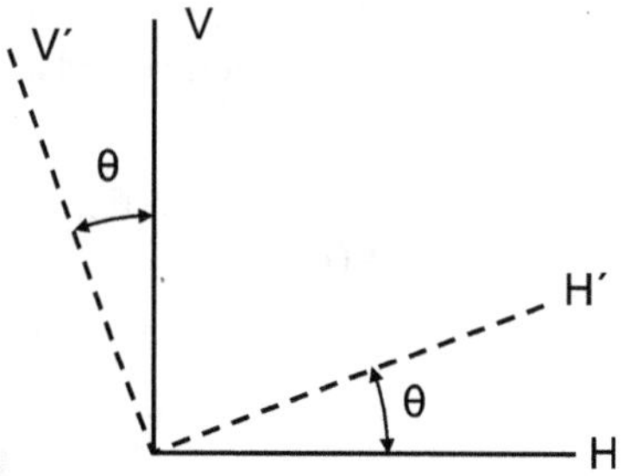

FIGURE 20.8 Two bases of state vectors rotated by an angle θ.

Express the state of Equation (20.51) in the rotated basis.

(b) Show that the photons in state of Equation (20.51) are always perpendicular to each other, regardless of the basis used.

4. A polarizer-oriented angle θ projects the light onto state

$$|\theta\rangle = |H'\rangle = \cos\theta|H\rangle + \sin\theta|V\rangle.$$

(a) Find an expression for the joint probability $P_\phi(\theta_1,\theta_2)$ for state $|\Psi\rangle$ of Equation (20.20).

(b) Derive the expression of Equation (20.43) for this state.

(c) Derive the correlation for mixed states of Equation (20.44).

5. Show that the inequality of Equation (20.46) is violated for state ψ when $\theta_1 = -\pi/4$, $\theta_1' = 0$, $\theta_2 = -\pi/8$, and $\theta_2' = \pi/8$

6. Show that the inequality of Equation (20.46) with E_{mix} of Equation (20.44) is satisfied when $\theta_1 = -\pi/4$, $\theta_1' = 0$, $\theta_2 = -\pi/8$, and $\theta_2' = \pi/8$

Lab 20C Single-Photon Interference and the Quantum Eraser

This experiment examines quantum interference in a vivid way: one photon at a time. The apparatus from the previous experiments has to be modified to include an interferometer. The experiment also serves to exercise the application of quantum algebra to an experiment. We will use the same down-conversion as Lab 20A (single crystal, no polarization entanglement).

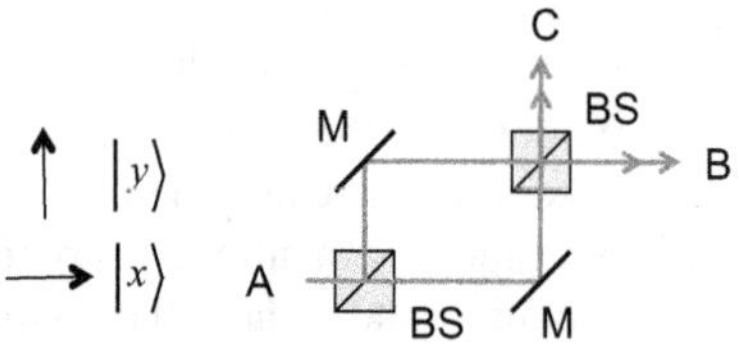

FIGURE 20.9 Schematic of a Mach-Zehnder interferometer. Optical elements are mirrors (M) and beam splitters (BS). Photons go from A to either B or C. The propagation directions are denoted by corresponding kets.

20C.1 Theory

Interference

We begin with a Mach-Zehnder interferometer. This is a rectangular interferometer that has two symmetric beam splitters and two mirrors arranged as shown in Figure 20.9. We use the Dirac notation to represent the states. To facilitate the algebra, we will use the matrix notation. Since the light can only go in two directions, we label those directions by x and y. The matrix representation of the state vectors is then

$$|x\rangle = \begin{pmatrix} 1 \\ 0 \end{pmatrix} \tag{20.54}$$

and

$$|y\rangle = \begin{pmatrix} 0 \\ 1 \end{pmatrix}, \tag{20.55}$$

The beam splitters mentioned earlier either reflect or transmit a photon. They can be represented by a unitary operator. The operator for a symmetric beam splitter applies to the state vectors:

$$\hat{B}|x\rangle = r|y\rangle + t|x\rangle \tag{20.56}$$

$$\hat{B}|y\rangle = t|y\rangle + r|x\rangle. \tag{20.57}$$

From the above relations, we can construct the matrix representation for the beam splitter operator:

$$\hat{B} = \begin{pmatrix} t & r \\ r & t \end{pmatrix}, \tag{20.58}$$

The 90° mirror is represented by the operator $\hat{M}$, which transforms one state into the other. It acts as the NOT operator for our basis (see also Question 1 in Section 20C.3 below):

$$\hat{M}|x\rangle = |y\rangle \tag{20.59}$$

$$\hat{M}|y\rangle = |x\rangle \tag{20.60}$$

The arms of an interferometer may not have the same length, and so the light accumulates a phase that depends on the path in which it travels. We can account for the effect of the phase by using a unitary phase-shift operator

$$\hat{G} = \begin{pmatrix} e^{i\delta_1} & 0 \\ 0 & e^{i\delta_2} \end{pmatrix}, \tag{20.61}$$

where $\delta_i = 2\pi \ell_i / \lambda$ with ℓ_i being the length of the path traveled in arm i. The phase operator $\hat{G}$ is diagonal, so the state vectors are eigenstates of the phase-propagating operator. The state of the light after the interferometer is

$$|\phi_0'\rangle = \hat{Z}|\phi_0\rangle, \tag{20.62}$$

where

$$\hat{Z} = \hat{B}\hat{G}\hat{M}\hat{B}. \tag{20.63}$$

The probability of a photon entering the interferometer through port A (in state $|x\rangle$) and exiting the interferometer through port B (also in state $|x\rangle$), as shown in Figure 20.9, is

$$\mathcal{P} = \left| \langle x | \hat{Z} | x \rangle \right|^2. \tag{20.64}$$

It can be shown (see Question 3 in Section 20C.3 below) that the probability is given by

$$\mathcal{P}_{AB} = \frac{1}{2}\left(1 + \cos\delta\right), \tag{20.65}$$

where $\delta = \delta_1 - \delta_2 = 2\pi\Delta L/\lambda$, with $\Delta L = \ell_1 - \ell_2$. The intuitive way to understand this result is Feynman's approach:[*] "if an event can occur in two (or more) indistinguishable ways, then the probability of the outcome is the square of the sum of the amplitudes for each case." In this case, the probability amplitudes have a phase due to the traveled paths, so the sum of probability amplitudes gives rise to interference: quantum interference.

Distinguishability

In the quantum eraser, we use the polarization degree of freedom in addition to the direction of propagation. The polarization space is two-dimensional. A convenient basis is the one where photons are either horizontally or vertically polarized, which we denote by $|H\rangle$ and $|V\rangle$, respectively. There are two important operators for polarization: The polarizer and the waveplate. In particular, the half-wave plate can rotate the orientation of the polarization. Let us focus on the case where the optic axis of the waveplate forms 45° with the horizontal. In this case, it acts as a NOT gate for the horizontal and vertical polarization eigenstates:

$$\hat{H}_{45°}|H\rangle = |V\rangle$$

$$\hat{H}_{45°}|V\rangle = |H\rangle$$

We send a photon through the interferometer, but past the second mirror we put in one of the arms an element that flips the polarization (the half-wave plate set to 45°). To account for the two degrees of freedom we need to combine the two spaces. Algebraically this is represented by the tensor product of the spaces. The input state is now

$$|\psi_i\rangle = |x\rangle \otimes |V\rangle, \tag{20.66}$$

where we have set the polarization of the input state to be vertical, and symbol "$\otimes$" represents the tensor operation for the two states that live in different Hilbert subspaces. After the first beam splitter, the state of the photon is

$$|\psi^i\rangle = \left(\hat{B} \otimes \hat{I}\right)\left(|x\rangle \otimes |V\rangle\right) = r|y\rangle|V\rangle + t|x\rangle|V\rangle, \tag{20.67}$$

and where in the last step we have dropped the tensor-product symbol for simplicity, as done with the polarization-entangled photon pairs. After the mirrors, the state is

$$|\psi^{ii}\rangle = \hat{M} \otimes \hat{I}|\psi^i\rangle = r|x\rangle|V\rangle + t|y\rangle|V\rangle. \tag{20.68}$$

After the half-wave plate in one of the arms, and including the phase due to travel, the state is:

$$|\psi^{iii}\rangle = r\hat{G}|x\rangle\,\hat{H}_{45°}|V\rangle + t\hat{G}|y\rangle|V\rangle = r|x\rangle|H\rangle e^{i\delta_1} + t|y\rangle|V\rangle e^{i\delta_2} \tag{20.69}$$

Past the second beam splitter, the state is then

$$|\psi^{iv}\rangle = r^2|y\rangle|H\rangle e^{i\delta_1} + rt|x\rangle|H\rangle e^{i\delta_1} + t^2|y\rangle|V\rangle e^{i\delta_2} + tr|x\rangle|V\rangle e^{i\delta_2}. \tag{20.70}$$

---

[*] R.P. Feynman, R.B. Leighton, and M. Sands, *Lectures in Physics* V. 3 (1963).

If we put a detector on the port B, the detection amounts to a projection onto state $|x\rangle$, or

$$|x\rangle\langle x|\psi^{iv}\rangle = |x\rangle\left(rt\langle H|e^{i\delta_1} + tr\langle V|e^{i\delta_2}\right). \tag{20.71}$$

The probability is then

$$\mathcal{P}_x = \left\||x\rangle\langle x|\psi^{iv}\rangle\right\|^2 = \frac{1}{2}. \tag{20.72}$$

Because the probability is constant there is no interference. In Feynman's words: "if the experiment is capable of determining whether one or another alternative path is actually taken, the probability of the event is the sum of the probabilities for each alternative." That is, the polarization labels the two paths or alternatives, and so the probability amplitudes are squared before adding, and there is no cancellation. Our algebra agrees with this concept because the states $|H\rangle$ and $|V\rangle$, conveyors of the path information, are orthogonal.

Erasure

If after the interferometer we place a polarizer forming 45° with the horizontal, then the polarizer projects the polarization states to state $|D\rangle$ of Equation (20.31). After this projection, the probability of detecting the photon is

$$\mathcal{P}_x = \left\||x\rangle\left(rt|D\rangle\langle D|H\rangle e^{i\delta_1} + tr|D\rangle\langle D|V\rangle e^{i\delta_2}\right)\right\|^2 = \frac{1}{4}\left(1 + \cos\delta\right). \tag{20.73}$$

Interference reappears! In Feynman's language, the polarizer erased the path information, so the alternatives could interfere again. Mathematically, the polarizer projected equally the two orthogonal states onto the state $|x\rangle|D\rangle$, retaining their phases, and leading to interference.

Interference and non-interference are often attached the labels of "wave" and "particle." Niels Bohr argued for the principle of complementarity, that wave and particle aspects of matter are complementary to each other. Thus, in interference we see the wave aspect, and in non-interference we see the particle aspect, but we cannot see the two aspects simultaneously. What is somewhat puzzling in this experiment is that we get to decide whether we see wave or particle aspects *after* the photon went through the interferometer.

Single Photons

If you recall Lab 20A, we did a measurement where twin photons take distinct paths, with one photon going toward a detector and the other photon going toward a beam splitter. One way to refer to the photons heading to the beam splitter is to say that they are heralded. That is, that their presence is announced by their twin partner. Coincidence detection ensures that we only consider this situation. The results are then interpreted as single photons heading toward a beam splitter. The results are consistent with this: either the detector after the transmitted path or the one after the reflected path clicks, but not both. If you recall a previous discussion, photons from a laser do not show this result because we never know if one or more than one photons are traveling at the same time, and so both detectors are likely to click. In this experiment, we do the same with heralded photons: they head to the interferometer, and their announcement by the presence of the twin partner ensures that the experiment is one where single photons travel through the interferometer. Any interference seen will then be due to the interference of a single photon with itself.

Coherence Length

The photons that are detected first pass through band-pass filters. That is, their energy is uncertain by the bandwidth of the filters, specified by $\delta\lambda$. Because the energy of the photons is indistinguishable within the bandwidth of the filter, we can then think that the photon is a wavepacket: a superposition of energy states within the bandwidth of the filter. The frequency of the light ν is related to the wavelength by $\nu = c/\lambda$, where c is the speed of light. Thus, the uncertainty in the frequency is related to the uncertainty in the wavelength by $\delta\nu = c\lambda^2/\delta\lambda$. Consider now that the photon is an energy wavepacket. By the Fourier theorem, this corresponds to a temporal wavepacket – a pulse in time, with a temporal width $\delta t = 1/\delta\nu$. This is a property of waves, although some argue that it is fundamentally tied to the Heisenberg uncertainty principle. The length of the wavepacket, known as the coherence length is then given by

$$\ell_c = c\delta t = \frac{\lambda^2}{\delta\lambda}. \tag{20.74}$$

The value of the coherence length is relevant to this experiment because for interference to occur, the interferometer paths have to be indistinguishable. If the path-length difference is greater than the coherence length, then the paths are distinguishable because a timing measurement could distinguish the path taken by the photon. In other words, the photon traveling through the shorter path will arrive distinguishably sooner than when traveling through the longer path. Thus, for interference to occur, the path-length difference must satisfy

$$\Delta L < \ell_c. \tag{20.75}$$

20C.2 Procedure

This setup adds to the setup of Lab 20A in which we use a single crystal, but if this follows Lab 20B, you need to switch the crystal to the one used previously. The added components (Table 20.4) are those that constitute an interferometer. The alignment is delicate but feasible,[*] so instructions must be followed closely. This is because the down-converted photons have a wide bandwidth: $\delta\lambda = 40$ nm. This corresponds to a coherence length $\ell_c = 16$ μm. So, to see interference, we must align the interferometer to have arm lengths equal to within 16 μm!

The first steps in the set up involve placing each element of the interferometer and aligning it, as shown in Figure 20.10a. The interferometer must be as small as possible, making room only for a waveplate in

TABLE 20.4

Parts for Lab 20C

Qty	Part	Description/Comments
2	Non-polarizing beam splitters on pedestal (or similar) mounts	(G) They must be identical; purchased together.
1	Mirror with pedestal (or similar) mount	(I) Pedestals are very stable.
1	Mirror on a translation stage	(J) Identical to the previous mirror.
1	Piezo-electric	To be used as spacer between the translation stage and its micrometer.
1	×10–15 voltage amplifier	For driving the piezo-electric.
1	Fiber-coupled spectrometer	One that displays the spectrum on a screen.
1	Small incandescent bulb	6-V bulb will suffice.
1	Diverging lens and mount	(L) To inspect the interference pattern produced by the interferometer.
2	Half-wave plates in rotational mounts	(H) 810 nm, zero order.
1	Polarizer	(P) Aligned for transmission of diagonally polarized light.

---

[*] E.J. Galvez et al., *Am. J. Phys.* **73**, 127–140 (2005).

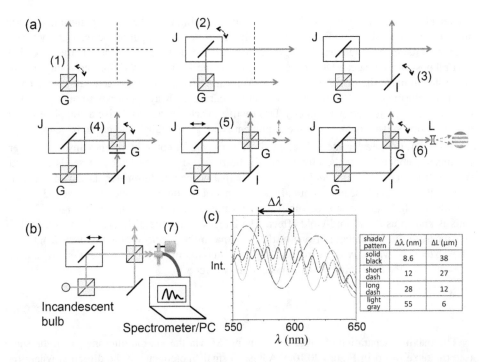

FIGURE 20.10 (a) Step-by-step schematic of the set up and alignment of the interferometer. (b) Schematic of the final adjustment of the interferometer using a spectrometer. Numbers within parenthesis refer to steps in the procedure. (c) Modulated spectrum of continuous light passing through the interferometer with different path lengths ΔL.

each arm. An alignment laser beam will be used for the alignment of the components, one at a time. A preliminary step is to make sure that this laser beam is parallel to the rows of holes of the breadboard.

Alignment

1. Insert first beam splitter (G) as shown in Figure 20.10a. Adjust its position so that the alignment laser beam is centered on the beam splitter. Align the 90° reflection of the alignment beam to be parallel to the holes of the breadboard.

2. Insert the mirror on the linear stage (J) and align the 90° reflection. The piezo-electric spacer should be inserted before aligning the mirror.

3. Insert the mirror on mount (I) and align the 90° reflection. After this step, the two beams should cross in free space at the same height. Verify this.

4. Insert second cube beam splitter (G). Position it so that the two beams cross at the cube's diagonal beam-splitting layer. Faint scattering off this layer helps with this. Block one of the beams and align the 90° reflection off the beam splitter. After this alignment, the beams coming from the two arms are parallel to each other and at the same height, but perhaps not overlapping.

5. Adjust the micrometer of the translation stage, which displaces one of the beams. Do this adjustment so that the two beams fully overlap on a screen placed after the interferometer.

6. Add a diverging lens (L) after the interferometer to expand the beam. Fringes should be seen. The tilt of the beam splitter about a vertical axis controls the vertical tilt of the fringes. That is, when the two beams form an angle relative to each other in a horizontal plane, the fringes are vertical. By tilting the beam splitter about the vertical axis in the wrong direction increases the density of fringes. Tilting the beam splitter in the opposite (correct) direction makes the vertical fringes broader and farther apart. Do so until you see perfectly horizontal fringes. Tilting the beam splitter about a horizontal axis changes the density of the horizontal fringes. Adjust this tilt so that these fringes get broad and further apart from each other. You may have to iterate

between horizontal and vertical tilts but continue adjusting until the whole pattern is uniform and has no fringes. Lean on the breadboard slightly to see the output of the interferometer get brighter or darker as a sign that the degree of interference (constructive or destructive) is varied. If at all possible, connect a DC power supply to the piezo-electric. Varying the voltage on the piezo should make the interference change. You can also use a function generator set to triangular waveform (with a DC offset set so that the output is always positive), to make sure that the piezo-electric is working properly. Destructive interference should give a completely dark output. A bull's-eye indicates that the alignment is not good enough and has to be redone.

7. At this point, the interferometer is aligned. Be very careful not to bump into any of the interferometer components or the alignment will be ruined. Place the fiber that sends light into the spectrometer so that it is looking into the output of the interferometer, as shown in Figure 20.10b. Place a small incandescent bulb at the input of the interferometer. Observe the fringes modulating the continuous spectrum of the incandescent light, as shown in Figure 20.10c. This is known as the Alford-Gold effect. The fringes reveal the difference in path length. If you measure the wavelength of one of the maxima in the spectrum, then its wavelength satisfies $\Delta L = m\lambda_1$, where m is an integer. The next maximum with higher wavelength satisfies $\Delta L = (m-1)\lambda_2$. By combining these two equations we get

$$\Delta L = \frac{\lambda_1 \lambda_2}{\lambda_2 - \lambda_1}. \tag{20.76}$$

The maxima separation changes by changing ΔL via the via the micrometer on the stage (see the table insert in Figure 20.10c). Adjust in small increments in the direction where the separation between maxima increases. Do so until the entire spectrum is one giant maximum or minimum. Remove the bulb and spectrometer and allow the alignment beam to go through the interferometer. Check that the interferometer is still aligned, adjusting the second beam splitter if not.

8. With the alignment beam still on, place the collimator so that the beam enters the fiber, as done in previous experiments. Once this is done, place the filter on the collimator and turn the alignment laser off.

Interference

9. Turn off the lights and turn on the detectors. Adjust the collimator to maximize the signal after the interferometer and the coincidences. It is useful to block one of the arms of the interferometer so that changes in the alignment are not confused with interference. Do a scan of the piezo to see oscillations in the coincidences AB and the singles from the detector after the interferometer (B). Best fringes are those where the minimum in the interference is as close to zero as possible. This is single-photon interference!

10. An interesting variation to this experiment is to add a third collimator (C, as in Lab 20A) and locate it to receive the light from the second port of the interferometer. Then when the voltage on the piezo is scanned, the coincidences AC should be the opposite of those of AB.

11. We could take advantage of this opportunity and do a Hanbury-Brown-Twiss measurement, similar to Lab 20A, but now with the interferometer serving as a beam splitter. Thus, we should not see triple coincidences that are not accidental. We see fringes in AB and in AC, but they are not created by the same photon because each photon is detected whole in either B or C. This result also gives one of the most intriguing aspects of the experiment: photons interfere like waves but are yet detected whole.

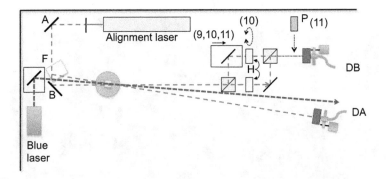

FIGURE 20.11 Schematic of the setup to recreate the quantum eraser. The numbers in parenthesis refer to the numbered steps in the procedure. Additional optical elements include half-wave plates (H) and a polarizer (P).

Erasure

12. Place half-wave plates in each arm of the interferometer and set both to 0°. We use two half-wave plates because one alone would imbalance the optical path length of the two arms. One of them serves only to compensate for the added path length introduced by the other one. The full setup should look like the one in Figure 20.11.

13. Do a piezo scan. It should show prominent fringes.

14. Rotate one of the waveplates by 45°. This will rotate the polarization of the photon going through that arm by 90° to become horizontal. Redo the scan. It should show no fringes. In reality, making the two polarizations exactly orthogonal is very difficult, so one may see slight oscillations. Comment on the results.

15. Place the polarizer (H) tilted 45° after the interferometer. This is the eraser. Redo the scan. Explain your observations.

20C.3 Questions

1. Use Equations (20.59–20.60) to obtain a matrix for the mirror operator $\hat{M}$.

2. Find an expression for $\hat{Z}$.

3. Derive the results of Equations (20.65, 20.72 and 20.73).

4. When $\delta = n\pi$, with n odd, the probability in Equation (20.65) is 0. Where does the energy go? It must be to C. Find the probability of the photon going from A to C.

5. Derive Equation (20.76).

FIGURE 20.12   Enrique Galvez working with a student.

Dr. Enrique J. "Kiko" Galvez is a Professor of Physics and Astronomy at Colgate University, where he has been for the last 30 plus years. His interests straddle between fundamental research and educational-laboratory development. He studies light, from either a quantum perspective, investigating the intricacies of quantum physics, or from a classical picture, where light's description contains singularities and other mathematical topologies. He is coauthor of three textbooks and numerous journal articles with student coauthors. In 2010, he received the American Physical Society (APS) Prize for a Faculty Member for Research in an Undergraduate Institution. In 2020, he received the APS Jonathan F. Reichert and Barbara Wolff-Reichert Award for Excellence in Advanced Laboratory Instruction (Figure 20.12).

21

Nuclear and Particle Physics

Brett Fadem

CONTENTS

Who ordered that?

– I.I. Rabi* (in response to the discovery of the muon)

21.1   The Standard Model

Particle physics is the study of the fundamental interactions that govern the behavior of nature's most elementary particles. The term "elementary particle" in this context is meant to denote entities without substructure and out of which all other composite systems are made. As a counterexample, even though protons are often referred to as "particles," they are themselves extremely complicated systems made of elementary particles.

The underpinning of our current models is quantum field theory, which itself is a merging of quantum mechanics and special relativity. The "standard model" consists of a set of quantum field theories that describe all the known fundamental forces except gravity. The elementary particles that form the players in these theories are often listed in a table, organized according to the interactions in which they take part, as shown in Figure 21.1.

The particles are arranged into broad categories. Entries in the first four columns are considered "matter particles," while the fifth column represents "force carriers" that mediate interactions between the matter particles. We think of the force carriers as messengers exchanged between the matter particles (on the other hand, force carriers are particles in their own right – the photon being the most famous example).

The matter particles fall into two categories: quarks and leptons. The term used to distinguish the different types of quarks is "flavor." The quark flavors are up, down, charm, strange, top, and bottom. While leptons are observed in isolation, with the exception of quarks and gluons in a quark-gluon plasma, quarks are always bound together to form composite entities called hadrons.

Hadrons, in turn, come in two varieties: baryons and mesons. More on these below.

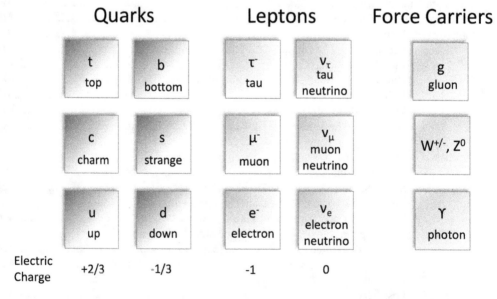

FIGURE 21.1   A table listing the known elementary particles. For every particle in the table except the neutrinos, the masses are known but not listed. The recently discovered Higgs boson is also not listed, but belongs among the force carriers in the final column. The intrinsic angular momentum (or "spin") of the particles is also not listed. For every particle above, there is a corresponding antiparticle with opposite charge. In some cases, such as the photon, particles are their own antiparticles.

---

* Rabi won the Nobel Prize in 1944 for nuclear magnetic resonance. He also was involved in many other important physics breakthroughs, both on the scientific and political side, including the creation of Brookhaven National Lab and CERN.

The Electric Charge

Underneath each of the first four columns, the electric charge of each particle in that column is listed. The only force carriers with a non-zero electric charge are the W+ and the W−. Incidentally, mass can be considered the "charge" of the gravitational field (remember, though, that gravity is not part of the standard model, and the presumed force carrier associated with it, the graviton, is not listed on the table). However, since the electromagnetic force is but one of those covered in the standard model and there are other forces and associated fields, the particles in the table have other types of charge. More to come on this when we discuss the fundamental forces below.

The Particle Masses

The masses of the elementary particles are also not listed in the table. In the case of the neutrinos, only upper limits on the masses are known. For decades they were treated as massless, but one of the major accomplishments in recent experimental particle physics was the discovery that the probability of obtaining a certain type of neutrino upon measurement varies as the neutrino travels along its way. This is called "neutrino oscillation" and would be impossible for massless neutrinos given current theory.

While the masses of the particles aren't listed in the table, in the case of the quarks and leptons, the rows have special significance. The bottom row is termed the first generation. These are the least massive (or "lightest") quarks and leptons: the up and down quarks, electron, and (presumably) electron neutrino. The middle row contains the second generation or mid-mass quarks and leptons, and the top row contains the third generation or greatest mass quarks and leptons. These masses vary by orders of magnitude between the first and third generations.

The Intrinsic Angular Momenta or "Spins"

Elementary particles carry an intrinsic angular momentum ("spin"). Elementary matter particles possess so-called ½ integer spin, while force carriers possess integer spin. More on this to come.

"Color" and "Weak" Charge

In addition to electric charge, quarks have "color" charge. The term "color" here is used metaphorically and has nothing to do with the actual colors of visible light. For that matter, the so-called "weak" force can act on any quark or lepton. While students are taught in introductory courses that forces are pushes and pulls, when acting on quarks, the weak force can change one type of quark into another and is thus sometimes referred to as the flavor changing force.

The Fundamental Forces

It is the underlying interactions or forces that make sense of the entries in the table of elementary particles. There are four known fundamental forces: electromagnetic, weak, strong, and gravitational. The theories of the standard model were developed to characterize the first three of these forces. In fact, the first two are now described by a single theory.

Just as there was a time, centuries ago, that electric and magnetic phenomena were considered distinct, electromagnetic and weak forces are typically listed separately. Electric and magnetic phenomena were unified under Maxwell's theory of classical electromagnetism in the 19th century, and more recently under the quantum field theory called quantum electrodynamics (or "QED"). QED remains the archetype of a quantum field theory. After its development, a quantum field theory with a unified explanation for both electromagnetic and weak phenomena was developed, called "electroweak" theory. Considered from this perspective, the separate electromagnetic and weak forces have been combined under a single electroweak description.

The so-called nuclear force is now understood to be a manifestation of the strong force that governs interactions between quarks. Quarks can have any one of three charges, and antiquarks can have any one of three anti-charges. Because of the way composite particles arise out of combinations of these quarks, we label the three charges "red," "green," and "blue," and refer to the strong force as the color force. The associated quantum field theory is called quantum chromodynamics (or "QCD").

Our list of fundamental forces and the theories that describe them takes the form shown in Figure 21.2.

There is a glaring inconsistency in our treatment of the first three forces (electromagnetic, weak, and strong) and the fourth, gravity. While the theories used to describe the first three forces are quantum field theories (based on the principles of both quantum mechanics and special relativity) the premier theory of gravity is not a quantum mechanical theory. There are no probability amplitudes in general relativity, no uncertainty principle, nor any other distinguishing feature of quantum mechanics. While electroweak theory and quantum chromodynamics (or "QCD") are both quantum field theories, there is no quantum field theory for gravity. The dream of a "final theory" unifying all four forces remains an active front of contemporary physics, but to this point, no serious contender has been validated by experiment. So, just as there is a seeming rift between our treatment of macroscopic (classical) phenomena and microscopic (quantum mechanical) phenomena, there is an actual rift between the theory of general relativity and the best experimentally corroborated quantum field theories.

Force	Theory
1. Electromagnetic	Electroweak
2. Weak	
3. Strong	QCD
4. Gravitational	General Relativity

FIGURE 21.2 Table of fundamental forces and the theories that describe them.

Thus, electroweak theory and QCD together (or perhaps QED, electroweak theory, and QCD) form the standard model of particle physics, while general relativity stands apart as a separate jewel in the crown of contemporary physics.

Unification of the strong force and the electroweak interaction under a single theory would itself constitute an extension to the standard model. Such theories are characterized as grand unified theories (or "GUTs") while theories that aim to explain all four forces have earned the impressive title "theories of everything" (or "TOEs").

There are currently a number of phenomena that are not explained by the standard model. For example, we neither understand the nature of dark matter nor the source of the accelerating expansion of the universe, termed "dark energy." So, the vast majority of stuff in the universe does not appear to have a place on our table of elementary particles!

Hadrons: Particles Made Out of Quarks

Quantum chromodynamics describes the interactions of quarks and yet, under ordinary conditions, it isn't the quarks that are observed. Instead, quarks combine in ways that can be described based on their "color" charges. The rule is that the composite particles formed by quarks must be colorless. There are two ways to do this, as shown in Figure 21.3: red, green, and blue quarks can combine to make colorless objects called baryons, or a given color and its anti-color (red and anti-red, blue and anti-blue, or green and anti-green) can combine to again make colorless objects called mesons. Anti-colored quarks are the antimatter versions of the corresponding colored quark objects. One can also form an antimatter baryon, or anti-baryon, from the combination of an anti-red, anti-green, and anti-blue quark.

The most famous hadron is the proton. A member of the baryon family, and the only stable hadron, the so-called "valence" quarks in the proton are two up quarks and one down quark (thus providing the +1 charge of the proton). Note again that by combining three quarks (red, green, and blue) a colorless object is formed. A full

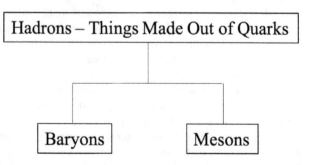

FIGURE 21.3 Quarks combine to make either baryons or mesons. Baryons consist of combinations of three quarks (red, green, and blue). The most famous examples are protons and neutrons. Mesons consist of quark–antiquark pairs. The most famous example is the pion.

description of the proton, however, goes beyond its valence structure. The proton also consists of gluons and virtual quarks popping in an out of existence. The detailed structure of the proton is an active area of research.

The neutron is also a baryon consisting of two down quarks and one up quark, forming a colorless object with a net electric charge of zero. While neutrons are not stable in isolation (precise measurements of their lifetime have also been an active area of research) they are, obviously, stable in some atomic nuclei.

The mesons are less well-known because none are stable and they don't form the underpinning of atoms in the same way as protons and neutrons. The most famous example of a meson is the pion. The pion comes in three varieties (π^+, π^0, and π^-) and consist of quark–antiquark pairs using up and down quarks.

In addition to three quarks coming together to form baryons and quark–antiquark pairs forming mesons, combinations consisting of four quarks and one antiquark can satisfy the rule that composite particles are colorless. The LHCb collaboration, for example, has observed such exotic baryon states.[*]

Interestingly, the muons that form the subject of this chapter's experiment are not made of quarks at all! They can be found in the leftmost lepton column of Figure 21.1. They appear in the row above the electron and, indeed, have properties that are very similar to the electron, except they are about 200 times more massive. More on them to come.

21.2 Active Areas of Research in Nuclear and Particle Physics

Active Areas in Nuclear Physics

With regard to nuclear physics, the 2015 Nuclear Science Advisory Committee created a document to guide the progress of nuclear physics over the coming decade. In their document, they cite the science questions that "animate nuclear physics today" as outlined by a preceding committee:

1. How did visible matter come into being and how does it evolve?
2. How does subatomic matter organize itself and what phenomena emerge?
3. Are the fundamental interactions that are basic to the structure of matter fully understood?
4. How can the knowledge and technical progress provided by nuclear physics best be used to benefit society?

The full recommendations are available via this section's page at ExpPhys.com.

The nuclear physics and particle physics communities are distinct. Nuclear physicists, true to their name, study the properties of nuclear matter. Although their principle preoccupation is with the theory of quantum chromodynamics, they frequently use nuclear systems to investigate physics beyond the standard model.

Active Areas of Particle Physics

Particle physicists are concerned with the standard model and physics beyond the standard model. Whereas nuclear physicists occupy themselves with the composite systems formed by quarks and gluons, particle physicists focus on investigations of the individual subatomic particles and their interactions.

The Particle Physics Project Prioritization Panel (P5) has developed a long-range plan similar to that of the nuclear physics community.

In the executive summary of the report, the following priorities are laid out:

- Use the Higgs boson as a new tool for discovery
- Pursue the physics associated with neutrino mass
- Identify the new physics of dark matter
- Understand cosmic acceleration: dark energy and inflation
- Explore the unknown: new particles, interactions, and physical principles

The full recommendations are available via this section's page at ExpPhys.com.

---

[*] R. Aaij et al. (LHCb Collaboration), *Phys. Rev. Lett.* 115, 072001 (2015).

Opportunities for Student Research

A great deal of research in nuclear and particle physics is undertaken at large laboratories such as the national laboratories in the United States and CERN in Switzerland where large particle accelerators are used to probe nuclear and particle interactions. However, there are a large number of nuclear and particle physics experiments deployed in settings apart from human-made particle accelerators. For example, a variety of cosmic ray detectors are deployed on the International Space Station, and a number of ground and underground experiments probe cosmic rays, cosmic ray showers, and neutrinos.

In most cases, the scale necessary to mount and conduct these experiments requires collaborations of scientists from multiple institutions. These collaborations vary from just a few scientists on smaller experiments, to thousands of scientists on the larger experiments. If one's local institution doesn't have physicists involved in this research, there are still a large number of summer REU (Research Experience for Undergraduates) positions that provide valuable research opportunities for undergraduates at cutting-edge facilities. Students often get to spend summers at national laboratories in the United States, at CERN in Switzerland, or at premier research universities in the United States.

From the above considerations, it should be clear that progress in nuclear and particle physics is not limited to the typical tools of the particle physics experimentalist. While the particle accelerator has been the most common location to work out the details of our theories, astrophysical phenomena and cosmological observations are also key to understanding particle physics. For that matter, smaller experimental setups, such as those on which gamma-ray spectroscopy are performed, have played a pivotal role in the development of nuclear and particle physics.

21.3   Cosmic Rays and Cosmic Ray Muons

The experiment described in this chapter involves the measurement of muon detection rates as a function of their direction relative to the zenith. One of the major sources of natural radiation on Earth is from muons. These come in two varieties, μ^+ and μ^-; the positive charged version is the antimatter partner of the negatively charged version. They are typically created in the upper atmosphere. As a glance back at Figure 21.1 will show, the muon is a lepton. The properties of a muon are similar to the electron, except that, at a mass of 105 MeV/c^2 it is about 200 times more massive. Another difference is that while the electron is stable, the muon is not. On average, a muon at rest will decay after 2.2 μs. However, since the decay of a muon is governed by quantum mechanics, this lifetime is only an average. Some muons decay sooner and others decay later.

Cosmic ray muons are secondary particles that are created by the collision of primary cosmic rays (often high energy protons) with atoms in the Earth's upper atmosphere, creating showers of secondary particles, as shown in Figure 21.4.

Among the particles created by the primary collision are pions and muons. The mean lifetime of a pion is less than that of a muon (in fact, nearly all charged pions decay into muons) and the interaction cross section between the atmosphere and the pions is greater than that of the muons because pions interact both through the

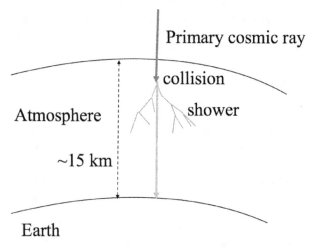

FIGURE 21.4   A cartoon depicting the creation of cosmic ray muons. The cosmic rays consist of the primary particles (often protons). Cosmic ray muons are secondary particles created by the collision of the primary cosmic ray with an atom in the atmosphere. Muons take longer to decay than other many other particles, and it is this fact, combined with the effects of time dilation, that lead to the prevalence of muons at sea level.

strong force and the electromagnetic force while the muons do not interact via the strong force. Primarily as a result of their longer lifetimes, muons are detected at the surface of the Earth in much greater numbers than other elementary particles created in these showers (all high-speed particles gain the advantage of time dilation). Neutrinos are also incident on the surface of the Earth, but detecting these particles is exceedingly difficult.

A very important use of cosmic ray muons is in the commissioning of nuclear and particle physics detectors. For example, while waiting for a new particle accelerator to come online, collaborations can determine whether their detectors are working using cosmic ray muons. They are ubiquitous in the development and testing of new detectors.

21.4 Modeling Muon Rates

We will explore the claim made by the Particle Data Group (PDG) document* on cosmic rays, that cosmic ray muon rates are proportional to $(\cos \theta)^2$, where θ is the angle that the muons make with the vertical. We are going to tackle this problem in stages. Initially, we will make as many simplifying assumptions as possible in order to build the simplest model possible. If our simplest model doesn't explain the angular dependence, we will relax assumptions that seem to be the most out of line with reality and compare our more sophisticated model with the claim. Let us see if, in the end, our model displays the dependence quoted by the PDG.

What causes the rate to vary by angle in this way? Perhaps the most obvious guess is that it has something to do with the decay of the muon. Muons decay by the reaction

$$\mu^- \rightarrow e^- + \nu_\mu + \overline{\nu}_e \text{ for the muon}$$

$$\mu^+ \rightarrow e^+ + \overline{\nu}_\mu + \nu_e \text{ for the anti-muon}$$

Since path lengths depend on θ, perhaps the $(\cos \theta)^2$ dependence is simply due to decay of muons as they travel through the atmosphere.

The decay of muons at rest follows the radioactive decay law $N = N_0 e^{-\frac{t}{\tau}}$, where N is the number of muons at time t, N_0 is the initial number of muons, and the time constant $\tau = 2.20\,\mu s$, is the mean life of a muon.

Exercise: (a) By setting $\dfrac{N}{N_0} = \dfrac{1}{2}$, determine the relationship between mean life and half-life. **(b)** Use your result to determine the half-life of a muon at rest.

However, the muons we are modeling are moving at a speed v in the Earth reference frame. So, their mean life is extended due to time dilation: $\tau_{\text{Earth}} = \gamma\tau$, where τ_{Earth} is the mean life in the Earth frame, τ is the mean life in the frame of the muon, and $\gamma = 1/\sqrt{1 - v^2/c^2}$.

So, for moving muons we have:

$$N = N_0 e^{-\frac{t}{\gamma\tau}} \tag{21.1}$$

In the actual experiment we are going to perform, we will not observe a collection of muons all moving at the same speed and simply note the time at which each decays, regardless of its location. Instead, our collection of muons will be built up, count by count, by a detector at rest on Earth. Each muon will be characterized by a different γ.

---

* M. Tanabashi et al. (Particle Data Group), *Phys. Rev. D* **98**, 030001 (2018), also available at http://pdg.lbl.gov/2018/reviews/rpp2018-rev-cosmic-rays.pdf.

Muon Path Lengths

The PDG suggests that muons are typically created at altitudes of about 15 km. For this back of the envelope initial stab at an analysis, let's assume that all cosmic ray muons are created at this altitude (even though this could not possibly be true). Let us further use a flat Earth approximation to determine the path length between the creation point of a muon, and the location of its detection.

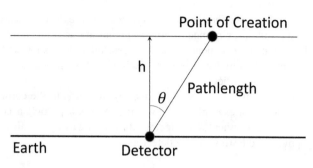

FIGURE 21.5   Calculating path length

From Figure 21.5, it is clear that the path length is given by

$$\text{path length} = \frac{h}{\cos\theta}$$

where $h = 15$ km is the altitude at which the muons are created.

Exercise: Given the above model, what is the path length of a muon detected at an angle of 60 degrees? (Answer below.*)

For a spherical Earth, you can show in Homework Problem 1 that

$$d_s = -R_E \cos\theta + \sqrt{\left(R_E \cos\theta\right)^2 + \left[\left(R_E + h\right)^2 - R_E^2\right]}.$$

For an angle $\theta = 60°$, we obtain $d_s = 29.89$ km. There is only a discrepancy of about 0.35% between the flat Earth result and the spherical Earth result, so it seems that we are safe with a flat Earth approximation to this problem out to angles of about 60°, and we'll use $d = d_{\text{flat}}$ for the path lengths.

To summarize our progress so far, we assume that all muons are created at a height of 15 km above the Earth and we apply a flat Earth approximation.

Predicting the Detection Rate

We can express the time in terms of the path length $t = d/v = h/(v \cos\theta)$

Using Equation (21.1), we then have

$$N = N_0 e^{-\frac{h}{v\gamma\tau\cos\theta}}$$

This is the total number of particles at the angle θ; the detector in our experiment detects a fixed fraction of these. Therefore, the rate should be proportional to the exponential above.

$$R \propto e^{-\frac{h}{v\gamma\tau\cos\theta}} \tag{21.2}$$

This does not match $(\cos\theta)^2$ dependence claimed by the PDG. While there is a $\cos\theta$ in the expression, we find it in the denominator of an exponential. You can make a more detailed comparison in Homework Problem 2.

For a more sophisticated model, we need to include effects such as energy loss by the muons as they travel through the atmosphere. You can explore this modeling further with the set of exercises under this section at ExpPhys.com

---

* The path length of the muon approaching at a polar angle of 60° should be (under a flat Earth approximation)

$$d_{\text{flat}} = \frac{h}{\cos\theta} = \frac{15\,\text{km}}{\cos 60°} = 30.00\,\text{km}$$

21.5   Homework Problems

1. The geometry for muon creation and travel through the atmosphere for a spherical Earth is as shown in Figure 21.6. Show that the muon path length is $d_s = -R_E \cos\theta + \sqrt{(R_E \cos\theta)^2 + \left[(R_E + h)^2 - R_E^2\right]}$.

   Hint: start by applying the law of cosines to the triangle whose longest side is formed by the line from the center of the Earth to the point of muon creation.

2. Create graphs comparing the $R \propto (\cos\theta)^2$ dependence found by PDG with the model given by Equation (21.2) for three kinetic energies: 1 GeV, 4 GeV, and 6 GeV. (The range of kinetic energies for muons as they travel through the atmosphere is 4 to 6 GeV.) Comment briefly on the degree of agreement between the two models, including whether there are ranges of angles for which they agree, and the extent to which agreement depends on energy. Hint: recall that $E_{\text{tot}} = KE + mc^2 = \gamma mc^2$.

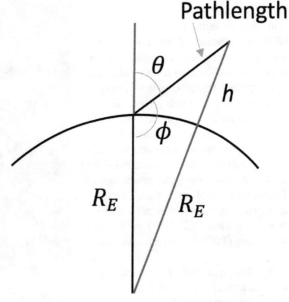

FIGURE 21.6   Geometry for Homework Problem 1.

3. The observation of muons at the surface of the Earth provided an early confirmation of the time dilation predicted by special relativity. To explore this effect with an example, assume that the kinetic energy of a particular muon is 6 GeV. If it is created at 15 km above the surface of the Earth and moves directly toward the surface, how far will it have traveled after a time equal to the mean life of the muon as measured by an observer at rest on the surface of the Earth?

Lab 21A Muon Telescope

Pre-lab reading: Sections 6.8, 7.4, 13.2.2, 21.3, and 21.4.

Labs you are assumed to have completed: 6G (or equivalent), 8A (8B and 8C recommended)

Safety precautions: If soldering the SiPM PCB, use a fan to disperse fumes, and take care not to get burnt by the soldering tip. If using lead-based solder, properly dispose of excess solder, and wash your hands when you're done.

21A.1 Introduction

In this lab, you will assemble a muon "telescope," i.e., an instrument which measures the rate of atmospheric muons that reach the telescope as a function of their angle from the zenith. There are many possible extensions but in the basic experiment you will use a novel mechanical design for the muon telescope consisting of a mailing tube and 3D printed parts, a combination of existing electronic circuit designs* and standard

---

* The choice of silicon photomultiplier, design of the silicon photomultiplier printed circuit board, and design of the amplification and signal stretching stages of the electronics are taken from the article "The desktop muon detector: A simple, physics-motivated machine- and electronics-shop project for university students" by S.N. Axani, J.M. Conrad, and C. Kirby, *Am. J. Phys.* 85, 948 (2017). In this chapter, this article will be referred to as Axani.

combinations of circuit elements presented in many undergraduate electronics classes to process the signals from silicon photo-multipliers (SiPMs), and a data acquisition system that includes an Arduino (or Arduino clone) and a computer. Alternatively, if time permits, you may develop your own electronics, mechanical designs, and data acquisition system. The completed result is shown in Figure 21.7.

In this experiment, raw signals from SiPMs are "conditioned" for subsequent processing. The voltages, for example, are amplified to make the signal more robust and easier to manipulate. This process is necessary for nearly all modern particle physics detectors in research settings around the world. In the case of the experiment described here, the subsequent electronics is not able to process very short signal pulses, so an additional step is taken to widen the signal pulses in time. The manipulation of analog signals for subsequent processing, say to prepare for analog to digital conversions (ADC) or time to digital conversions (TDC) is extremely common.

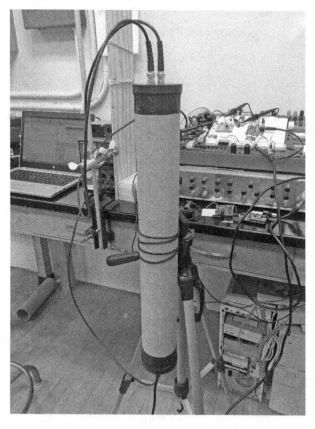

FIGURE 21.7   Picture of a completed muon telescope on a tripod.

Pre-lab question 1: If one has a detecting medium that consists of a thin horizontal disk 3 inches (7.62 cm) in diameter, what muon rate would you expect? (Assume all muons that strike the disk are detected. The muon flux at sea level is 1 cm^{-2} min^{-1}.)

How Do Two Muon Detectors Constitute a Telescope?

The apparatus you will build must distinguish muons arriving from a specific direction from other muons. The drawing below reveals how the geometry of the telescope constrains the directions of the muons.

Figure 21.8 illustrates the angular acceptance of the muon telescope. The optical axis is tilted at an angle, θ, from vertical. A horizontal strip of sky is also shown both to guide the eye and illustrate a region of sky in which muons originate. While this slice of sky is more accurately described by a piece of a spherical shell, over the range of sky positions that can put a muon into the telescope, the use of a plane is a reasonable approximation, as shown in Section 21.4.

By tilting the telescope at an angle θ from the zenith, muons arriving at angles between $\theta - \alpha$ to $\theta + \alpha$ have a chance of firing both detectors (shown as cylinders in Figure 21.8) almost simultaneously (since the muons move very close to the speed of light). Two of the dashed lines in the drawing determine a half angle α. These "extreme" rays denote the minimum and maximum angles that can put a muon through the light producing material of each detector (or "counter"). In this case, each cylinder represents a 3-inch (7.62 cm) diameter and 1.5-inch (3.81 cm) long piece of plastic scintillator (described briefly below).

21A.2 Building the Muon Telescope: Overview of Mechanical Structure

Two independent detectors are encased in a 3-inch inner diameter tube. We use a thick, inexpensive mailing tube. The tube is cut to a length that constrains the half opening angle, α, to the design specification.

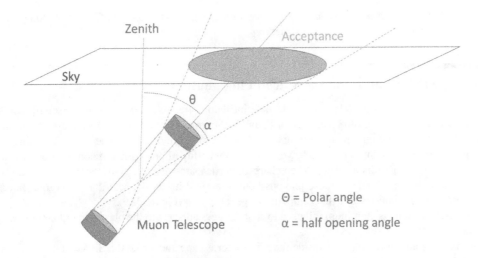

FIGURE 21.8 Diagram illustrating the geometry of the muon telescope.

Pre-lab question 2: Determine the length of the tube necessary for $\alpha = 10°$, assuming that the scintillators have diameter 7.62 cm and are much shorter than the tube length.

Figure 21.9 depicts the mechanical assembly of each detector in the telescope. From bottom to top, the assembly consists of an end cap and part of a spacer mechanism, the top of the spacer mechanism, a plastic disk with holes through which four wires are fed (and on which aluminum foil is taped or glued on one side), an SiPM holder, an SiPM and SiPM PCB, a cylinder of plastic scintillator, and an additional piece of plastic (again with foil on one side). The trickiest part of the procedure is soldering the four wires

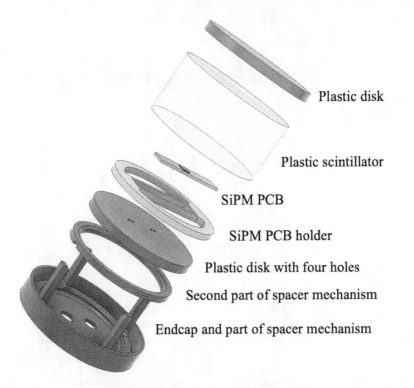

FIGURE 21.9 Exploded view of one of the two detectors that combine to make a muon telescope.

first to the SiPM PCB and then to bulkhead connectors attached to the end cap. This is the final step in the assembly process of the mechanical structure.

21A.3 The Scintillator and the Silicon Photomultiplier

The detecting medium of the telescope is plastic scintillator, often referred to as organic scintillator. We have used Saint-Gobain BC-408 plastic scintillator. At the time of publication, 3-inch diameter, 1.5-inch thick cylinders polished on top and bottom cost about US$60 each on ebay. (The reason for using such thick pieces of plastic scintillator is to leave open the possibility of performing muon lifetime measurements with the scintillators as well.) When charged, particles traverse the scintillator and thousands of photons are produced. Some of these photons bounce around inside the scintillator via total internal reflection. Other photons escape the scintillator or get absorbed (Figure 21.9).

Concept test: Why is aluminum foil used to coat surfaces adjacent to the plastic scintillator? (Answer below.*)

Ultimately, some of the photons inside the scintillator escape and strike the SiPM. Optical grease is used at the interface of the scintillator and the SiPM. An SiPM consists of an array of 1000s of avalanche photodiodes all wired in parallel.

As with all diodes, each avalanche photodiode is characterized by a junction between an "n-type" material and a "p-type" material. The n type material consists of a base substance, such as silicon, which has four outermost electrons, doped with atoms that contain five outermost electrons such as phosphorus. Four of the five outermost electrons in a dopant atom participate in covalent bonding, placing the phosphorus atom in the lattice, while the fifth becomes a conduction electron (see Figure 21.10).

Note that the overall charge of the n type material is neutral, since each conduction electron is balanced by a proton in the nucleus of the dopant atom.

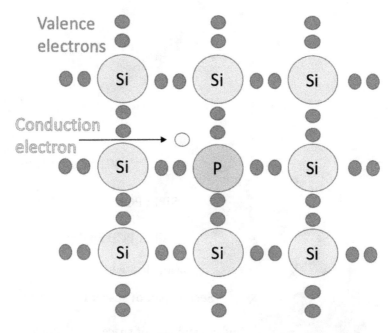

FIGURE 21.10 Visualization of an n type material depicting the lattice structure and doping that gives rise to electron charge carriers.

---

* For photons that escape the plastic scintillator, aluminum foil will sometimes reflect them back into it. This is the same reason why the SiPM printed circuit board is white, and why you should 3D print the SiPM holder in white.

Similarly, the p type material is characterized by doping with atoms that contain three valence electrons, such as boron. The three outermost electrons in the dopant atoms participate in covalent bonding but there is a "hole" where one of the electrons taking part in the covalent bonding should be (see Figure 21.11). Nearby electrons can fill such holes, causing the hole to move around like a free charge carrier, and it can be modeled as a positive charge. The overall substance has a net zero charge because for every hole there is a dopant atom with one less proton than the surrounding atoms in the lattice.

At the junction of the n and p type materials, electrons in the n type material diffuse over to the p type material and fill holes. Similarly, holes in the p type material diffuse into the n type material (see Figure 21.12). This

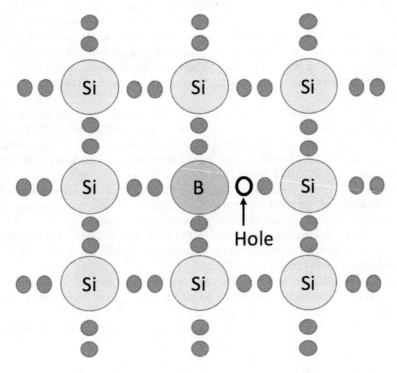

FIGURE 21.11 Visualization of a p type material depicting the lattice structure and doping that gives rise to "hole" charge carriers.

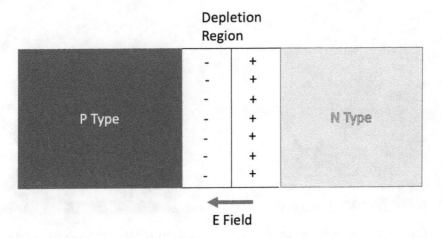

FIGURE 21.12 Visualization of a PN junction before the application of an external field.

process, however, leaves a net negative charge at the lattice sites near the junction on the p side of the material, and a net positive charge at the lattice sites on the n type material. As this diffusion proceeds, an electric field develops which tends to reduce the number of additional conduction electrons that diffuse from the n type material to the p type material and the number of holes that diffuse from the p type material to the n type material.

However, placing an external potential across the junction can shrink the depletion region until the material can conduct again. This is referred to as forward bias. Under reverse bias voltage, however, the depletion region grows, and the only way to generate a current is if charge carriers are created in the depletion region. This can occur when photons strike atoms in the depletion region, creating electron/hole pairs. Under high reverse bias voltage these electron/hole pairs are accelerated in opposite directions, smashing into other atoms and creating additional electron/hole pairs. An avalanche occurs and a detectable current is generated. If light strikes a number of microcells in the SiPM, a significant current can be generated. The SiPMs used in this experiment are 6 mm×6 mm ON Semiconductor C series MICROFC-60035-SMTs. The best reverse bias voltage at which to operate these SiPMs is 29.4 volts. SiPMs detect single photons, but the size of current produced by the device is proportional to the number of microcells triggered.

Each plastic scintillating cylinder is mated to an SiPM with optical grease and each SiPM is soldered to a printed circuit board (PCB). Only SiPM pins 1 and 3 are soldered to the SiPM PCB. Pin 1 can be located using the fact that the wire configuration on the edge near its corner is different than the other corners. See the SiPM datasheet for more information. Images of the SiPM, SiPM PCB, and SiPM PCB holder are shown in Figure 21.13 and Figure 21.14.

The circuit on the SiPM PCB, taken from Axani is shown in Figure 21.15. The resistors and capacitors connected to the cathode of the SiPM function as low pass filters to even out the power supply voltage over time. Resistor R3 converts the signal current pulse to a voltage pulse.

One wire connects the reverse bias voltage to the SiPM PCB and a second wire carries the signal from the SiPM PCB to a bulkhead connector on the telescope end cap. In addition, there are two ground wires that are connected to the SiPM PCB.

Figure 21.16 shows a detector assembly, both before and after installation. Care should be taken not to allow any exposed wire or solder pads to contact the aluminum foil. In addition, optical grease is placed on the SiPM before it is placed in contact with the scintillator.

Three-dimensional printer files are provided on ExpPhys.com to produce the end cap, spacer mechanism, and SiPM holder.

FIGURE 21.13   SiPM, SiPM PCB, and SiPM PCB holder. The SiPM is soldered to the front.

FIGURE 21.14   SiPM PCB back side. SiPM PCB circuit elements are soldered on the back.

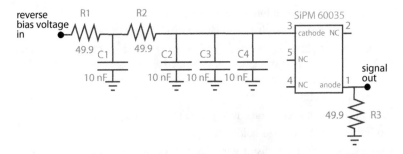

FIGURE 21.15 SiPM printed circuit board circuit design taken (adapted from the Axani paper).

FIGURE 21.16 At the center front is an example of a disk with aluminum foil attached. On the left is a completed detector assembly, and on the right is a detector assembly installed in a tube.

21A.4 Assembling the Mechanical Structure and Initial Testing

One of the primary requirements of the mechanical structure is that it should be light tight (let no external light in). The reason for this is that the means of detection are flashing lights in the scintillator. If external light gets into the telescope, then the individual flashes of light from the scintillator will be overwhelmed by the external light. The light signal would be buried in light noise.

- Prepare plastic disk pieces:
  - Cut pieces of aluminum foil to glue to one side of each of the two plastic disks. Leave holes where wire is fed through the plastic disk holes, and areas that will be close to solder pads of SiPM PCB.

- Tape or glue the aluminum foil to the plastic disk pieces.
- Glue the end cap, spacer mechanism, and plastic disk as necessary:
  - In newer versions of the 3D print design, one piece of the spacer is printed attached to the endcap and the other part of the spacer is printed attached to the disk with four holes. The parts that require glue should be obvious.
- Attach the bulkhead connectors to the end cap
- Label each bulkhead connectors on outside of end cap (signal, bias)
- Connect scintillator, SiPM, SiPM holder, wires:
  - Calculate nearly exact lengths of wire that will connect the SiPM PCB and the bulkhead connectors. A little slack can be provided, but long wires degrade fast signals.
  - Cut wires to size
    - Choose reasonable colors for each wire
  - Solder the four wires to the SiPM PCB. It might be wise to wear a grounding strap when building the electronics.
  - Snap the SiPM PCB into the SiPM PCB holder.
  - Put optical grease on the SiPM, connect it to the scintillator and tape around circumference to hold the SiPM mechanism to the scintillator
- Connect end cap assembly to SiPM assembly:
  - Feed the wires through the plastic disk.
  - Tape around the circumference to secure.
- Repeat process to create a second counter assembly
- Insert first assembly into one end of mailing tube
- Tape around outer circumference of mailing tube to secure end cap to tube
- Place packing peanuts in tube so that just enough room is provided inside tube for second counter assembly
  - After second assembly is inserted and taped, the two counters, packing peanuts, and tape on the outside of the tube should hold everything together under slight compression; this is required to fully stabilize the scintillators.

Attach the telescope to a tripod (see Figure 21.7). Four coaxial cables are used to provide two channels of reverse bias voltage and feed two signals to an electronic breadboard. As stated above, the SiPMs require 29.4 volts reverse bias voltage, so either one dual channel 30 volt power supply or two 30 volt power supplies is/are required.

Examining the Raw Signal

After applying the reverse bias voltage, it is time to check that muon signals are produced. Use an oscilloscope: the amplitude of the raw signal should be about 10–100 mV, and the length of the raw signal should be roughly ½ μs. It should have a steep rise followed by a long decaying tail (as shown in Figure 21.17).

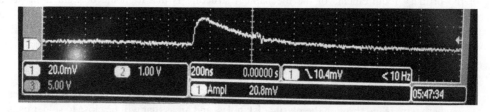

FIGURE 21.17   Scope trace of the raw signal from the SiPM and SiPM PCB.

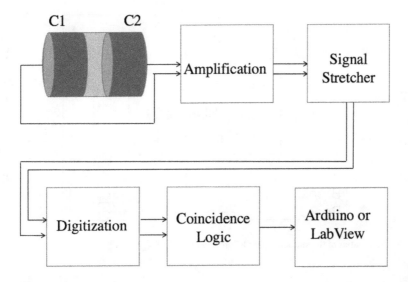

FIGURE 21.18 Flow chart for the muon telescope electronics that process the signal after its initial generation using the SiPM PCB.

21A.5 Designing and Assembling the Signal-Conditioning Electronics

A flow chart showing the electronics that processes the signals from the muon telescope is shown in Figure 21.18.

The output of the SiPM PCBs are voltage signals. These signals must first be amplified. *Please review Section 6.9 before continuing.* You can use a surface mount LT6201 dual op amp to both amplify and later stretch the signal in time (note that in the Axani article, the signal stretching circuit fragment is referred to as a peak detecting circuit). Alternatively, a through hole IC, the OPA350 has been used successfully to amplify the signal. The SiPM PCB and the electronic designs for the amplification and signal stretching stages are taken from the Axani article. The pinout for the LT6201 dual op amp recommended in the article can be found on its data sheet.

As mentioned above, the LT6201 is a surface mount IC, so to use it with a solderless breadboard, one option is to solder the LT6201 to a "breakout" board, like the one shown in Figures 21.19 and 21.20 from Adafruit.* You can easily find the pinout for the LT6201 on the web.

Amplification

A non-inverting op amp amplifier circuit is employed to amplify the signal, and is followed by a high-pass filter to remove any DC offset, as shown in Figure 21.21 (non-inverting amplifiers are described in Section 6.9.)

FIGURE 21.19 An LT6201 dual op amp soldered to a surface mount IC breakout board.

---

* www.adafruit.com/product/1212.

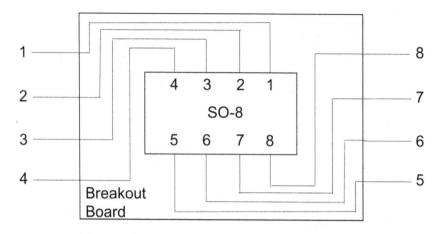

FIGURE 21.20   Schematics of the surface mount IC breakout board.

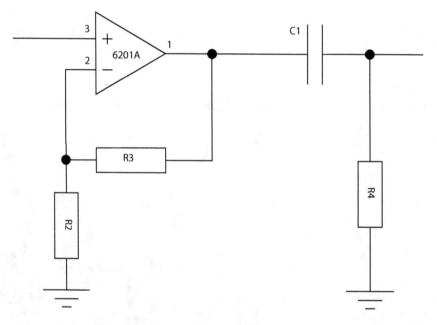

FIGURE 21.21   Circuit diagram for the non-inverting op amp amplifier. It is followed by a low pass RC filter to remove any DC offset produced by the amplifier. Figure drawn by Moira Ferrer.

   Pre-lab question 3: Choose R3 and R2 resistors that would produce a gain of 101 for an ideal op
      amp. Let's assume that you choose R2 = 100 ohms.

   To determine a design value for the gain of your own amplification stage, use an oscilloscope to
examine the raw signals. Estimate the minimum pulse height you want to accept as a legitimate count
by limiting the rate to less than the expected rate calculated in Pre-lab question 1. Once you have
determined a scope trigger that limits the rate of triggers to a number less than the expected rate calculated in Pre-lab question 1, choose a gain that will amplify this minimum raw signal to somewhere
between 2 and 3 volts.
   Once you have designed and built your non-inverting amplifier, examine the output on a scope.
   In-class exercise: Measure the gain between the raw signal and the amplified signal. An example of
doing so is shown in Figure 21.22.

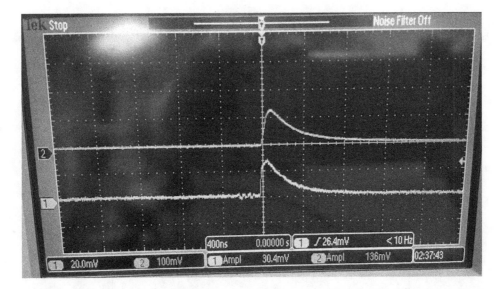

FIGURE 21.22 Signal from the SiPM PCB (bottom trace) and amplified signal (top trace). This amplification is much smaller than you should expect to obtain when following the instructions in the text.

The signal shown in Figure 21.22 is only amplified by about a factor of 5. Your amplification should be much larger (although your actual gain will probably not be as large as your design value). Consider the gain bandwidth product for this op amp when trying to understand why a non-inverting amplifier designed for a gain of 101 might wind up with a somewhat smaller gain. Also, when attempting to amplify by this large factor, you may notice a significant DC offset. This is the motivation for following this amplification stage with a high-pass filter.

In-class exercise: Choose values for the capacitor and R4 that will remove any DC offset introduced by the op amp without significantly attenuating the output signal of the op amp. This amounts to choosing component values that will yield reasonable high-pass filter parameters. To help you determine reasonable values for these components, consider the fact that the width of the signals is about ½ μs.

Signal Stretching (or "Peak Detecting") Circuit

You can implement the signal stretcher, referred to as a peak detecting circuit in Axani, with the other half of the dual op amp. The purpose of this part of the electronics is to make it easier to process the signal with inexpensive and somewhat slow electronics in the subsequent stages. Making the signal longer will allow looser criteria for the performance of the comparator that will be used to digitize the signal, and the longer digital signal will be easier for the subsequent electronics and Arduino to process (Figure 21.23).

You probably have not previously thought about op amp circuits that include diodes, so only think about each of the following questions for about 30 seconds before looking at the answer.

(1) If you omitted D1, D2, and the capacitor, then what would this circuit do?

Answer to (1): it is a gain 2 non-inverting amplifier.

(2) Now add D1, but continue to omit D2 and the capacitor. Use a simple model for the diodes in which they completely block current for reverse voltage and also for any forward voltage below a threshold value (0.45 V for the Schottky diodes shown), but allow current to flow easily for forward voltage higher than the threshold. This means that, when current is flowing through the diode, there is a 0.45 V drop across it.

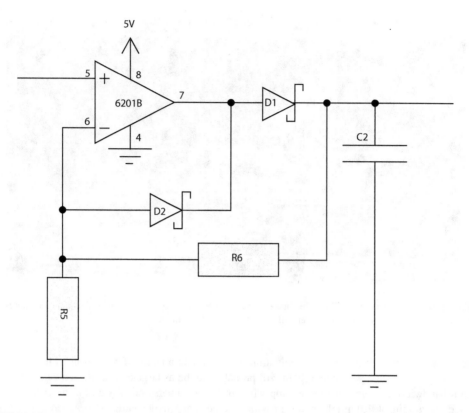

FIGURE 21.23 Signal stretching circuit. Axani specifies 100 kΩ resistors for R5 and R6, and 10 nF for the capacitor C2. D1 and D2 can be implemented with 1N5148 Schottky diodes. Figure drawn by Moira Ferrer.

If the input is a linear voltage ramp from 0 V to 2 V as shown in Figure 21.24, sketch the output of the circuit, i.e., the voltage at the right side of D1. (The power supplies to the op amp are +5 V and 0 V. The op amp we're using is "rail to rail," meaning that the output can go all the way to the positive and negative supply voltages.)

Answer to (2): the circuit still acts as a gain 2 amplifier, because the diode is inside the feedback loop; by the "golden rules," V_- must equal V_+, so (because of the voltage divider formed by the two resistors), the circuit output must be twice V_{in}, as shown in Figure 21.25. (Since there is a 0.45 V drop across D1, the output of the op amp itself is 0.45 V higher than the output of the circuit.)

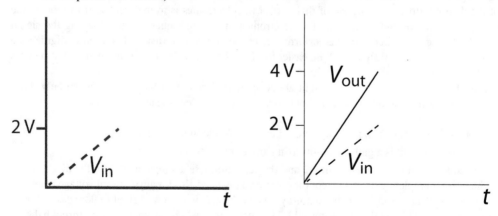

FIGURE 21.24 Prompt for question 2.   **FIGURE 21.25** Answer to question 2.

(3) Now add the capacitor and D2. Again, the input voltage is ramped from 0 V to 2 V, but then is ramped back to 0 V, as shown in Figure 21.26. Because of D1, the op amp can add charge to the capacitor, but cannot remove charge. The capacitor can discharge through R6, but because this is a large resistor, this discharge happens relatively slowly. Therefore, as soon as V_{in} starts to drop, V_+ becomes lower than V_-, causing the output of the op amp to go as low as it can (0 V). Use these ideas to complete the sketch the output of the circuit for the time shown in the graph.

The answer is shown in Figure 21.27.

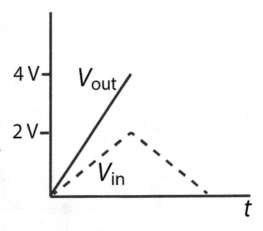

FIGURE 21.26   Prompt for question 3.

(4) In terms of the capacitor C2, and resistors R5 and R6, what is the RC time constant for the decay shown in the previous part?

Answer to (4): since the output of the op amp is at 0 V, there is more than a 0.45 V difference across D2, so the current discharging C2 can flow through R6, then through D2 to the op amp output. Therefore, the RC time is $R6 \times C2$.

(5) Once the output of the circuit falls below 0.9 V, the RC time constant for the discharge of C2 changes. What is the new value?

Answer to (5): when the output of the circuit falls below 0.9 V, the voltage drop across D2 falls below 0.45 V, so current can no longer flow through it. So, the current discharging C2 must flow through both R5 and R4 to ground. So, the RC time is $(R5+R6) \times C2$.

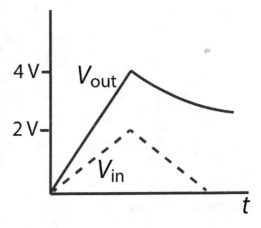

FIGURE 21.27   Answer to question 3.

The Schmitt Trigger – Digitizing the Signal with a LM311

A so-called Schmitt trigger discriminates and digitizes the amplified and stretched signal. Only signals with a large enough amplitude will create an output pulse and that output pulse will be digital (~5 v or ~0 v). The Schmitt Trigger is a comparator circuit that operates under positive feedback (see Figure 21.28). Only pulses that satisfy some minimum pulse height requirement will send the output from high voltage to low voltage. Furthermore, Schmitt triggers exhibit the interesting feature of hysteresis. When the input voltage is low and increasing, the switching voltage is higher than when the input pulse is high and getting lower. This stops the output from oscillating due to noisy input close to the switching voltage.

To understand the operation of the Schmitt trigger, imagine that instead of resistor R7, we substitute an open circuit, i.e., we remove the feedback. If R8 and R9 are both 5K resistors, then the voltage at the non-inverting input of the comparator will be half of the total voltage or 2.5 volts. The output of the comparator is the best the comparator can do to amplify the difference between the non-inverting and inverting inputs. This means that when the voltage on the inverting input is less than 2.5 volts, the output will be the voltage on the high side of the pull-up resistor, 5 volts. On the other hand, when the voltage on the inverting input is greater than 2.5 volts, the difference between the non-inverting input and the inverting input is negative, and the output is driven to the lowest possible voltage, or 0 volts, as shown in Figure 21.29.

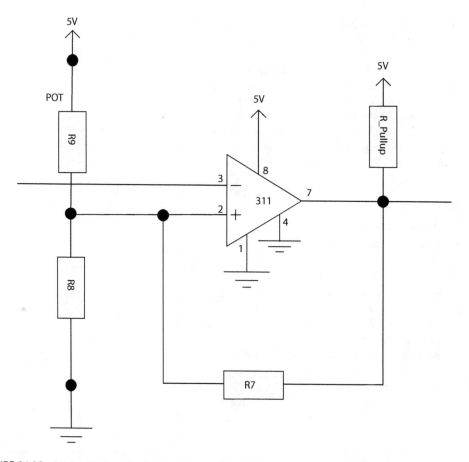

FIGURE 21.28 Implementation of a Schmitt Trigger with a 311 Comparator. The dashed line that encloses resistances R8 and R9 is actually implemented with a 5K or 10K potentiometer, and the feedback resistor should be chosen to provide the hysteresis desired. Figure drawn by Moira Ferrer.

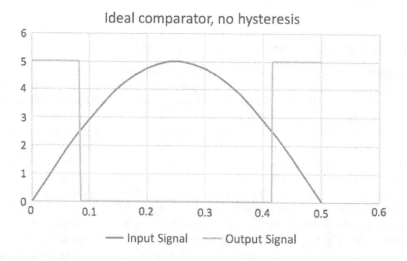

FIGURE 21.29 Expected behavior of an ideal comparator with no hysteresis.

On the other hand, if the input signal is noisy, the output can oscillate between high and low several times, thus the need for hysteresis.

When the input is low and the output is high, the switching voltage is given by

$$V_{\text{switch, input low}} = 5\text{ volts} \times \frac{R_8}{R_8 + (R_7 \parallel R_9)}$$

where

$$(R_7 \parallel R_9) = \frac{R_7 \times R_9}{R_7 + R_9}$$

On the other hand, when the input is high and the output is low, the switching voltage is given by

$$V_{\text{switch, input high}} = 5\text{ volts} \times \frac{(R_7 \parallel R_8)}{R_9 + (R_7 \parallel R_8)}$$

with $(R_7 \parallel R_8)$ defined similarly.

If, for example, R8 and R9 are both set to 2500 Ω, while R7 is set to 20 kΩ, a hysteresis of ± 0.147 volts. When waiting for the input signal, the switching point is 2.65 volts, and when waiting for the input signal to subside, the switching point is 2.35 volts. A hysteresis of this level is reasonable.

The switching point when waiting for a signal should be set to give a reasonable expectation that when a large enough signal comes in to trigger the comparator, one is reasonably confident the signal is real. One way to estimate this minimum pulse height requirement is to take your answer for the maximum expected rate for a single scintillator and raise the voltage threshold until the rate of triggers above the threshold is less than your expected rate under ideal conditions.

Pre-lab question 4: R8 and R9 are set by a potentiometer. If it is a 5k potentiometer, then the sum of R8 and R9 must be 5k. Determine the resistances to use for R8 and R9 so that the middle tap of the potentiometer is at 2.5 volts, and choose R7 so that a hysteresis of about 91 mV is achieved.

Once you have completed building the non-inverting op amp amplifier, signal stretching circuit, and Schmitt trigger, you should examine the signal at each of these stages of processing. If you have a four-channel oscilloscope available, your results might look something like those of the first three channels in Figure 21.30.

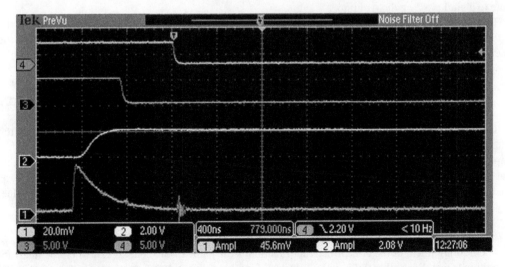

FIGURE 21.30 Scope traces from various points in the signal processing chain. The bottom trace is the "raw" pulse from the SiPM PCB, the next-to-bottom trace is the signal after it has been amplified and stretched, the next-to-top trace is the signal after the Schmitt trigger, and the top trace is the coincidence signal that results after completing the electronics of the next section.

The Coincidence Trigger Using a 74LS00 Quad NAND Gate

Of course, two copies of all the preceding electronics must be implemented, one for each of the two scintillators (or two channels). Having completed this task, you now have to combine the digital outputs of the two Schmitt triggers to determine coincidences.

In-class exercise: Please design the digital electronics that will take as inputs the signals from each of the two Schmitt triggers and produce a high (5 volts) output when there is no signal, and a low (0 volt) output when there is a coincidence signal. This can be done, for example, with three of the four gates on a quad NAND IC. See the digital electronics chapter for the basics of what a NAND gate does.

21A.6 Counting the Number of Coincidence Signals Using an Arduino Uno/SparkFun Redboard

The output of the QUAD NAND gate (the coincidence signal) is sent to an Arduino microcontroller. The experiment has been performed with a variety of data acquisition devices, including both the Arduino Uno and the Arduino Nano. Arduino Nano clones can be obtained at a remarkably low cost, but we found it to be a bit of a hassle to get the CH340 drivers that are used with the inexpensive versions of the Nano to work under all software/hardware scenarios, we advocate using the ubiquitous Arduino Uno R3. More precisely, we use the SparkFun Redboard which is nearly a match to the Arduino Uno R3.[*] For future reference, whenever the term RedBoard is used in subsequent text, the Arduino Uno R3 could just as easily be used.

The Arduino is going to monitor the status of the output of the NAND gate. When no signal is present, the output of the NAND gate is high, as shown in the truth table of the previous section. The output of the NAND gate only goes low when there is a coincidence between the two muon detectors in the muon telescope. Thus, you need to program the Arduino to monitor the output of the NAND gate and produce a count every time the output of the NAND gate goes from high to low.

An ideal mechanism for this task is the "hardware interrupt." Interrupts allow programs/sketches to monitor a pin, waiting for an event to happen, without locking the microprocessor in an explicit loop for this purpose. When an interrupt is used, it doesn't matter where the program/sketch is in its execution, control is handed over to the interrupt code which handles the event and then returns the program to where it left off. The Arduino Uno is based on the ATmega328 microcontroller, and this microcontroller allows for two so-called "external" interrupts that are implemented on digital pin 2 and digital pin 3. For your purpose, plan to use digital pin 2 to implement the interrupt. Wire the output of the final NAND gate to digital pin 2 on the Arduino.

To program the Arduino hardware interrupt, you are going to need to place a line of code in the setup() function. (The setup() function contains commands that are only executed once at the start of the sketch). One of the lines of code that should appear in the setup() function should look something like the following:

```
AttachInterrupt(digitalPinToInterrupt(interruptPin), addone, FALLING);
```

The AttachInterrupt() routine sets up the interrupt. The first argument of this routine requires the interrupt number. It turns out that interrupt 0 maps to digital pin 2. The beauty of the digitalPinToInterrupt() function is that it will determine the appropriate interrupt number for a given interrupt pin, in this case, it will return 0 when the interruptPin = 2. Thus, we could call the function with the argument 2, as in digitalPinToInterrupt(2), but in our case, we have defined a global constant variable equal to this value:

```
const byte interruptPin = 2;
```

The next argument of the AttachInterrupt function is the name of the code to call when the interrupt condition is met. We will call this code "addone." Finally, we must define the scenario that triggers the

---

[*] Information about the SparkFun Redboard can be found at www.sparkfun.com/products/13975.

interrupt. The argument "FALLING" causes the Arduino to wait for the digital pin to go from high to low before triggering the interrupt. This explains the details of the AttachInterrupt() function.

As for the code that is executed when the interrupt is triggered, you will write an addone function that simply adds one to a variable every time it is enacted. A little care should be taken with the type of variable that holds the count. In our case, we chose the type: "volatile int." This is an integer variable that is stored in a place in the microcontroller (in RAM) in which the presence of an interrupt won't mess up the value of the variable. This global variable might be declared as follows:

```
volatile int counts = 0;
```

The other thing the Arduino microcontroller must do is send the count tally to the computer. It uses serial communication via the USB cable that connects the Arduino to the computer to do this. This turns out to be particularly convenient because the Arduino Integrated Development Environment sports a serial monitor that enables communication between the Arduino and the computer.

To summarize, each time the interrupt pin is brought from high to low, the interrupt function, "addone," is called. This function adds one to the current tally. Meanwhile, in the main body of the program, the current tally is compared with the tally manipulated by the interrupt routine. Every time the main program determines that the interrupt tally is higher than the main program tally, the main program tally is incremented (bringing it into alignment) and the new count tally is sent to the computer on a virtual serial port that runs through the USB cable between the computer and the Arduino. Code to do this is shown in Figure 21.31.

You might consider altering the code shown in Figure 21.31 slightly so that in addition to the count tally, the time elapsed, in milliseconds, since the Arduino started running is listed. This is accomplished by using the millis() function. Each time the Arduino receives a new coincidence count, in addition to sending the new count tally from the Arduino to the PC, it can send the time elapsed, in milliseconds, since the sketch started running.

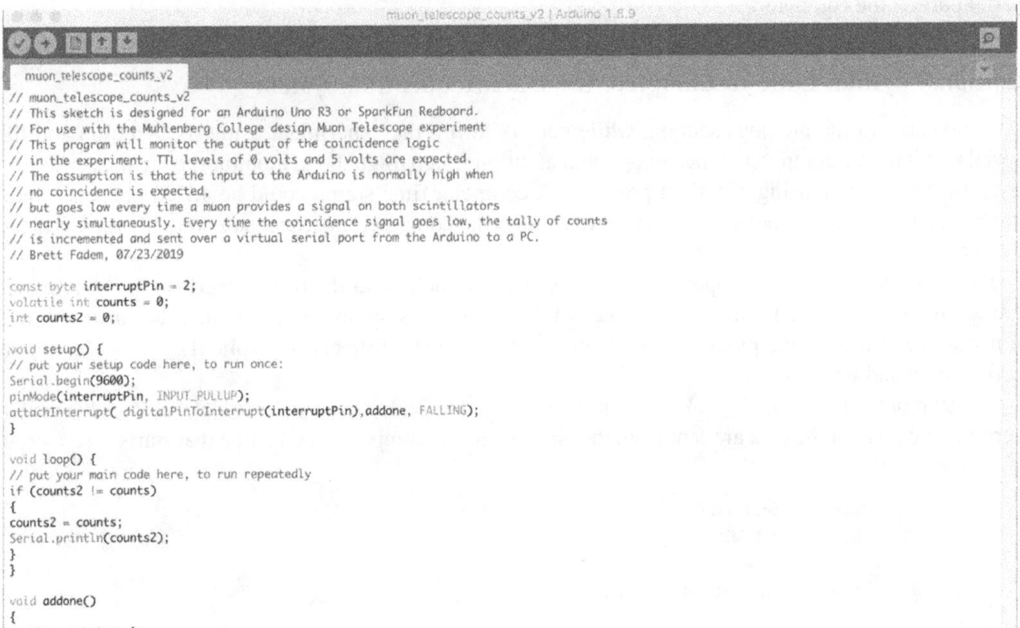

```
muon_telescope_counts_v2 | Arduino 1.8.9

muon_telescope_counts_v2
// muon_telescope_counts_v2
// This sketch is designed for an Arduino Uno R3 or SparkFun Redboard.
// For use with the Muhlenberg College design Muon Telescope experiment
// This program will monitor the output of the coincidence logic
// in the experiment. TTL levels of 0 volts and 5 volts are expected.
// The assumption is that the input to the Arduino is normally high when
// no coincidence is expected,
// but goes low every time a muon provides a signal on both scintillators
// nearly simultaneously. Every time the coincidence signal goes low, the tally of counts
// is incremented and sent over a virtual serial port from the Arduino to a PC.
// Brett Fadem, 07/23/2019

const byte interruptPin = 2;
volatile int counts = 0;
int counts2 = 0;

void setup() {
  // put your setup code here, to run once:
  Serial.begin(9600);
  pinMode(interruptPin, INPUT_PULLUP);
  attachInterrupt( digitalPinToInterrupt(interruptPin),addone, FALLING);
}

void loop() {
  // put your main code here, to run repeatedly
  if (counts2 != counts)
  {
    counts2 = counts;
    Serial.println(counts2);
  }
}

void addone()
{
  counts = counts + 1;
}
```

FIGURE 21.31   Screenshot of an Arduino sketch that should send data to the computer via serial communication through the USB cable.

A check of the discrepancy between keeping time this way with the Arduino millis() function and keeping track of the time elapsed using the system time on the computer that the Arduino is communicating with revealed a discrepancy of about 0.03%. For the purpose of these measurements, this systematic error is much less than other uncertainties in the experiment. While this method of timing should work, as a matter of principle, it should be pointed out that it introduces a tiny amount of "dead time" into the measurement. Every time a coincidence triggers an interrupt, for the time it takes the interrupt function to execute, the millis() timer ceases to time. While we can argue that this error is negligible, we should acknowledge that it is there.

For more information about interrupts and serial communication, please see Monk (2013).[*]

If you add code so that both the elapsed time on the Arduino and the count tally are printed out each time a count is received, you might consider ending development of the data acquisition system, calling it complete and declaring victory. If the time elapsed, as measured by the Arduino, is sent to the serial monitor along with the count tally, the number of counts received within a certain time interval, say between one and four hours, can be determined by examining the count tally and time elapsed listing on the serial monitor. Alternatively, you could run until a specified number of counts, say 100, is received at each angle and note how long it took to obtain that number of counts at each angle. For a closer examination of the distribution in time of the various counts, the time and count tally rows can be copied to a spreadsheet or otherwise exported for subsequent analysis.

In-class exercise: Alter the code provided so that upon detection of each coincidence, both the count tally and time elapsed are sent to the serial monitor on the computer.

On the other hand, you might find it convenient to automate the data taking so that a run is ended either after a certain amount of time has elapsed or after a certain number of counts are obtained. Furthermore, one might automatically write a file containing the count tally, Arduino time elapsed, and/or PC timestamp for each count obtained. Creating such files obviates the need to cut and paste information by hand. In the case of experiments with large data sets, the difficulty of using a cut and paste method becomes prohibitive, but that won't be a problem for this experiment. The optional discussion that follows outlines how one might perform these steps using a Python program to monitor the virtual serial port on the computer.

Optional Python Code to Complete the Data Acquisition System

As mentioned in the previous section, while one could run the experiment using the Arduino and the serial monitor on a computer alone, an optional additional step might be to monitor the virtual serial port used by the Arduino using a Python program. A computer time stamp could be added to each coincidence count tally (and time) sent from the Arduino to the computer and the result could be accumulated in a computer file.

The "import serial" and "import time" commands can help with the tasks described in the previous paragraph by importing the relevant modules for this task. Using an Anaconda distribution, we found it necessary to import the pyserial module into our Anaconda distribution explicitly before the import serial command worked.

The start of your code might look like that shown in Figure 21.32. The Python script monitors the virtual serial port on which the data are sent from the Arduino to the computer via the line that starts "ser=serial."

```
import serial
import time

file = open("muon_counter_data_071818.txt", "w")

ser = serial.Serial('/dev/cu.wchusbserial1410',9600)
```

FIGURE 21.32 A snippet of Python code from a program that could form the end point of the data acquisition system.

---

[*] S. Monk (2013). *Programming Arduino Next Steps: Going Further with Sketches.* McGraw-Hill Education TAB.

Serial(…)." Macs and PCs have different conventions for naming these virtual serial ports. In the code snippet shown, which ran on a Mac, the name was '/dev/cu.wchusbserial1410'; this name can be obtained from the Arduino IDE used to upload the Arduino code, under "Tools…Port." The virtual serial port designation on a PC usually looks something like "COM4" or some other string that starts with "COM" and ends with a number. The particular strings in the single quotes will vary in each case but will follow these formats.

One can write the Python code so that the program runs until a desired count tally is obtained. We have used 100 counts at each angle.

For the computer timestamp, we used the "time.time()" command. Each time a coincidence count (and the time at which the count was received according to the Arduino millis() function) is received from the Arduino over the virtual serial port, a string is created containing this information, and an additional string is concatenated with it containing the time stamp from the computer running the Python program. The string is written to a text file which will wind up containing one line of information for each coincidence count received.

In-class exercise (optional): Complete a Python script that accomplishes the goals stated above. Please refer to the Python coding chapter.

21A.7 Data Taking

The experiment involves pointing the muon telescope at a variety of angles and measuring the count rate at each angle.

In-class exercise: Assume that you run the experiment at each angle until 100 counts are obtained. What is the statistical uncertainty in the number of counts for each angle [hint: Poisson statistics govern this counting experiment]? What is the fractional statistical error in each measurement under these circumstances? Please refer to the uncertainty analysis section of this book.

You might run the experiment longer for the data point at 0° so as to shrink that particular statistical error bar as much as possible.

In-class exercise: Use graphing software to plot the rate of the muons as a function of angle. Be sure to title the plot, label the axes (including units), and provide statistical error bars on the data points. Overlay the plot of a scaled cosine squared function for comparison.

21A.8 Post-lab Questions

1. Determine the goodness of fit or chi squared per degree of freedom between your measurements and the scaled cosine squared prediction for the rate (using the scale factor that minimizes this chi squared per degree of freedom). Comment on the agreement of the cosine squared parameterization from this perspective.

2. [Challenging] You may find that the count rates you measure from each of the angles in the experiment do not lead to entirely independent data points in the sense that muons coming in at some particular angles will be measured by two distinct orientations of the muon telescope. For example, consider the case in which data were taken at 15° and 30°. Since the half opening angle of the muon telescope is about 10°, some muons from as much as 25° were included in the 15° data point. For that matter, some muons from as little as 20° were included in the 30° data point. One way to approach this issue would be to determine the theoretical prediction for these data points under the assumption that the rate does indeed vary as the square of the cosine but taking into account the actual acceptance of the muon telescope. How big is this correction?

3. By taking data for an extended period of time at 0°, determine if there is any systematic variation in the rate as a function of time of day.

4. Explain how one might go about correcting the data for accidental coincidences between detectors. To do this, consider the expected rate of accidental coincidences given the widths of the digital signals in time from each of the two detectors.

Brett Fadem is a Professor of Physics at Muhlenberg College. In addition to teaching courses across the undergraduate physics curriculum and physics classes for non-science majors, he and his students perform research with the PHENIX collaboration, a group of scientists who have built a large detector on the Relativistic Heavy Ion Collider ("RHIC") ring at Brookhaven National Laboratory. The quark-gluon plasma, a state of matter in which protons and neutrons melt into their constituents (and the state of "nuclear" matter in the universe microseconds after the big bang) was discovered at RHIC in 2005. In recent years, he has focused his attention on developing low cost particle detectors for the undergraduate lab that are based on the same principles as the cutting-edge detectors at RHIC and the LHC. For nearly ten years between high school and college, Fadem pursued an acting career in New York City, where he was trained by Uta Hagen and Herbert Berghof. He performed a number of plays at the HB Playwrights Foundation, and later started a small company called the Stillwaters Theatre Company with other young actors. He also enjoys playing violin and discussing philosophical questions with his family during moments when they wish he wouldn't (Figure 21.33).

FIGURE 21.33   Brett Fadem and a student with the muon telescope.

Index

Note: "f" means "in footnote"

10X probe, 108

A/D, 133
Absolute tolerance, 42
Absorption coefficient, beta particle, 275
Absorption spectroscopy, 345
Abstracts, 23
Active matter, 328
Activity, 268
ADC, 133
Addition in quadrature, 100
Affirmative action, 63
Airy pattern, 259
ALARA, 270
Algorithm, 316
    for 1D random walk, 316
    for 2D random walk, 318
    for chemotaxis, 321
    for jumping, 324
Aliasing, 139
Alignment, optical, 239, 243
Allyship, 62f
Alpha decay, 268
Alpha particle, 268
    interaction with matter, 275
Amplifier
    bandwidth, 111
    definition, 86
    differential, 90
    lock-in, 124
    noise, 103
    oscillations, 87, 112
    single-ended, 90
    zero offset, 120
Anaconda prompt, 205
Analog input/output common, 90, 118
Analog to digital converter, 133
AND, 146
Angular momentum of photons, 347
Annotated bibliography, 292
Aperture, 336
Apparaus, modeling of, 12
Arduino, 142, 170
    basic I/O, 172
    motor control, 188
    Muon telescope data acquisition, 426
    Muon telescope serial port, 427
    pulse width modulation, 178, 186, 189
    thermostat, 185
Arithmetic Logic, 155
Article database, 9
Aspheric lenses, 233

Authorship criteria, 60
Automation, 196
Avalanche photodiode (APD), 274

Bad ground box, 117
Baltimore Case, 56f
Bandwidth
    amplifier, 111
    definition, 86
    equivalent noise (ENBW), 101, 103, 300
    optical filter, 397
    photon, 397
Battery, ideal, 75
Bayh-Dole act, 65
BBO crystal, 379, 391
Beam expanders, 238
Beamsplitters, 235
Becquerel (Bq), 268
Behavior, 312
    chemotaxis, 308, 314
    jumping, 323
    thermotaxis, 315
Bell inequality, 390
Belmont Report, 69
Berg, Robbie, biography, 369
Beta decay, 268
Beta particle, 268
    absorption coefficient, 275
    interactions with matter, 275
Biblical-sounding mnemonic, 138
Biomechanics, 323
Block diagrams, 295
Boltzmann's constant, 302
Boolean algebra, 148–151
Boolean literature searches, 8
Breadboards, optical, 229
Brewsters angle, 255
Brick wall filter, 101
Buffer, perfect, 78
Bull's eye pattern, 246
Bush, Vannevar, 68

Capacitive voltage divider, 87
Capillary length, 333
Challenger disaster, 66
Chaos, 340
Chaotic pendulum, 294
Chemotaxis, 308
    combined with temperature gradient, 315
    ethyl acetate gradient, 310
    trajectory analysis, 314
Chirality, 252
Chi-square 47, 283
Circular dichroism, 343, 346–347

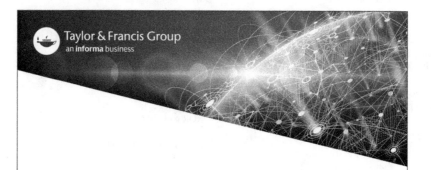

CPSIA information can be obtained
at www.ICGtesting.com
Printed in the USA
LVHW061547131221
706071LV00003B/351